AF388975

Geothermal Energy Utilization and Technologies 2020

Geothermal Energy Utilization and Technologies 2020

Editors

Carlo Roselli
Maurizio Sasso

MDPI • Basel • Beijing • Wuhan • Barcelona • Belgrade • Manchester • Tokyo • Cluj • Tianjin

Editors

Carlo Roselli
Università degli Studi del Sannio
Italy

Maurizio Sasso
Università degli Studi del Sannio
Italy

Editorial Office
MDPI
St. Alban-Anlage 66
4052 Basel, Switzerland

This is a reprint of articles from the Special Issue published online in the open access journal *Energies* (ISSN 1996-1073) (available at: https://www.mdpi.com/journal/energies/special_issues/GEUT2020).

For citation purposes, cite each article independently as indicated on the article page online and as indicated below:

LastName, A.A.; LastName, B.B.; LastName, C.C. Article Title. *Journal Name* **Year**, *Volume Number*, Page Range.

ISBN 978-3-0365-0704-0 (Hbk)
ISBN 978-3-0365-0705-7 (PDF)

Contents

About the Editors

Carlo Roselli is an associate professor in Engineering Thermodynamics and Energetics at the Department of Engineering of the Università degli Studi del Sannio (Benevento, Italia), who received his M.Sc. in Mechanical Engineering at the University of Naples Federico II in Energy Engineering and his Ph.D. in Energy Engineering at the Università degli Studi del Sannio. His main interests are focused on the dynamic simulations of energy systems, energy efficiency, microcogeneration, micropolygeneration, ground source heat pump, desiccant cooling, geothermal energy, energy efficiency, renewable energy as well as solar heating and cooling. He has been involved in different national and international research projects. Among them, two have been promoted by the International Energy Agency (Annex 42, Annex 54), and another one (GEOGRID: Innovative Technologies and Systems for the Sustainable Use of Geothermal Energy), especially focused on the geothermal sector, has been supported by a local Italian Institution (Regione Campania). He has co-authored more than 110 papers (international journals, chapters' books, conference proceedings) (https://orcid.org/0000-0002-4423-8147).

Maurizio Sasso is a full professor of Engineering Thermodynamics and Renewable Energy Technology at the Department of Engineering of the Università degli Studi del Sannio (Benevento, Italia), who received his M.Sc. in Mechanical Engineering at the University of Naples Federico II. He received a research grant from ENI (ENERGETICA '87) on the topic: "Study of the applicative possibility and design of a medium-power geothermal convector" 1987–1988. He has been involved in different national and international research projects. Among them, two have been promoted by the International Energy Agency (Annex 42, Annex 54 as a subtask B leader). He was a member of a research group supporting Campania Region for developing the 2001 Regional Energy Plan (PER). His activity has been focused on the guidelines, planning and programming of geothermal energy in the Campania region. He had the scientific responsibility of a research group involved in the project "GEOGRID: Innovative technologies and systems for the sustainable use of geothermal energy", PO FESR 2014-2020. His main interests are focused on: applied thermodynamics, thermoeconomic analysis and the optimization of thermal systems, energy efficiency, microcogeneration, micropolygeneration, ground source heat pump, gas engine-driven heat pump, desiccant-based HVAC systems and geothermal energy. He has co-authored more than 190 scientific papers and two books (https://www.unisannio.it/it/user/396/curriculum).

Preface to "Geothermal Energy Utilization and Technologies 2020"

The increasing interest in renewable energy sources is leading the geothermal sector towards playing an important role in meeting the final energy demand of different countries. This can help reduce dependency on energy imports and can vary the energy mix. Possible applications can vary from small scale (residential) to large scale (city). Geothermal energy, unlike other renewable sources, has the significant benefit of being a programmable source, therefore it is not dependent on the time of day or on particular weather conditions.

Geothermal energy could be considered for direct use in space heating and cooling, greenhouse heating, aquaculture, bathing, district heating networks and industrial uses. Where specific conditions are available (high-temperature hydrothermal resources, aquifer systems medium temperatures, and hot dry rock), geothermal energy could be converted into electricity by means of different technologies (flash steam, dry steam, binary). The cogeneration and polygeneration systems could be also considered to exploit geothermal sources. A smart energy community being supported by geothermal sources could be a sustainable and interesting option.

In 2017, the global geothermal power generation was 84.8 TWh, while the cumulative capacity reached 14 GW. The capacity is expected to reach 17 GW by 2023 as estimated by the International Energy Agency. Despite its significant potential, the contribution of geothermal energy to global heating, cooling, and power demand is relatively low. This book aims to advance the contribution towards focusing the attention on technologies to guarantee the faster spread of geothermal energy and on the policies supporting the exploitation of this source.

Carlo Roselli, Maurizio Sasso
Editors

Article

Assessment of Geothermal Resources in the North Jiangsu Basin, East China, Using Monte Carlo Simulation

Yibo Wang [1,2], Lijuan Wang [3], Yang Bai [1,2,4], Zhuting Wang [5], Jie Hu [1,2,4,*], Di Hu [6], Yaqi Wang [1,2,4] and Shengbiao Hu [1,2,4,*]

1 State Key Laboratory of Lithospheric Evolution, Institute of Geology and Geophysics, Chinese Academy of Sciences, Beijing 100029, China; ybwang@mail.iggcas.ac.cn (Y.W.); baiyang16@mails.ucas.ac.cn (Y.B.); wangyaqi@mail.iggcas.ac.cn (Y.W.)
2 Innovation Academy for Earth Science, Chinese Academy of Sciences, Beijing 100864, China
3 Key Laboratory of Earth Fissures Geological Disaster, Ministry of Land and Resources, Geological Survey of Jiangsu Province, Nanjing 210018, China; wang_lijuan@126.com
4 College of Earth and Planetary Sciences, University of Chinese Academy of Sciences, Beijing 100864, China
5 School of Mines, China University of Mining and Technology, Xuzhou 221116, China; wangzhuting123@163.com
6 Key Laboratory of Exploration Technologies for Oil and Gas Resources, Ministry of Education, Yangtze University, Wuhan 434023, China; cugdeehus@gmail.com
* Correspondence: hujie161@mails.ucas.ac.cn or hujie802@163.com (J.H.); sbhu@mail.iggcas.ac.cn (S.H.); Tel.: +86-010-8299-8519 (J.H.); +86-010-8299-8533 (S.H.)

Abstract: Geothermal energy has been recognized as an important clean renewable energy. Accurate assessment of geothermal resources is an essential foundation for their development and utilization. The North Jiangsu Basin (NJB), located in the Lower Yangtze Craton, is shaped like a wedge block of an ancient plate boundary and large-scale carbonate thermal reservoirs are developed in the deep NJB. moreover, the NJB exhibits a high heat flow background because of its extensive extension since the Late mesozoic. In this study, we used the Monte Carlo method to evaluate the geothermal resources of the main reservoir shallower than 10 km in the NJB. Compared with the volumetric method, the Monte Carlo method takes into account the variation mode and uncertainties of the input parameters. The simulation results show that the geothermal resources of the sandstone thermal reservoir in the shallow NJB are very rich, with capacities of $(6.6–12) \times 10^{20}$ J (mean 8.6×10^{20} J), $(5.1–16) \times 10^{20}$ J (mean 9.1×10^{20} J), and $(3.2–11) \times 10^{20}$ J (mean 6.6×10^{20} J) for the Yancheng, Sanduo and Dai'nan sandstone reservoir, respectively. In addition, the capacity of the geothermal resource of the carbonate thermal reservoir in the deep NJB is far greater than the former, reaching $(9.9–15) \times 10^{21}$ J (mean 12×10^{21} J). The results indicate capacities of a range value of $(1.2–1.7) \times 10^{21}$ J (mean 1.4×10^{22} J) for the whole NJB (<10 km).

Keywords: geothermal resource; Monte Carlo simulation; assessment; thermal reservoir; North Jiangsu Basin

Citation: Wang, Y.; Wang, L.; Bai, Y.; Wang, Z.; Hu, J.; Hu, D.; Wang, Y.; Hu, S. Assessment of Geothermal Resources in the North Jiangsu Basin, East China, Using Monte Carlo Simulation. *Energies* **2021**, *14*, 259. https://doi.org/10.3390/en14020259

Received: 4 December 2020
Accepted: 29 December 2020
Published: 6 January 2021

Publisher's Note: MDPI stays neutral with regard to jurisdictional claims in published maps and institutional affiliations.

1. Introduction

In the past two centuries, the extensive use of non-renewable energy (e.g., nuclear energy, coal, and so on) in the world has caused many environmental disasters [1–3]. Among them, the consumption of fossil fuels has led to a sharp increase in the concentration of greenhouse gases in the atmosphere, which has led to an increase in global warming. Therefore, green and renewable energy has been highly regarded by more and more countries. Among those types of energies, geothermal energy is now receiving wide attention because of its unique characteristics: wide distribution, resource-richeness, safety and stability, cleanliness and low-carbon nature.

In the four decades since China's reform and opening up, rapid economic growth and the use of fossil fuels have made China the world's largest carbon emitter (more than

100 million tons of CO_2 per year). The direct use of geothermal energy has increased rapidly with the increasing demand for energy efficiency in buildings and the reduction of CO_2 emissions, and has maintained its position as the world's number one in recent years [1–3]. The North Jiangsu Basin is located in Jiangsu Province, the most developed province in (Eastern) China, and is a typical 'hot in summer and cold in winter' region. Existing detailed geological data, abundant geothermal resources [4], and an important economic basis are an important basis for the efficient utilization of geothermal resources in the North Jiangsu Basin and, at the same time, provide an important possible model for the future use of geothermal resources in the Yangtze River Delta.

The assessment of geothermal resources is the estimation of the amount of geothermal energy which might be extracted from the inner Earth and economically used in the future. A regional resource assessment can, on the one hand, provide a framework for the government or industry with a long-term energy strategy and policy, and on the other hand, help us to ensure rational planning and exploitation of geothermal resources. A geothermal resources assessment has been carried out using the volumetric method in the NJB [5–7] and the Jiangsu Province [8–11]. However, there are currently three main problems: (1) the low amount of geological and geothermal data constrains the accurate geothermal resources assessment; (2) carbonate thermal reservoirs are currently the most important and favorable reservoirs for geothermal resources, yet they have not attracted enough attention from researchers; (3) the volumetric method does not take into account the uncertainty of parameters involved, but rather assigns a specific value to the thermal storage geometry/physical property parameters.

This paper updates and summarizes high-quality temperature profiles measured in geothermal wells, oil wells, and national hydrological observation wells representative of the major structural units in the NJB. We established the whole density and porosity column of the strata (or rocks) in different geologic ages in the NJB. Specific heats of representative rocks from different thermal reservoirs were also determined. Based on these parameters, we evaluated the uncertainties in parameters such as reservoir area and temperature through monte Carlo simulations and finally calculated the geothermal resource potential of the NJB (less than 10 km). We focused our study on considering parameter uncertainties (Monte Carlo method) and the evaluation of geothermal resources in deep carbonate thermal reservoirs. This research can provide reliable primary data for further study of the geothermal resource planning and development, long-term energy strategy policy, environmental protection, etc.

2. Geologic Setting

The NJB, situated in the northeastern Lower Yangtze Craton, is a mesozoic-Cenozoic sedimentary basin. The basin is bounded by the Su'nan and the Sulu uplifts to the south and north, respectively, and bounded to the west by the Tancheng-Lujiang Fault Zone, with an area of approximately 35,000 km^2. Controlled by NE-NNE faults, the NJB can be divided into the Jianhu uplift, the northern Yanfu depression, and the southern Dongtai depression, and the two depressions contain 22 highs and sags (Figure 1).

2.1. Tectonic Evolution and Stratigraphy

The basement of the NJB mainly consists of three parts, namely metamorphic rocks in the Proterozoic, the Yangtze marine clastic and carbonates in the Paleozoic-Mesozoic, and volcanic rocks forming in the middle Triassic to Early Cretaceous (Qiu et al., 2006). Carbonate rocks in the NJB are mainly found in the Upper Ordovician-Ordovician (Huangxu Formation, Dengying Formation, Hetang Formation, mufushan Formation, Paotaishan Formation, Guanyinshan Formation, Lunshan Formation, Honghuayuan Formation, Dawan Formation, and Tangtou Formation), Carboniferous (Jinling Formation, Laohudong Formation, Huanglong Formation, and Chuanshan Formation) and Permian (Qixia Formation, Gufeng Formation, Longtan Formation, and Dalong Formation) strata. Extensive drilling and geophysical data show that the sedimentary thickness since Paleozoic exceeds 11,000 m, with Cenozoic strata over 7000 m thick (Figure 2).

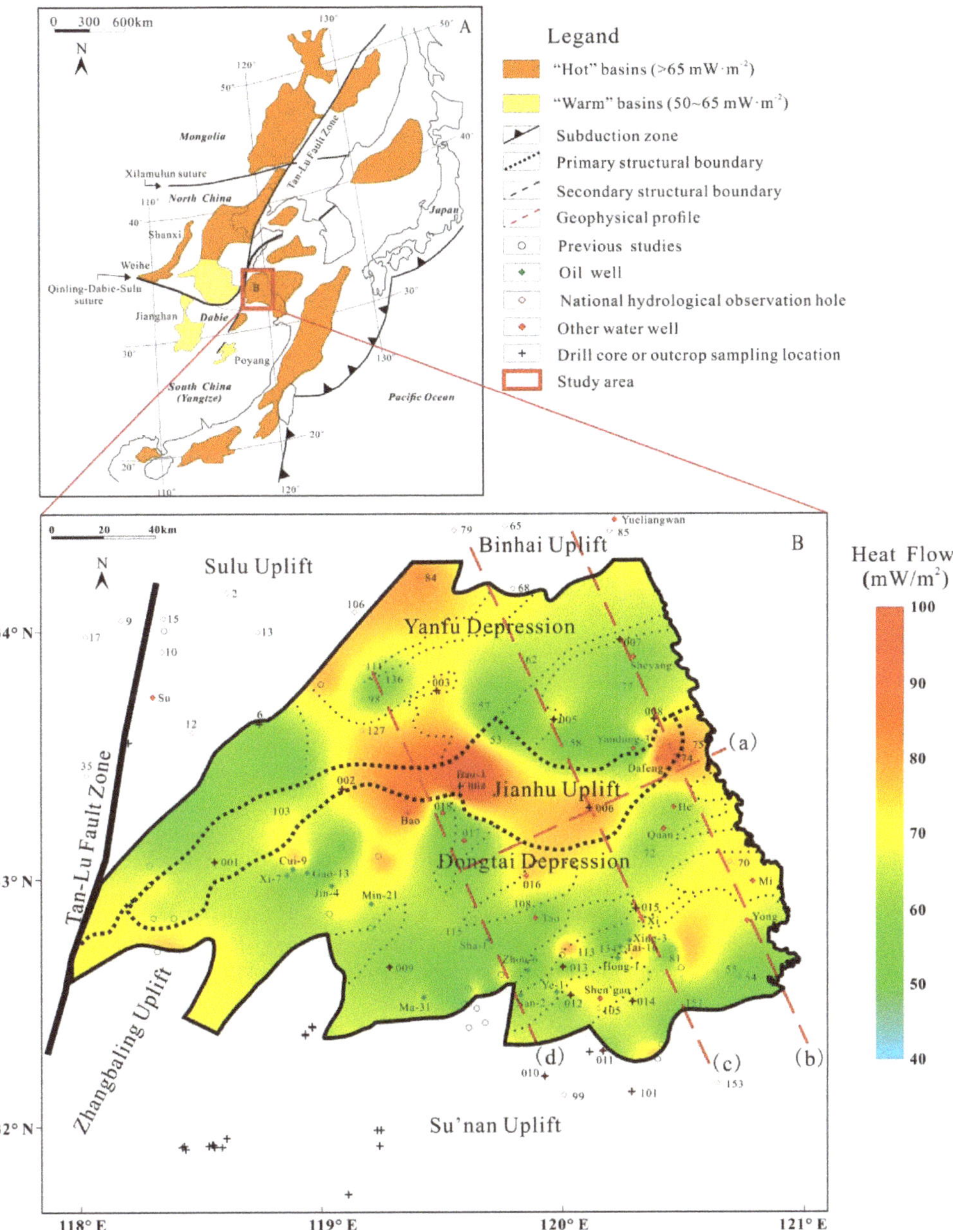

Figure 1. Geothermal background in the North Jiangsu Basin. (**A**): Schematic geological map of East Asia (modified after Grimmer, et al. [12]). (**B**): Schematic geological structural map of the North Jiangsu Basin (modified after Wang, et al. [4]).

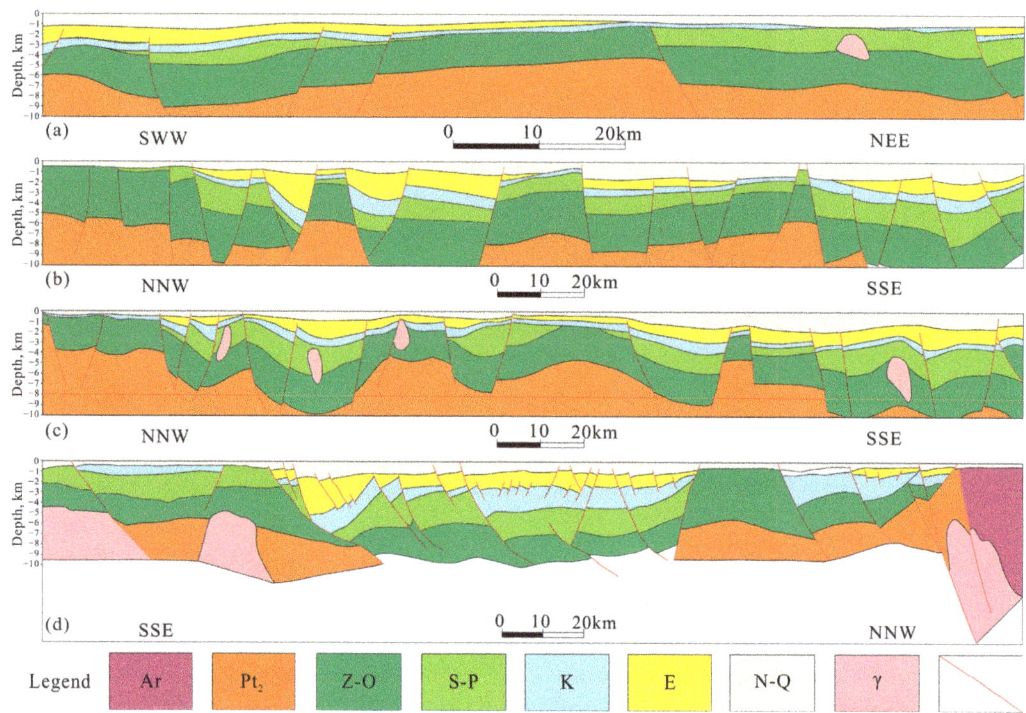

Figure 2. Geological interpretation sections in the North Jiangsu Basin ((**a,c**) are modified from Wang, et al. [4] (**b**) is revised from oil company reports, (**d**) is modified from Chen [13], Qiao, et al. [14]). The four sections are shown in Figure 1B. Ar, Archean; Pt$_2$, mesoproterozoic; Z-O, Ediacaran to Ordovician; S-P, Silurian to Permian; K, Cretaceous; E, Paleogene; N-Q, Neogene to Quaternary; γ, magmatic rock.

In the Triassic period, the NJB was uplifted due to the collision between North China Craton and Yangtze Craton, and the possible southward squeeze of the Siberian plate [15–17]. The Lower Yangtze craton thus entered the terrestrial sedimentary environment. Since then, the NJB has been primarily characterized by three tectonic evolution episodes: (1) Yizheng movement with fluvial facies and lacustrine facies, resulting in the Cretaceous Taizhou Formation and the Paleocene Fu'ning Formation; (2) Wubao movement with lake delta fluvial-lacustrine facies, forming the Eocene Dai'nan Formation and Sanduo Formation; and (3) Sanduo movement with the Neocene-Quaternary fluvial facies [18–21].

2.2. Geothermal Background

Thermal history research in the Lower Yangtze Craton suggests that the last tectono-thermal event in the NJB probably occurred in the early Palaeogene and that back-arc extension caused by the subduction of the Izanagi plate is likely to be the main reason. The thermal history inversions from the Yancan-1 and An-1 wells show that the peak heat flow value could be around 85 mW·m^{-2}, after which it gradually decreased and stabilized [22]. The present-day high mantle heat flow (35–43 mW·m^{-2}) indicates that the higher geothermal background in the NJB is probably mainly influenced by mantle activity [23].

A heat flow map is an important representation of the geothermal background of a structural unit and is an important basis for regional geothermal resources assessment. Based on previous work [4], we updated and produced an up-to-date heat flow map of the NJB: a total of 78 high-quality heat flow values from within the basin and 165 published heat flow data from around the basin were used (Figure 1). The distribution density of heat flow sites and the quality of heat flow data in the NJB are extremely high, both nationally and globally. The measured minimum and maximum heat flow values in the NJB are 46 mW·m^{-2} (Yandong-3) and 110 mW·m^{-2} (site 004), respectively, with an average value of 67 mW·m^{-2}, which is higher than the mean heat flow in Continental China and line with the average global continental value [24,25]. As shown in Figure 1, the middle and eastern Jianhu Uplift, the northern Jinhu sag, and the eastern Dongtai sag in the NJB are relatively high heat flow regions (>70 mW·m^{-2}). The Jianhu Uplift and Huai'an high have the highest geothermal heat flow, with an average geothermal value of over 72 mW·m^{-2}, followed by the Sujiazui high and Xiaohai high, with average geothermal values of 71 mW·m^{-2}. In general, heat flow values in the Jianghu uplift are higher than those in the depressions, and the heat flow in some highs of the depressions is also higher, with some even exceeding that in the Jianhu Uplift.

2.3. Types of Geothermal Reservoir

Thermal reservoirs in the NJB contains two primary types: Cenozoic sandstone thermal reservoirs, and Ediacaran-Paleozoic carbonate thermal reservoirs. Cenozoic thermal reservoirs can be divided vertically by lithology, porosity, temperature, etc., into the Palaeogene Dai'nan Formation (E_2d), the Sanduo Formation (E_3s), and the Neogene Yancheng Formation (Ny). The Dai'nan reservoir with its depth varying from 270 to 2800 m is deeper than the Sanduo reservoir (65–1900 m) and Yancheng reservoir (45–1700 m) and therefore has a higher temperature. The large porosity (>20%) of the Cenozoic sandstone thermal reservoirs results in good water-rich conditions and is an important basis for the efficient use of geothermal resources. The distribution of the Cenozoic sandstone thermal reservoirs in the NJB is shown in Figure 3.

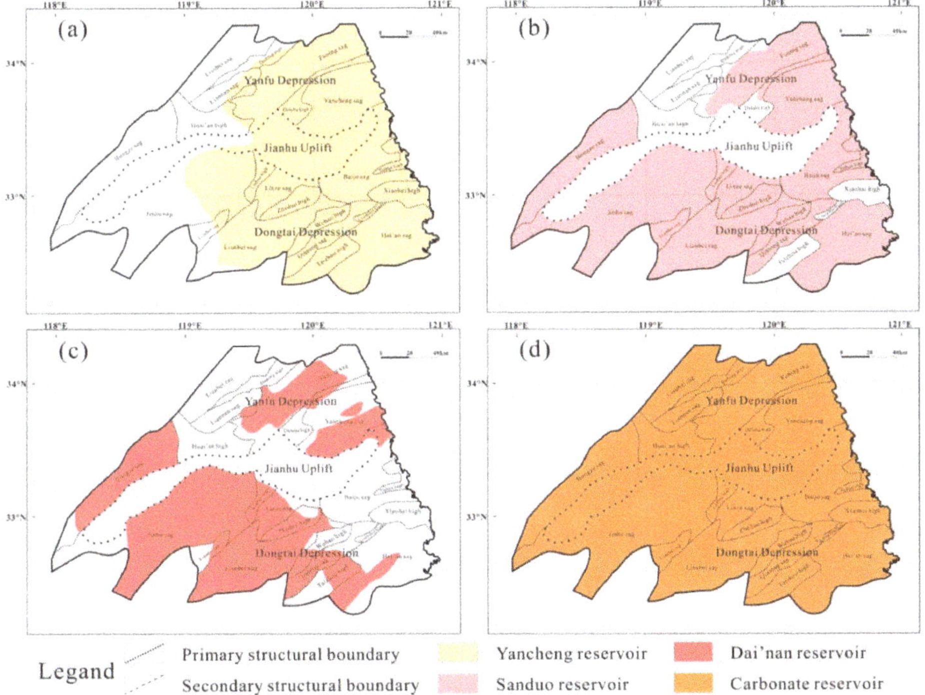

Figure 3. Distribution of Cenozoic sandstone thermal reservoirs and carbonate thermal reservoirs in the North Jiangsu Basin. (**a**) Yancheng reservoir; (**b**) Sanduo reservoir; (**c**) Dai'nan reservoir; (**d**) Carbonate reservoir.

Numerous geophysical profiles and sedimentological evidence show that the NJB has a large-scale carbonate thermal reservoir at depth, characterized by a wide range and large scale [4,13,14,26–28]. Carbonate rocks in the NJB are much thicker and hotter than the Cenozoic sandstone thermal reservoirs, suggesting a very rich geothermal resource potential. The carbonate reservoirs in the NJB act as medium-high temperature reservoirs and provide optimum conditions for geothermal resource development. Due to the development of possible karst fissures inside the thermal reservoirs, which provide good access for groundwater activity, the heat energy from the deep can be rapidly transferred in a convective manner to the bottom of the sedimentary cover, giving the thermal reservoirs a high temperature. In addition, the unique water-rich nature and permeability of the carbonate karst zone contribute to the efficient extraction and recharge of groundwater. Although there is some variation in the stratigraphic thickness of the Ediacaran-Ordovician rocks, carbonate rocks develop at depth throughout the North Jiangsu Basin, as shown in Figure 3. Considering the accuracy of the profiles and the difficulty of extracting geothermal resources, we only evaluated carbonate thermal reserves above 10 km.

3. Methodology

3.1. Assessment method

There are several methods (e.g., surface heat flux method, heat storage modeling, statistical analysis, analogy method) available for geothermal resources assessment, the most widely used being the volumetric method proposed by muffler and Cataldi [29]. The volumetric method calculates the total geothermal energy in the fluids in the rock masses and pores of a study area, i.e., the geothermal energy accumulation. We use the volumetric method in combination with the monte Carlo method to minimize parameter uncertainty. If the porosity of the reservoir and the thermophysical properties of the water and rock are known, then the total geothermal resources can be calculated by the following three equations (Equations (1)–(3)).

$$Q_{Total} = Q_R + Q_W \tag{1}$$

$$Q_R = A \, D \, (1 - \varphi) \, \rho_R \, c_R \, (T - T_0) \tag{2}$$

$$Q_W = A \, D \, \varphi \, \rho_W \, c_W \, (T - T_0) \tag{3}$$

where Q_{Total}, Q_W, and Q_R represent the total geothermal energy and that stored in pore water and rock, J; A (m^2) and D (m) are the area and thickness of the thermal reservoirs; ρ_R and ρ_W denote the density of the geothermal reservoir and geothermal water, kg·m^{-3}; c_R and c_W are the specific heat of the geothermal reservoir rock and geothermal water, J·kg^{-1}·°C^{-1}; φ is rock porosity, %; T and T_0 represent the average temperature of geothermal reservoir and surface, °C.

3.2. Monte Carlo Simulation

The monte Carlo simulation is a numerical simulation method in which probabilistic phenomena are studied as objects. It is a calculation method for inferring unknown quantities of characteristics by taking statistical values from sample surveys. The main steps involved in carrying out a geothermal resource evaluation using the monte Carlo method are: (1) Defining the input parameter set; (2) Setting different distribution models for the input parameters; (3) Defining the relationship between input and output parameters according to the mathematical model; (4) Defining the output parameters; (5) Setting the number of iterations; (6) Simulating and analyzing the results of the calculations and provide a probability distribution function for each result.

In the monte Carlo simulation process, different distribution models can be selected for the input parameters, e.g., uniform, pert, triangular, and lognormal distributions, with each model having its unique frequency distribution. Taking into account the range and pattern of variation of the different parameters, we give different distribution models for the input parameters. The distribution of the measured porosity in the formations suggests that the triangular distribution may be more consistent with the actual porosity pattern; two-dimensional profile and deep borehole data constrain the variation of the thermal reservoir thickness as the triangular distribution; the overall linear increasing of the temperature in thermal reservoirs indicates that a triangular distribution of reservoir temperature should be chosen; pert distribution models were chosen by the specific heat and rock density based on the skewness of distribution characteristics and continuity of parameter variation. Taking the Jinhu sag as an example, the following models are chosen for the input parameters (Figure 4): the annual average temperature (T_0) and water density (ρ_W) are given a constant uniform model; the geothermal reservoir area (A) is set to a range of uniform distribution model; the variation of geothermal reservoir temperature (T), porosity (φ) and thickness (D) are given a triangular distribution model; for the specific heat of the geothermal reservoir rock (c_R) and water (c_W), and the rock density (ρ_R), the pert distribution model is chosen. Taking into account the accuracy of the simulation results, the number of iterations in this study was set to be 10,000.

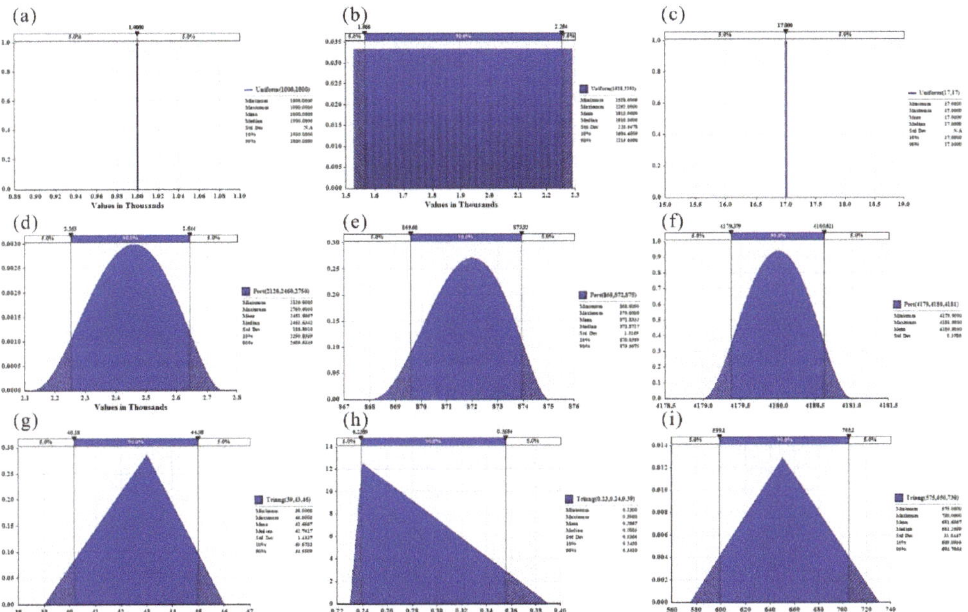

Figure 4. Distribution models of input parameters in the Jinhu sag ((**a**) water density; (**b**) the geothermal reservoir area; (**c**) the annual average temperature; (**d**) the geothermal reservoir rock density; (**e**) the specific heat of rocks; (**f**) the specific heat of water; (**g**) geothermal reservoir temperature; (**h**) the geothermal reservoir rock porosity; (**i**) the thickness of the geothermal reservoir)).

4. Database

4.1. Temperature Logs

The determination of parameter ranges and distribution models is key for the assessment of geothermal resources. Temperature is one of the most important parameters for geothermal resources assessment and steady-state temperature measurement in boreholes is the most direct and effective way to obtain the true temperature in deeper formations. From 2018 to 2019, temperature profiles of 99 boreholes were acquired [4]. The temperature and depth data of the boreholes was measured using a consecutive logging system of a 5000 m long cable and a Platinum thermal resistance sensor. This field work is the first systematic steady-state temperature measurements in the whole NJB, with the same instruments and researchers. To date, we have obtained high-quality temperature logs from a total of 110 boreholes (Figure 1) and selected 28 representative temperature logs of each sub-structural unit in the NJB for display, as shown in Figure 5.

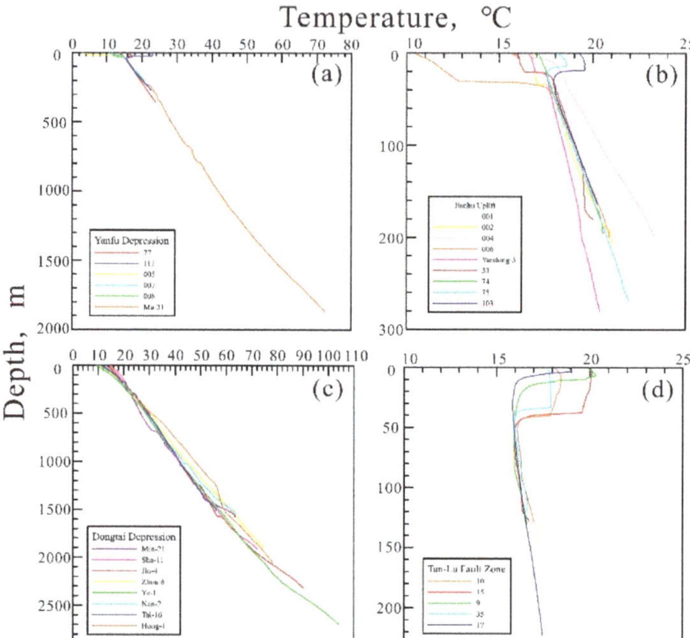

Figure 5. The representative temperature-depth profiles for each sub-structural unit in the North Jiangsu Basi (**a**) Yanfu Depression; (**b**) Jianhu Uplift; (**c**) Dongtai Depression; (**d**) Tan-Lu Fault Zone. (modified from Wang, et al. [4].

The water level of each borehole (generally less than 40 m) can be determined according to the temperature logs (Figure 5). The temperature increases (nearly) linearly with depth more than 50 m, showing a conductive type [30,31]. Based on the depth and temperature data, the temperature gradients (the portion above (including) the Yancheng Formation) of each sub-structural unit in the NJB were calculated. Regionally, the middle-eastern Jianhu Uplift and the eastern Dongtai Depression is a zone of the high temperature gradient, with the average temperature gradient value higher than 35 °C·km^{-1}. Vertically, the temperature gradient in the NJB increases with depth, and the temperature gradients of Ny and E$_2$s-E$_2$d are generally lower than that of E$_1$f-K$_2$t. Formations of the Neogene and above are characterized by a low temperature gradient, generally between 15–30 °C·km^{-1}, which may be the result of heat redistribution and groundwater activity in the shallow. The high sand content in the Ny, E$_2$s, and E$_2$d formations, resulting in high thermal conductivity and low temperature gradients; whereas, the Palaeocene and Cretaceous formations have a relatively high mud content and thus a high temperature gradient of 30–40 °C·km^{-1}.

4.2. Thermal Conductivity

Thermalphysical properties, particularly thermal conductivity measurement, are fundamental to the study of regional deep geothermal field. We carried out thermal conductivity test on 264 samples (Table 1; sample locations, see Figure 1). moreover, we collected the thermal conductivity data tested by Wang and Shi, and Wang et al. [32,33] (Table 1). According to the data above, we established the thermal conductivity column of different formations (Erathem) in the NJB (Table 1).

Table 1. Thermal conductivity of the formations in the North Jiangsu Basin.

Period	Formation	Abbreviation	Thermal Conductivity (This Work)		Thermal Conductivity (Previous Work *)		Thermal Conductivity $(W \cdot m^{-1} \cdot K^{-1})$	Number
			$W \cdot m^{-1} \cdot K^{-1}$	Number	$W \cdot m^{-1} \cdot K^{-1}$	Number		
Q	Dongtai	Q	1.6	87			1.6	87
N	Yancheng	N	1.7	16			1.7	16
E	Sanduo	E_3s	1.4	6	1.5	5	1.5	11
	Dai'nan	E_2d	2.4	3	2.4	7	2.4	10
	Fu'ning	E_4f			1.9	5	1.9	5
		E_3f	2.6	4	2.6	4	2.6	8
		E_2f	2.4	5	2.3	3	2.4	8
		E_1f	2.5	2	2.5	6	2.5	8
K	Taizhou	K_2t	3.0	1	2.4	4	2.6	5
	Chishan	K_2c			1.5	6	1.5	6
	Pukou	K_2p	2.3	8	2.4	10	2.4	18
		K_1			2.7	2	2.7	2
J		J	4.4	4	2.2	1	3.9	5
T		T	2.8	2	3.1	21	3.1	23
P	Dalong	P_2d			2.9	10	2.9	10
	Longtan	P_2l	2.8	10	2.5	6	2.7	16
	Qixia	P_1q	3.3	3	3.2	20	3.2	23
C	Chuanshan	C_3c	3.3	3	3.3	6	3.3	9
	Huanglong	C_2h	2.8	2	3.4	11	3.4	13
	Hezhou	C_1h	3.2	2	3.2	2	3.2	4
	Laohudong	C_1l			5.7	3	5.7	3
	Gaolishan	C_1g	4.0	3	3.8	3	3.9	6
D	Wutong	D_3w	3.7	5	5.2	6	4.5	11
		D_{1+2}			4.1	4	4.1	4
S	maoshan	S_2m^1	8.0	1			8.0	1
	Fentou	S_2f	3.9	8	4.2	10	4.1	18
	Gaojiabian	S_1g	3.3	17	3.3	3	3.3	20
O	-	O	3.9	1	3.6	17	3.6	18
$\in$	Guanyintai	$\in_3g$	3.3	2			3.3	2
	Paotaishan	$\in_3p$			4.7	13	4.7	13
	mufushan	$\in_3m$	4.6	2			4.6	2
Z	Dengying	Z_2d	6.1	6			6.1	6
	Huangxu	Z_2h	2.6	4			2.6	4
	Liantuo	Z_1l			3.8	1	3.8	1
Pt	Jinping	Pt_3j	3.2	39			3.2	39
	Pichengyan	Pt_2p	3.4	18			3.4	18

* Previous work is from Wang and Shi, and Wang et al. [32,33].

The results show that the average thermal conductivity of the strata in the region fluctuates considerably, with the smallest value being the clay and sand of the Quaternary with a thermal conductivity of 0.6 $W \cdot m^{-1} \cdot K^{-1}$ and the largest value being the Silurian $S_2 m^1$ with a thermal conductivity of 8.0 $W \cdot m^{-1} \cdot K^{-1}$. In general, the thermal conductivity shows a gradual decrease with the strata from old to new. The thermal conductivity of the Palaeozoic strata is generally high, greater than 3.0 $W \cdot m^{-1} \cdot K^{-1}$; the mesozoic strata have the second highest thermal conductivity, mostly 2.0–3.0 $W \cdot m^{-1} \cdot K^{-1}$; and the Cenozoic strata have the lowest thermal conductivity, between 1.5 and 2.5 $W \cdot m^{-1} \cdot K^{-1}$. The stratigraphic thermal conductivity of the NJB can be roughly divided into two parts, namely the shallow low thermal conductivity section and the deep high thermal conductivity section. The shallow Cretaceous-Quaternary thermal conductivity is low (<2.7 $W \cdot m^{-1} \cdot K^{-1}$), especially the Late Paleozoic-Quaternary thermal conductivity is mostly less than 2.0 $W \cdot m^{-1} \cdot K^{-1}$, which can provide a good cover for heat preservation. The thermal conductivity of the deeper Ediacaran-Jurassic is relatively high, generally greater than 3.0 $W \cdot m^{-1} \cdot K^{-1}$, especially the thermal conductivity of the Silurian-Devonian and Ediacaran-Cambrian can be more than 4.0 $W \cdot m^{-1} \cdot K^{-1}$, and the thermal conductivity of the Dengying Formation can be more than 6.0 $W \cdot m^{-1} \cdot K^{-1}$, all of which can be used as good thermal storage layers.

4.3. Input Parameters

The temperature range of the geothermal reservoirs is an important parameter for geothermal resource assessment. In combination with the geophysical profiles, we used Equation (4) to calculate the temperature range corresponding to each thermal reservoir at depth (z), based on the average heat flow (Q_s) in the various sub-structural units in the NJB. The temperature at a given depth for heat flow in each sub-structural unit in the NJB with a constant surface heat flow Q_s, surface temperature T_0, thermal conductivity λ, and radiogenic heat production A is defined as

$$T(z) = T_0 + z \, Q_s / \lambda - z^2 \, A / (2\lambda) \tag{4}$$

where λ ($W \cdot m^{-1} \cdot K^{-1}$) and A ($\mu W \cdot m^{-3}$) represent the measured thermal conductivity and heat production of the rock, respectively.

The main parameters, such as porosity and heat production, are measured data, and detailed explanations are given in Wang, et al. [4]. In addition, we carried out measurements for specific heat, porosity, and density for the main lithologies of each geothermal reservoirs, other parameters were mainly set regarding the literature: the annual average temperature of each sub-structural units [34] and water density (1000 $kg \cdot m^{-3}$) were taken as constant values; the specific heat of the water in the geothermal reservoir varied with temperature [35]; the variation in thickness and area of the geothermal reservoir was mainly based on actual geophysical profiles and sedimentological evidence [4,13,14,28,36]. The parameters for the Yancheng Formation, the Sanduo Formation, the Dai'nai Formation, and the carbonate thermal reservoirs are reported in Tables 2–5, respectively (the most likely values are shown in parentheses).

Table 2. Parameters for the Yancheng Formation thermal reservoir.

The Tectonic Units	Area (km²)	Thickness (m)	The Temperature of the Thermal Reservoir (°C)	Porosity (%)	The Density of the Rock (kg·m⁻³)	Specific Heat of Rock (J·kg⁻¹·°C⁻¹)	Water Density (kg·m⁻³)	Specific Heat of the Water (J·kg⁻¹·°C⁻¹)	The Annual Average Temperature (°C)
Hongze sag	960~1400 (1200)	30~240 (140)	20~27 (23)	23~39 (24)	2100~2800 (2500)	790~820 (790)	1000	4200	15
Liannan sag	240~360 (300)	100~300 (190)	20~29 (24)	23~39 (24)	2100~2800 (2500)	790~830 (800)	1000	4200	16

Table 2. *Cont.*

The Tectonic Units	Area (km^2)	Thickness (m)	The Temperature of the Thermal Reservoir (°C)	Porosity (%)	The Density of the Rock (kg·m^{-3})	Specific Heat of Rock (J·kg^{-1}·°C^{-1})	Water Density (kg·m^{-3})	Specific Heat of the Water (J·kg^{-1}·°C^{-1})	The Annual Average Temperature (°C)
Lianbei sag	240~360 (300)	120~290 (200)	20~27 (24)	23~39 (24)	2100~2800 (2500)	790~820 (800)	1000	4200	15
Fu'ning sag	1700~2200 (2200)	100~470 (320)	20~36 (29)	23~39 (24)	2100~2800 (2500)	790~860 (830)	1000	4200	15
Yancheng sag	1700~2100 (2100)	320~600 (460)	27~38 (33)	23~39 (24)	2100~2800 (2500)	820~860 (850)	1000	4200	15
Huai'an high	800~1000 (1000)	45~120 (85)	19~23 (21)	23~39 (24)	2100~2800 (2500)	780~790 (790)	1000	4200	17
Dadong high	40~60 (50)	90~140 (120)	19~21 (20)	23~39 (24)	2100~2800 (2500)	780~790 (790)	1000	4200	15
Dalaba high	320~480 (400)	340~400 (370)	29~32 (30)	23~39 (24)	2100~2800 (2500)	730~850 (840)	1000	4200	16
Jianhu Uplift	2100~3200 (2600)	190~600 (300)	25~43 (30)	23~39 (24)	2100~2800 (2500)	800~870 (840)	1000	4200	17
Jinhu sag	1500~2300 (1900)	580~730 (650)	39~46 (43)	23~39 (24)	2100~2800 (2500)	870~880 (870)	1000	4200	17
Linze sag	280~350 (350)	250~450 (350)	27~50 (38)	23~39 (24)	2100~2800 (2500)	820~880 (860)	1000	4200	17
Gaoyou sag	2100~2700 (2600)	500~740 (620)	43~73 (58)	23~39 (24)	2100~2800 (2500)	870~900 (880)	1000	4200	17
Baiju sag	1300~1600 (1600)	700~960 (830)	44~55 (49)	23~39 (24)	2100~2800 (2500)	870~880 (880)	1000	4200	16
Qintong sag	910~1100 (1100)	390~610 (500)	30~61 (46)	23~39 (24)	2100~2800 (2500)	840~890 (880)	1000	4200	17
Hai'an sag	3000~3800 (3800)	550~1000 (810)	39~57 (50)	23~39 (24)	2100~2800 (2500)	870~890 (880)	1000	4200	17
Lingtangqiao high	120~180 (150)	370~620 (500)	31~40 (36)	23~39 (24)	2100~2800 (2500)	840~870 (860)	1000	4200	17
Liubao high	240~300 (300)	320~490 (405)	27~52 (40)	23~39 (24)	2100~2800 (2500)	820~880 (870)	1000	4200	17
Zheduo high	720~900 (1200)	390~600 (500)	30~64 (47)	23~39 (24)	2100~2800 (2500)	840~890 (880)	1000	4200	17
Wubao high	320~400 (400)	580~650 (610)	38~67 (53)	23~39 (24)	2100~2800 (2500)	860~890 (880)	1000	4200	17
Taizhou high	560~730 (700)	390~450 (420)	28~49 (39)	23~39 (24)	2100~2800 (2500)	820~880 (870)	1000	4200	17
Yuhua high	240~300 (300)	630~740 (680)	42~47 (44)	23~39 (24)	2100~2800 (2500)	870~880 (870)	1000	4200	17
Xiaohai high	500~620 (620)	310~730 (520)	30~49 (40)	23~39 (24)	2100~2800 (2500)	840~880 (870)	1000	4200	17

Table 3. Parameters for the Sanduo Formation thermal reservoir.

The Tectonic Units	Area (km^2)	Thickness (m)	The Temperature of the Thermal Reservoir (°C)	Porosity(%)	The Density of the Rock (kg·m^{-3})	Specific Heat of Rock (J·kg^{-1}·°C^{-1})	Water Density (kg·m^{-3})	Specific Heat of the Water (J·kg^{-1}·°C^{-1})	The Annual Average Temperature (°C)
Hongze sag	1400~1800 (1800)	250~980 (610)	38~72 (55)	23~28 (27)	2300~2800 (2500)	870~900 (880)	1000	4200	15
Lianbei sag	240~360 (300)	400~600 (500)	43~53 (48)	23~28 (27)	2300~2800 (2500)	870~880 (880)	1000	4200	15
Fu'ning sag	750~1130 (940)	0~500 (260)	25~50 (43)	23~28 (27)	2300~2800 (2500)	840~880 (870)	1000	4200	15
Yancheng sag	1600~2100 (1900)	360~400 (380)	48~51 (50)	23~28 (27)	2300~2800 (2500)	870~880 (880)	1000	4200	15
Jinhu sag	4000~5000 (5000)	300~1100 (500)	36~87 (72)	23~28 (27)	2300~2800 (2500)	860~920 (900)	1000	4200	17
Linze sag	240~350 (300)	0~350 (100)	50~55 (52)	23~28 (27)	2300~2800 (2500)	880~890 (880)	1000	4200	17
Gaoyou sag	160~280 (250)	160~280 (250)	70~89 (80)	23~28 (27)	2300~2800 (2500)	900~920 (910)	1000	4200	17
Baiju sag	230~340 (280)	0~350 (100)	55~57 (56)	23~28 (27)	2300~2800 (2500)	880~890 (890)	1000	4200	16
Qintong sag	910~1100 (1100)	100~350 (300)	49~74 (60)	23~28 (27)	2300~2800 (2500)	880~910 (890)	1000	4200	17
Hai'an sag	3000~3800 (3800)	50~800 (200)	59~63 (61)	23~28 (27)	2300~2800 (2500)	890	1000	4200	17
Lingtangqiao high	360~450 (450)	150~450 (250)	42~47 (45)	23~28 (27)	2300~2800 (2500)	870~880 (870)	1000	4200	17
Liubao high	240~300 (300)	0~300 (100)	42~54 (45)	23~28 (27)	2300~2800 (2500)	870~880 (870)	1000	4200	17
Zheduo high	680~900 (850)	50~300 (100)	55~65 (58)	23~28 (27)	2300~2800 (2500)	880~890 (890)	1000	4200	17
Wubao high	320~400 (400)	200~280 (240)	62~81 (72)	23~28 (27)	2300~2800 (2500)	890~910 (900)	1000	4200	17

Table 4. Parameters for the Dai'nan Formation thermal reservoir.

The Tectonic Units	Area (km^2)	Thickness (m)	The Temperature of the Thermal Reservoir (°C)	Porosity (%)	The Density of the Rock (kg·m^{-3})	Specific Heat of Rock (J·kg^{-1}·°C^{-1})	Water Density (kg·m^{-3})	Specific Heat of the Water (J·kg^{-1}·°C^{-1})	The Annual Average Temperature (°C)
Hongze sag	1400~1800 (1800)	250~540 (390)	63~74 (67)	16~29 (24)	2700~2800 (2700)	890~910 (900)	1000	4200	15
Lianbei sag	200~300 (250)	200~400 (300)	54~61 (58)	16~29 (24)	2700~2800 (2700)	880~890 (890)	1000	4200	15
Fu'ning sag	200~500 (400)	0~340 (270)	45~55 (52)	16~29 (24)	2700~2800 (2700)	870~880 (880)	1000	4200	15
Yancheng sag	280~420 (350)	250~470 (360)	57~63 (60)	16~29 (24)	2700~2800 (2700)	890	1000	4200	15
Jinhu sag	3300~4200 (4000)	100~600 (350)	106~120 (113)	16~29 (24)	2700~2800 (2700)	950~990 (970)	1000	4200	17
Gaoyou sag	1800~2200 (2600)	190~250 (220)	74~90 (81)	16~29 (24)	2700~2800 (2700)	910~930 (910)	1000	4200	17
Qintong sag	340~430 (520)	0~330 (250)	60~80 (67)	16~29 (24)	2700~2800 (2700)	890~900 (900)	1000	4200	17
Hai'an sag	190~290 (240)	50~300 (100)	60~66 (63)	16~29 (24)	2700~2800 (2700)	890~900 (890)	1000	4200	17
Wubao high	40~60 (50)	150~200 (170)	70~78 (74)	16~29 (24)	2700~2800 (2700)	900~910 (910)	1000	4200	17

Table 5. Parameters for the carbonate thermal reservoir.

The Tectonic Units	Area (km^2)	Thickness (m)	The Temperature of the Thermal Reservoir ($^\circ$C)	Porosity (%)	The Density of the Rock (kg·m^{-3})	Specific Heat of Rock (J·kg^{-1}·$^\circ$C^{-1})	Water Density (kg·m^{-3})	Specific Heat of the Water (J·kg^{-1}·$^\circ$C^{-1})	The Annual Average Temperature ($^\circ$C)
Hongze sag	900~1400 (1100)	1000~2000 (1500)	130~170 (160)	0.5~6 (2)	2700~2900 (2800)	950~970 (960)	1000	4300	15
Liannan sag	550~880 (660)	1000~1900 (1500)	92~120 (110)	0.5~6 (2)	2700~2900 (2800)	920~940 (930)	1000	4200	16
Lianbei sag	400~640 (480)	1400~1900 (1700)	110~140 (120)	0.5~6 (2)	2700~2900 (2800)	930~950 (940)	1000	4200	15
Fu'ning sag	1100~1700 (1300)	900~2400 (1800)	76~220 (160)	0.5~6 (2)	2700~2900 (2800)	900~990 (960)	1000	4300	15
Yancheng sag	1100~1700 (1300)	1500~2500 (2000)	170~210 (190)	0.5~6 (2)	2700~2900 (2800)	970~980 (980)	1000	4400	15
Huai'an high	500~800 (600)	620~1600 (1100)	62~96 (79)	0.5~6 (2)	2700~2900 (2800)	890~920 (910)	1000	4200	17
Dadong high	130~200 (150)	1800~2200 (2000)	54~99 (77)	0.5~6 (2)	2700~2900 (2800)	880~920 (900)	1000	4200	15
Dalaba high	200~320 (240)	2300~2500 (2400)	62~100 (83)	0.5~6 (2)	2700~2900 (2800)	890~930 (910)	1000	4200	16
Jianhu Uplift	2200~3500 (2600)	1000~2100 (1600)	25~160 (94)	0.5~6 (2)	2700~2900 (2800)	840~960 (920)	1000	4200	17
Jinhu sag	2500~4000 (3000)	1500~1800 (1700)	140~180 (160)	0.5~6 (2)	2700~2900 (2800)	950~970 (960)	1000	4300	17
Linze sag	180~280 (210)	1700~2700 (2200)	100~160 (130)	0.5~6 (2)	2700~2900 (2800)	930~970 (950)	1000	4300	17
Gaoyou sag	1300~2100 (1600)	1400~1800 (1600)	130~160 (140)	0.5~6 (2)	2700~2900 (2800)	940~970 (960)	1000	4300	17
Baiju sag	800~1300 (960)	1500~1800 (1700)	130~180 (150)	0.5~6 (2)	2700~2900 (2800)	950~970 (960)	1000	4300	16
Qintong sag	570~910 (680)	1400~2300 (1900)	100~180 (140)	0.5~6 (2)	2700~2900 (2800)	930~970 (960)	1000	4300	17
Hai'an sag	1900~3000 (2300)	1600~2100 (1900)	180~230 (200)	0.5~6 (2)	2700~2900 (2800)	970~990 (980)	1000	4500	17
Lingtangqiao high	230~360 (270)	1000~1800 (1500)	120~170 (150)	0.5~6 (2)	2700~2900 (2800)	940~970 (960)	1000	4300	17
Liubao high	150~240 (180)	1600~2300 (2000)	88~140 (110)	0.5~6 (2)	2700~2900 (2800)	910~950 (930)	1000	4200	17
Zheduo high	450~720 (540)	760~1700 (1200)	84~120 (100)	0.5~6 (2)	2700~2900 (2800)	910~940 (920)	1000	4200	17
Wubao high	200~320 (240)	1300~1500 (1400)	110~150 (130)	0.5~6 (2)	2700~2900 (2800)	930~960 (950)	1000	4300	17
Taizhou high	370~590 (440)	1300~1500 (1400)	81~120 (100)	0.5~6 (2)	2700~2900 (2800)	910~940 (930)	1000	4200	17
Yuhua high	150~240 (180)	1600~2300 (1900)	130~170 (150)	0.5~6 (2)	2700~2900 (2800)	950~970 (960)	1000	4300	17
Xiaohai high	310~500 (370)	1500~1700 (1600)	92~140 (120)	0.5~6 (2)	2700~2900 (2800)	920~950 (940)	1000	4200	17

5. Results

5.1. Simulation of the Cenozoic Sandstone Thermal Reservoirs

According to the parameters in Table 2, we used the monte Carlo method to calculate the geothermal resource base for the thermal reservoir of the Neogene Yancheng Formation in the NJB. Figure 6a indicates the geothermal resource base in which the value varies from 6.6×10^{20} J to 1.2×10^{21} J (mean value $(8.6 \pm 0.6) \times 10^{20}$ J), with the highest probability being 0.86×10^{21} J (probability > 8 %) and a 90 % probability of a range of $(0.76–0.96) \times 10^{21}$ J. It shows a geothermal resource potential of 3.5×10^{16} J·km^{-2} by dividing the average value by the Yancheng thermal reservoir area.

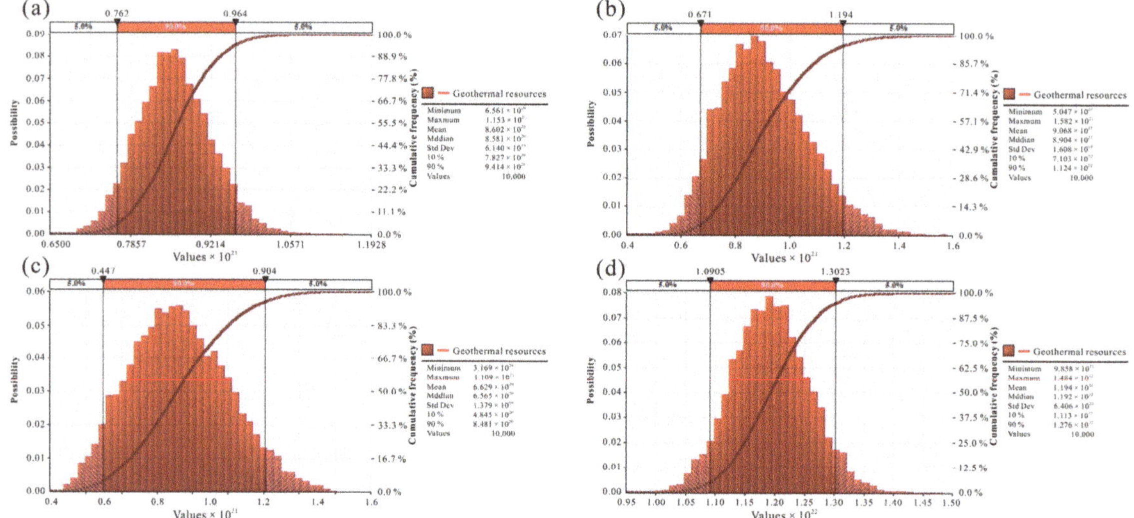

Figure 6. Simulation results for the primary thermal reservoirs in the North Jiangsu Basin ((**a**) Yancheng thermal reservoir; (**b**) Sanduo thermal reservoir; (**c**) Dai'nan thermal reservoir; (**d**) carbonate thermal reservoir).

Figure 6b shows the potential of the geothermal resource in which the value changes from 5.0×10^{20} J to 1.6×10^{21} J with a mean value of $(9.1 \pm 0.2) \times 10^{20}$ J), with nearly 7% probability of the most likely value of 0.86×10^{21} J, and a probability of < 5% that the geothermal resources > 1.2×10^{21} J or < 0.67×10^{21} J. Dividing by area, it indicates a geothermal potential of 4.5×10^{16} J·km^{-2}.

Figure 6c presents a geothermal resource potential for the Dai'nan thermal reservoir of 3.2×10^{20} J to 1.1×10^{21} J with a mean value of $(6.6 \pm 0.1) \times 10^{20}$ J), with a 90 % probability that the energy potential is $(0.45–0.90) \times 10^{21}$ J. Considering the geothermal resource potential per unit of thermal storage area, it indicates a potential of 6.7×10^{16} J·km^{-2}.

5.2. Simulation of the Carbonate Thermal Reservoirs

Figure 6d indicates that the geothermal resource potential for the carbonate thermal reservoir sums to 9.9×10^{21} J to 1.5×10^{22} J with a mean value of $(1.2 \pm 0.1) \times 10^{22}$ J). From Figure 6d, we can get that the geothermal potential is above 1.3×10^{22} J or below 1.1×10^{22} J in which the probability is less than 5 %, and the most likely value of the whole is 1.2×10^{22} J (nearly 8%). The geothermal resource potential per unit of the thermal storage area is 3.7×10^{17} J·km^{-2}, which is much larger than that of the Cenozoic sandstone thermal reservoirs. The geothermal resource base of the carbonate thermal reservoir in the NJB is extremely large, indicating a great potential for geothermal resources.

5.3. Monte Carlo Simulation Results for the Total Geothermal Resources in the North Jiangsu Basin

The total geothermal resource base of the Cenozoic sandstone and carbonate thermal reservoirs were estimated through running the monte Carlo simulation for iterations, with the results being shown in Figure 7. It indicates the geothermal resource base in which the value ranges from 1.2×10^{22} J to 1.7×10^{22} J (mean value $(1.4 \pm 0.1) \times 10^{22}$ J), with the highest probability being $1.4–1.5 \times 10^{22}$ J (probability nearly 7%), and a 90 % probability of a range of $(1.3–1.6) \times 10^{22}$ J. The Yancheng, Sanduo, Dai'nan, and carbonate thermal reservoirs account for about 6%, 6%, 5%, and 83% of the total geothermal resource base, respectively.

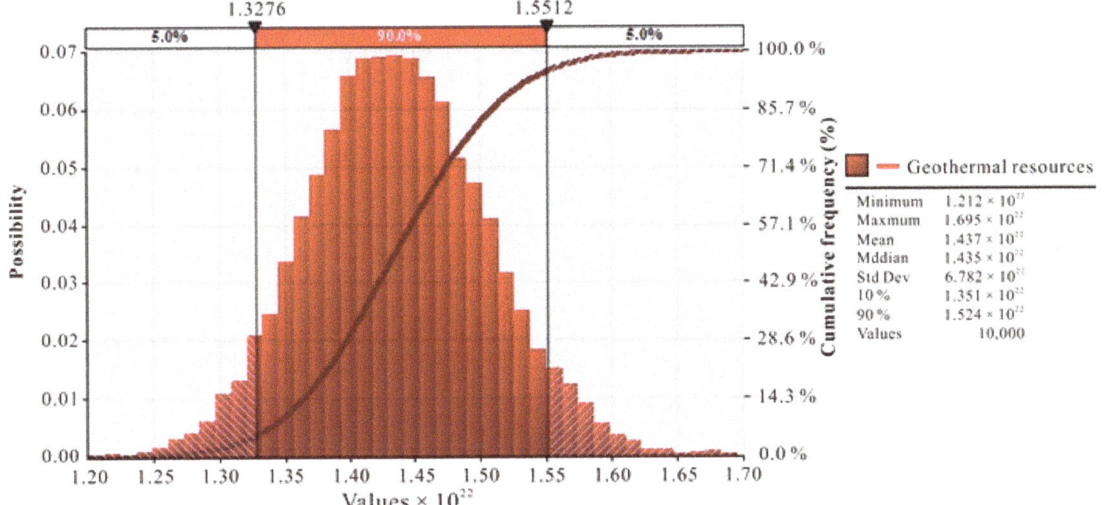

Figure 7. Geothermal resource potential simulation results for the sum of the four thermal reservoirs.

6. Conclusions

This study used the monte Carlo method to perform the detailed assessment of geothermal resources in the North Jiangsu Basin. Geothermal resources within the Cenozoic sandstone and carbonate (<10 km) thermal reservoirs total $(1.2–1.7) \times 10^{22}$ J, with a mean value of $(1.4 \pm 0.1) \times 10^{22}$ J and a most likely value of 1.4×10^{22} J, which is equivalent to 4.9×10^{13} tons of standard coal heat content. moreover, the most promising carbonate thermal reservoir owns an average geothermal resource potential of $(1.2 \pm 0.1) \times 10^{22}$ J, about five times that of the Cenozoic thermal reservoirs. The abundant geothermal energy of the North Jiangsu Basin may alleviate energy shortages to a certain extent and promote exemplary geothermal heating in the 'hot in summer and cold in winter' region of the Yangtze River Delta.

Author Contributions: Conceptualization, Y.W. (Yibo Wang) and Y.B.; methodology, Z.W.; software, J.H.; validation, J.H. and S.H.; formal analysis, Y.W. (Yibo Wang) and Y.W. (Yaqi Wang); investigation, Y.W. (Yibo Wang), Z.W., D.H. and Y.B.; data curation, L.W. and Z.W.; writing—original draft preparation, Y.W. (Yibo Wang); writing—review and editing, Y.B.; visualization, Z.W.; supervision, S.H.; project administration, S.H.; funding acquisition, L.W. All authors have read and agreed to the published version of the manuscript.

Funding: This research was accomplished under the support of the Jiangsu Fund Project (Special Study on the Investigation of Geothermal Field and Crustal Thermal Structure in the North Jiangsu Basin, JITC-1902AW2919; JITC-1802AW2292).

Institutional Review Board Statement: Not applicable.

Informed Consent Statement: Not applicable.

Data Availability Statement: The data presented in this study are available on request from the corresponding author. The data are not publicly available due to [Privacy].

Acknowledgments: We thank the Geological Survey of Jiangsu Province very much, who provided enormous amount of help on the borehole selection and sample collection that are significant for our temperature logging work and geothermal resources assessment.

Conflicts of Interest: The authors declare no conflict of interest.

References

1. Lund, J.W.; Boyd, T.L. Direct Utilization of Geothermal Energy 2015 Worldwide Review. *Geothermics* **2016**, *60*, 66–93. [CrossRef]
2. Lund, J.W.; Freestod, D.H.; Boyd, T.L. Direct Utilization of Geothermal Energy. *Geothermics* **2011**, *40*, 159–180. [CrossRef]
3. Lund, J.W.; Toth, A.N. Direct utilization of geothermal energy 2020 worldwide review. *Geothermics* **2021**, *90*, 101915. [CrossRef]
4. Wang, Y.; Hu, D.; Wang, L.; Guan, J.; Bai, Y.; Wang, Z.; Jiang, G.; Hu, J.; Tang, B.; Zhu, C.; et al. The present-day geothermal regime of the North Jiangsu Basin, East China. *Geothermics* **2020**, *88*, 101829. [CrossRef]
5. Min, W.; Yu, Y.; Lu, Y.; Gu, X. Assessment and zoning of geothermal resources in the northern Jiangsu Basin. *Shanghai Land Resour.* **2015**, *36*, 90–94.
6. Wang, S.; Hu, J.; Yan, J.; Li, F.; Chen, N.; Tang, Q.; Guo, B.; Zhan, L. Assessment of Geothermal Resources in Petroliferous Basins in China. *Math. Geol.* **2019**, *51*, 271–293. [CrossRef]
7. Zhang, W.; Wang, G.; Liu, F.; Xing, L.; Li, m. Characteristics of geothermal resources in sedimentary basins. *Geol. China* **2019**, *46*, 255–268.
8. Wang, Y.; Liu, Y.; Dou, J.; Li, m.; Zeng, m. Geothermal energy in China: Status, challenges, and policy recommendations. *Util. Policy* **2020**, *64*, 101020. [CrossRef]
9. Hou, J.; Cao, m.; Liu, P. Development and utilization of geothermal energy in China: Current practices and future strategies. *Renew. Energy* **2018**, *125*, 401–412. [CrossRef]
10. Xu, L.; Wang, L.; Yang, Q. An Estimation of Hot-Dry-Rock (HDR) Resources in Jiangsu Province. *Geol. J. China Univ.* **2014**, *20*, 464–469.
11. Zhu, J.; Hu, K.; Lu, X.; Huang, X.; Liu, K.; Wu, X. A review of geothermal energy resources, development, and applications in China: Current status and prospects. *Energy* **2015**, *93*, 466–483. [CrossRef]
12. Grimmer, J.C.; Jonckheere, R.; Enkelmann, E.; Ratschbacher, L.; Hacker, B.R.; Blythe, A.E.; Wagner, G.A.; Wu, Q.; Liu, S.; Dong, S. Cretaceous−Cenozoic history of the southern Tan-Lu fault zone: Apatite fission-track and structural constraints from the Dabie Shan (eastern China). *Tectonophysics* **2002**, *359*, 225–253. [CrossRef]
13. Chen, A. Tectonic features of the Subei Basin and the forming mechanism of its dustpan-shaped fault depression. *Oil Gas Geol.* **2010**, *31*, 140–150.
14. Qiao, X.; Li, G.; Ming, L.; Wang, Z. CO_2 storage capacity assessment of deep saline aquifers in the Subei Basin, East China. *Int. J. Greenh. Gas Control* **2012**, *11*, 52–63. [CrossRef]
15. Guo, X.; Encarnacion, J.; Xu, X.; Deino, A.; Li, Z.; Tian, X. Collision and rotation of the South China block and their role in the formation and exhumation of ultrahigh pressure rocks in the Dabie Shan orogen. *Terra Nova* **2012**, *24*, 339–350. [CrossRef]
16. Yin, A.; Nie, S. An indentation model for the North and South China collision and the development of the Tan-Lu and Honam Fault Systems, eastern Asia. *Tectonics* **1993**, *12*, 801–813. [CrossRef]
17. Zhang, R.Y.; Liou, J.G.; Ernst, W.G. The Dabie–Sulu continental collision zone: A comprehensive review. *Gondwana Res.* **2009**, *16*, 1–26. [CrossRef]
18. Chen, A. Dynamic mechanism of formation of dustpan subsidence, Northern Jiangsu. *Geol. J. China Univ.* **2001**, *07*, 408–418.
19. Lu, Y.; Xu, Y.; Liu, M.; Yan, M.; Xu, S.; Lu, W. Tectonic features and evolution of the north margin of Northern Jiangsu Basin revealed by seismic reflection profile. *J. Geol.* **2017**, *41*, 624–630.
20. Qian, J. Oil and gas fields formation and distribution of Subei Basin-research compared to Bohai Bay Basin. *Acta Pet. Sin.* **2001**, *22*, 12–16.
21. Qiu, H.; Xu, Z.; Qiao, D. Progress in the study of the tectonic evolution of the Subei basin, Jiangsu, China. *Geol. Bull. China* **2006**, *Z2*, 1117–1120.
22. Zeng, P. *The Application of the Thermometric Indicators to the Study of Thermal Evolution in the Lower-Yangtze Region*; China University of Geosciences: Beijing, China, 2005.
23. Wang, Y. *Differences of Thermal Regime of the Cratons in Eastern China and Discussion of Its Deep Dynamic mechanism*; University of Chinese Academy of Sciences: Beijing, China, 2020.
24. Jiang, G.; Hu, S.; Shi, Y.; Zhang, C.; Wang, Z.; Hu, D. Terrestrial heat flow of continental China: Updated dataset and tectonic implications. *Tectonophysics* **2019**, *753*, 36–48. [CrossRef]
25. Lucazeau, F. Analysis and mapping of an Updated Terrestrial Heat Flow Data Set. *Geochem. Geophys. Geosyst.* **2019**, *20*, 1–24. [CrossRef]
26. Zeng, P. Comprehensive Interpretation of Large Section of G78 Area and Structural Features of Subei Basin. *J. Oil Gas Technol.* **2007**, *29*, 82–86.
27. Fang, C.; Huang, Z.; Teng, L.; Xu, F.; Zhou, D.; Yin, Q.; Shao, W.S. Lithofacies palaeogeography of the Late Ordovician Kaitian Stage-the early Silurian Rhuddanian Stage in Lower Yangtze region and its petroleum geological significance. *Geol. China* **2020**, *47*, 144–160.
28. Zhao, T. Structural Characteristics of Marine Mesozoic and Paleozoic and Prediction of Distribution of Their Main Assemblage in the Northern Jiangsu Area. Master's Thesis, China University of Petroleum, Qingdao, China, 2017.
29. Muffler, P.; Cataldi, R. Methods for regional assessment of geothermal resources. *Geothermics* **1977**, *7*, 53–89. [CrossRef]
30. Hu, S.; He, L.; Wang, J. Heat flow in the continental area of China: A new data set. *Earth Planet. Sci. Lett.* **2000**, *179*, 407–419. [CrossRef]

31. Pasquale, V.; Verdoya, M.; Chiozzi, P. *Geothermics: Heat Flow in the Lithosphere*; Springer: Berlin/Heidelberg, Germany, 2014; pp. 15–49.
32. Wang, L.; Li, C.; Shi, Y.; Wang, Y. Distribution of geotemperature and terrestrial heat flow density in lower Yangtze area. *Chin. J. Geophys.* **1995**, *38*, 469–476.
33. Wang, L.; Shi, Y. *Geothermal Study on the Oil and Gas Basin*; Nanjing University Press: Nanjing, China, 1989.
34. Liu, J.Y.; Zhuang, D.F.; Luo, D.; Xiao, X. Land-cover classification of China: Integrated analysis of AVHRR imagery and geophysical data. *Int. J. Remote Sens.* **2003**, *24*, 2485–2500. [CrossRef]
35. Engineering ToolBox. Water-Specific Heat. 2004. Available online: https://www.engineeringtoolbox.com/specific-heat-capacity-water-d_660.html (accessed on 30 December 2004).
36. Yang, L. The Structural Evolution in the Cenozoic of Subei Basin and Its Relationship with Oil and Gas. Master's Thesis, Yangtze University, Wuhan, China, 2015.

Article

Techno-Economic Assessment of Mobilized Thermal Energy Storage System Using Geothermal Source in Polish Conditions

Dominika Matuszewska [1], Marta Kuta [1] and Piotr Olczak [2,*]

[1] Department of Energy and Fuels, AGH University of Science and Technology, 30 Mickiewicza Ave., 30-059 Cracow, Poland; dommat@agh.edu.pl (D.M.); marta.kuta@agh.edu.pl (P.O.)
[2] Mineral and Energy Economy Research Institute, Polish Academy of Sciences, 7A Wybickiego St., 31-261 Cracow, Poland
* Correspondence: olczak@min-pan.krakow.pl

Received: 21 May 2020; Accepted: 28 June 2020; Published: 2 July 2020

Abstract: The paper considers technical and economic possibilities to provide geothermal heat to individual recipients using a mobile thermal storage system (M-TES) in Polish conditions. The heat availability, temperature and heat cost influence the choice of location—Bańska Niżna, near Zakopane in the southern part of the Poland. The indirect contact energy storage container was selected with phase change material characterized by a melting temperature of 70 °C and a heat storage capacity of 250 kJ/kg, in the amount of 800 kg. The economic profitability of the M-TES system (with a price per warehouse of 6000 EUR, i.e., a total of 12,000 EUR—two containers are needed) can be achieved for a heat demand of 5000 kWh/year with the price of a replaced heat source at the level of 0.21 EUR/kWh and a distance between the charging station and building (heat recipient) of 0.5 km. For the heat demand of 15,000 kWh/year, the price for the replaced heat reached EUR 0.11/kWh, and the same distance. In turn, for a demand of 25,000 kWh/year, the price of the replaced heat source reached 0.085 EUR/kWh. The distance significantly affected the economic profitability of the M-TES system—for the analyzed case, a distance around 3–4 km from the heat source should be considered.

Keywords: geothermal energy; mobile thermal energy storage (M-TES); phase change material (PCM); LCOH; heat transport; renewable energy source

---

1. Introduction

For several years, a global energy transformation due to the growing awareness of greenhouse gas (GHG) emissions and fossil fuel depletion has been observed. In this context, the use of renewable energy sources, including geothermal energy, became one of the main strategies of this transformation. Geothermal energy is produced mainly as a result of the decay of potassium, uranium and thorium radioisotopes [1] in the planetary interior as thermal energy [2]. Compared with other renewable energy sources, it has many advantages, among others, large reserves of geothermal sources, wide distribution and scope of applications, good stability, and high utilization efficiency of use. Additionally, once stabilized, they can be used for a long period of time [3,4]. In addition, geothermal sources are the only renewable energy source that is not affected by solar radiation and gravitational attraction between the sun and the moon [5]. However, the main problem with geothermal resources is that the speed of geothermal reservoir exploitations can be faster than heat replacement, and nowadays this is considered as the challenge connected with geothermal heat utilization. It largely depends on such factors as geological time scale, geothermal applications and used-heat reinjection methods [6].

The great potential of geothermal resources used for electric power generation has lately been recognized more and more [7]. However, geothermal heat has been widely used directly for centuries by humans and animals, and can be found in various regions of the world [8]. Heat appears on the earth's surface in various forms that affect the way it is used. The wide spectrum of geofluid temperature determines the way it is used, and the range of applications as a temperature function is shown in Figure 1, which is based on the Lindal diagram [9,10]. Currently, geothermal heat can be utilized in two forms: direct use of hot springs for bathing, aquaculture, geothermal heating and indirect in electric power generation [11,12]. Global geothermal power capacity by the end of 2019 totaled 13.93 GW, with annual electricity generation reaching 85.98 TWh in 2017 (most recent data) [13]. Geothermal power plants can be categorized by the technologies exploiting geothermal resources, according to power plant configuration: dry steam plant (23%), single flash (41%), double flash (19%), binary (14%) and advanced geothermal energy conversion systems such as tripe flash, hybrid flash-binary systems, hybrid fossil-geothermal systems, hybrid back pressure system (3%) [14,15]. A two-group classification (binary cycles for lower well enthalpies and steam cycles for higher enthalpies) has been suggested by Valdimarsson [16].

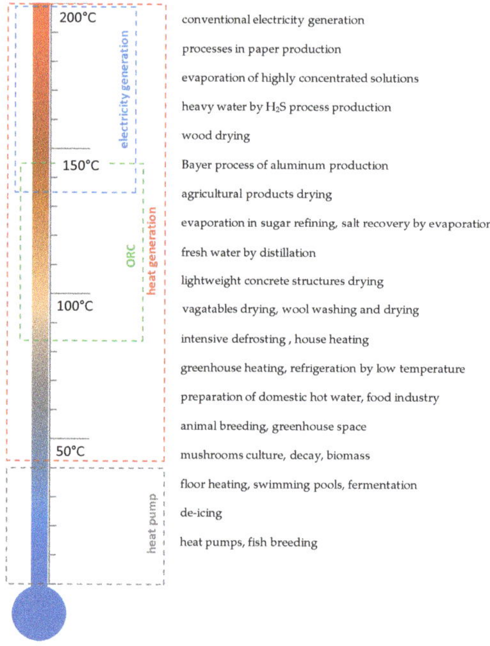

Figure 1. Modified Lindal diagram. Source: own study, based on [9,10].

The geothermal heat can be used directly in many applications depending on a temperature range (mostly used reservoirs with temperature between 20 °C and 150 °C) [17,18]. In 2018, the direct utilization of geothermal energy was identified and reported for 82 countries with an estimated global installed thermal capacity at 70.33 GW$_{th}$ (with 0.265 capacity factor and a growth rate 7.7% annually) [19,20]. Lund and Boyd [21] show that heat energy utilization reaches 163,287 GWh and identify nine utilization categories: geothermal heat pump (90,297 GWh), bathing and swimming (33,147 GWh), space heating (24,493 GWh), greenhouse heating (7347 GWh), aquaculture pond heating (3266 GWh), industrial uses (2939 GWh), cooling and snow melting (653 GWh), agricultural drying (653 GWh) and others (490 GWh). As it can be seen from these categories, the most popular are geothermal heat pumps, which can be used in many applications under different conditions [22,23]. Figure 1 shows that the direct use (DU) of geothermal energy varies in range from high-temperature industrial applications to very low-temperature aquaculture. Additionally, in some ranges, the geothermal heat utilized in DU can not only meet thermal requirements for some applications, but also can be used in combined heat and power configurations (CHP) [24]. Matuszewska and Olczak [25] show the modified binary system to provide heat and electricity simultaneously from low-enthalpy geothermal reservoirs (normally used for providing district heating in Polish conditions). Carotenuto et al. [26] analyze novel solar–geothermal low-temperature district heating and cooling and domestic hot water systems. The case study has been developed for the district area of Monterusciello in Southern Italy, where the geothermal source obtains 55 °C. Hepbasli and Canakci [27] show, in the example of geothermal district heating in Izmir-Balcova, that geothermal energy is cheaper than other energy resources in Turkey (including fossil fuels) and thus it can make a contribution in GHG emissions. Sander's [28] study challenges opportunities of geothermal district heating systems based on the analysis of China, Germany, Iceland, and the U.S. country assessments.

One of the ways to deliver geothermal heat to the recipient in a less conventional way than with the help of standard heat pipelines is to use a mobile heat storage, which will be the subject of the following analysis.

Mobilized thermal energy storage systems are based on the use of heat in a place other than its generation, collection, and storage. The basic assumption is storing waste heat, excess heat or heat produced from renewable energy sources in an amount larger than local demand. Heat transported using mobile heat containers can be stored using phase change materials based on the use of latent heat or using zeolites that use thermochemical transformations. The article cites examples of both possibilities; however, the analysis focuses on mobile thermal energy storage (M-TES) based on phase change materials.

The stored heat can be transported by road, rail, water, or a combination of these. Due to the greatest universality, road transport is the most popular. It can take place using trucks (with large sizes of the container) or cars equipped with a trailer (with smaller container sizes). The heat is transported to recipients, where the M-TES is discharged.

Potential heat sources can be selected from: renewable heat sources, including geothermal heat, industrial plants and biogas plants producing large amounts of waste heat. A number of requirements are set for heat sources: the possibility of effective heat collection and transfer to the phase change material (PCM)-container, an appropriate heat temperature enabling material full phase transformation, constant heat availability, an adequate amount of available heat, high thermal efficiency. Additionally, the heat transportation process from heat source to the recipient location must meet the relevant requirements; the distance cannot be too large. Distance depends on the parameters of the PCM-container and the heat demand at the recipient. The literature shows that it is not advisable for the distance to be more than 30 km [29]. The M-TES needs to be thermally protected to minimize heat losses in transport and during unloading. The M-TES system assumes the use of two heat containers working in parallel and enabling constant access to heat. When the first one is unloaded at the recipient location, the second one is loaded and waits for the delivery to the user.

The place of heat reception should be characterized by heat demand in the right range, suitable heat parameters and operating mode. Potential recipients can be buildings like single and multi-family houses, public buildings, and industry (transport within the same industrial plant).

The mobile thermal energy storage system's work cycle is presented in Figure 2.

Due to the fact that the M-TES parameters can be flexibly adapted to the parameters of the heat source and the recipient, it is possible to select the appropriate PCM filling the heat container, where parameters will be adapted to the temperature of the heat source, and at the same time, will also be adapted to the needs of the recipients. At the stage of M-TES system design, parameters such as the frequency of deliveries or the size of the container dedicated to a given solution should be considered. These parameters are connected to the size and seasonality of the heat demand. The heat demand (at every hour) is effected by a number of parameters, such as: building size and its condition (statistical [30] and methodical approach [31]), number of residents [31], building usage hours and characteristics [32], and requirements for maintaining the temperature at a certain level, related to weather conditions [33].

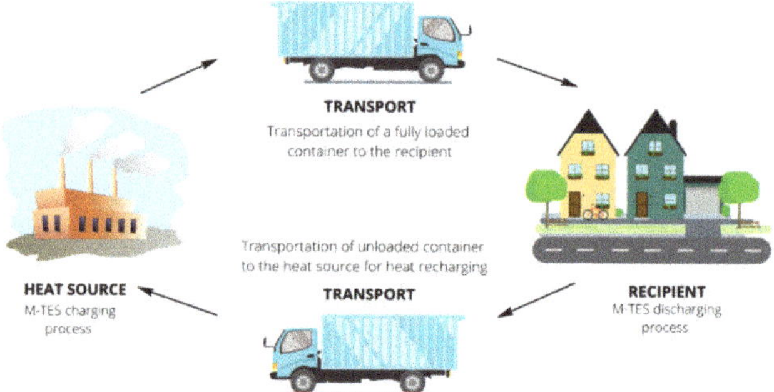

Figure 2. Work cycle of the mobile thermal energy storage system.

In recent years, projects dedicated to mobile heat transport technology have been conducted, as well as the first projects completed with commercial implementations. Selected examples of such solutions are presented below.

Yan Wang et al. [34] designed and tested an M-TES system dedicated to renewable energy and recovered industrial waste heat. The storage unit contained a heat exchanger in the form of compact storage tubes filled with PCM. The system uses 215 kg sodium acetate trihydrate ($CH_3COONa \cdot 3H_2O$), which is characterized by: melting temperature: 331 K, density 1330 kg/m^3, specific heat 3200 J/(kg·K), thermal conductivity 0.8 W/(m·°C) and enthalpy 260 kJ/kg. The tests showed that 1200 s would be enough to melt 215 kg of PCM filling 80 tubes. The tested system has been able to store 125,576 kJ of heat.

Weilong Wang et al. [35] built and tested lab-scale direct and indirect thermal energy storage containers filled with the phase change material erythritol, which is characterized by melting temperature 118 °C and heat density 330 kJ/kg. The system has been dedicated for the supply of heat recovered from the industry. Two options of M-TES were compared: direct- and indirect-contact containers, and it was concluded that the direct-contact M-TES can be charged and discharged faster than the indirect-contact container (for tested systems).

Marco Deckert et al. [36] optimized and tested an M-TES system based on the system commercially available. The system consists of two 20-foot-long tanks filled with sodium acetate trihydrate, which melts at 58 °C. The heat exchanger is built of 24 tubes. To enhance the charging and discharging process, tubes in one part of the storage were extended with graphite structures. The tested M-TES has been able to store up to 2 MWh of the heat, including 1.3 MWh of latent heat. The authors concluded that the technical and economic feasibility depends on the M-TES storage capacity but also on heat demand of the recipients. The tests indicated areas of improvement and demonstrated the possibility of commercial use for M-TES.

Xuelai Zhang et al. [37] proposed an M-TES system in the form of container with dimensions: 4.1 m × 2 m × 1.25 m (inner tank: 3.6 m × 1.8 m × 1.1 m). The container was filled with stainless steel balls (radius of 60 mm, 80 mm and 100 mm) with new phase change material consisting of 0.4% nanocopper + 99.6% erythritol, which is characterized by phase change temperature 118 °C and latent heat 362.2 kJ/kg. The authors concluded that the biggest impact on the results had the temperature of the oil bath (higher speed obtained in lower initial temperature) and the size of the balls (easier melting for smaller diameter, but heat dissipation is faster for larger sizes).

Andreas Krönauer et al. [38] present the results of one-year long tests where the M-TES was charged to 130 °C, using extraction steam from a waste incineration plant. The heat was transported to the recipient located 7 km from the heat source and used in the drying processes. The tank was filled with 14 tons of Zeolite and had a storage capacity of 2.3 MWh. The authors conducted analysis and identified areas for improvement to increase CO_2 reduction and improve efficiency.

This paper focuses on the economical assessment of M-TES using geothermal heat in Polish conditions. There is a lack of information about the profitability of using M-TES for geothermal reservoir utilizations. Additionally, most research focuses on determination of the operating conditions of the geothermal reservoir, and on the amount of energy that can be extracted, but they do not consider the problems with large dispersion of recipients (which may cause an increase in the costs of building a heating network), as well as topography unfavorable for this type of solution. Considering this, the application of M-TES for geothermal heat is justified primarily in places where it is impossible or unprofitable to build a heating network.

In Poland, the temperature of geothermal water does not exceed 100 °C, and most reservoirs have a depth of 3000 m [39]. The most promising and interesting areas from a geothermal point of view are in the southern part of the country, the Podhale Basin. It has sufficient operation conditions for district heating, recreation, balneotherapy and other applications. Additionally, the heating system located there and operated by PEC Geotermia Podhalańska S.A. is currently the largest geothermal installation in Poland with a geothermal heating system capacity ca. 40.8 MW$_{th}$ (the total installed capacity of the heating plant is 80.5 MW$_{th}$) [40]. The analyzed geothermal heating system is among the largest in Europe and is steadily upgraded [41].

The example of PEC Geotermia Podhalańska S.A. was considered in further analysis due to the location in a mountainous area, which may prevent the extension of a heating network, or significantly increase construction costs. In addition, the price of heat generation using geothermal sources is competitive in relation to other sources and amounts to approx. 0.0322 EUR/kWh, at the same time, the cost of building a heating network is high. Due to the large dispersion of recipients, as well as the landform, the price of heat from geothermal energy, after inclusion of the heat network cost, is high. PEC Geotermia Podhalańska S.A. currently has a heating network with a length of over 100 km, in order to connect new customers not located on the existing network, it is necessary to consider the expansion of the network in terms of the economics of the solution. Due to the high dispersion and differences in the foundation height of buildings, the heating network development costs are high. Therefore, based on global examples, it seems worthwhile to consider the use of M-TES. Figure 3 presents that the price for heat from geothermal energy is lowest compared with other geothermal district heating systems in Poland. This means that, if for PEC Geotermia Podhalańska S.A. the solution is economically justified, it can be further considered for facilities with a higher price for geothermal heat.

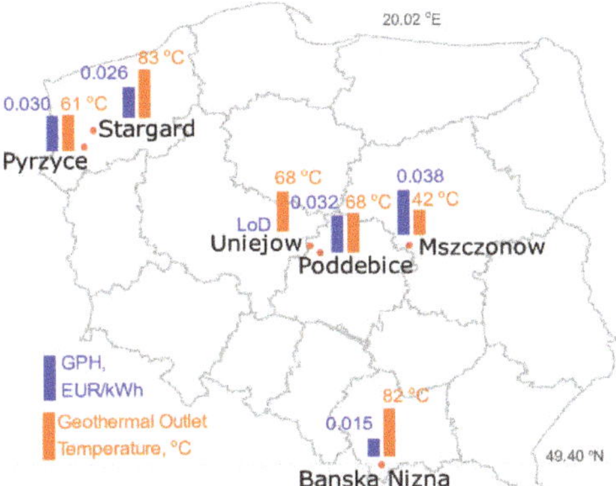

Figure 3. Map of Poland with geothermal sources and heat price (without distribution and charge for power). Source: own study based on [42,43].

Based on the references above, the M-TES for the use of geothermal heat form Bańska Niżna well was analyzed. The economic assessment of these problem was conducted. PEC Geotermia Podhalańska S.A. was selected due to the lowest geothermal heat prices (with no additional fee) and highest geothermal temperature from the production well. There are enough studies about M-TES for geothermal heat utilization, especially in Polish conditions. The novelty of this paper is in analyzing the economical profitability of applying such solutions in mentioned conditions. The paper is structured as follows: Section 2 presents used methods and assumptions; in Section 3, the results are presented; and finally, conclusions are provided in Section 4.

2. Methods and Assumptions

The main purpose of this analysis is to determine the economic profitability of the solution focused on replacing the current heat source in the building with geothermal heat transported by a M-TES. The analysis began by determining the relationship between the M-TES heat capacity and total building heat demand (THD). The next step was to determine the impact of various parameters on Economical Profitability. Finally, the impact of using M-TES transport systems on the reduction of carbon dioxide emission was estimated.

The calculations presented in this article were carried out for M-TES, which uses 800 kg of phase change material for heat storage (maximum amount of PCM that can be safely transported by car with a trailer with a permissible total weight of 3.5 Mg). The material that could potentially be used in this application was selected, taking into account the parameters of the heat source, requirements resulting from the characteristics of the place of heat reception and the phase change materials requirements typical for this type of applications (e.g., appropriate phase transition temperature, high value of heat storage capacity, the highest possible value of the thermal conductivity coefficient, stability of properties after many working cycles, nontoxic, non-harmful for human or environment, nonflammable).

Materials suitable for use in this application include: PureTemp 68 (melting temperature: 68 °C, heat storage capacity: 213 kJ/kg, thermal conductivity liquid/solid: 0.15/0.25 W/(m·°C) [44]; ATP 70 (melting temperature: 70 °C, heat storage capacity: 250 kJ/kg, thermal conductivity 0.2 W/(m·°C) [45]; GAIA HS PCM78 (melting temperature: 78 °C, heat storage capacity: 260 kJ/kg, thermal conductivity liquid/solid: 0.5/0.98 W/(m·°C) [46]; RT 70 HC (melting temperature: 70 °C, heat storage capacity: 260 kJ/kg, thermal conductivity 0.2 W/(m·°C) [47]. For calculation purposes it has been selected PCM characterized by following properties: temperature: 70 °C, heat storage capacity: 250 kJ/kg. The amount of PCM as well as the weight of the entire M-TES was selected in a way which it could be transported by car with a trailer with a permissible total weight of 3.5 Mg. The use of this type of solution was adapted to various transport requirements related to the landform around the heat source (Bańska Nizna).

2.1. Total Heat Demand

The total heat demand (THD) in the range of 5000–25,000 kWh was adopted for the analysis. The limiting values for the above-mentioned range has been determined on the basis of:

1. Heat demand for households using coal-based heat sources. In the first place, these sources should be replaced by renewable energy, due to carbon dioxide and pollution emission to the air [48]. For coal-based heat sources, the average heat demand per m^2 of usable floor space is 222 kWh/(m^2·year) for insulated buildings and 253 kWh/(m^2·year) for non-insulated buildings [30]. The average usable floor space of households (in rural areas) is 108.3 m^2 and the total energy demand is 216 kWh/(m^2·year) [48]. At the same time, the area up to several km from Bańska Niżna is characterized by a large diversity of buildings (from small traditional one- and two-room houses to multi-story residential houses) as well as their energy efficiency. DHW demand was determined on the basis of average consumption of DHW per person 35 dm^3/(person·day) [31] and the number of households in Lesser Poland Voivodeship [49], as well as on the basis of efficiency of heat production and transport for coal sources. The results are shown in Figure 4.

2. The ratio of heat loss to THD (lower limit of THD range)—discussed in detail later in the article.

3. Possibilities for the number of M.TES exchanges during the day—also discussed in detail later in the article.

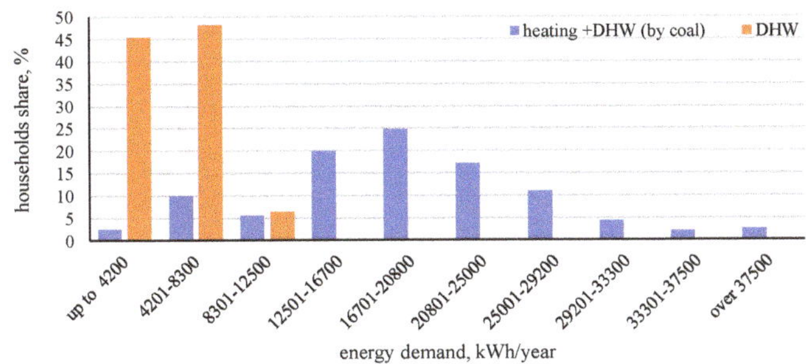

Figure 4. Demand for thermal energy in households in Poland. Data limited to coal sources. Source: own study, based on [30,31,50].

This value applies to heat for central heating (CH) and domestic hot water (DHW). Lower demand values (from 5000 kWh/year) were not taken into account due to the high share of heat losses (calculated according to formula no. 1) [51] from M-TES which is HL(M-TES, year) = 2300 kWh/year to the total amount of heat delivered to the building from M-TES (5000 kWh/year). It was assumed that M-TES thermal insulation will provide a two-time lower heat loss stream than in heat storage solutions typically used inside buildings.

$$HL(M-TES, \tau) = \frac{1h}{1000} \cdot (T(M-TES, \tau) - Ta(\tau)) \cdot ni \cdot chl \cdot \sqrt{V(M-TES)}, \ kWh \tag{1}$$

where:

HL(M-TES,τ)—heat losses from M-TES at the hour τ, kWh

T(M-TES,τ)—M-TES temperature at the hour τ, °C

Ta(τ)—outside temperature, °C

ni—number of M-TES for each house: 2 (one close to house, second for loading)

chl—coefficient for heat losses. It has been assumed from (as $\frac{1}{2}$ of value from [51]: 0.08), W/(K·dm$^{3/2}$)

V(M-TES)—volume of M-TES, dm^3

The calculations assume T(M-TES,τ) at the level of 70 °C. For the determination of outside temperature (Ta), data from the Typical Reference Year (Figure 3) [52] were used. On this basis, the annual sum of heat losses was calculated—Equation (2).

$$HL(M-TES, \ year) = \sum_{\tau=1}^{8760} HL(M-TES, \tau), \frac{kWh}{year} \tag{2}$$

The 25,000 is a defined upper safe limit for the use of two heat stores due to the rate of heat distribution at the recipient and its charging at the base station, e.g., in Geotermia Podhalańska.

The distribution of total heat demand THD for heating Q(CH) and for domestic hot water preparation Q(DHW) was assumed in a 3:2 ratio [30]. This ratio matters regarding the frequency of M-TES exchange (winter/summer relationship) and for heat losses during the transportation: a greater share of DHW means that, in warmer months, there are more trips (and exchanges) than in the case of a smaller share. However, the analysis does not include heat losses during transport, due to the fact that its time is a maximum of several minutes. In summary, it was considered that the ratio Q(CH)/Q(DHW) according to specific THDs was separated, and has no significant impact on further calculations.

The heat demand for heating, taken from the range of 3000–15,000 kWh, was divided into individual hours of the year according to the hourly temperature data from the Typical Reference Year for Zakopane (10 km from Bańska Niżna) (Figure 5).

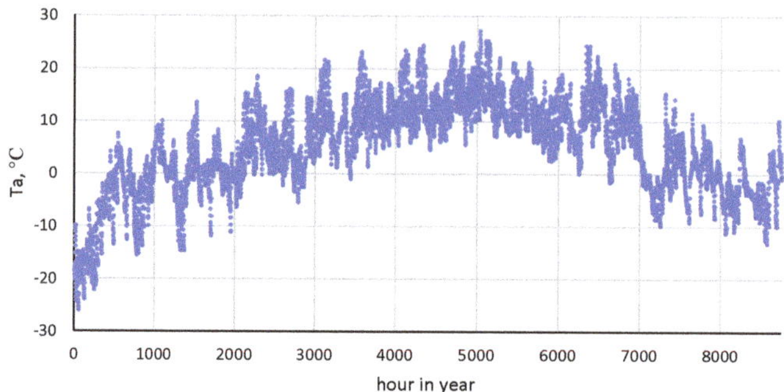

Figure 5. Outside temperature in Zakopane according to Typical Reference Year. Source: own study, based on [52].

Data on the heat demand for the preparation of domestic hot water were divided according to hours based on data from work [53].

The change in the amount of heat accumulated in M-TES in each hour was determined from the formula:

$$Q(M - TES, \tau) = Q(M - TES, \tau - 1) - HL(M - TES, \tau) - Q(CH, \tau) - Q(DHW, \tau), \text{ kWh} \qquad (3)$$

where:

$Q(M\text{-}TES,\tau)$—heat accumulated inside the M-TES at the hour: τ, kWh
$Q(M\text{-}TES,\tau - 1)$—heat accumulated inside M-TES at the hour: $\tau - 1$, kWh
$HL(M\text{-}TES,\tau)$—heat losses from M-TES at the hour: τ, kWh
$Q(CH,\tau)$—heat needed for the central heating purposes at the hour: τ, kWh
$Q(DHW,\tau)$—heat needed for the domestic hot water purposes at the hour: τ, kWh
τ—hour (time)

When $Q(M\text{-}TES, \tau)$ would reach zero in a given hour of computing simulation, it is considered in the calculations that the M-TES was replaced with a fully charged one. In this case, $Q(M\text{-}TES, \tau-1)$ is replaced by a value equal to the heat capacity M-TES, which is 55 kWh.

2.2. Economical Profitability and NPV

The following assumptions were made in the analysis:

- Euro exchange rate 1 EUR = 4.5 PLN,
- The total costs of the driver's work (and also the person servicing the M-TES exchange in the heat source location and at the buildings) for the employer, TWC = 5 Euro/hour,
- The total cost of transportation by the car with a trailer with a total permissible weight up to 3.5 Mg, TTC: 0.2 EUR/km. This cost includes the purchase of the fuel, service costs, inspections, and depreciation of the car,
- Unit transport time, UTT: 2 min/km,
- Container exchanging time at the heat source and at the recipient, together with time for changing magazines, TfCMs: 12 min,
- Geothermal heat price without distribution, GHP: 0.0322 EUR/kWh [42],

Fixed prices were assumed in the analysis due to the assumed proportional increase in the prices of heat (geothermal and other), transport costs (including labor costs).

Economical profitability (EP) M-TES solutions were calculated in comparison to other technologies for heat production in buildings on the basis of estimating the costs of their use over 20 years. EP values were presented as the difference between the cost of the current solution (e.g., coal boiler) and planned installation, which is M-TES. This difference was calculated for a 20-year perspective and referred to one year of use—Equation (4). Due to the fact that in both cases (base and M-TES), investment costs refer to the total heat demand, and often the service costs are not linearly dependent on the amount of heat demand, the presentation of the comparison results in a unitary way was considered as too high a simplification.

$$EP(dist., THD, LCOH.OS, M-TES.P) = THD \cdot (LCOH.OS(THD) - LCOH.G(dist., M-TES.P, THD)), \frac{EUR}{year} \quad (4)$$

where:

EP—economical profitability, EUR/year
dist.—distance between building and Geothermal Base Station, km
THD—total heat demand for building, kWh/year
LCOH.OS—Levelized Cost of Heating (for 20 years) for other than Geothermal source, equation 10 (based on [54]), EUR/kWh
LCOH.G—Levelized Cost of Heating (for 20 years) for Geothermal source, Equation (5), EUR/kWh
MTES.P—M-TES price, EUR

$$LCOH.G(dist., M-TES.P, THD) = \frac{n \cdot (CGH(THD) + TC(dist., THD)) + ni \cdot M-TES.P}{n \cdot THD}, \frac{EUR}{kWh} \quad (5)$$

where:

LCOH.G—Levelized Cost of Heating (for n years) for Geothermal source, EUR/kWh
n—lifetime of M-TES, year
CGH—Cost of Geothermal Heat (calculated by Equation (6)), EUR/year
TC—transport and work cost (calculated by Equation (7)), EUR/year
ni—number of M-TES for each house: 2 (one close to house, second for loading in base station)
M-TES.P—M-TES price, EUR
THD—Total heat demand for the building, kWh/year
dist.—distance between building and Geothermal Base Station, km

$$CGH(THD) = (THD + HL(M-TES, year)) \cdot GHP, \frac{EUR}{year} \quad (6)$$

where:

CGH—Cost of Geothermal Heat, EUR/year
THD—Total heat demand for the building, kWh/year
HL(M-TES, year)—heat losses from M-TES in whole year, kWh/year
GHP—unit geothermal heat price, EUR/kWh

$$TC(dist., THD) = M-TES.EX(THD) \cdot \left(TWC \cdot \frac{1}{60} \cdot (TfCMs + UTT \cdot dist. \cdot 2) + dist. \cdot 2 \cdot TTC \right), \frac{EUR}{year} \quad (7)$$

where:

TC—transport and work cost, EUR/year
M-TES.EX—number of M-TES exchanging per year
TWC—total work cost, EUR/h
TfCMs—time for M-TES exchanging, min
UTT—unit transport time, min/km

TTC—total transport cost, EUR/km

dist.—distance between building and Geothermal Base Station, km

THD—Total heat demand for the building, kWh/year

To use geothermal heat with the use of M-TES, every household needs to have two heat containers; while one of them is used to supply heat to the building, the other one is charged at the charging station in Geotermia Podhalańska S.A. (in the Geothermal Base Station). The cost of a single container is determined from the following relationship:

$$M - TES.P = \text{Phase change material cost} + \text{container cost} + \text{trailer cost} \tag{8}$$

LCOH.OS in the calculations was adopted in the interval from 0.05 to 0.25 EUR/kWh. LCOH.OS was not clearly defined because it can differ for every building. Theoretically, the upper limit of LCOH.OS should be slightly higher than the price of electricity in Poland, which is currently around 0.15 EUR/kWh for domestic consumers [30]. However, the adopted range (up to 0.25 EUR/kWh) also considers the possible scenario of an increase in energy prices or the possibility of scaling the proposed solution outside the Poland. An example of the LCOH.OS calculation for a building (THD = 10,000 kWh/year) heated with a coal boiler was made using Equation (9).

$$LCOH.OS = \frac{n \cdot (FC.OS(THD) + WC(THD)) + nl \cdot OS.P}{n \cdot THD}, \frac{EUR}{kWh} \tag{9}$$

where:

LCOH.OS—Levelized Cost of Heating (for n years) for other than Geothermal source, EUR/kWh

n—lifetime of M-TES, year

FC.OS—total cost of fuel for other than Geothermal heat source, EUR/year

THD—Total heat demand for the building, kWh/year

WC—work cost, EUR/year

nl—correction coefficient of lifetime for other source = n/lifetime of other source

OS.P—other source price, EUR

The analysis of the example also includes: the cost of fuel 200 EUR/Mg of coal with a calorific value of 24 MJ/kg, efficiency 80% [31]. It displays the cost of 0.0375 EUR/kWh. FC.OS = 375 EUR/year. The cost of service was associated with the working time, and it was estimated as about $\frac{1}{2}$ hours per day in the heating season. In total, it is about 100 h/year for the town of Bańska Niżna. The cost of service, due to the work mainly being the owners (no additional costs), was valued as $\frac{1}{2}$ TWC. The total work cost was obtained: WC = 100 h × $\frac{1}{2}$ × 5 EUR/h = 250 EUR. For a 10,000 kWh/year demand, a unit cost of 0.025 EUR/kWh is implied. The cost of the heat source, i.e., in this case, a coal boiler, was estimated at EUR 1500 (OS.P). Assuming a lifetime of 10 years and THD = 10,000 kWh/year, the unit investment cost is 0.015 EUR/kWh. The cost of auxiliary energy (consumed in the building) was not included in the analysis since it is needed in virtually every solution, including the application of the M-TES system. The total for the above example with a coal boiler in the conditions of Bańska Niżna, Zakopane and the surrounding area of LCOH.OS is 0.0775 EUR/kWh.

In addition, Net Present Value (NPV) was calculated, according to Formula No. 10, for the following assumptions:

- lifetime of M-TES, n = 20 years
- discount rate (r) from 0% to 6% [55]
- M-TES.P = 6000 EUR
- initial investment cost IO = 2·M-TES.P

$$NPV = \left[\sum_{t=1}^{n} \frac{CF_t}{(1+r)^t} \right] - IO \tag{10}$$

where:

CF_t is the cash flow in the year t, EUR

t—year of the analysis

n—lifetime of M-TES, year

I0 = ni·M-TES.P initial investment value, EUR

r—discount rate, %

$$CF_t = (CGH(THD) + TC(dist., THD)) - THD·LCOH.OS, \frac{EUR}{year} \qquad (11)$$

where:

CF_t is the cash flow in the year t, EUR

CGH—Cost of Geothermal Heat, EUR/year

THD—Total heat demand for the building, kWh/year

TC—transport and work cost, EUR/year

dist.—distance between building and Geothermal Base Station, km

LCOH.OS—Levelized Cost of Heating (for n = 20 years) for other than Geothermal source, EUR/kWh

2.3. Carbon Dioxide Emission

The following specific emission values were assumed for the calculations:

- Transportation of the M-TES storage by car, on the trailer, EM.CO2.KM = 0.240 kg CO_2/km [56]
- emission related to the current heat source (EM.CO2.OS) [57]:

  - coal boiler: 0.375 kg CO_2/kWh [53,58],
  - gas boiler: 0.200 kg CO_2/kWh [58,59],

- associated with auxiliary energy for the service of pumps and devices in a geothermal heating plant (1% of THD + HL(M-TES)) in a geothermal source (EM.CO2.G): 0.0078 kg CO_2/kWh (it is about kWh of energy produced from a geothermal source) [60].

The reduction of CO_2 emissions caused by replacing the current heat source with a geothermal source was calculated according to Formula No. 12:

$$RE.CO2(THD, dist.) = THD·(EM.CO2.OS - EM.CO2.G) - EM.CO2.T(M - TES.EX(THD), dist.), \frac{kgCO_2}{year} \qquad (12)$$

where:

RE.CO2—reduction of CO_2 emissions, kgCO2/year

THD—Total heat demand for the building, kWh/year

dist.—distance between building and Geothermal Base Station, km

EM.CO2.OS—emission from other than Geothermal source, kg CO_2/kWh

EM.CO2.G—Emission connected with Geothermal source, kg CO_2/kWh

EM.CO2.T—annual CO_2 emissions related to the M-TES transportation, determined from the formula no. 13, kg CO_2/year

M-TES.EX—Mobile Thermal Energy Storage exchanged

$$EM.CO2.T(M - TES.EX(THD), dist.) = 2·dist.·M - TES.EX(THD)·EM.CO2.KM \qquad (13)$$

where:

EM.CO2.T—annual CO_2 emissions related to the M-TES transportation, kg CO_2/year

M-TES.EX—Mobile Thermal Energy Storage exchanged

THD—Total heat demand for the building, kWh/year

dist.—distance between building and Geothermal Base Station, km

EM.CO2.OS—emission from other than Geothermal source, kg CO_2/kWh

EM.CO2.KM—emission from car per km, kg CO_2/km

Emissions associated with transport are linearly dependent on the distance. For example, for a distance of 5 km between Geothermal Base Station and Building and M-TES.EX = 200/year, transport related emissions would 480 kg CO_2/year. A total of 2000 km would be driven.

3. Results

3.1. Analysis of M-TES System Operation in Relation to the Total Heat Demand Size for the Building

For different levels of total heat demand for the building (THD), the hourly heat demand (Q(CH) and Q(DHW)) as well as heat losses were calculated. The results for THD = 25,000 kWh/year in the winter period are shown in Figure 6, and in the summer period in Figure 7. The M-TES charge level (initial Q(M-TES)) 55 kWh in a given hour means that M-TES exchanged occurred in this hour.

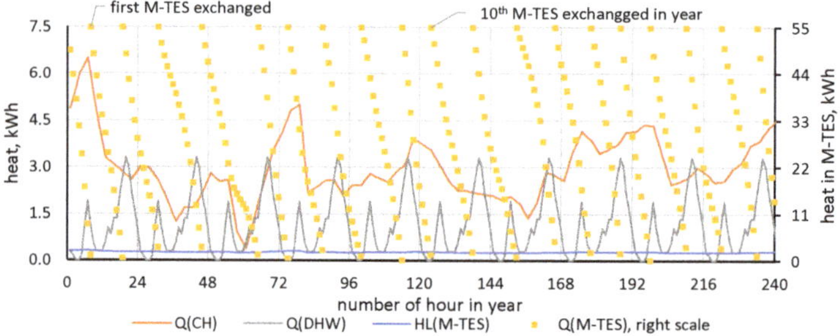

Figure 6. Heat demand and heat losses and heat accumulated in mobile thermal storage system (M-TES) for each hour in the 1–240-h period of the year (January).

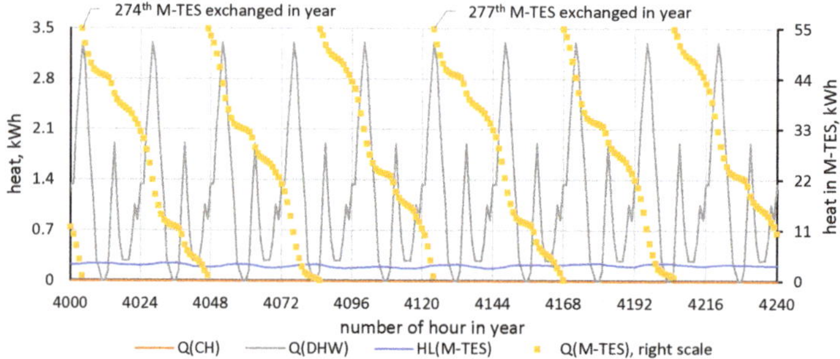

Figure 7. Heat demand and heat losses and heat accumulated in M-TES for each hour in the 4000–4240-h period of the year (June).

As can be seen in Figure 6, on the second day of the year, M-TES should be changed twice (for THD = 25,000 kWh/year). In turn, during the warmer period (June), i.e., between 4000–4240 h of the year, the tank should be replaced much less frequently (six times in 10 days; in the example for the winter, the sum was 10), and the heat losses are also lower. On aggregate for months, the number of M-TES exchanges is shown in Figure 8.

For THD = 5000 kWh/year, the maximum number of exchanges per month is 19, for THD = 15,000 it is 48, and for 25,000 it amounts to 78 (December). A feasibility analysis of the M-TES replacement in the context of various types of restrictions was carried out. The results presented below show the example, based on the calculations for the last value (78). The first of the analyzed restrictions is the M-TES charging and discharging rate: 78 exchanges mean almost three exchanges per day—the average charging power is approx. 7 kW. The maximum estimated charging power of 14 kW is realized with a flow temperature of 86 °C and a return temperature of 72 °C. Under these conditions, the flow rate would be 0.25 dm^3/s. For THD = 25,000, the simulation received nine days with four M-TES exchanges per day (this means the average charging power of 9 kW). A similar situation is found for the M-TES discharging process. Further research includes work related to the irregularity of the M-TES maximum charging/discharging rate.

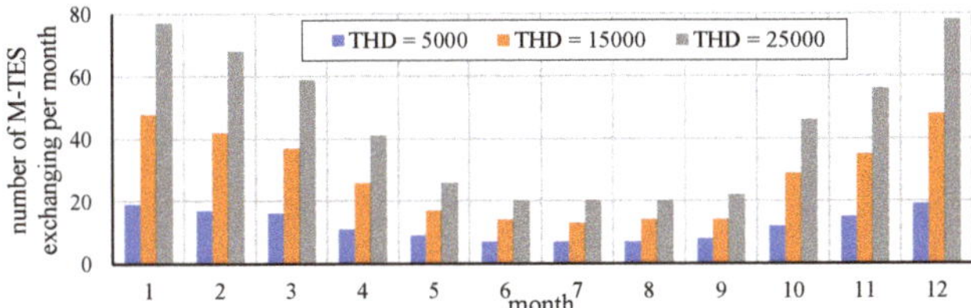

Figure 8. Number of M-TES exchanges (M-TES.EX) during each month of the year for different value of total heat demand (THD): 5000, 15,000 and 25,000 kWh/year.

In total, the number of exchanges in the year M-TES (M-TES.EX) is 147, 337 and 533, respectively. The number of exchanges needed to ensure the demand of THD = 15,000 kWh/year is 2.3 times higher than for THD = 5000 kWh/year, while the ratio of demand is three times higher. The reason for this phenomenon is the practically constant level of heat losses (HL (M-TES)) independent of the THD value in the adopted range.

The second possible restriction affecting the number of exchanges is the level of transport complexity in the mountainous area and—resulting from this—the transport time of one of the two M-TES dedicated for the building using this system. The longer the transportation time, the more time should be deducted for the M-TES charging process. The fully charged M-TES is transported to households and replaced with the empty M-TES (Q(M-TES) $\approx$ 0). Therefore, the discharging process is interrupted only during the M-TES replacements. The remaining time is spent on the M-TES charging process, excluding the transport to Geothermal Base Station and connection to the charging spot.

The use of the proposed M-TES system (with a heat capacity of 55 kWh) for a greater heat demand (more than THD = 25,000 kWh/year) may result in the M-TES dual storage system being insufficient due to the speed of heat use (recipient) or M-TES charging. Therefore, the maximum heat demand assumed for the building was checked. For this purpose, weather data from the last 20 years 2000–2019 were used (Figure 9). ERA5-Land data were analyzed for the location of Bańska Niżna (Figure 3.) [61,62].

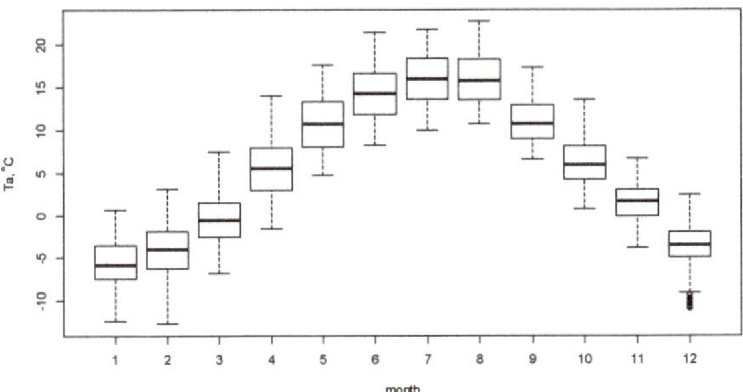

Figure 9. Outside temperature (Ta) in Banska Nizna: average temperature in each day hour in each month in 2000–2019.

The lowest average monthly temperature in the period 2000–2019 was recorded in February 2012. For this month, detailed weather data were obtained from ogimet.com. Then, the number of M-TES exchanges during the following days was determined. The highest accepted value of THD = 25,000 was taken into account. The points of exchanges are shown in Figure 10 in combination with the outside temperature. For the first three days of February 2012, the estimated number of M-TES exchanges would be five per day. This number of exchanges means a minimum charging or discharging time of four hours, which means a necessary charging power of 14 kW.

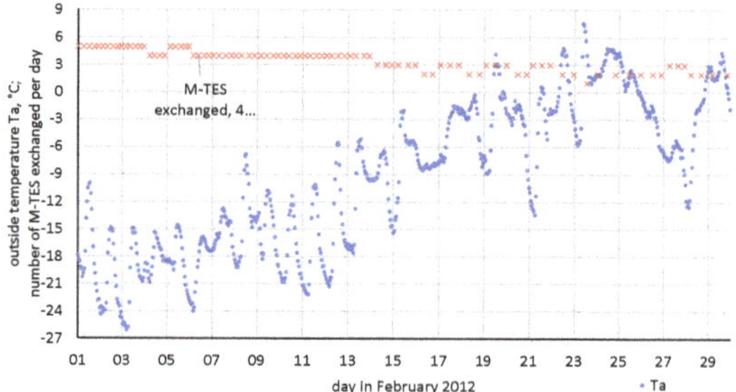

Figure 10. Outside temperature (Ta) in Zakopane and number of M-TES exchanges (×) calculated for February 2012. Source: own study, based on [63].

3.2. Economical Profiltability

The results in the form of Economical Profitability for THD = 5000, 10,000, 15,000, 20,000 and 25,000 kWh/year are presented in the distance, along with the Levelized Cost of Heating (LCOH) function of the current heat source (LCOH.OS): Figures 11–15: (a) for M-TES price = EUR 2000, (b) for M-TES price = EUR 6000 (c) for M-TES price = EUR 10,000.

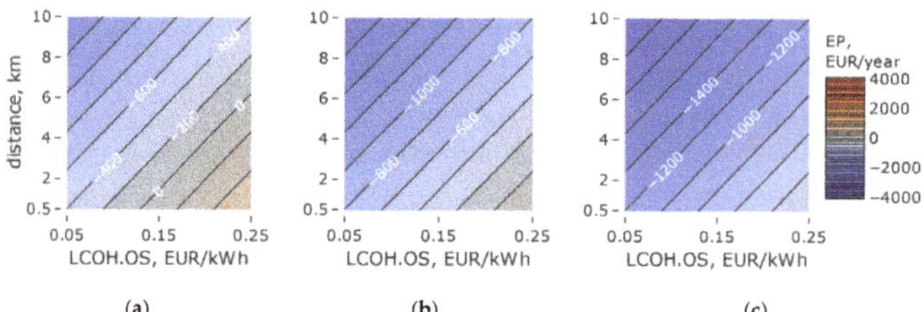

(a) (b) (c)

Figure 11. The Yearly Economical Profitable as a function of Levelized Cost of Heating from non-geothermal source (LCOH.OS) and distance between Geothermal Base Station and House, for Total Heat Demand (THD) = 5000 kWh/year and for: (**a**) M-TES price = 2000 EUR; (**b**) M-TES price = 6000 EUR; (**c**) M-TES price = 10,000 EUR.

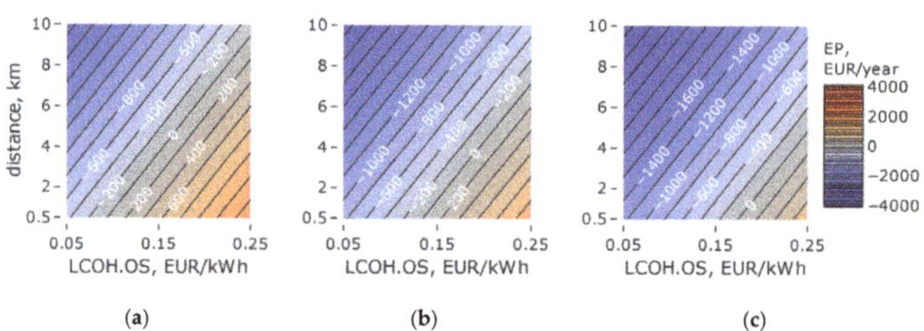

(a) (b) (c)

Figure 12. The Yearly Economical Profitable as a function of Levelized Cost of Heating from non-geothermal source (LCOH.OS) and distance between Geothermal Base Station and House, for Total Heat Demand (THD) = 10,000 kWh/year and for: (**a**) M-TES price = 2000 EUR; (**b**) M-TES price = 6000 EUR; (**c**) M-TES price = 10,000 EUR.

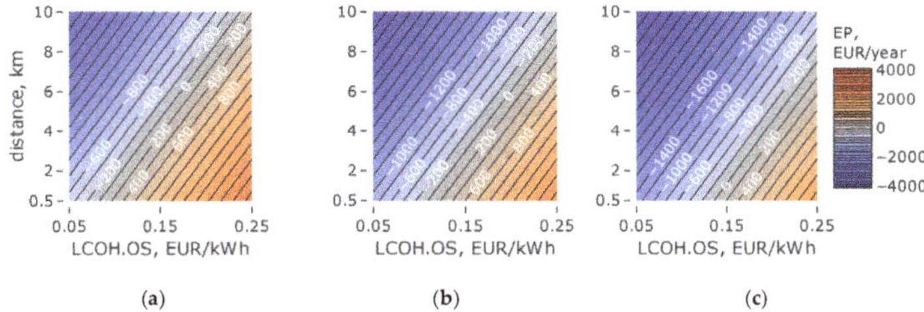

Figure 13. The Yearly Economical Profitable as a function of Levelized Cost of Heating from non-geothermal source (LCOH.OS) and distance between Geothermal Base Station and House, for Total Heat Demand (THD) = 15,000 kWh/year and for: (**a**) M-TES price = 2000 EUR; (**b**) M-TES price = 6000 EUR; (**c**) M-TES price = 10,000 EUR.

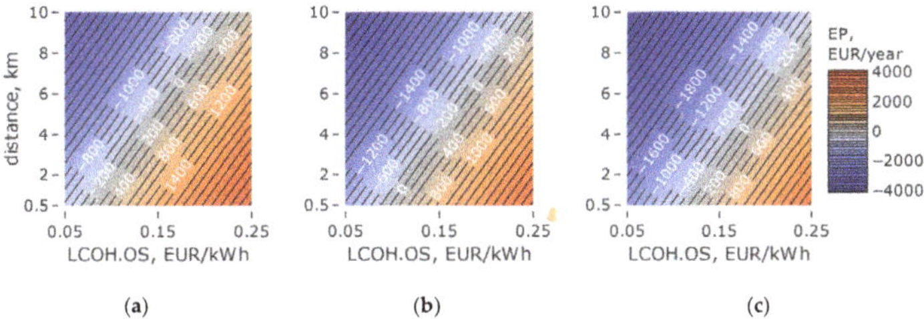

Figure 14. The Yearly Economical Profitable as a function of Levelized Cost of Heating from non-geothermal source (LCOH.OS) and distance between Geothermal Base Station and House, for Total Heat Demand (THD) = 20,000 kWh/year and for: (**a**) M-TES price = 2000 EUR; (**b**) M-TES price = 6000 EUR; (**c**) M-TES price = 10,000 EUR.

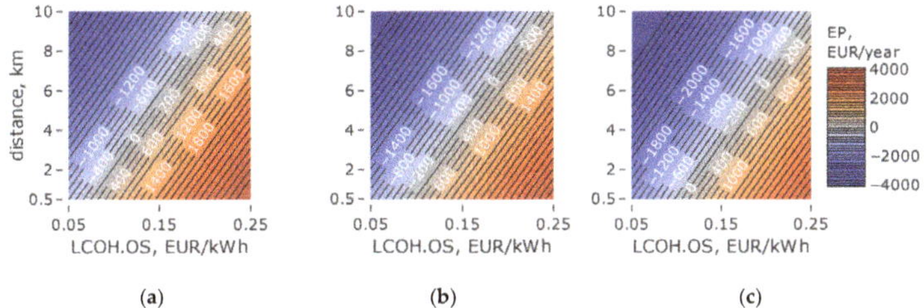

Figure 15. The Yearly Economical Profitable as a function of Levelized Cost of Heating from non-geothermal source (LCOH.OS) and distance between Geothermal Base Station and House, for Total Heat Demand (THD) = 25,000 kWh/year and for: (**a**) M-TES price = 2000 EUR; (**b**) M-TES price = 6000 EUR; (**c**) M-TES price = 10,000 EUR.

A profitability of at least zero will be achieved at M-TES price = 6000 EUR for the total heat demand (THD):

- 5000 kWh/year; at the price LCOH.OS = 0.21 EUR/kWh and distance = 0.5 km;
- 10,000 kWh/year; at the price LCOH.OS = 0.135 EUR/kWh and distance = 0.5 km; at the price 0.235 EUR/kWh and distance = 6 km;
- 15,000 kWh/year; at the price LCOH.OS = 0.11 EUR/kWh and distance = 0.5 km; at the price 0.205 EUR/kWh and distance = 6 km;
- 20,000 kWh/year; at the price LCOH.OS = 0.095 EUR/kWh and distance = 0.5 km; at the price 0.19 EUR/kWh and distance = 6 km;

- 25,000 kWh/year; at the price LCOH.OS = 0.085 EUR/kWh and distance = 0.5 km; at the price 0.18 EUR/kWh and distance = 6 km.

At higher demand values (THD) from the analyzed range, the results in the form of EP assume higher values, because the unit cost of M-TES decreases. This relationship is also influenced by the fact that M-TES price is usually higher than half of the cost of another heat source converted into 20 years of lifetime ($\frac{1}{2}\cdot$nl$\cdot$OS.P). In addition, the above-mentioned dependence (EP on M-TES.P) is affected by the HL (M-TES)/THD ratio, which is lower at higher THD values, because the HL (M-TES) value is practically constant. HL (M-TES) depends, among other things, on temperature in M-TES and the outside temperature.

An example of the impact of THD on EP is illustrated by a comparison of Figures 11b and 15b (for M-TES.P = EUR 6000). For the same LCOH.OS price equal to 0.15 EUR/kWh, five times more heat demand is added to the increase of EP from −300 EUR/ear to 1500 EUR/year, at distance = 0.5 km. When LCOH.OS is 0.1 EUR/kWh, EP achieves a result of −550 EUR/ear and 200 EUR/year respectively. Additionally, in Figure 15b, there is a higher slope of the isoline compared to Figure 11b. This demonstrates the relatively smaller impact of distance on EP compared to the effect of LCOH.OS on EP. A similar comparison (THD = 5000 and 25,000 kWh/year) was made for LCOH.OS equal to 0.1 EUR/kWh and variable M-TES.P—Figure 16. The impact of M-TES price clearly becomes smaller with higher heat demand. In turn, a greater impact on EP is identified for distance.

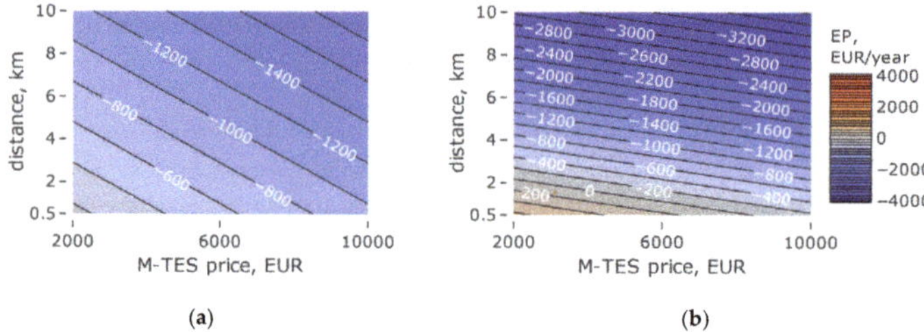

(a) (b)

Figure 16. The Yearly Economical Profitable as a function of M-TES price and distance between Geothermal Base Station and Building, for LCOH.OS = 0.1 EUR/kWh and for: (**a**) Total Heat Demand (THD) = 5000 kWh/year; (**b**) THD = 25,000 kWh/year.

3.3. NPV Analysis

The Net Present Value (NPV) was calculated (for 20 years and distance = 0.5 km) for M-TES.P = 6000 EUR and THD = 15,000 EUR/year (LCOH.G without M-TES.P = 0.045 EUR/kWh). The results for various discount rates are shown in Figure 17. At a discount rate of 6%, NPV reaches zero for LCOH.OS = 0.115 EUR/kWh, and for a discount rate of 0% at 0.082 EUR/kWh.

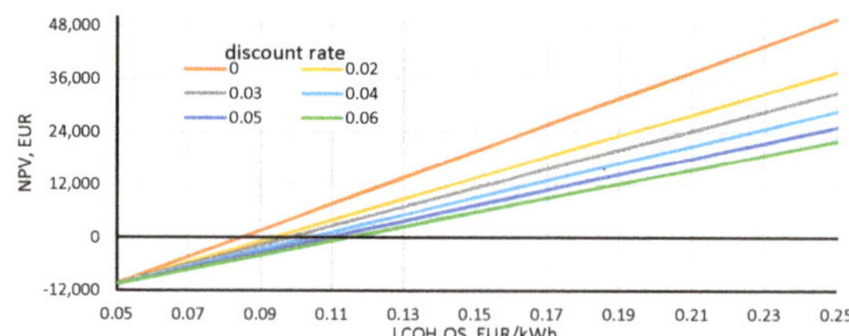

Figure 17. Net Present Value (NPV) as a function of LCOH.OS for different discount rate value, for distance = 0.5 km.

To verify the impact of changing the value of selected parameters (+/−20%) on the NPV change, a sensitivity analysis was performed—Table 1. The results of the analysis were sorted from the largest to the smallest impact, but this is related to the base level of the parameters—a different base level would have a different impact on NPV (example distance 0.5 km and 5 km—Table 1). Significantly, the ratio of heat demand for heating and domestic hot water preparation (with the same THD) does not play any role in the NPV value, but only as shown in Section 2 for different frequencies of M-TES replacement in individual months. At the same time, the authors are aware that the results of the sensitivity analysis are affected by the adopted baseline values, e.g., as shown in Table 1, the assumed baseline distance will have a different impact on NPV change.

Table 1. NPV sensitivity analysis.

| Parameter | Base Value | Unit | Change of Parameter Base Value | |
| | | | −20% | 20% |
			NPV Change	
LCOH.OS	0.0775	EUR/kWh	−114.7%	60.3%
M.TES.P	6000	EUR	45.8%	−45.7%
THD	15000	kWh/year	−32.4%	32.4%
LCOH.G	0.0322	EUR/kWh	30.9%	−30.9%
distance*	5	km	13.6%	−13.6%
distance	0.5	km	3.5%	−3.5%
r	4	%	8.9%	−7.8%
Q(H)/Q(DHW)	1.5	-	0.0%	0.0%

distance*—only this example is calculated for base value 5 km, others for 0.5 km.

3.4. Carbon Dioxide Emission Reduction

The results in the form of quantitative CO_2 emissions related to M-TES transport are presented in Table 2. The results in the form of a quantitative reduction of CO_2 emissions in the event of the replacement of a coal or gas source by a geothermal source by a mobile heat storage are presented in Table 2. The results presented in Tables 2 and 3 were estimated for the assumptions presented in Section 2.3.

Table 2. Carbon dioxide emission of M-TES transport as a function of distance between building and M-TES load base, and THD. Source: own study.

| Distance, km | Carbon Dioxide Emission of M-TES Transport, kg/year for THD: | | |
	5000	15,000	25,000
0.5	38	87	137
1	76	174	275
2	152	347	550
3	227	521	824
4	303	695	1099
5	379	869	1374
6	455	1042	1649
7	531	1216	1924
8	606	1390	2198
9	682	1564	2473
10	758	1737	2748

Journeys related to transportation M-TES for a year to cover the heat demand for a building THD = 5000 kWh/year for a distance of 0.5 km are associated with CO_2 emissions from the car at the level of 38 kg CO_2/year, a distance of 10 km will result in 20x greater emissions. At THD = 25,000 kWh/year carbon dioxide emission is about 3.6 higher than in THD = 5000 kWh/year.

For a building with THD = 25,000 kWh/year and a distance of 0.5 km from the M-TES charging station, the use of geothermal source will bring:

- A total of 9 Mg CO_2 emission reduction when replacing a coal-fueled heat source,
- A total of 4.8 Mg CO_2 emission reduction when replacing a natural gas heat source. For a distance of 10 km and THD = 5000 kWh/year, and when replacing a source powered by natural gas, the reduction will be only 0.2 Mg CO_2/year.

Table 3. Carbon dioxide emission reduction as a function of distance, THD and replacement heat source. Source: own study.

| Distance, km | Reduction of Carbon Dioxide Emission, kg/year | | | | | |
| | for Replacement of Coal Source, for THD: | | | for Replacement of Natural Gas Source, for THD: | | |
	5000	15,000	25,000	5000	15,000	25,000
0.5	1774	5394	9009	928	2854	4777
1	1736	5307	8872	890	2767	4639
2	1660	5133	8597	814	2594	4365
3	1585	4959	8322	738	2420	4090
4	1509	4785	8047	662	2246	3815
5	1433	4612	7772	587	2072	3540
6	1357	4438	7498	511	1899	3265
7	1281	4264	7223	435	1725	2991
8	1206	4090	6948	359	1551	2716
9	1130	3917	6673	283	1377	2441
10	1054	3743	6398	208	1204	2166

4. Conclusions

This article considered, for Polish conditions, the technical and economic possibilities of providing geothermal heat to individual recipients using an M-TES system. For this purpose, heat storage, PCM and the best location regarding geothermal heat availability (temperature and the cost) were selected. The chosen location is Bańska Niżna near Zakopane (southern part of the Poland). An indirect contact energy storage container filled with PCM characterized by melting temperature 70 °C and heat storage capacity 250 kJ/kg was selected, in the amount of 800 kg. The total M-TES heat capacity is 55 kWh.

The discussed PCM-container is satisfactory to supply a building with a total heat demand (central heating and domestic hot water) up to 25,000 kWh/year. This was established based on the data for the month with the lowest average temperature of the last 20 years (February 2012). Above this demand value, consideration should be given to the use of M-TES with larger heat capacity. The economic profitability of the M-TES system (with a price per warehouse of 6,000 EUR, i.e., a total of 12,000 EUR—two containers are needed) can be achieved for a heat demand of 5000 kWh/year with the price of a replaced heat source at the level of 0.21 EUR/kWh and distance between charging station and building (heat recipient) 0.5 km. For the heat demand of 15,000 kWh/year, the price for replaced heat reached EUR 0.11/kWh, and the same distance. In turn, for a demand of 25,000 kWh/year, the price of the replaced heat source reached 0.085 EUR/kWh.

The NPV value for the M-TES system was also determined. With a heat demand of 15,000 kWh/year, the NPV was calculated for 20 years of operation and a 6% discount rate that reached zero for a levelized cost of heating 0.115 EUR/kWh, and for a discount rate of 0% at 0.082 EUR/kWh. The levelized cost of heating in the case of geothermal energy without M-TES costs (2 × 6000 EUR) amounted to 0.045 EUR/kWh.

The economic profitability is significantly affected by the distance. For the adopted assumptions and at current prices, there are no arguments for transporting geothermal heat over distances bigger than 3–4 km away from the heat source, even in the case of the replaced heat price at the level of the current electricity price in Poland. This also concerns heat losses during the transportation, and ecological aspects. Considering the building with 25,000 kWh/year of total heat demand, which is located 0.5 km from the M-TES charging station, the use of a geothermal source can bring 9 Mg CO_2 emission reduction in the case of replacing a coal-fueled heat source or 4.8 Mg CO_2 emission reduction in the case of replacing a natural gas heat source. By increasing the distance to 10 km, a decrease in emission reduction of up to 6.4 Mg CO_2 and 2.2 Mg CO_2, respectively, can be observed.

Author Contributions: All the authors have contributed toward developing and implementing the ideas and concepts presented in the paper. All the authors have collaborated to obtain the results and have been involved in preparing the manuscript. All authors have read and agreed to the published version of the manuscript.

Funding: This work was supported by AGH—University of Science and Technology number 16.16.210.476.

Abbreviation

CGH	Cost of Geothermal Heat, EUR/year
chl	coefficient for heat losses, $W/(K{\cdot}dm^{3/2})$
CF	Cash Flow, EUR
CH	Central heating
DHW	Domestic hot water
dist.	distance between building and Geothermal Base Station, km
DU	direct use
EM.CO2.G	Emission connected with Geothermal source, kg CO_2/kWh
EM.CO2.KM	Emission from car per km, kg CO_2/km
EM.CO2.OS	Emission from other than Geothermal source, kg CO_2/kWh
EP	Economical profitability, EUR/year
EUR	Euro
FC.OS	total cost of fuel for other than Geothermal heat source, EUR/year
GHG	greenhouse gases
GHP	unit geothermal heat price, EUR/kWh
HL	Heat loss, kWh
IO	initial investment value, EUR
LCOH	Levelized Cost of Heating, EUR/kWh
LCOH.G	Levelized Cost of Heating (for n years) for Geothermal source
LCOH.OS	Levelized Cost of Heating (for n years) for other than Geothermal source
M-TES	Mobile Thermal Energy Storage
M-TES.EX	Mobile Thermal Energy Storage exchanged
M-TES.P	M-TES price, EUR
n	lifetime of M-TES, year
ni	number of M-TES for each house
nl	correction coefficient of lifetime for other source
NPV	Net present value, EUR
OS.P	other source price, EUR
Q	heat, kWh
r	discount rate, %
RE.CO2	reduction of CO_2 emissions, kgCO_2/year
T	Temperature, °C
t	subsequent year
Ta	Outside temperature, °C
TC	Transport and work cost, EUR/kWh
TfCMs	Time for M-TES exchanging, min
th	thermal
THD	Total heat demand for the building, kWh/year
TPEH_GEO	Total price of energy for heating—geothermal source, EUR/kWh
TPEH_OS	Total price of energy for heating—other source (not geothermal), EUR/kWh
TTC	Total transport cost, EUR/km
TWC	Total work cost, EUR/h
V	volume, dm^3
UTT	unit transport time, min/km
WC	work cost, EUR/year
τ	time

References

1. McCay, A.T.; Harley, T.L.; Younger, P.L.; Sanderson, D.C.W.; Cresswell, A.J. Gamma-ray spectrometry in geothermal exploration: State of the art techniques. *Energies* **2014**, *7*, 4757–4780. [CrossRef]
2. Stober, I.; Bucher, K. Geothermal energy: From theoretical models to exploration and development. In *Geothermal Energy: From Theoretical Models to Exploration and Development*; Springer: Berlin/Heidelberg, Germany, 2013; ISBN 9783642133527.
3. Li, K.; Bian, H.; Liu, C.; Zhang, D.; Yang, Y. Comparison of geothermal with solar and wind power generation systems. *Renew. Sustain. Energy Rev.* **2015**, *42*, 1464–1474. [CrossRef]
4. Zhu, J.; Hu, K.; Lu, X.; Huang, X.; Liu, K.; Wu, X. A review of geothermal energy resources, development, and applications in China: Current status and prospects. *Energy* **2015**, *93*, 466–483. [CrossRef]

5. Younger, P.L. *Energy: All That Matters Paperback*; Hodder and Stoughton/John Murray: London, UK, 2014.

6. Moya, D.; Aldás, C.; Kaparaju, P. Geothermal energy: Power plant technology and direct heat applications. *Renew. Sustain. Energy Rev.* **2018**, *94*, 889–901. [CrossRef]

7. Xia, L.; Zhang, Y. An overview of world geothermal power generation and a case study on China—The resource and market perspective. *Renew. Sustain. Energy Rev.* **2019**, *112*, 411–423. [CrossRef]

8. Redko, A.; Redko, O.; DiPippo, R. Heating with geothermal systems. In *Low-Temperature Energy Systems with Applications of Renewable Energy*; Academic Press: London, UK, 2020; pp. 177–224.

9. Gudmundsson, J.S.; Freeston, D.H.; Lienau, P.J. Lindal diagram. *Trans. Geotherm. Resour. Counc.* **1985**, *9*, 15–17.

10. Kaczmarczyk, M.; Tomaszewska, B.; Operacz, A. Sustainable utilization of low enthalpy geothermal resources to electricity generation through a cascade system. *Energies* **2020**, *13*, 2495. [CrossRef]

11. Dickson, M.H.; Fanelli, M. *Geothermal Energy: Utilization and Technology*; Earthscan: London, UK, 2013; ISBN 9781315065786.

12. Younger, P.L. Geothermal energy: Delivering on the global potential. *Energies* **2015**, *8*, 11737–11754. [CrossRef]

13. IRENA Geothermal Energy. Available online: https://www.irena.org/geothermal (accessed on 15 May 2020).

14. Bertani, R. Geothermal power generation in the world 2010–2014 update report. *Geothermics* **2016**, *60*, 31–43. [CrossRef]

15. DiPippo, R. *Geothermal Power Plants: Principles, Applications, Case Studies and Environmental Impact*, 3rd ed.; Butterworth-Heinemann: London, UK, 2012; ISBN 978-0-08-098206-9.

16. Valdimarsson, P. Geothermal power plants and main components. In Proceedings of the Short Course on Geothermal Drilling, Resource Development and Power Plants, Santa Tecla, El Salvador, 16–22 January 2011.

17. Anderson, D.N.; Lund, J.W. *Direct Utilization of Geothermal Energy: A Technical Handbook*; Geothermal Resources Council: Davis, CA, USA, 1979.

18. Glassley, W.E. *Geothermal Energy—Renewable Energy and the Environment*, 2nd ed.; CRC Press/Taylor&Francis: Boca Raton, FL, USA, 2014; ISBN 9780429161988.

19. Lund, J.W.; Freeston, D.H.; Boyd, T.L. Direct application of geothermal energy: 2005 Worldwide review. *Geothermics* **2005**, *34*, 691–727. [CrossRef]

20. Lund, J.W.; Freeston, D.H.; Boyd, T.L. Direct utilization of geothermal energy 2010 worldwide review. *Geothermics* **2011**, *40*, 159–180. [CrossRef]

21. Lund, J.W.; Boyd, T.L. Direct utilization of geothermal energy 2015 worldwide review. *Geothermics* **2016**, *60*, 66–93. [CrossRef]

22. Marrasso, E.; Roselli, C.; Sasso, M.; Tariello, F. Global and local environmental and energy advantages of a geothermal heat pump interacting with a low temperature thermal micro grid. *Energy Convers. Manag.* **2018**, *172*, 540–553. [CrossRef]

23. Alkhwildi, A.; Elhashmi, R.; Chiasson, A. Parametric modeling and simulation of low temperature energy storage for cold-climate multi-family residences using a geothermal heat pump system with integrated phase change material storage tank. *Geothermics* **2020**, *86*, 101864. [CrossRef]

24. Mangi, P. Geothermal direct use application: A case of geothermal Spa and demonstration centre at olkaria geothermal project, Kenya. In Proceedings of the Short Course IX on Exploration for Geothermal Resources, Naivasha, Kenya, 2–23 November 2014.

25. Matuszewska, D.; Olczak, P. Evaluation of using gas turbine to increase efficiency of the Organic Rankine Cycle (ORC). *Energies* **2020**, *13*, 1499. [CrossRef]

26. Carotenuto, A.; Figaj, R.D.; Vanoli, L. A novel solar-geothermal district heating, cooling and domestic hot water system: Dynamic simulation and energy-economic analysis. *Energy* **2017**, *141*, 2652–2669. [CrossRef]

27. Hepbasli, A.; Canakci, C. Geothermal district heating applications in Turkey: A case study of Izmir-Balcova. *Energy Convers. Manag.* **2003**, *44*, 1285–1301. [CrossRef]

28. Sander, M. Geothermal district heating systems: Country case studies from China, Germany, Iceland, and United States of america, and schemes to overcome the gaps. *Trans. Geotherm. Resour. Counc.* **2016**, *40*, 769–776.

29. Miró, L.; Gasia, J.; Cabeza, L.F. Thermal energy storage (TES) for industrial waste heat (IWH) recovery: A review. *Appl. Energy* **2016**, *179*, 284–301. [CrossRef]

30. CSO. *Energy Consumption in Households in 2018*; CSO: Warsaw, Poland, 2019.

31. Ministry of Development Regulation of the Minister of Infrastructure and Development of 27 February 2015 on the Methodology for Determining the Energy Performance of a Building or Part of a Building, and Energy Performance Certificates. Available online: http://prawo.sejm.gov.pl/isap.nsf/download.xsp/WDU20150000376/O/D20150376.pdf (accessed on 22 March 2020).

32. Kryzia, D.; Kuta, M.; Matuszewska, D.; Olczak, P. Analysis of the potential for gas micro-cogeneration development in Poland using the Monte Carlo method. *Energies* **2020**, *13*, 3140. [CrossRef]

33. Specjał, A.; Lipczynska, A.; Hurnik, M.; Król, M.; Palmowska, A.; Popiołek, Z. Case study of thermal diagnostics of single-family house in temperate climate. *Energies* **2019**, *12*, 4549. [CrossRef]

34. Wang, Y.; Yu, K.; Ling, X. Experimental study on thermal performance of a mobilized thermal energy storage system: A case study of hydrated salt latent heat storage. *Energy Build.* **2020**, *210*, 109744. [CrossRef]

35. Wang, W.; Guo, S.; Li, H.; Yan, J.; Zhao, J.; Li, X.; Ding, J. Experimental study on the direct/indirect contact energy storage container in mobilized thermal energy system (M-TES). *Appl. Energy* **2014**, *119*, 181–189. [CrossRef]

36. Deckert, M.; Scholz, R.; Binder, S.; Hornung, A. Economic efficiency of mobile latent heat storages. *Energy Procedia* **2014**, *46*, 171–177. [CrossRef]

37. Zhang, X.; Chen, X.; Han, Z.; Xu, W. Study on phase change interface for erythritol with nano-copper in spherical container during heat transport. *Int. J. Heat Mass Transf.* **2016**, *92*, 490–496. [CrossRef]

38. Krönauer, A.; Lävemann, E.; Brückner, S.; Hauer, A. Mobile sorption heat storage in industrial waste heat recovery. *Energy Procedia* **2015**, *73*, 272–280. [CrossRef]

39. Igliński, B.; Buczkowski, R.; Kujawski, W.; Cichosz, M.; Piechota, G. Geoenergy in Poland. *Renew. Sustain. Energy Rev.* **2012**, *16*, 2545–2557. [CrossRef]

40. Bujakowski, W.; Tomaszewska, B.; Miecznik, M. The Podhale geothermal reservoir simulation for long-term sustainable production. *Renew. Energy* **2016**, *99*, 420–430. [CrossRef]

41. Tomaszewska, B.; Pająk, L. Cooled and desalinated thermal water utilization in the Podhale heating system. *Miner. Resour. Manag.* **2013**, *29*, 127–139. (In Polish) [CrossRef]

42. URE Tariffs Published in 2019. (In Polish). Available online: https://bip.ure.gov.pl/bip/taryfy-i-inne-decyzje-b/cieplo/3784,Taryfy-opublikowane-w-2019-r.html (accessed on 22 April 2020).

43. Kępińska, B. A review of geothermal energy uses in Poland in 2016–2018. *Tech. Pozyskiwań Geol. Geoterm. Zrównoważony Rozw.* **2018**, *1*, 11–27.

44. Puretemp Global Authority on Phase Change Material. Available online: http://www.puretemp.com/ (accessed on 8 April 2020).

45. Axiotherm Axiotherm PCM. Available online: www.axiotherm.de/en/produkte/axiotherm-pcm/ (accessed on 10 April 2020).

46. GlobalESystem Phase Change Materials. Available online: www.global-e-systems.com/en/phase-change-materials/ (accessed on 9 April 2020).

47. Rubitherm PCM RT Line. Available online: www.rubitherm.eu/en/index.php/productcategory/organische-pcm-rt (accessed on 10 April 2020).

48. Kaczmarczyk, M.; Sowizdżał, A.; Tomaszewska, B. Energetic and environmental aspects of individual heat generation for sustainable development at a local scale—A case study from Poland. *Energies* **2020**, *13*, 454. [CrossRef]

49. CSO. *National Census*; Central Statistical Office: Warsaw, Poland, 2014.

50. CSO. Household Forecast for 2016–2050. Available online: https://stat.gov.pl/obszary-tematyczne/ludnosc/prognoza-ludnosci/prognoza-gospodarstw-domowych-na-lata-2016-2050,9,4.html (accessed on 19 April 2020).

51. En, B.S. Heating systems in buildings—Method for calculation of system energy requirements and system efficiencies. *Management* **2007**. [CrossRef]

52. Ministry of Development Typical Reference Year. Available online: https://www.gov.pl/web/fundusze-regiony/dane-do-obliczen-energetycznych-budynkow (accessed on 18 December 2019).

53. Olczak, P.; Zabagło, J.; Kandefer, S.; Dziedzic, J. Influence of solar installation with flat-plate collectors in a detached house on pollutants emission and waste stream. In *Between Evolution and Revolution—In Search of an energy Strategy*; WAT: Poznań, Poland, 2015; pp. 739–752.

54. Doračić, B.; Novosel, T.; Pukšec, T.; Duić, N. Evaluation of excess heat utilization in district heating systems by implementing levelized cost of excess heat. *Energies* **2018**, *11*, 575. [CrossRef]

55. Kryzia, D.; Kopacz, M.; Kryzia, K. The valuation of the operational flexibility of the energy investment project based on a gas-fired power plant. *Energies* **2020**, *13*, 1567. [CrossRef]

56. Kryzia, D.; Kopacz, M.; Orzechowska, M. Estimation of carbon dioxide emissions and diesel consumption in passenger cars. *Bull. Miner. Energy Econ. Res. Inst. Polish Acad. Sci.* **2015**, *90*, 79–92.

57. Pająk, L.; Tomaszewska, B.; Bujakowski, W.; Bielec, B.; Dendys, M. Review of the low-enthalpy lower cretaceous geothermal energy resources in Poland as an environmentally friendly source of heat for urban district heating systems. *Energies* **2020**, *13*, 1302. [CrossRef]

58. IoEP-NRI. *Inventory of Greenhouse Gases in Poland for 1988–2017*; The National Centre for Emissions Management: Warsaw, Poland, 2019.

59. Olczak, P.; Olek, M.; Kryzia, D. The ecological impact of using photothermal and photovoltaic installations for DHW preparation. *Polityka Energ. Energy Policy J.* **2020**, *23*, 65–74. [CrossRef]

60. WSKAŹNIKI EMISYJNOŚCI CO2, SO2, NOx, CO i pyłu całkowitego DLA ENERGII ELEKTRYCZNEJ. Available online: https://www.kobize.pl/uploads/materialy/materialy_do_pobrania/wskazniki_emisyjnosci/Wskazniki_emisyjnosci_grudzien_2019.pdf (accessed on 7 April 2020).

61. Copernicus Climate Change Service (C3S). ERA5: Fifth Generation of ECMWF Atmospheric Reanalyses of the Global Climate. Copernicus Climate Change Service Climate Data Store (CDS). Available online: https://climate.copernicus.eu/climate-data-store (accessed on 24 March 2020).

62. European Centre for Medium-Range Weather Forecasts (ECMWF) ERA5. Available online: https://cds.climate.copernicus.eu/cdsapp#!/dataset/reanalysis-era5-land-monthly-means?tab=form (accessed on 24 March 2020).

63. Ogimet Synop Based Summary by States form 2019. Available online: http://ogimet.com/resynops.phtml.en (accessed on 24 April 2020).

Article

Effects of Cyclic Heating and Water Cooling on the Physical Characteristics of Granite

Xiangchao Shi [1,2,*], Leiyu Gao [1], Jie Wu [1], Cheng Zhu [3], Shuai Chen [1] and Xiao Zhuo [1]

[1] State Key Lab of Oil and Gas Reservoir Geology and Exploitation, Southwest Petroleum University, Chengdu 610500, China; 201921000633@stu.swpu.edu.cn (L.G.); 201821000577@stu.swpu.edu.cn (J.W.); 201921000608@stu.swpu.edu.cn (S.C.); 201721000594@stu.swpu.edu.cn (X.Z.)

[2] Key Laboratory Deep Underground Science and Engineering (Ministry of Education), Sichuan University, Chengdu 610065, China

[3] Department of Civil & Environmental Engineering, Rowan University, Glassboro, NJ 08028, USA; zhuc@rowan.edu

* Correspondence: sxcdream@swpu.edu.cn

Received: 21 January 2020; Accepted: 27 April 2020; Published: 29 April 2020

Abstract: This paper aims to study the effect of cyclic heating and flowing-water cooling conditions on the physical properties of granite. Ultrasonic tests, gas measured porosity, permeability, and microscope observations were conducted on granite after thermal treatment. The results showed that the velocity of P- and S-waves decreased as the number of thermal cycles increased. The porosity increased with the number of the thermal cycles attained at 600 °C, while no apparent changes were observed at 200 and 400 °C. The permeability increased with the increasing number of thermal cycles. Furthermore, microscope observations showed that degradation of the granite after thermal treatment was attributed to a large network of microcracks induced by thermal stress. As the number of thermal cycles increased, the number of transgranular microcracks gradually increased, as well as their length and width. The quantification of microcracks from cast thin section (CTS) images supported the visual observation.

Keywords: granite; physical characteristics; cyclic; thermal treatment; water cooling

---

1. Introduction

Geothermal energy is an important component of renewable energy, and most of the deep geothermal resource is stored in hot dry rock (HDR) [1]. HDR is defined as a hot and almost waterless geothermal system. Common HDR systems include granite or other crystalline basement rocks. Rock temperature varies from 150 to 500 °C at maximum depths of 5–6 km [2,3]. The use of deep HDR resources will contribute to the mitigation of the environmental pollution caused by traditional fossil energy [4]. Enhanced geothermal system (EGS) is an effective engineering method to exploit HDR resources [5,6]. To enhance the permeability of HDR, typical methods include increasing the heat transfer area and improving the efficiency of the hydraulic fracturing system. During the creation of the fracture network and the subsequent thermal energy exploitation, water is injected into the bedrock [7]. Studying the variation of rock physical properties after cyclic thermal treatment with flowing water cooling can provide a theoretical foundation for the EGS system.

A number of previous studies have conducted experiments to investigate the evolution of the physical and mechanical properties of the bedrock under temperature variations. Typical factors to consider include the heating temperature, heating rate, and cooling condition. With increasing temperature, rock experiences more significant deterioration. Hydraulic properties, such as porosity [8,9] and permeability [10,11], increase with a rise of the temperature, whereas mechanical properties, such as the P-wave velocity [11,12], unconfined compressive strength [7,12–15], tensile strength [11,16–18], and Young's modulus [9], decrease with increasing temperature [19]. The different types of microcracking also have been studied using 2D and 3D observations [20–23]. For example, thermal damage in Beishan granite subjected to high temperature treatment (from 100 to 800 °C at different heating rates, ranging from 1 to 15 °C/min) was studied in order to assess thermal effects on physical and mechanical properties. Results of acoustic emission (AE) monitoring, mechanical and physical properties measurements all indicated that heating rates had a significant impact on thermal damage. 5 °C/min was recognized as the critical heating rate for standard samples. Thermal stress induced by temperature gradients plays an important role in governing the damage in samples treated at a heating rate above 5 °C/min, at higher heating rates, thermal cracking is dominated by the stress concentrations caused by high thermal gradients [24,25]. Additionally, these physical and mechanical characteristics depend on different cooling conditions. Rock degradation caused by rapid cooling with water cooling or liquid nitrogen cooling was more severe compared with those under slower cooling such as air cooling, or cooling in a furnace [26–28]. Other rock types encountered in EGS projects, such as sandstone, have shown a similar trend [29–32]. The initial permeability of sandstone under certain pressure conditions was found to increased nonlinearly with the increase in temperature. Moreover, unconfined compressive strength and elastic parameters (i.e., elastic modulus and Poisson's ratio) of calcarenite decreased as the temperature was increased from 105 to 600 °C [33].

Many previous studies have investigated rock damage after a single heating and cooling treatment, however, few have considered the effect of cyclic treatments. Gräf studied the effects of cyclic thermal-heating treatment on the thermal expansion behavior of granite, however, the heating temperature and the number of treatment cycles were limited [34]. Mahmutoglu investigated the effect of thermal cycles on the mechanical behavior of marble and Buchberger sandstone. The results of unconfined compression, Brazilian and "Continuous Failure State" triaxial tests, pointed out that all of the mechanical parameters decreased gradually with an increasing number of heating cycles [35]. Additionally, Rong et al. studied the effect of thermal cycles on marble and granite subjected to air cooling on P-waves, stress–strain relationships, and acoustic emissions. The results revealed that thermal cyclic loading weakens the mechanical properties of the rock [36]. Furthermore, Wu et al. studied the effects of thermal cycles on the density, permeability, and unconfined compressive strength of granite subjected to liquid nitrogen cooling. Liquid nitrogen cooling was found to have a greater effect on the physical and mechanical properties of granite than air cooling [37]. Previous literature has also revealed that microcracks influence the physical and mechanical characteristics of the rock. Research has also quantified the microscopic responses of rock, particularly microcracking [38], to such processes.

Considering that water is the most common fluid used to extract thermal energy from HDR, in this experimental study, we investigate the effects of cyclic heating and water cooling on the physical and mechanical properties of granite, including a quantitative analysis of the resulting microcracks.

2. Materials and Methodology

2.1. Rock Samples

Granite was selected as the experimental material in our study. The samples were fine-grained granite with a grain size ranging from 0.5 to 1.5 mm (Chinese granite G655), which were collected from an outcrop located in Zhangzhou, Fujian, China (117.86° E, 24.83° N). No fissures were observed in the original rock. Two shapes of granite specimens were used: cylinders with the dimensions of 25 mm in diameter and 50 mm in length and discs with the dimensions of 25 mm in diameter and 10 mm in thickness (Figure 1). The mineralogy of the granite is described in Section 3.4.

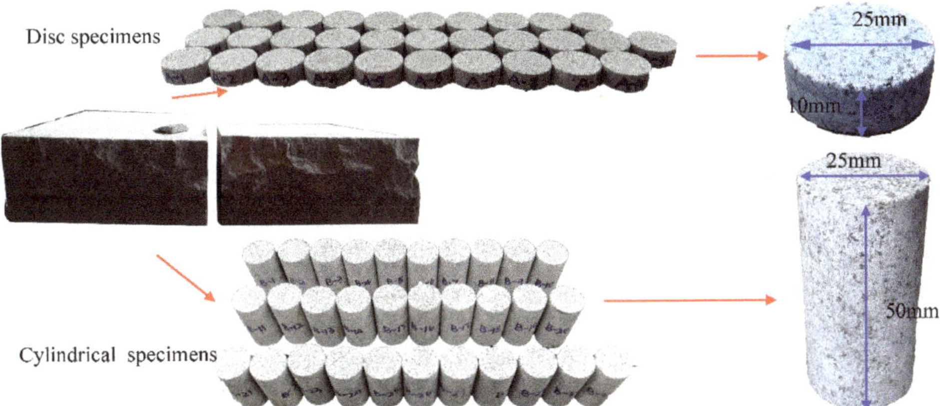

Figure 1. Photographs of the rock specimens.

2.2. Methodology of Heating and Cooling

The granite cylinders were used for the measurement of porosity, permeability, P- and S-wave velocity measurements, whereas the granite disc samples were used for microstructural characterization. The P-wave and S-wave velocity of the cylindrical specimens were measured prior to thermal treatment. Thirty-one cylinders and thirty-one discs with similar P-wave velocities were divided into five groups with six specimens in each group (see Table 1). A SX-G04123 box-type electric furnace was used in heating (power 2.5 kW, maximum temperature 1200 °C). The process of heating a rock specimen to a pre-determined temperature and then cooling it down to room temperature with water was regarded as one thermal cycle. A heating rate of 1.5 °C/min was used to avoid the influence of thermal shock [39]. The pre-determined temperature, once reached, was kept constant for 5 minutes in the furnace to avoid the influence of subsequent heating time at a pre-determined temperature. The specimens were then taken out of the furnace and cooled down to room temperature (20 °C) with flowing water (shown in Figure 2). For each group, the pre-determined temperatures were set to 100–600 °C to mimic a high-temperature condition of deep bedrocks in an EGS system. The numbers of thermal cycles for each group was either 0 (i.e., no thermal treatment), 1, 2, 4, 8, and 16.

Table 1. Thermal treatment conditions of granite employed in this study. (A: disc specimens, B: cylindrical specimens).

Temperature (°C)		100	200	300	400	500	600	
Number of cycles	1	A1, B1	A2, B2	A3, B3	A4, B4	A5, B5	A6, B6	A31, B31 (No thermal treatment)
	2	A7, B7	A8, B8	A9, B9	A10, B10	A11, B11	A12, B12	
	4	A13, B13	A14, B14	A15, B15	A16, B16	A17, B17	A18, B18	
	8	A19, B19	A20, B20	A21, B21	A22, B22	A23, B23	A24, B24	
	16	A25, B25	A26, B26	A27, B27	A28, B28	A29, B29	A30, B30	

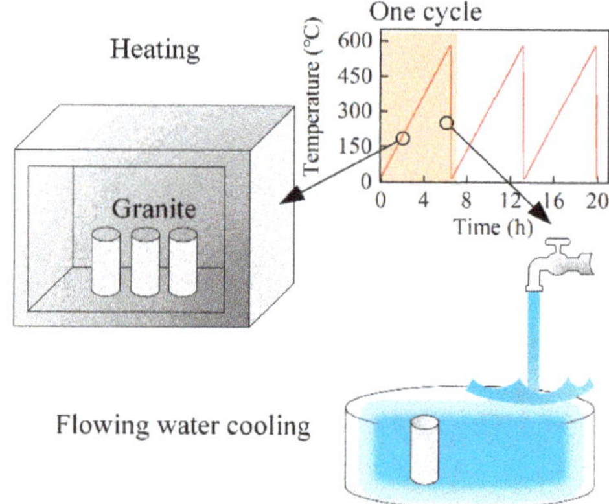

Figure 2. Schematic diagram of the heating and cooling process.

2.3. Ultrasonic Wave Velocity Measurements

Before measuring the porosity and permeability, P and S-wave velocities of the rock specimens were measured after thermal treatment for different temperatures and numbers of thermal cycles were measured using an ultrasonic apparatus (HS-YS2A Type). Two transducers fixed by a holder applying the same force were positioned at the ends of the specimen. A pressure gauge and hand wheel were used to apply an equal force with transducers during every test (see Figure 3). Each specimen was tested twice to ensure repeatability.

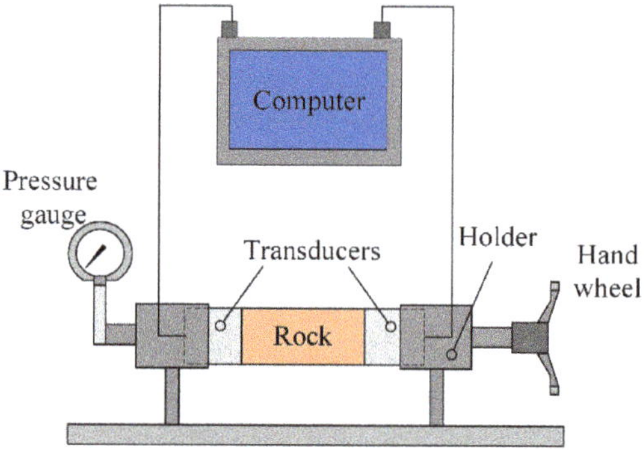

Figure 3. Schematic of the ultrasonic test.

2.4. Porosity and Gas Permeability Measurements

We measured the porosity and steady-state permeability of the three temperature groups (200, 400, and 600 °C) using an SCMS-E high-temperature, high-pressure, multi-parameter core measurement system. The test procedure was performed in accordance with the methods suggested by the Practices for Core Analysis (GB/T 29172-2012) in Chinese. The measurement gas was nitrogen, while the test temperature was room temperature (10–15 °C), and the confining pressure was 3.5 MPa (Figure 4).

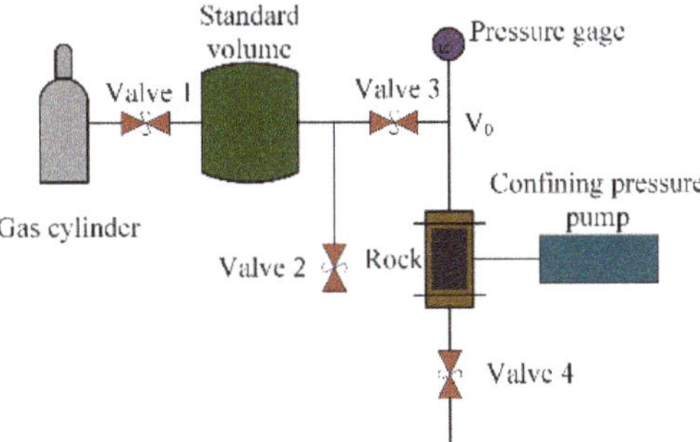

Figure 4. Schematic of the porosity and permeability measurement system.

2.5. Microscopic Observation

Cast thin sections (CTS), i.e., thin sections of rock impregnated with colored epoxy, were used for highlighting the microcracks. The disc rock samples were saturated with blue epoxy to distinguish pores and fracture from rock matrices. The procedure of obtaining CTS is shown in Figure 5. All disc rock specimens were impregnated with colored epoxy after thermal treatment. After leveling, lapping and polishing, a thin section of size 25 × 0.03 mm was obtained. The impregnated sample was bonded to the surface of a piece of glass for further processing. The thin section image was then photographed under plane polarized light and cross polarized light by using a polarizing microscope (Zeiss Scope A1) with an attached digital camera.

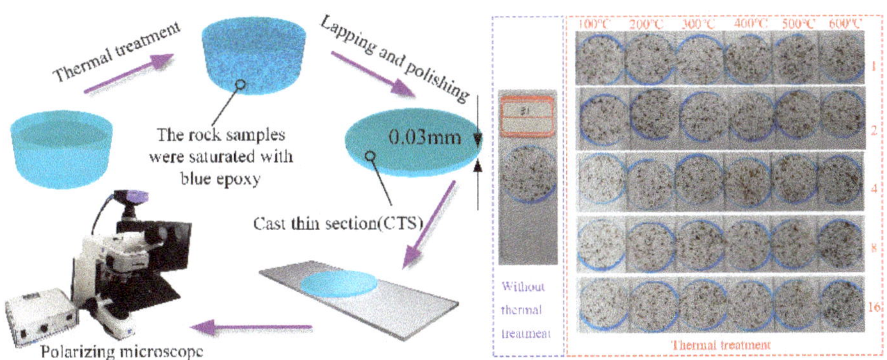

Figure 5. Schematic of polarizing microscope observation process.

2.6. Image Processing

Microcracks appear blue under plane-polarized light because the blue epoxy filled the microcracks. We selected the color of blue epoxy (R: 70, G: 132, B: 171). Usually, it is not consistent in every CTS image. Then, we set the tolerance to 70 to ensure that the blue parts of the image were successfully selected. The results are shown in Figure 6b. Manual interactive thresholding segmentation was used for the segmentation process. The thresholding of 40 was applied to the intermediate image based on the image histogram (see Figure 7). The binarized image result is shown in Figure 6d, which was further used for quantitative analysis (Figure 6c).

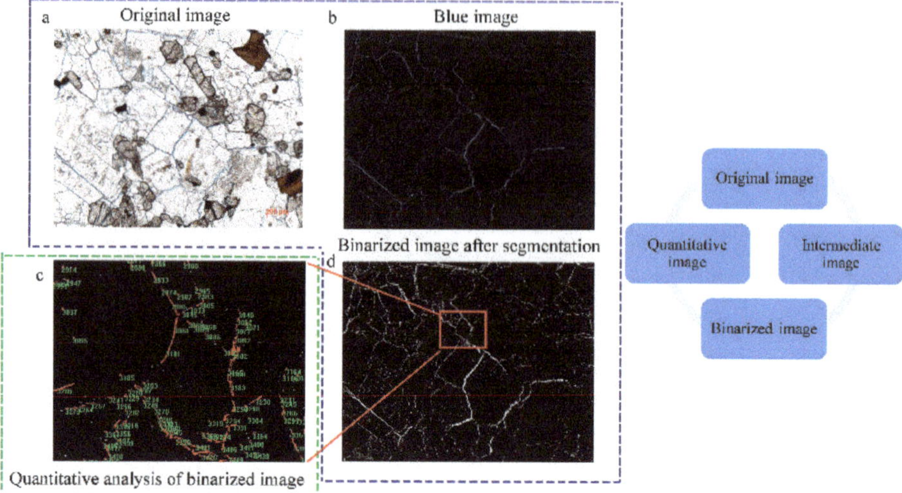

Figure 6. Schematic view of image processing. (The blue part is the preparation process, while the green part is the measurement process).

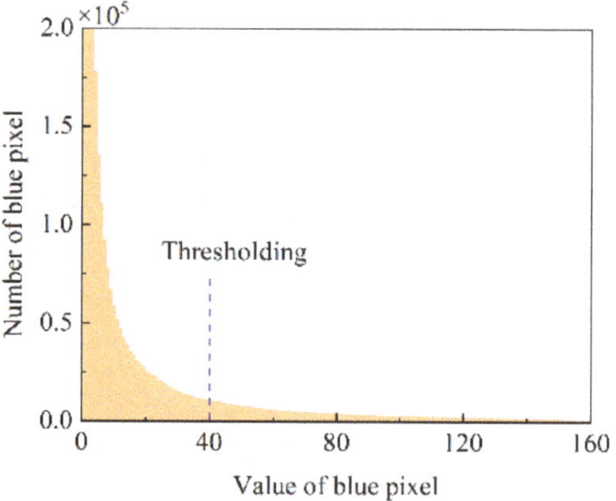

Figure 7. Segmentation via image thresholding.

We then measured four different parameters to measure: area, size (width), size (length), dendrites—one pixel thick open branches (Figure 8). The number of isolated elements was automatically counted (Figure 6d). It should be noted that the size width was not the actual width of the fracture. Finally, the image statistical data were exported to a worksheet for post processing. The image porosity was calculated according to Equation (1).

$$Porosity = \frac{White\ pixels}{All\ pixels} \times 100\% \tag{1}$$

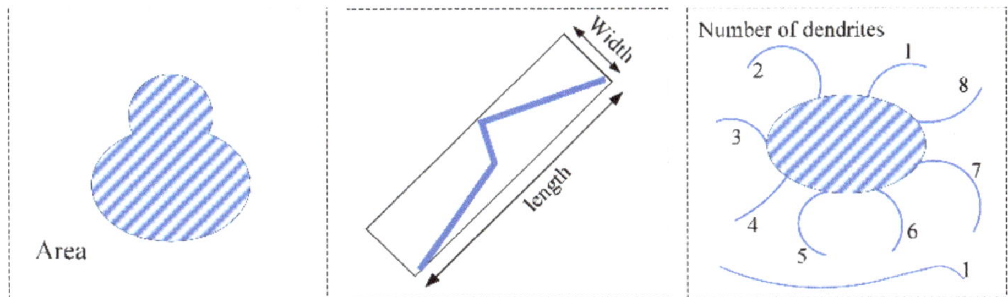

Figure 8. Three different types of parameters for measurements.

3. Results

3.1. Porosity

As shown in Figure 9, high temperatures (600 °C) had a greater effect on porosity than low heating temperatures (200 and 400 °C). Moreover, at 600 °C, porosity increased substantially as the cycle number increased from zero to two. At 200 and 400 °C, the influence of the number of thermal cycles on the porosity and permeability is negligible.

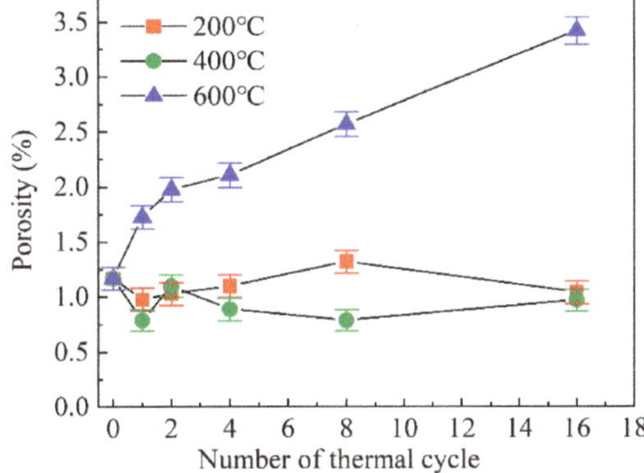

Figure 9. Effect of the number of thermal cycles and temperature on granite porosity.

3.2. Gas Permeability

The permeability variations with the number of thermal cycles at different heating are plotted in Figure 10. The trends were similar for all three different heating temperatures. A positive correlation was observed between thermal cycling and permeability increase at 400 and 600 °C, while there was an irregular rise against thermal cycles at 200 °C. In addition, the permeability of granite at 600 °C was significantly higher than that at 400 °C and 200 °C. At 600 °C, the permeability of granite increased from 0.0001 to 4.7770 mD after 16 thermal cycles.

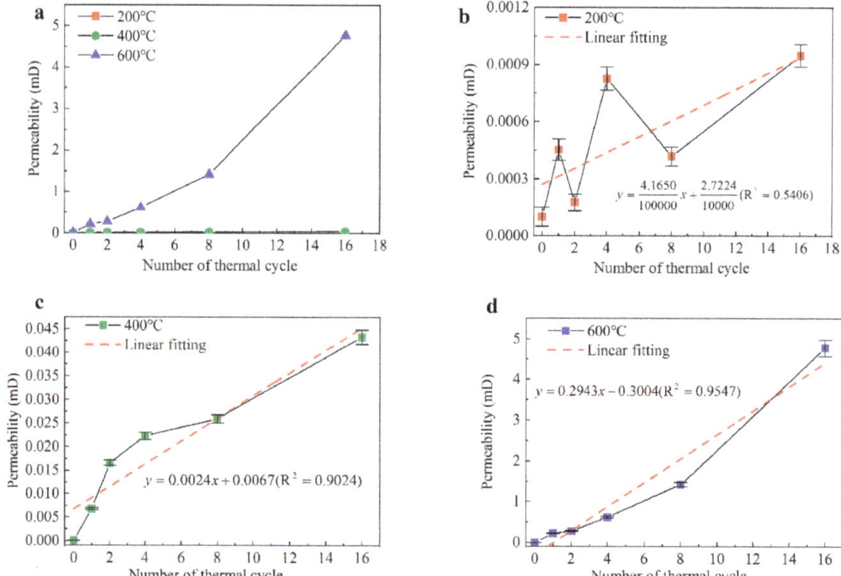

Figure 10. Permeability characteristics after thermal treatment. (**a**) Three different temperatures using same y axis. (**b**) The pre-determined temperature was 200 °C. (**c**) The pre-determined temperature was 400 °C. (**d**) The pre-determined temperature was 600 °C.

3.3. P- and S-Wave Velocity

Ultrasonic wave test is commonly used to detect the interior failure in a rock because of its simple and non-destructive characteristics. The typical P- and S-wave forms recorded and used to determine the velocities are presented in Figure 11 and the velocity results are shown in Figure 12. It appears that both P-wave velocity and S-wave velocity exhibited a similar negative correlation with heating temperature and the number of thermal cycles. The gradient of both P-wave and S-wave velocity decreased as the number of cycles increased. At 600 °C and one cycle of the thermal treatment, P- and S-wave velocities decreased by 73.6% and 58.6%, respectively. At 600 °C, after 16 cycles of the thermal treatment, the velocity reduced by 84.3% and 82.4%, respectively.

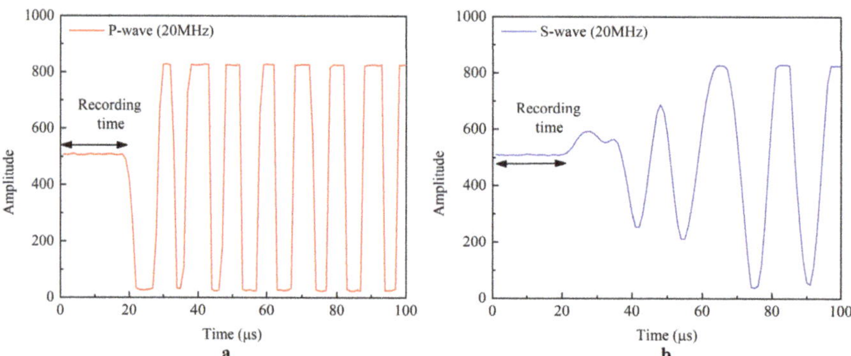

Figure 11. Typical P- and S wave forms recorded and used to determine velocities. (**a**) P-wave form. (**b**) S-wave form.

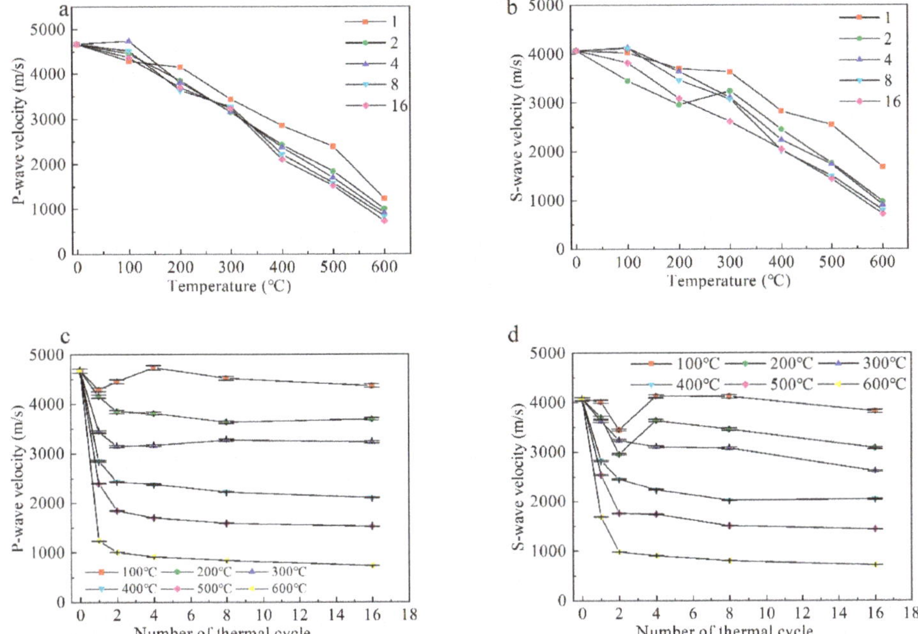

Figure 12. Variation of P-wave velocity and S-wave velocity changes with the number of thermal cycles and temperature. (**a**) Variation of P-wave velocity with temperature. (**b**) Variation of S-wave velocity with temperature. (**c**) Variation of P-wave velocity with the number of thermal cycles. (**d**) Variation of S-wave velocity with the numbers of thermal cycles.

3.4. Rock Microstructure

The microscopic observations of A31 (no thermal treatment) are shown in Figure 13. The granite is mainly composed of feldspar, quartz, and biotite with a small amount of pyroxene and magnetite. Anorthosite is 578 μm in length and 251 μm in width. Quartz is 458 μm in length and 243 μm in width. Clear boundaries were observed between mineral grains (Figure 13b). No blue epoxy was not observed in the plane-polarized image (Figure 13a), indicating that the granite has negligible porosity.

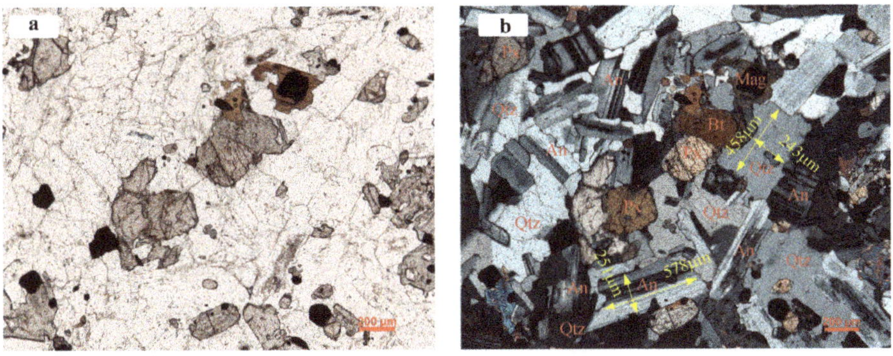

Figure 13. Polarized micrographs of granite without thermal treatment. (**a**) Plane polarized image; (**b**) cross polarized image. Qtz—Quartz; An—Anorthite; Bt—Biotite; Px—Pyroxene; Mag—Magnetite.

Figure 14 presents the CTS observation results for the granite after one thermal cycle at 600 °C. The mineral grains and the associated microcracks are clearly affected by thermal treatment. Two types of microcracks were observed: grain boundary microcracks, transgranular microcracks (including intracrystalline microcracks) [12]. Grain boundary microcracks describe the microcracks developed at the boundary between different minerals, whereas transgranular microcracks are those developed in the interior of the mineral grains. These microcracks were predominantly confined within minerals, with some passing through multiple grains. Occasionally, transgranular microcracks and grain boundary microcracks developed connectivity at an angle of approximately 90°. Furthermore, the location of the microcracks was related to mineral species. Transgranular microcracks were typically developed in feldspar, and the development direction was almost perpendicular to the optical twin crystal direction of feldspar. Grain boundary microcracks were typically between feldspar and quartz, and between feldspar and feldspar [25,40]. Almost no transgranular microcracks were observed in biotite.

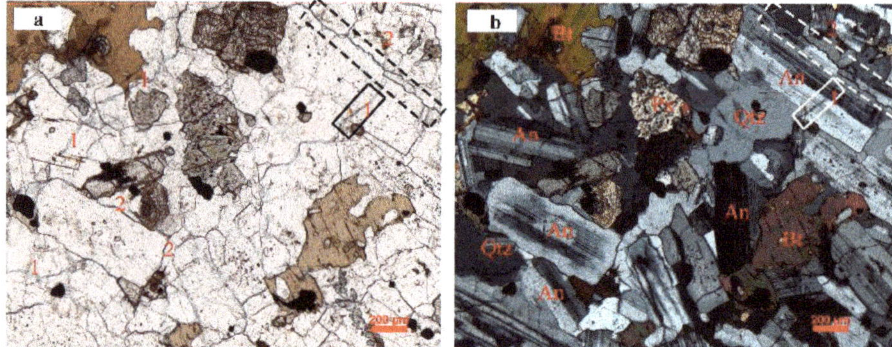

Figure 14. The types and location of granite fracture. (**a**) Plane polarized image; (**b**) cross polarized image. Qt—Quartz; An—Anorthose; Bt—Biotite; Px—Pyroxene. 1—grain boundary microcracks; 2—transgranular microcracks. (600 °C, 1 cycle).

Plane-polarized and cross-polarized CTS images of the granite specimens with different heating temperatures (100–600 °C, one cycle) are presented in Figures 15 and 16, respectively. Heating temperature affected the microcracks development behaviors. In the range of 100–300 °C, minimal blue epoxy was observed in the images. Grain boundary microcracks were sparsely distributed in a specimen heated to 400 °C (see Figure 15d), however, as the temperature exceeded 400 °C, more grain microcracks could be observed in the granite. For the specimen treated at 500 °C heating temperature (see Figures 15e and 16e), transgranular microcracks were observed in feldspar. Grain boundary microcracks also existed but in a smaller proportion than those transgranular microcracks in the granite subjected to 600 °C. According to the plane-polarized microscope observation, as shown in Figure 15f (600 °C), abundant grain boundary microcracks were found along feldspar and quartz grains. Moreover, the transgranular microcracks and grain boundary microcracks occasionally coalesced.

Thermal cycles also had a significant influence on microcrack evolution, especially that of transgranular. The plane-polarized and cross-polarized photographs of CTS for granite specimens treated at 600 °C with different numbers of thermal cycles are presented in Figures 17 and 18. With the increase of thermal cycles, both the length and width of transgranular microcracks increased, as well as the number of grain boundary microcracks and transgranular microcracks. Figures 17f and 18f present sample A30, which was subject to 600 °C heating and 16 thermal cycles. The maximal width of transgranular reached approximately 20 μm. Microcracks were well developed and most mineral boundaries had grain boundary microcracks. The width of the transgranular microcracks was larger than that of grain boundary microcracks.

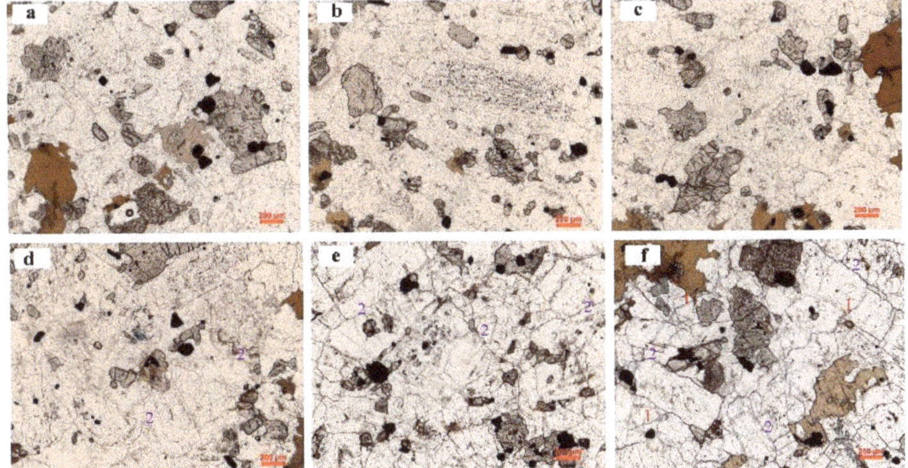

Figure 15. Plane polarized images of microcracks in granites subjected to different heating temperatures. (**a**) 100 °C; (**b**) 100 °C; (**c**) 200 °C; (**d**) 400 °C; (**e**) 500 °C; (**f**) 600 °C; 1—grain boundary microcracks; 2—transgranular microcracks.

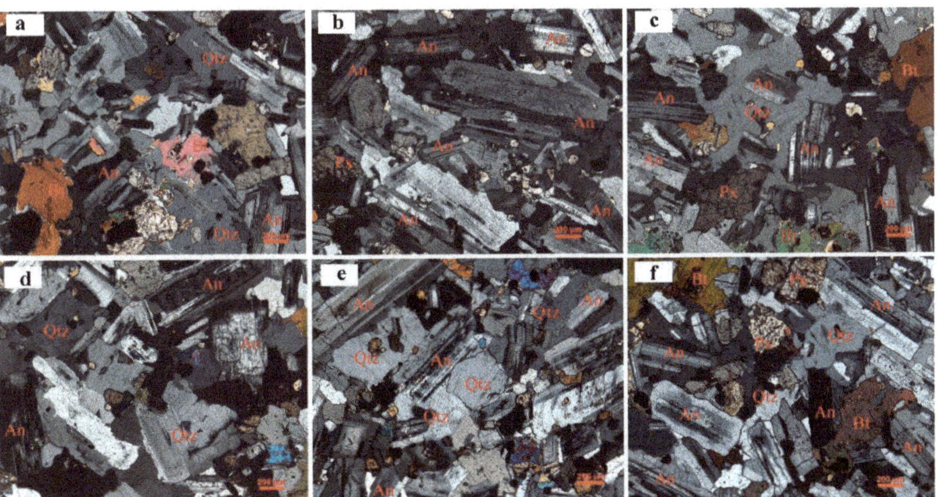

Figure 16. Cross polarized images of microcracks for granites subjected to different heating temperatures. (**a**) 100 °C; (**b**) 100 °C; (**c**) 200 °C; (**d**) 400 °C; (**e**) 500 °C; (**f**) 600 °C. Qtz—Quartz; An—Anorthose; Bt—Biotite; Px—Pyroxene.

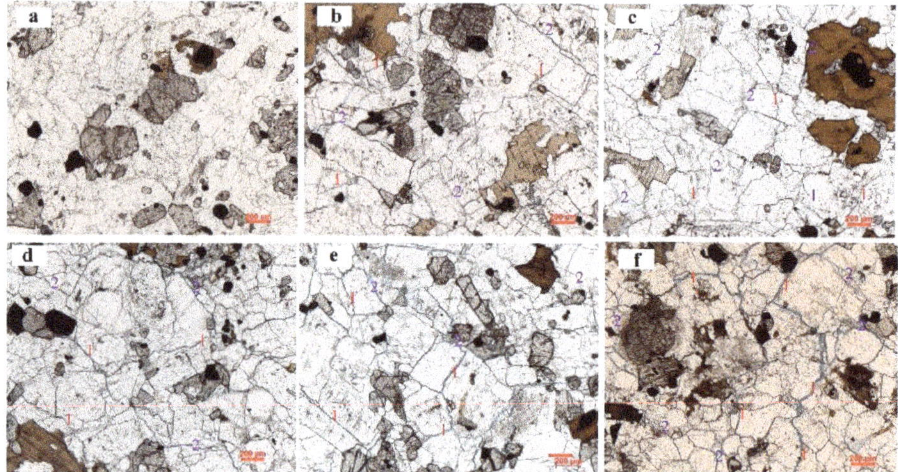

Figure 17. Plane polarized images of microcracks for granites with 600 °C heating temperature subjected to different numbers of the thermal cycle. (**a**) No thermal treatment; (**b**) 1 cycle; (**c**) 2 cycles; (**d**) 4 cycles; (**e**) 8 cycles; (**f**) 16 cycles. 1—grain boundary microcracks; 2—transgranular microcracks.

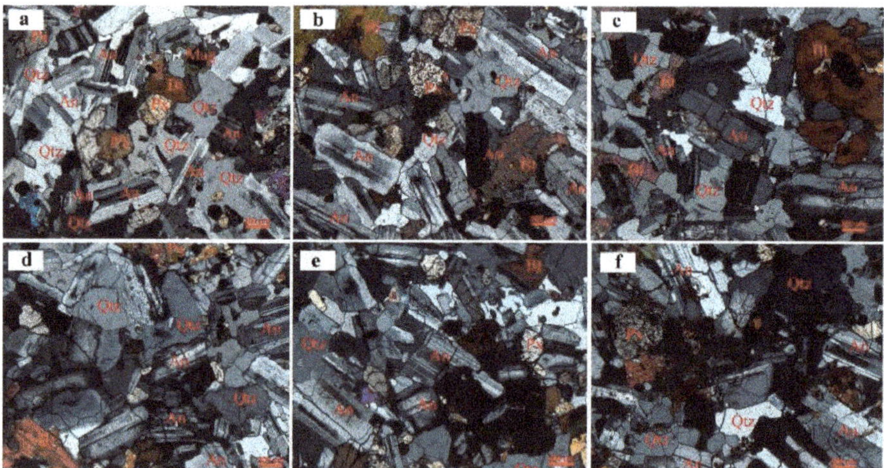

Figure 18. Cross polarized images of microcracks for granites with 600 °C heating temperature subjected to different numbers of the thermal cycle. (**a**) No thermal treatment; (**b**) 1 cycle; (**c**) 2 cycles; (**d**) 4 cycles; (**e**) 8 cycles; (**f**) 16 cycles. Qtz—Quartz; An—Anorthose; Bt—Biotite; Px—Pyroxene.

3.5. Microcrack Morphology

The granite morphology was analyzed through the CTS images (Figure 19), which were sorted according to the microcrack area and arranged vertically. The length of microcracks and the number of inflexion points increased with the number of thermal cycles (see Figure 19). This implies that the development and cross-cutting of grain-boundary microcracks and transgranular microcracks developed and crossed together. Hence, thermal cycles had a substantial influence on high-temperature granite subjected to water cooling.

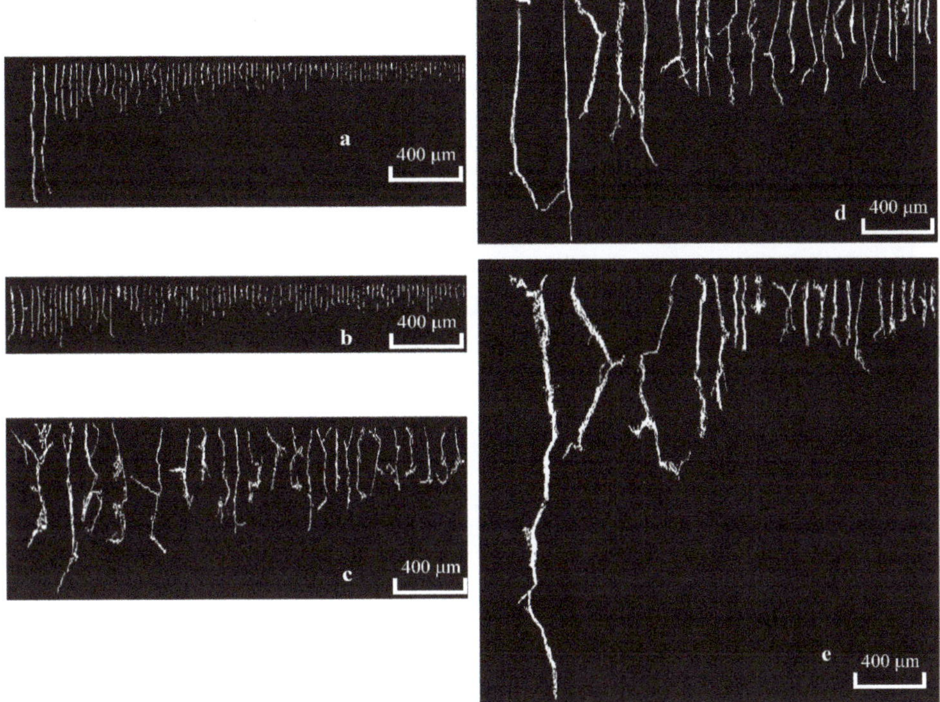

Figure 19. The morphology for granite with 600 °C heating temperature and different thermal cycles. (**a**) 1 cycle; (**b**) 2 cycles; (**c**) 4 cycles; (**d**) 8 cycles; (**e**) 16 cycles.

We conducted a statistical analysis of the microcracks area to explore the distribution of pores size. As shown in Figure 20, relative frequency of microcracks descended rapidly from 0.35 to 0.05. 80% of pores contained less than 21 pixels or 17.01 μm^2 (1 pixel $\approx$ 0.81 μm^2). Tiny pores were the most numerous among all microcrack sizes.

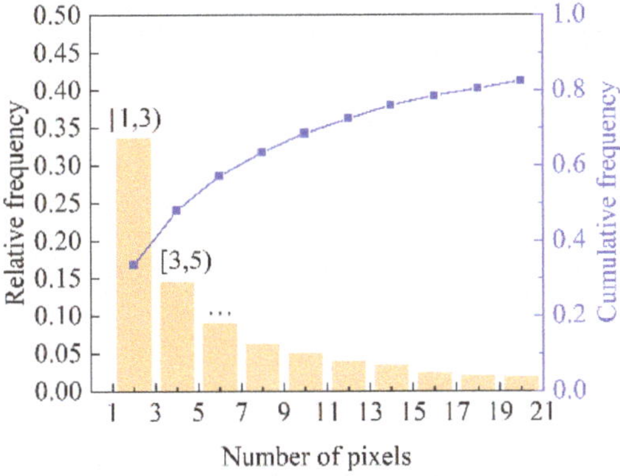

Figure 20. Distribution of microcracks with 600 °C heating temperature and 8 thermal cycles.

The results of the CTS image processing are presented in Figure 21. Generally, differences were observed between the porosity measured via gas permeability measurement and microscope observations. The latter increased by 3.5 times from one cycle to 16 cycles, which rose from 1.61% to 5.67%. Conversely, the former only increased from 1.73% to 3.42% (Figure 21a). Measurements of maximum porosity, maximum length, and maximum width tended to increase with a greater number of thermal cycles (see Figure 21b–d). The line charts revealed that the development of microcracks induced by thermal stress increased with the rise of thermal cycles. The number of measured dendrites is shown in Figure 22, which also increased with the number of thermal cycles, by almost 13 times to 63. This clearly illustrates the expansion of microcracks with the number of thermal cycles at high heating temperatures.

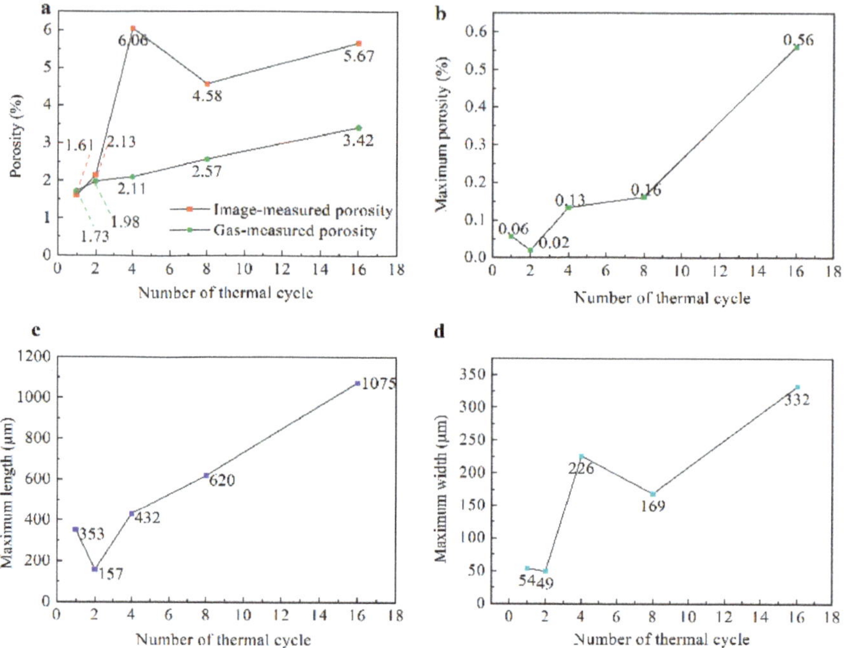

Figure 21. Results of the cast thin section (CTS) image processing versus the numbers of the thermal cycle. (**a**) Variation of image porosity. (**b**) Variation maximum image porosity. (**c**) Variation maximum fracture length. (**d**) Variation maximum fracture width.

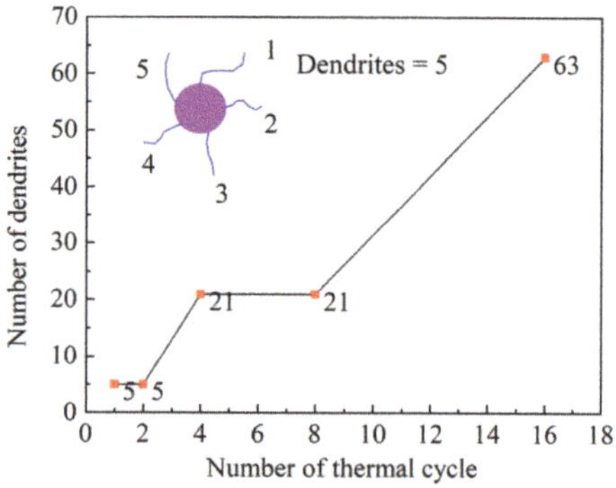

Figure 22. Effect of the number of thermal cycles on the number of dendrites.

4. Discussion

Porosity and crack density are the most important properties, playing a major role in the structural integrity of a rock [30]. They are related to the type and arrangement of mineral grains, internal pores, and microcracks. High temperature thermal treatment could increase the porosity of granite. Some researchers reported that 400 °C is a threshold temperature for granite to change its structure [9,30,41]. Below the threshold temperature, no significant relationship was found between the porosity of granite and the number of thermal cycles, suggesting that the increasing number of thermal cycles does not contribute to the propagation of microcracks. However, beyond a temperature of 600 °C, the increase of cyclic thermal treatment is associated with the increase of rock porosity.

Permeability is in relation to pores and especially fractures. Generally, fractures play an important role in the percolation capacity. Obviously, the thermal treatments can enhance the permeability of granite. The permeability is significantly improved as heating temperature rises [11,17,42]. Interestingly, the porosity at 400 °C was lower than that at 200 °C except after two thermal cycles, whereas the permeability exhibited the exact opposite trend. This reflected that the enhanced connectivity between granite microcracks was enhanced. Pore and crack networks facilitated fluid flow through the specimens, leading to the increased permeability. As a result, it demonstrated reversely that propagation of microcracks increased with the number of thermal cycles. The number of thermal cycles also had significant effects on the permeability of granite.

The CTS image processing employed in this study exhibits some limitations. First of all, the images cannot reveal all microcracks in the sample because of their finite resolution and size. Secondly, the images are 2D images rather than 3D images; thus, some microcracks cannot be observed for technical reasons. These two limitations would result in an underestimation of the porosity. Third, due to the inconsistent blue color of epoxy, some noise pollution is inevitable in the CTS image, which would lead to an overestimated porosity. Nevertheless, these images still provide useful and quantitative descriptions of microcracks in the granite.

During the process of heating and cooling, a series of physical and chemical reactions have occurred in granite (see Figure 23). When the heating temperature exceeded 100 °C, water inclusions that originally existed in the granite pores due to wettability and capillary force-escaped rock in the form of gas [8]. As a result, the P-wave velocity decreased. The chemical formula for biotite is $K\{(Mg_{<0.67}, Fe_{>0.33})_3[AlSi_3O_{13}](OH)_2\}$. In the temperature range of 300–500 °C, due to the escape of crystal water and dissociation of the H^+ and OH^- (the existing form of constitution water in the mineral crystal lattice structure), the mineral framework was destroyed and microcracks developed in rock [17,42]. When the heating temperature reaches 573 °C, low-temperature α quartz turns into high-temperature β quartz, which is accompanied by a sudden volume expansion [7,43–45]. This transition results in severe derogation of the granite. During the heating process, due to differences in the thermal expansion coefficients of different minerals, thermal stresses accumulate at granular interfaces, even if the temperature field outside is uniform. This expansion mismatch primarily contributes to the development of grain boundary microcracks [24,46]. During the cooling process, granite specimens are placed in the flowing water, which induces the sudden variation of temperature and results in the gradual formation of grain boundary microcracks and transgranular microcracks among rock minerals. Water then invades the connected microcracks, resulting in new chemical reactions in the minerals. Because the transition of quartz to β quartz is a reciprocal reaction, the granite specimens experience repeated damage with an increasing number of thermal cycles.

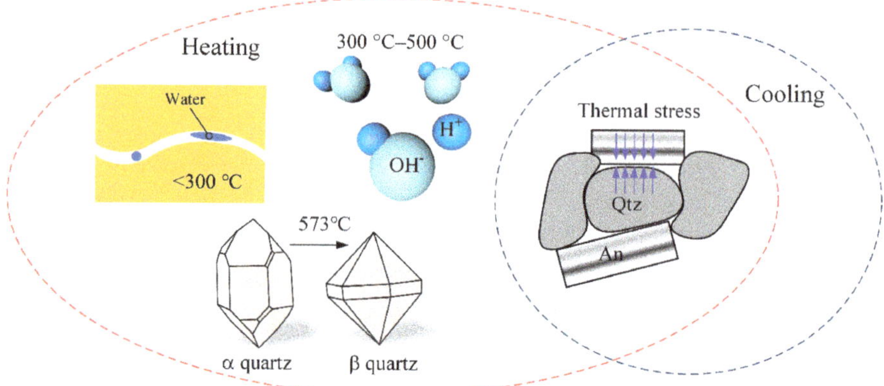

Figure 23. Mechanisms of thermal damage in heating and cooling.

5. Conclusions

We conducted a set of physical experiments to investigate the effect of cyclic thermal treatment and water cooling on the physical characteristics of granite. The results show that the thermal cycling has a significant influence on the physical characteristics (i.e., porosity, permeability, the seismic velocity). The results contribute to the fundamental understanding of the evolution of porosity and permeability in HDR geothermal systems. Qualitative and quantitative analyses of CTS images led to the following conclusions:

(1) Physical characteristics changed significantly after flowing water cooling at high heating temperatures versus the number of thermal cycles. P- and S-waves reduced with the increase of thermal cycles. Porosity did not change substantially at heating temperatures of less than 400 °C. The permeability increased by four orders of magnitude compared to the samples without thermal treatment, which is susceptible than porosity.

(2) Both grain boundary microcracks and transgranular microcracks were found. The primary effect of heating was grain boundary cracking during the first thermal cycle. Increasing the number of thermal cycles, transgranular microcracks also developed in the rock. Both types of grain boundary microcracks and transgranular microcracks coalesced to form a fracture network.

(3) Quantification of the crack morphology from CTS images indicated that the large number of microcracks that developed in the granite during high-temperature treatment changed the rocks physical properties. The length of microcracks increased by one order of magnitude.

Author Contributions: Conceptualization, X.S. and C.Z.; data curation, J.W. and L.G.; formal analysis, L.G. and S.C.; supervision, C.Z. and L.G.; validation, X.Z. and X.S.; writing—original draft, L.G., X.Z. and X.S. All authors have read and agreed to the published version of the manuscript.

Funding: This research was funded by the National Science Foundation of China, grant number 51774428, and Sichuan International Science and Technology Innovation Cooperation/Hong Kong, Macao and the Taiwan Science and Technology Innovation Cooperation, grant number 2019YFH0166, and the National Students' Innovation and Entrepreneurship Training Program of China, grant number 20180615013.

References

1. Huang, W.; Cao, W.; Jiang, F. A novel single-well geothermal system for hot dry rock geothermal energy exploitation. *Energy* **2018**, *162*, 630–644. [CrossRef]

2. Breede, K.; Dzebisashvili, K.; Liu, X.; Falcone, G. A systematic review of enhanced (or engineered) geothermal systems: Past, present and future. *Geotherm. Energy* **2013**, *1*, 4. [CrossRef]

3.   Tomac, I.; Sauter, M. A review on challenges in the assessment of geomechanical rock performance for deep geothermal reservoir development. *Renew. Sustain. Energy Rev.* **2018**, *82*, 3972–3980. [CrossRef]

4.   Pan, S.-Y.; Gao, M.; Shah, K.J.; Zheng, J.; Pei, S.-L.; Chiang, P.-C. Establishment of enhanced geothermal energy utilization plans: Barriers and strategies. *Renew. Energy* **2019**, *132*, 19–32. [CrossRef]

5.   Kelkar, S.; WoldeGabriel, G.; Rehfeldt, K. Lessons learned from the pioneering hot dry rock project at Fenton Hill, USA. *Geothermics* **2016**, *63*, 5–14. [CrossRef]

6.   Wan, Z.; Zhao, Y.; Kang, J. Forecast and evaluation of hot dry rock geothermal resource in China. *Renew. Energy* **2005**, *30*, 1831–1846. [CrossRef]

7.   Kumari, W.G.P.; Ranjith, P.G.; Perera, M.S.A.; Chen, B.K.; Abdulagatov, I.M. Temperature-dependent mechanical behaviour of Australian Strathbogie granite with different cooling treatments. *Eng. Geol.* **2017**, *229*, 31–44. [CrossRef]

8.   Wang, P.; Yin, T.; Li, X.; Zhang, S.; Bai, L. Dynamic Properties of Thermally Treated Granite Subjected to Cyclic Impact Loading. *Rock Mech. Rock Eng.* **2019**, *52*, 991–1010. [CrossRef]

9.   Zhang, F.; Zhao, J.; Hu, D.; Skoczylas, F.; Shao, J. Laboratory Investigation on Physical and Mechanical Properties of Granite After Heating and Water-Cooling Treatment. *Rock Mech. Rock Eng.* **2017**, *51*, 677–694. [CrossRef]

10. Feng, Z.; Zhao, Y.; Zhang, Y.; Wan, Z. Real-time permeability evolution of thermally cracked granite at triaxial stresses. *Appl. Therm. Eng.* **2018**, *133*, 194–200. [CrossRef]

11. Jin, P.; Hu, Y.; Shao, J.; Zhao, G.; Zhu, X.; Li, C. Influence of different thermal cycling treatments on the physical, mechanical and transport properties of granite. *Geothermics* **2019**, *78*, 118–128. [CrossRef]

12. Peng, J.; Rong, G.; Cai, M.; Yao, M.-D.; Zhou, C.-B. Physical and mechanical behaviors of a thermal-damaged coarse marble under uniaxial compression. *Eng. Geol.* **2016**, *200*, 88–93. [CrossRef]

13. Chen, Y.; Ni, J.; Shao, W.; Azzam, R. Experimental study on the influence of temperature on the mechanical properties of granite under uni-axial compression and fatigue loading. *Int. J. Rock Mech. Min. Sci.* **2012**, *56*, 62–66. [CrossRef]

14. Gautam, P.K.; Verma, A.K.; Jha, M.K.; Sharma, P.; Singh, T.N. Effect of high temperature on physical and mechanical properties of Jalore granite. *J. Appl. Geophys.* **2018**, *159*, 460–474. [CrossRef]

15. Yang, S.; Ranjith, P.G.; Jing, H.-W.; Tian, W.-L.; Ju, Y. An experimental investigation on thermal damage and failure mechanical behavior of granite after exposure to different high temperature treatments. *Geothermics* **2017**, *65*, 180–197. [CrossRef]

16. Hu, X.; Song, X.; Liu, Y.; Cheng, Z.; Ji, J.; Shen, Z. Experiment investigation of granite damage under the high-temperature and high-pressure supercritical water condition. *J. Pet. Sci. Eng.* **2019**, *180*, 289–297. [CrossRef]

17. Zhang, W.; Sun, Q.; Hao, S.; Geng, J.; Lv, C. Experimental study on the variation of physical and mechanical properties of rock after high temperature treatment. *Appl. Therm. Eng.* **2016**, *98*, 1297–1304. [CrossRef]

18. Yin, T.; Li, X.; Cao, W.; Xia, K. Effects of Thermal Treatment on Tensile Strength of Laurentian Granite Using Brazilian Test. *Rock Mech. Rock Eng.* **2015**, *48*, 2213–2223. [CrossRef]

19. Heuze, F.E. High-temperature mechanical, physical and Thermal properties of granitic rocks—A review. *Int. J. Rock Mech. Min. Sci. Geomech. Abstr.* **1983**, *20*, 3–10. [CrossRef]

20. Fredrich, J.T.; Wong, T.-F. Micromechanics of thermally induced cracking in three crustal rocks. *J. Geophys. Res. Solid Earth* **1986**, *91*, 12743–12764. [CrossRef]

21. Chen, Y.; Kobayashi, T.; Kuriki, Y.; Kusuda, H.; Mabuchi, M. Observation of microstructures in granite samples subjected to one cycle of heating and cooling. *J. Jpn. Soc. Eng. Geol.* **2008**, *49*, 217–226. [CrossRef]

22. Arena, A.; Delle Piane, C.; Sarout, J. A new computational approach to cracks quantification from 2D image analysis: Application to micro-cracks description in rocks. *Comput. Geosci.* **2014**, *66*, 106–120. [CrossRef]

23. Delle Piane, C.; Arena, A.; Sarout, J.; Esteban, L.; Cazes, E. Micro-crack enhanced permeability in tight rocks: An experimental and microstructural study. *Tectonophysics* **2015**, *665*, 149–156. [CrossRef]

24. Rossi, E.; Kant, M.A.; Madonna, C.; Saar, M.O.; Rudolf von Rohr, P. The Effects of High Heating Rate and High Temperature on the Rock Strength: Feasibility Study of a Thermally Assisted Drilling Method. *Rock Mech. Rock Eng.* **2018**, *51*, 2957–2964. [CrossRef]

25. Chen, S.; Yang, C.; Wang, G. Evolution of thermal damage and permeability of Beishan granite. *Appl. Therm. Eng.* **2017**, *110*, 1533–1542. [CrossRef]

26. Zhao, Z.; Liu, Z.; Pu, H.; Li, X. Effect of Thermal Treatment on Brazilian Tensile Strength of Granites with Different Grain Size Distributions. *Rock Mech. Rock Eng.* **2018**, *51*, 1293–1303. [CrossRef]

27. Isaka, B.; Gamage, R.; Rathnaweera, T.; Perera, M.; Chandrasekharam, D.; Kumari, W. An Influence of Thermally-Induced Micro-Cracking under Cooling Treatments: Mechanical Characteristics of Australian Granite. *Energies* **2018**, *11*, 1338. [CrossRef]

28. Sarout, J.; Cazes, E.; Delle Piane, C.; Arena, A.; Esteban, L. Stress-dependent permeability and wave dispersion in tight cracked rocks: Experimental validation of simple effective medium models. *J. Geophys. Res. Solid Earth* **2017**, *122*, 6180–6201. [CrossRef]

29. Liu, Q.; Qian, Z.; Wu, Z. Micro/macro physical and mechanical variation of red sandstone subjected to cyclic heating and cooling: An experimental study. *Bull. Eng. Geol. Environ.* **2017**, *78*, 1485–1499. [CrossRef]

30. Sirdesai, N.N.; Mahanta, B.; Ranjith, P.G.; Singh, T.N. Effects of thermal treatment on physico-morphological properties of Indian fine-grained sandstone. *Bull. Eng. Geol. Environ.* **2017**, *78*, 883–897. [CrossRef]

31. Yang, S.; Hu, B. Creep and Long-Term Permeability of a Red Sandstone Subjected to Cyclic Loading After Thermal Treatments. *Rock Mech. Rock Eng.* **2018**, *51*, 2981–3004. [CrossRef]

32. Yang, S.; Xu, P.; Li, Y.; Huang, Y. Experimental investigation on triaxial mechanical and permeability behavior of sandstone after exposure to different high temperature treatments. *Geothermics* **2017**, *69*, 93–109. [CrossRef]

33. Brotóns, V.; Tomás, R.; Ivorra, S.; Alarcón, J.C. Temperature influence on the physical and mechanical properties of a porous rock: San Julian's calcarenite. *Eng. Geol.* **2013**, *167*, 117–127. [CrossRef]

34. Gräf, V.; Jamek, M.; Rohatsch, A.; Tschegg, E. Effects of thermal-heating cycle treatment on thermal expansion behavior of different building stones. *Int. J. Rock Mech. Min. Sci.* **2013**, *64*, 228–235. [CrossRef]

35. Mahmutoglu, Y. Mechanical Behaviour of Cyclically Heated Fine Grained Rock. *Rock Mech. Rock Eng.* **1998**, *31*, 169–179. [CrossRef]

36. Rong, G.; Peng, J.; Cai, M.; Yao, M.; Zhou, C.; Sha, S. Experimental investigation of thermal cycling effect on physical and mechanical properties of bedrocks in geothermal fields. *Appl. Therm. Eng.* **2018**, *141*, 174–185. [CrossRef]

37. Wu, X.; Huang, Z.; Cheng, Z.; Zhang, S.; Song, H.; Zhao, X. Effects of cyclic heating and LN2-cooling on the physical and mechanical properties of granite. *Appl. Therm. Eng.* **2019**, *156*, 99–110. [CrossRef]

38. Freire-Lista, D.; Fort, R.; Varas-Muriel, M. Thermal stress-induced microcracking in building granite. *Eng. Geol.* **2016**, *206*, 83–93. [CrossRef]

39. Dwivedi, R.D.; Goel, R.K.; Prasad, V.V.R.; Sinha, A. Thermo-mechanical properties of Indian and other granites. *Int. J. Rock Mech. Min. Sci.* **2008**, *45*, 303–315. [CrossRef]

40. Isaka, B.L.A.; Ranjith, P.G.; Rathnaweera, T.D.; Perera, M.S.A.; Silva, V.R.S.D. Quantification of thermally-induced microcracks in granite using X-ray CT imaging and analysis. *Geothermics* **2019**, *81*, 152–167. [CrossRef]

41. Chaki, S.; Takarli, M.; Agbodjan, W.P. Influence of thermal damage on physical properties of a granite rock: Porosity, permeability and ultrasonic wave evolutions. *Constr. Build. Mater.* **2008**, *22*, 1456–1461. [CrossRef]

42. Wu, X.; Huang, Z.; Song, H.; Zhang, S.; Cheng, Z.; Li, R.; Wen, H.; Huang, P.; Dai, X. Variations of Physical and Mechanical Properties of Heated Granite After Rapid Cooling with Liquid Nitrogen. *Rock Mech. Rock Eng.* **2019**, *52*, 2123–2139. [CrossRef]

43. Cheng, Y.; Luo, N.; Xie, X.; Li, M.; Yu, S. α-β phase transition of quartz based on molecular dynamics simulations. In Proceedings of the 9th International Conference on Properties and Applications of Dielectric Materials, Harbin, China, 19–23 July 2009.

44. Duffrene, L. Molecular dynamic simulations of the alpha; -beta; phase transition in silica cristobalite. *J. Phys. Chem. Solids* **1998**, *59*, 1025–1037. [CrossRef]

45. Griffiths, L.; Heap, M.J.; Baud, P.; Schmittbuhl, J. Quantification of microcrack characteristics and implications for stiffness and strength of granite. *Int. J. Rock Mech. Min. Sci.* **2017**, *100*, 138–150. [CrossRef]

46. Wang, H.F.; Bonner, B.P.; Carlson, S.R.; Kowallis, B.J.; Heard, H.C. Thermal stress cracking in granite. *J. Geophys. Res.* **1989**, *94*. [CrossRef]

Article

Low Enthalpy Geothermal Systems in Structural Controlled Areas: A Sustainability Analysis of Geothermal Resource for Heating Plant (The Mondragone Case in Southern Appennines, Italy)

**Marina Iorio [1], Alberto Carotenuto [2], Alfonso Corniello [3], Simona Di Fraia [2,*],
Nicola Massarotti [2], Alessandro Mauro [2], Renato Somma [4,5] and Laura Vanoli [2]**

[1]   Istituto Scienze del Mare, CNR, 80133 Napoli, Italy; marina.iorio@cnr.it
[2]   Dipartimento di Ingegneria, Università degli Studi di Napoli "Parthenope", 80143 Napoli, Italy;
     alberto.carotenuto@uniparthenope.it (A.C.); massarotti@uniparthenope.it (N.M.);
     alessandro.mauro@uniparthenope.it (A.M.); laura.vanoli@uniparthenope.it (L.V.)
[3]   Dipartimento di Ingegneria Civile Edile ed Ambientale, Università degli Studi "Federico II",
     80125 Napoli, Italy; corniell@unina.it
[4]   Istituto Nazionale di Geofisica e Vulcanologia Sezione di Napoli, 80124 Napoli, Italy; renato.somma@ingv.it
[5]   Istituto di Ricerca su Innovazione e Servizi per lo Sviluppo, CNR, 80134 Napoli, Italy
*   Correspondence: simona.difraia@uniparthenope.it

Received: 16 January 2020; Accepted: 5 March 2020; Published: 6 March 2020

Abstract: In this study, the sustainability of low-temperature geothermal field exploitation in a carbonate reservoir near Mondragone (CE), Southern Italy, is analyzed. The Mondragone geothermal field has been extensively studied through the research project VIGOR (Valutazione del potenzIale Geotermico delle RegiOni della convergenza). From seismic, geo-electric, hydro-chemical and groundwater data, obtained through the experimental campaigns carried out, physiochemical features of the aquifers and characteristics of the reservoir have been determined. Within this project, a well-doublet open-loop district heating plant has been designed to feed two public schools in Mondragone town. The sustainability of this geothermal application is analyzed in this study. A new exploration well (about 300 m deep) is considered to obtain further stratigraphic and structural information about the reservoir. Using the derived hydrogeological model of the area, a numerical analysis of geothermal exploitation was carried out to assess the thermal perturbation of the reservoir and the sustainability of its exploitation. The effect of extraction and reinjection of fluids on the reservoir was evaluated for 60 years of the plant activity. The results are fundamental to develop a sustainable geothermal heat plant and represent a real case study for the exploitation of similar carbonate reservoir geothermal resources.

Keywords: carbonate geothermal reservoirs; sustainable geothermal energy exploitation; southern Italy; numerical simulation

---

1. Introduction

The total (electric and thermal) potential energy of a geothermal reservoir depends on many variables such as thermal regime, geological structure, and hydrogeological properties. Recently, the use of groundwater, extracted via open-loop systems, is increasingly being considered as an efficient means for heating and cooling of buildings, especially in the heat pump systems.

This research concerns the geothermal carbonate reservoir, which makes the bedrock in the alluvial Mondragone plain, Southern Italy. The site has been extensively studied through the VIGOR project (coordinated by the Italian National Council Research and sponsored by the Italian Ministry of

Economic Development) dedicated to the evaluation of geothermal potential in four Italian Regions: Puglia, Calabria, Campania, and Sicily. In this plain, a well-doublet open-loop plant has been designed for the district heating of seven public schools of Mondragone town [1].

Low-temperature geothermal fields have been exploited over decades for industrial applications in Iceland, Hungary, China, Turkey, France, Germany, Russia, and other countries [2], and in recent times many authors propose the use of geothermal energy standalone or coupled with other renewable energy sources [3–5]. As an example, a classical low-temperature geothermal field of meteoric origin, in Kamchatka, Russia, has extracted thermal water since 1966 mainly in the mode of artesian flow to supply numerous swimming pools, district heating of two villages, greenhouse farming, and fish breeding [6]. The long-term exploitation (20 years) has been assessed finding an insignificant temperature decrease in the geothermal reservoir (0.5 °C) [2]. The geothermal power potential in a hot spring close to the Municipality of Isa, in Japan, has been analyzed through the gravity data estimating a power density of 30.4 kW_e/km^2 for a 20 year exploitation [7]. Another interesting investigation has been carried out in a volcanic research site in the North German Basin, where a well-doublet with one injection and one production well has been explored for future geothermal energy production [8]. Alternating the injection of low and high flow rates of water a rapid increase in the water level of an adjacent well has been observed as well as an increase of the overall productivity of the treated well [9].

With the exception of active volcanic areas, carbonate aquifers are the most important geothermal reservoirs in which low-to-medium temperatures are usually reached at significant depths from the ground level (>1000 m) (e.g., [10–16]).

In many countries (i.e., Southern Italy), fault-controlled carbonate extensional domains may present temperatures in the range of low or middle enthalpy ($T < 150$ °C), at relatively shallow depths (<0.5 km) [14,17–21]. In fact, fault-controlled geothermal sites are accompanied by hydrothermal activity, hot fluid circulation, and mineralization processes [22,23]. Fracturing and chemical dissolution of carbonates increase the permeability of rocks, enhancing the advection of hot fluids and generating heat and mass transport [24–28]. The setting of carbonate bedrock in Mondragone plain corresponds to this hydrogeological model.

In recent times, the installation of innovative pilot geothermal electric and heating power plants has been incentivized in Italy. In this framework, prior to drilling activities and plant design, knowing scientific-technical data related to the potential of geothermal reservoirs and the sustainability of the utilization represents a crucial task to assess the economic feasibility of the geothermal exploitation and the development of future project plants [29,30].

Other information can be obtained in such a way, such as the relevance of system boundary conditions during long-term utilization, the interference of fluids extraction and reinjection during production, and the effectiveness of geothermal production systems. Numerical simulation is recognized as a fundamental tool for the elaboration and assessment of using geothermal energy [31], which is otherwise considered to be highly risky [32–34].

In recent years, different multiphase simulation tools have been widely used to model the effects induced by the exploitation of geothermal resources [35–37]. Among them, the code TOUGH2®(Transport Of Unsaturated Groundwater and Heat) has been firstly applied to Wairakei (New Zealand) geothermal field [38–40] and subsequently to several other geothermal fields (i.e., [41,42]). Additionally, the Los Alamos National Laboratory Finite Element Heat and Mass Transfer (FEHM®) code has a good record of numerical modeling studies, mainly related to enhanced geothermal systems (EGS) [43–46], together with the U.S. Geological Survey code HYDROTHERM® [47–49]. The commercial software COMSOL®, in particular, is a widely used simulator that has been adapted (as for example with the link to MATLAB®) [50] to applications including geothermal studies and, in several cases, the evaluation of fault influence [23,51,52]. As the heat and mass transfer in the reservoir depends on the internal properties of the reservoir and of the surrounding rocks, there has been an increased effort to model geothermal porous reservoirs. However up to now, few sustainability exploitation analyses on fractured carbonate rocks have been done [53,54] and they mainly focus on middle-high enthalpy and deep geothermal reservoirs, establishing several dynamic relations between the properties of the equivalent porous medium and fracture aperture. The exploitation of a low enthalpy geothermal system has been proposed by Cherubini et al. [55], who simulated a fractured geological system, composed of a single homogeneous layer with an inclined fault. They found that the width and permeability of the fault significantly affect fluid flow and thermal field, with a concentration of thermal perturbation within the fault plane.

2. Research Object Description

In the present paper, a sustainability analysis of the exploitation, by means of a well-doublet open-loop plant, of a low-temperature reservoir consisting of tectonically fractured and karstic limestone is carried out. This aquifer hosts an artesian groundwater body with a pressure larger than 1.70 bars, a wellhead temperature of 14 °C, and a natural water flow of 23.5 L/s. In the present work, two of the seven schools analyzed within the VIGOR Project are considered to be supplied. The characteristics of the system proposed to feed the two public schools are reported in Table 1.

Table 1. The proposed system for the feeding of two public schools.

Thermal power plant	Network	Users
- Drawn water flow rates 6 L/s - One Titanium heat exchanger - Two Geothermal heat pumps with $T_{evap.} = 25.0\,^\circ$C - Total thermal power 160 kW - COP 4.00	- Principal Loop DN 100; branches DN 80 - Water flow rates 4.00 L/s - Network flow temperature 55.0 °C	- School Name "II Circolo" - School Name "Leonardo da Vinci" - Gym of School Name "Leonardo da Vinci"

A water flow of 6.00 out of 23.5 L/s was computed as sufficient to heat the considered schools (about 160 kW). The remainder is supposed to be by-passed and then mixed again to reinject the entire flow rate through a reinjection well [56]. A scheme describing the operation of the system is shown in Figure 1.

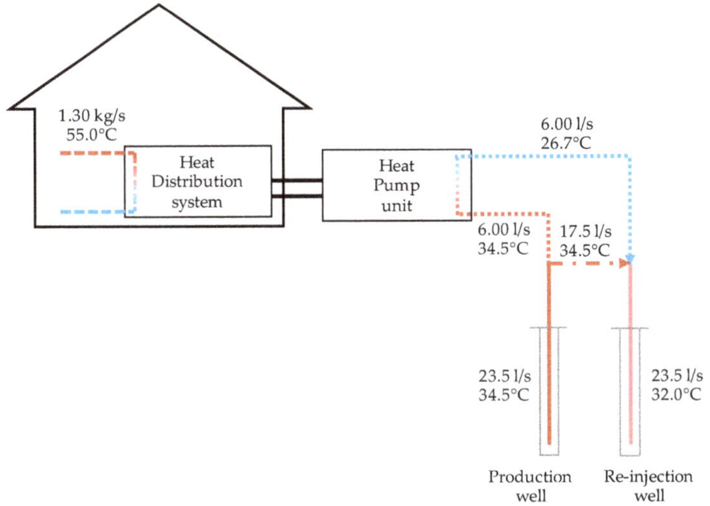

Figure 1. Scheme of the system operation.

The sustainability of such geothermal exploitation has been analyzed by a numerical simulation, comprising a one fault case and two faults case, using the commercial software COMSOL Multiphysics®to assess the reservoir thermal perturbation (for 60 years of system operation for heating needs) due to extraction and reinjection of fluids at a rate suitable to produce the required thermal power.

Due to low enthalpy values and high water pressure effects on the reservoir, mechanical deformations have been considered negligible. A further element of interest in the modeling is due to the position of the wells (of the extraction/injection) in the plant. In fact, for logistical constraints, the injection well is not downstream of the extraction well, but almost flanked by this with respect to the flow direction of the groundwater body. The numerical results obtained allow development of a solid conceptual model for the Mondragone geothermal reservoir exploitation and open the possibility to future applications to similar geothermal reservoirs which are widespread in the Mediterranean region.

3. Hydrogeological Setting and Conceptual Model of the Geothermal Area

The area of interest is located on the coastal plain of Mondragone town at the bottom of Mt. Petrino carbonate hill (Figure 2).

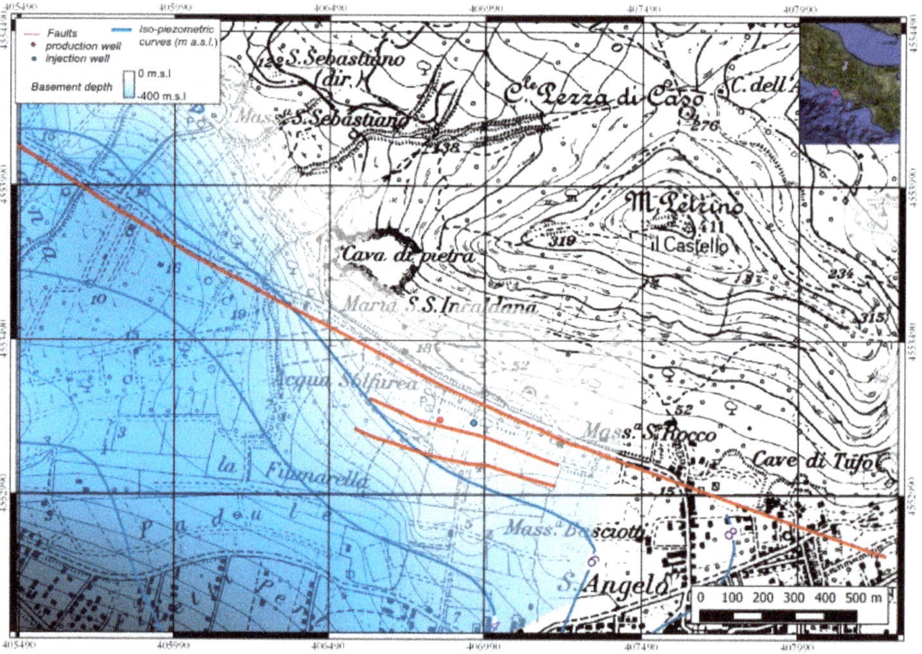

Figure 2. Localization and hydrogeological structural setting of the analyzed site. (Adapted from [1]).

This is at SW of Mt. Massico, which is a carbonate monocline trending SW bounded by high-angle normal faults along the margins of the Garigliano (NW), Volturno (SE), and Mondragone (SW) plains. These faults, several kilometers deep, have a throw estimated to be hundreds of meters [57–59]. Recent geophysical data have provided information about the bedrock of Mondragone plain. It is made by limestone and marly-arenaceous rocks and affected by several faults. The most important fault of this area crosses the whole plain from NW to SE and it is here coupled to other sub-parallel faults. The carbonate bedrock near Mt. Petrino has been found at a depth of −30 m below ground level (b.g.l.), under pyroclastic-alluvial deposits. From this zone, moving towards SE, the bedrock shows instead a regular depth towards the sea [1] (Figure 2).

Boreholes' data in the plain have shown the presence of a powerful bank of Campanian Ignimbrite tuff almost continuous across the plain. This tuff has low permeability and therefore separates the pyroclastic-alluvial aquifer (very poor) above the tuff, from the underlain alluvial deposits, deeper and more permeable, generating confined conditions in this lower aquifer. This last aquifer receives groundwater from lateral and vertical flows from the carbonate hill of Mt. Petrino, which is not hydro-geologically connected with the Mt. Massico, and from the carbonate bedrock, both shallow and deeper groundwater flow from NE to SW, toward the sea [22,60] (Figure 2).

Groundwater below the tuff presents interesting chemical characteristics: temperatures and average electrical conductivities are above 20 °C and >20,000 μS/cm, respectively, while, in the neighboring areas, these parameters show lower values (i.e., conductivity around 1000 μS/cm). These anomalies can be explained by the rise of deep mineralized waters (rich in gas, above all CO_2 of inorganic origin [61]) along the faults of the carbonate bedrock. These deep waters mix with groundwater below the tuff and increase their total dissolved solids and temperature, features that are distributed along the groundwater flow direction [22].

Additional data about the mineralized zone is derived from drilling, in the frame of VIGOR project, of a geothermal exploration well, 300 m deep, about 150 m from the base of Mt. Petrino (Figure 2). The first 167 meters of the well stratigraphy (Table 2) highlight that limestone starts under pyroclastic-alluvial deposits at about 30 m b.g.l., and that fracturing of crossed carbonate rocks reaches its maximum around 100–120 m b.g.l. (Table 3).

Table 2 reports the geothermal stratigraphy and the geophysical properties of the layers composing the analyzed domain. Density, thermal conductivity, and porosity were derived from literature, whereas the permeability was determined through "Lugeon tests" performed at different depths in the exploration well.

The experimental campaign has revealed the presence of artesian groundwater (under a maximum pressure of 1.70 bars) in the carbonate rocks, from which, in absence of confinement, about 23.5 L/s naturally flow [1]. After reaching the depth of 300 m, inside the well (and with a temporary casing) geophysical logs were performed (using a probe 2PFA-1000 / MATRIX converter) to measure temperature (Table 3) and verticality. During drilling, groundwater samples have been collected at different depths to analyze changes in chemical composition. The gases have been also sampled at two different sites, in correspondence of the geothermal exploration well and in its vicinity [60]. It was found that the chemical characteristics of groundwater along the depth are similar; the gases sampled are 98% CO_2, whose average concentration is 1380 mg/L [1,60].

Combining all data, a geothermal model that controls the mineralization of groundwater at the base of Mt. Petrino can be proposed [1,10,22,26,60,61]. According to this model, the geothermal reservoir corresponds to the carbonate bedrock of the Mondragone plain. Near Mt. Petrino, deep hot gases (mainly CO_2) rise along the faults of the bedrock, involving groundwater (sodium-bicarbonate type) typical of a reducing environment and of meteoric origin. From the carbonate bedrock, thermo-mineral groundwater spreads in the overlying pyroclastic-alluvial aquifer diluting gradually. Therefore, the geothermal process is closely linked to the faults, as shown by temperature increase, which occurs, as measured continuously along the well (Table 3), at the same levels of major fracturing degree. Therefore, it is very likely that these levels represent the preferential path along which the hot fluid rise occurs [60,61].

Table 2. Geothermal stratigraphy and geophysical properties of the layers composing the analyzed domain.

Layer Number	Depth from Ground Level (m)	Lithology	Density (kg/m^3)	Permeability (m^2)	Porosity (-)	Thermal Conductivity (W/(m K))
1	0–1.20	Silty-clayey vegetal soil	1600	1.02×10^{-12}	0.50	1.00
2	1.20–2.80	Gravel in silty-clayey matrix	1800	1.02×10^{-10}	0.30	2.20
3	2.80–5.50	Clays and silts, with sandy and peaty intercalations	1450	1.02×10^{-14}	0.60	1.50
4	5.50–15.0	Peat pits and clays with sandy and coarse interactions	1100	1.02×10^{-12}	0.65	2.00
5	15.0–25.0	Grey tuff	2000	1.02×10^{-13}	0.47	1.10
6	25.0–32.7	Sands alternated with clayey silts	1700	1.02×10^{-11}	0.35	2.00
7	32.7–34.8	Limestone from fractured to compact	2000	1.00×10^{-13}	0.15	2.80
8	34.8–51.4	Highly fractured and karst limestone	2200	1.00×10^{-12}	0.15	2.80
9	51.5–101	Highly fractured and karst limestone	2100	1.00×10^{-11}	0.30	3.00
10	101–123	Milonitic rock, sometimes cemented friction breccia	2500	1.00×10^{-10}	0.15	2.80
11	123–157	Very fractured limestone with karstic voids filled with loose material	2100	1.00×10^{-11}	0.15	2.80
12	157–167	Very fractured limestone, calcite and manganese concretions in the fractures, episodic empty karst. Strongly fractured and karstic	2200	1.00×10^{-12}	0.15	2.80

Table 3. Temperatures and lithology of the geothermal exploration well at different depths.

Depth from Ground Level (m)	Temperature (°C)	Lithology
0–32	38.5–40.0	Pyroclastic–alluvial deposits
32–124	40.0	Limestone passing to Milonitic rock
124–160	40.0–36.0	Milonitic rock passing to very fractured limestone
160–240	36.0–35.0	Limestone passing to very fractured limestone
240–300	33.0–30.0	Very fractured limestone passing to fractured limestone

4. Numerical Simulation of Geothermal Exploitation

In order to investigate the sustainability of geothermal exploitation, a numerical model was developed to analyze the coupled multiphase thermo-hydraulic processes due to the extraction and reinjection of groundwater. The model was implemented within the Finite Element commercial software COMSOL Multiphysics®, which is a powerful tool to reproduce coupled or multiphysics phenomena.

4.1. Governing Equations

Geothermal energy exploitation involves the interaction of different physical phenomena occurring in the ground, such as fluid flow, heat transport, chemical transport, and mechanical deformation [32,34,62]. In this work, only heat and fluid flow in fully saturated porous media were analyzed, considering the conservation laws of mass, momentum, and energy in a porous medium, that represents the ground.

The conservation of mass for a fluid flowing through a porous medium is expressed as:

$$\frac{\partial}{\partial t}(\rho \varepsilon) + \nabla \cdot (\rho \mathbf{u}) = 0, \tag{1}$$

where ρ is the fluid density, ε is the porosity, $\mathbf{u}$ is the seepage (Darcy) velocity vector. Therefore, the terms $\rho \varepsilon$ and $\rho \mathbf{u}$ represent the mass per unit volume within the porous matrix and the fluid mass flux, respectively. The flow velocity is low due to the low values of permeability and porosity of the domain. In this study, the porous medium flow was analyzed by using the well-known Darcy equation, which is the most common approach in geo-mechanic [63]. According to Darcy equation, the net flux across a face of porous surface, $\mathbf{u}$, is linearly related to the pressure gradient, ∇p, as

$$\mathbf{u} = -\frac{K}{\mu} \nabla p, \tag{2}$$

where K is the permeability of the porous medium and μ the geothermal fluid dynamic viscosity. Since the fluid flow is coupled to heat transfer, the dependence of fluid properties on temperature is taken into account.

The Darcy's flow model is coupled with heat transport in a porous medium to study the temperature distribution under geothermal exploitation. In the heat transport equation, the local thermal equilibrium between the solid and fluid phases is considered. Therefore, the solid temperature, T_s, is equal to the fluid temperature, T_f. Moreover, heat transport between solid and fluid phases is considered to be negligible. The heat transport in the subsurface is described as [64,65]

$$\left(\rho c_p\right)_{eq} \frac{\partial T}{\partial t} + \left(\rho c_p\right) \mathbf{u} \cdot \nabla T = \nabla \cdot \left(k_{eq} \nabla T\right), \tag{3}$$

where c_p is the specific heat, k is the thermal conductivity. The term $(\rho c_p)_{eq}$ represents the volumetric heat capacity of the porous medium. The subscript eq indicates that these are equivalent properties since they are spatially averaged to account for the porous matrix according to

$$\left(\rho c_p\right)_{eq} = \theta_p \rho_p c_{p,p} + \left(1 - \theta_p\right)\rho c_p, \tag{4}$$

$$k_{eq} = \theta_p k_p + \left(1 - \theta_p\right)k, \tag{5}$$

where θ_p is the porous matrix volume fraction and subscript p indicates quantities that are related to the solid porous matrix.

4.2. Boundary and Initial Conditions

The boundary conditions employed in the present model refer to the ground and the well domains. For logistical reasons, the position of the injection well is not the traditional one, which is downstream of the extraction well [66]. The injection well is in fact placed, along the direction of groundwater flow, parallel to the extraction well and not far behind this (Figure 2). The top and bottom surfaces of the ground domain are considered impermeable to mass ($u = 0$), such as the lateral surfaces parallel to the groundwater flow. The average hydraulic gradient, (i), in the groundwater layer was estimated, according to Corniello et al. [22], with a NE–SW direction of the flow, determined from the iso-piezometric map of the area (Figure 2).

A prescribed velocity u was assigned on the filtering lateral surfaces of the wells. This value was calculated considering the water flow rate of extraction and injection, $\dot{V}$, measured during the experimental campaign. As initial condition for the groundwater flow equation, the hydraulic head, H, derived from experimental measurements, was imposed on the whole domain. The bottom ground surface was assumed to be adiabatic, whereas a convective heat flux is imposed on the top surface of the domain. The external temperature was assumed to be equal to 16.0 °C, which is the yearly average temperature of the area. As concerns the wells, a fixed temperature, T, equal to that of flow extraction and injection, is imposed on the lateral surfaces. The water temperature values measured along the exploration well [1,56] are illustrated in Figure 3. The soil was assumed to be in local thermal equilibrium with the groundwater, then the measured temperature profile (Table 3) was considered as the initial condition on the ground domain. All boundary and initial conditions are summarized in Figure 3 and Table 4.

Table 4. Boundary and initial conditions.

Parameter	Symbol	Value	Unit
Hydraulic gradient	i_x	0.5%	m/m
Extraction flow rate	$\dot{V}_{out}$	84.6	m^3/h
Reinjection flow rate	$\dot{V}_{in}$	84.6	m^3/h
Hydraulic head	H	13.0	m
Extraction Temperature	T_{out}	34.5	°C
Injection Temperature	T_{in}	32.0	°C
External Temperature	T_{ext}	16.0	°C
Convective heat transfer coefficient	h	25.0	W/(m^2 K)

The experimental campaign revealed a sub-vertical fault zone, NW–SE oriented. The injection well is located at 135 m (Figure 2) from the production well, in the rearward direction. From the stratigraphical data of the exploration well, two different faults were found between 100–120 and 180–240 m b.g.l., according also to the results of the detailed geophysical survey. The deepest fault if projected towards the injection well may or may not meet this well at about 80 m depth. For this reason, two case studies were analyzed, one with the fault crossing the injection well and the other without. The fault zone presents the same characteristics of the layer identified by the number 10 in Table 2.

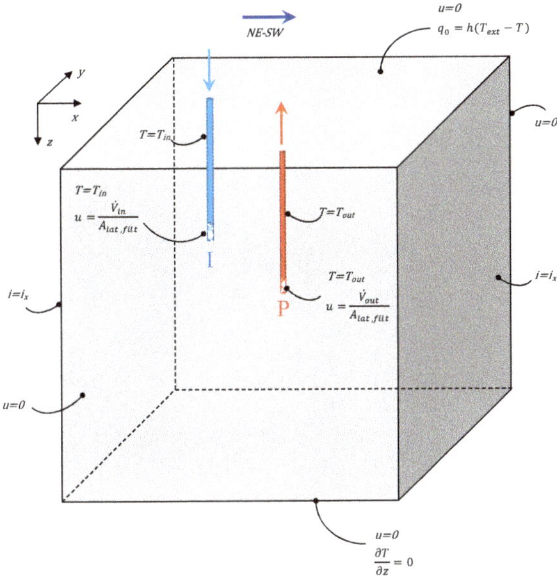

Figure 3. Sketch of boundary and domain initial conditions (for clarity the width of the wells is enlarged). P and I represent the production and injection wells, respectively.

5. Results and Discussion

The numerical model described in the previous section was used to study the sustainable use of the low-temperature geothermal field of Mondragone. In order to evaluate the sustainability of this use, the temperature field in the ground was modeled over a period of time in which the source was continuously used at the full capacity of the plant. As mentioned above, the analysis was carried out with and without the fault zone crossing the injection well, to assess its influence on heat transfer and fluid flow. The simulations that refer to a single fault are indicated as Case 1 from now on, whereas those with the domain that includes two fault zones are referred to as Case 2. Figure 4 shows a schematic representation of the domain considered for the analysis of Case 1 where a single fault, indicated as F1, is present. In Case 2 the domain is the same, with the addition of a fault zone crossing the injection well, as shown in Figure 5. For the sake of clarity, only the data concerning the fault crossing the injection well, indicated as F2, are reported in Figure 5.

The overall domain is defined as a rectangular parallelepiped of 163 m in depth with a basis equal to 250 × 200 m. Below this depth, the effect of the geothermal system is considered negligible due to the lower degree of fracturing in limestone. The fault zones have both a thickness of 20.0 m and a slope of 10° in relation to the vertical axis. The production and reinjection wells cross the fault cataclastic zone starting from 80.0 to 120 and 68.0 to 110 m b.g.l., respectively. Both the wells have a radius of 24.4 cm and a filtering section of 10.0 m in height from the top of the cataclastic zones.

The water properties, ρ, μ, c_p, and k are considered temperature-dependent. For the ground domain, literature values of thermal conductivity, density, porosity, and permeability of the eight layers found during the well drilling were considered. The parameters used in the numerical simulation are reported in Table 4.

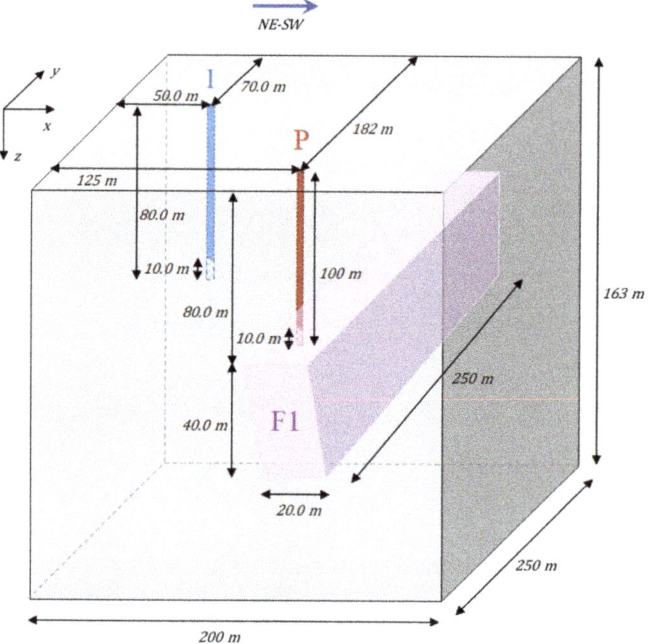

Figure 4. Computational domain in the case of a single fault (F1).

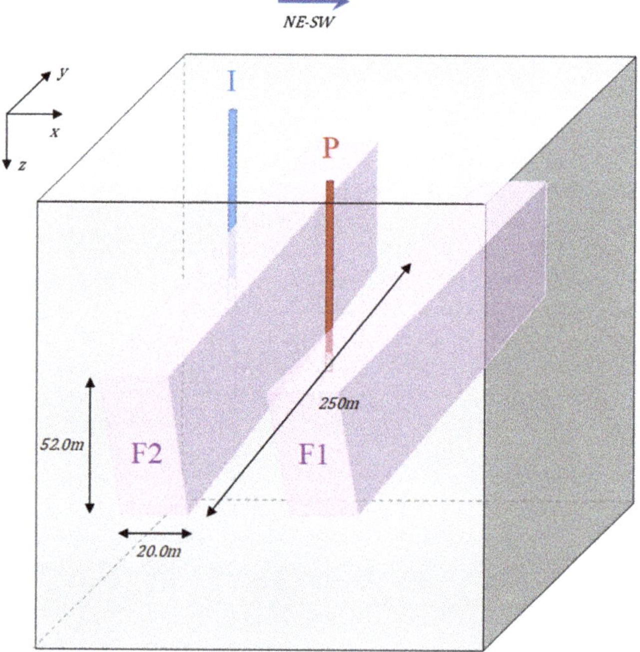

Figure 5. Computational domain in the case of double faults (F1 and F2).

A 3D unstructured mesh, refined near all the boundaries to capture the larger gradients of both flow velocity and temperature, was used. Indeed, the grid needs to be finer in the region of greatest change in the skin zone close to the well and can be coarser further out in the reservoir [67]. For the sake of clarity, the mesh used for the case of a single fault is reported in Figure 6.

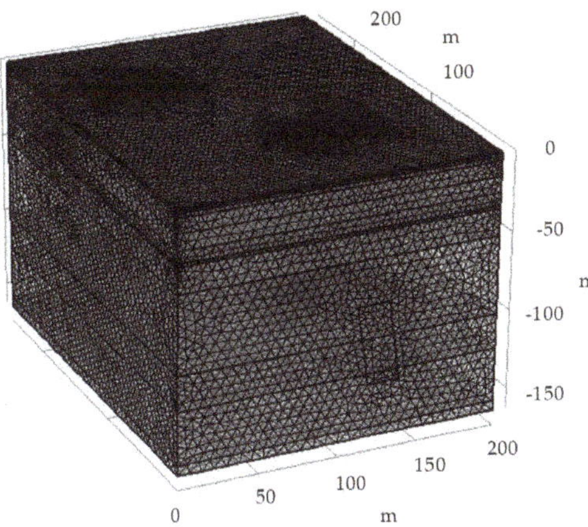

Figure 6. Mesh used to analyze the case of a single fault.

The details of the meshes used for the two cases are shown in Table 5. A mesh sensitivity study was carried out to obtain grid-independent results from the calculations.

Table 5. Mesh details for Case 1 (single fault) and Case 2 (two faults).

Parameter	Case 1	Case 2
Tetrahedral elements	3,373,929	3,671,990
Triangular elements	168,494	182,727
Edge elements	6134	6813
Vertex elements	160	180
Minimum element quality	0.03311	0.03311
Average element quality	0.6594	0.6601
Element volume ratio	3.95×10^{-6}	2.10×10^{-6}

In order to assess the impact of the geothermal exploitation on the ground domain, the groundwater velocity field and the ground temperature field were determined through the proposed model for a period of time of 60 years.

The temperature fields at different times in the xz plan cutting the domain in correspondence of the two wells' axes are reported in Figure 7 for the case of the single fault (Case 1) and in Figure 8 for the case of double (Case 2) fault zones. For the sake of clarity, the section cuts are also reported.

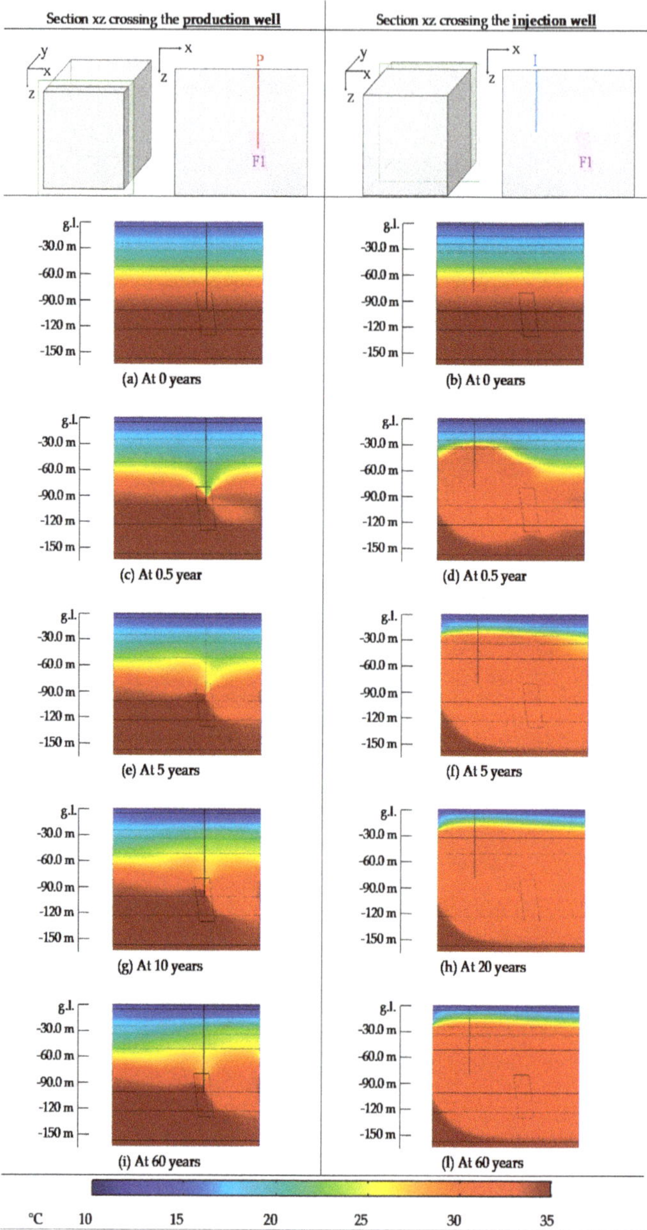

Figure 7. Temperature field of the sections *xz* crossing the wells, at different times, for Case 1 (single fault).

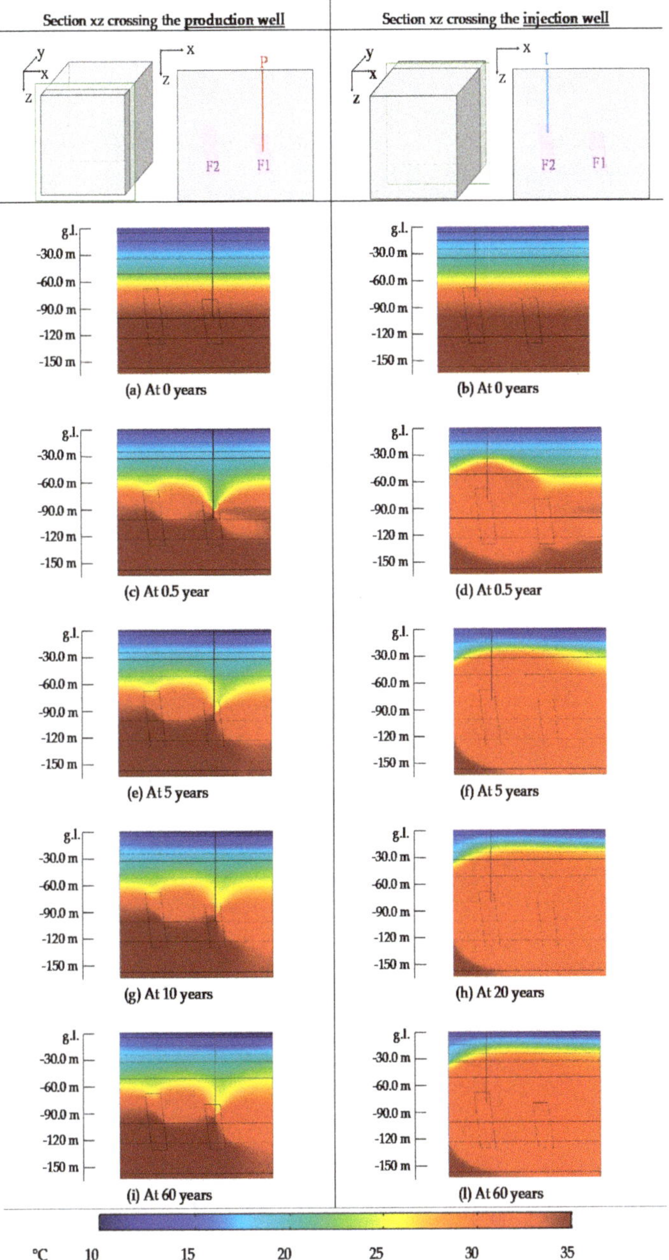

Figure 8. Temperature field of the sections *xz* crossing the wells, at different times, for Case 2 (double faults).

The effect of geothermal extraction is mainly contained in the area surrounding the production well. A moderate decrease in the ground temperature can be observed in the downstream region: the variation is similar in both Cases 1 and 2.

In the upstream section the temperature field remains almost unchanged in Case 1, whereas in Case 2 a temperature decrease can be observed in the region delimited by the two faults. The difference is due to the permeability and the thermal conductivity of the fault zone, which are larger than those of the surrounding material. However, in both cases after 10 years of operation the temperature field does not change any more (reaching steady state operation of the reservoir). The effect of injection is more pronounced than that of extraction in both cases. However, the thermal disturbance does not reach the superficial layers, since it is limited by the 5th layer which has lower values of permeability and thermal conductivity with respect to the other layers (Table 2). The injection in the fault slightly modifies the temperature distribution within the domain: the propagation of the thermal disturbance is lower in layers over the fault and higher towards the bottom of the domain. In both cases, the temperature field does not change any more after 20 years of operation, therefore the period of time needed to reach a steady condition is longer in the region surrounding the injection well than in the area of the production well. This is probably due to the higher difference in temperature between groundwater surrounding the injection well and injected water.

The temperature field, at different times, in the yz plan cutting the domain in correspondence of the axis of two wells is shown in Figures 9 and 10 for Cases 1 and 2, respectively. For the sake of clarity, the section cuts are also reported. In these plots, it can be noticed that the temperature decrease caused by the injection well is localized in the region of the domain behind the production well and it is less pronounced in Case 1. In Case 1 the temperature propagation is more pronounced towards the upper layers, whereas in Case 2 it is larger in the bottom of the domain.

The temperature distribution and the streamlines indicating the velocity field, at different depths and times, reported in Figure 11, clearly show that the direction of the thermal disturbance is significantly influenced by the groundwater flow. A stable temperature distribution is reached after 25 years at the lowest depths, whereas only 5 years are needed at highest depths. The thermal disturbance in the layer close to the top surface of the domain is concentrated in the region surrounding the injection well.

The fault located in correspondence of the injection well, F2, does not significantly affect the temperature distribution, except for the upstream region of the production well, where the temperature is slightly lower than in Case 1. Conversely, the fault influences the velocity field, due to the permeability, which is larger than in the adjacent layers. This variation in the velocity field is mainly responsible for the extension of the region affected by the thermal disturbance in Case 2.

At depths below 75.0 m from ground level the temperature and velocity fields are constant over time, therefore only their variation with depth was taken into account. Analyzing the plots in Figure 12, which are all related to the time of 60 years, it can be observed that the thermal disturbance is more extended in Case 2. In addition, at 155 m b.g.l. the thermal disturbance can be neglected in Case 1, whereas in Case 2 the region downstream of the injection well is still affected by the geothermal exploitation.

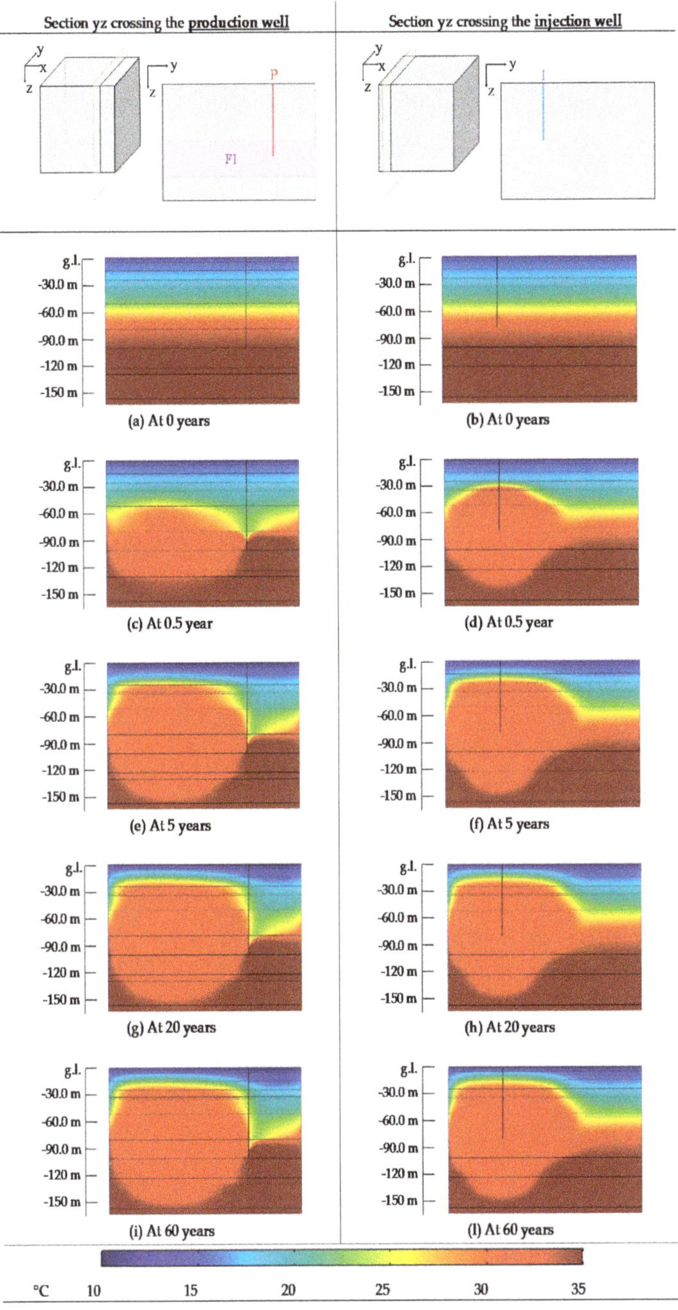

Figure 9. Temperature field of the sections yz crossing the wells, at different times, for Case 1 (single fault).

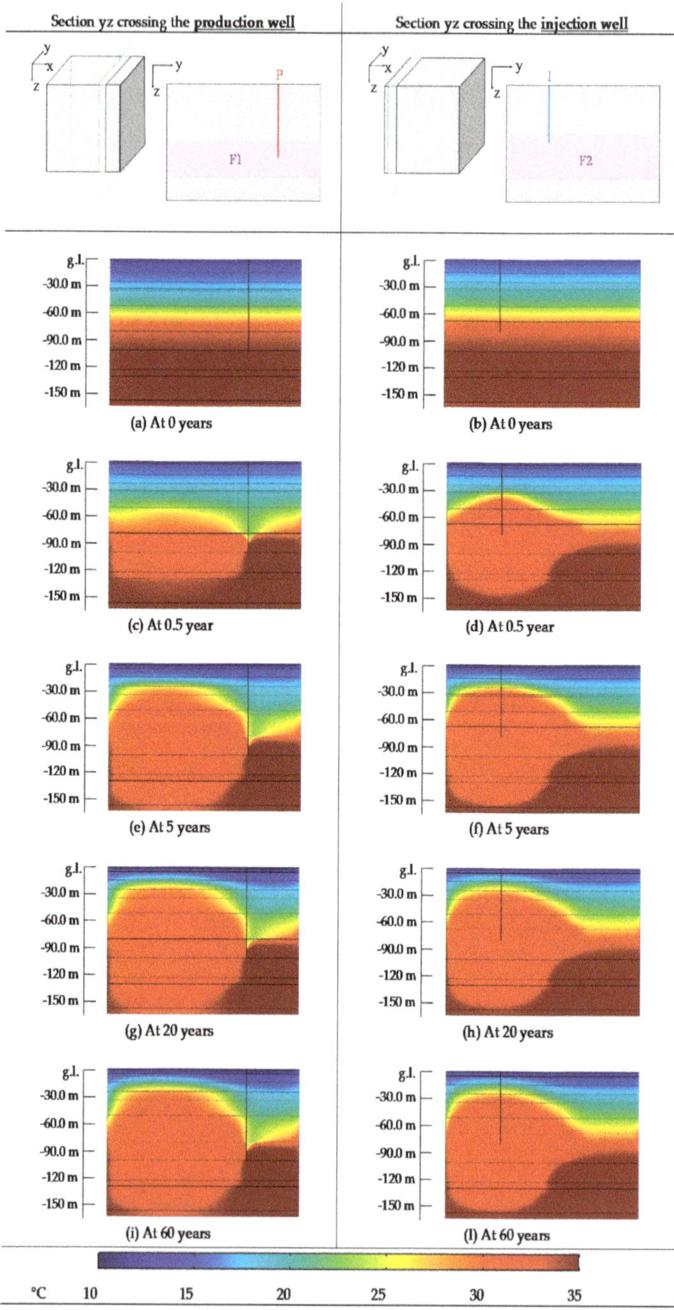

Figure 10. Temperature field of the sections yz crossing the wells, at different times, for Case 2 (double faults).

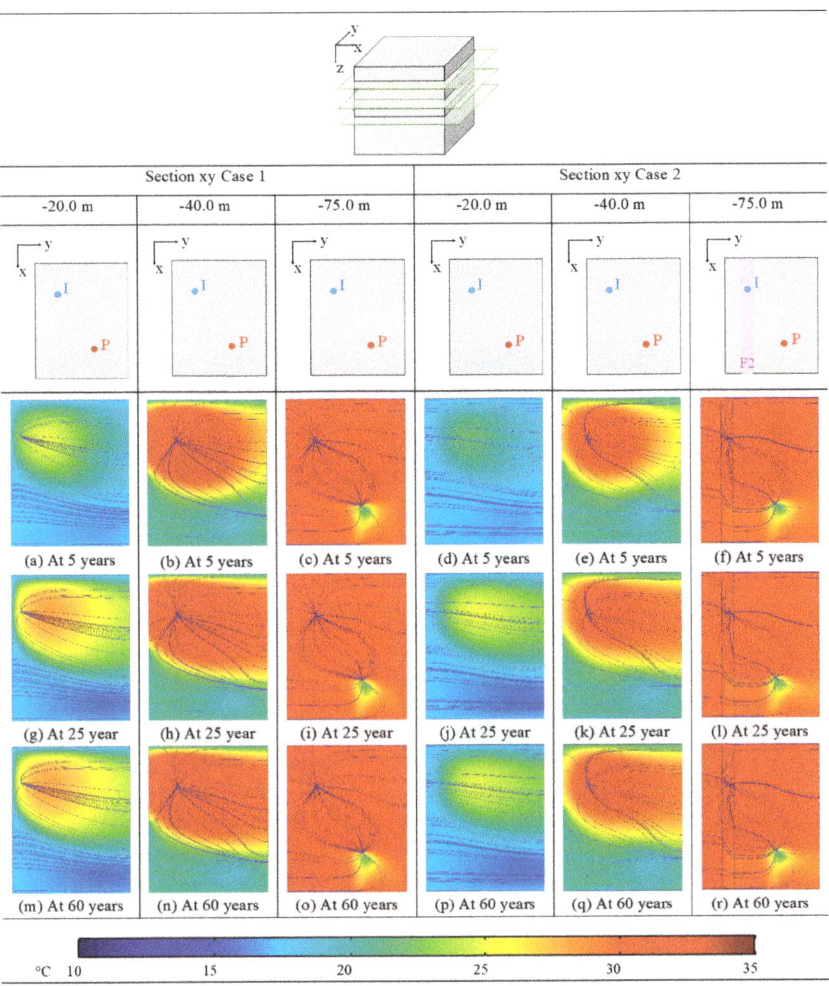

Figure 11. Temperature field at different depths and times, for Case 1 (single fault) and Case 2 (two faults).

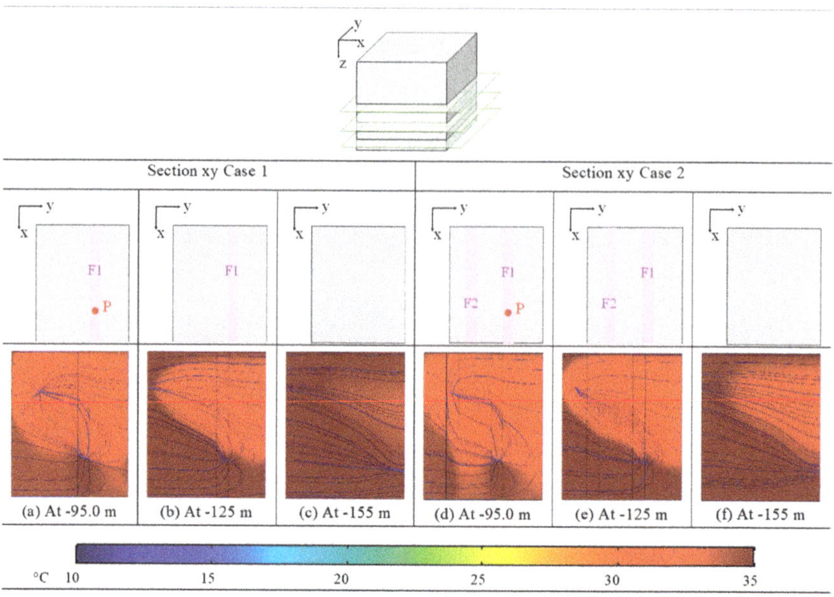

Figure 12. Temperature field at different depths, for Case 1 (single fault) and Case 2 (two faults).

The Darcy velocity field and magnitude at different depths for Cases 1 and 2 are reported in Figures 13 and 14, respectively. The *xy* sections reported in these figures are related to different layers of the domain, in order to analyze the influence of their properties on the groundwater flow. In the first two layers of the domain, (refer to Figure 13a,b and Figure 14a,b), both the distribution and the magnitude of Darcy velocity do not change from Case 1 to Case 2, and result as constant in the entire *xy* plans considered. This is due to the low permeability of the third layer, which avoids the effect of the geothermal exploitation on the groundwater flow reaching the top of the domain. Indeed, the groundwater flow in the lower layers of the domain, which are characterized by higher values of permeability, varies in the regions where the production and injection wells are located (refer to Figure 13c,d and Figure 14c,d).

Considering the injection well, the velocity magnitude decreases upstream and increases downstream, whereas the opposite trend can be observed in the region surrounding the production well. In Case 2, a lower propagation of the fluid flow disturbance can be observed: this is due to higher permeability of the material in the correspondence of the fault, which facilitates the groundwater flow, which is more concentrated in the tectonic region. In addition, considering Case 2, the velocity magnitude varies differently in the zones of injection and production, probably due to the different vertical extension of the faults. The fault located in the injection area, F2, is more extended and its effect on the upper layers is larger.

In the sections that cut the wells (refer to Figure 13f,g and Figure 14f,g), the Darcy velocity field and magnitude are similar in Case 1 and Case 2. Obviously, the magnitude is significantly higher than at other depths due to the flow extraction and injection. A different trend can be observed in sections 10.0 m distant from the bottom of the wells in the vertical direction (refer to Figure 13e,h and Figure 14e,h), mainly in the areas surrounding the injection well. The fault affects the groundwater flow, whose magnitude increases in the area surrounding the injection well and decreases upstream of the production well. However, a vertical distance of 10.0 m is sufficient to observe a significant decrease in the Darcy velocity, which tends to the undisturbed value as the horizontal distance from the wells increases. At higher depths, the velocity magnitude continues to decrease (refer to Figure 13i,l and Figure 14i,l). The velocity magnitude is higher in the central region of the considered section, where the groundwater has the same direction of the undisturbed fluid flow in both cases, injection and extraction.

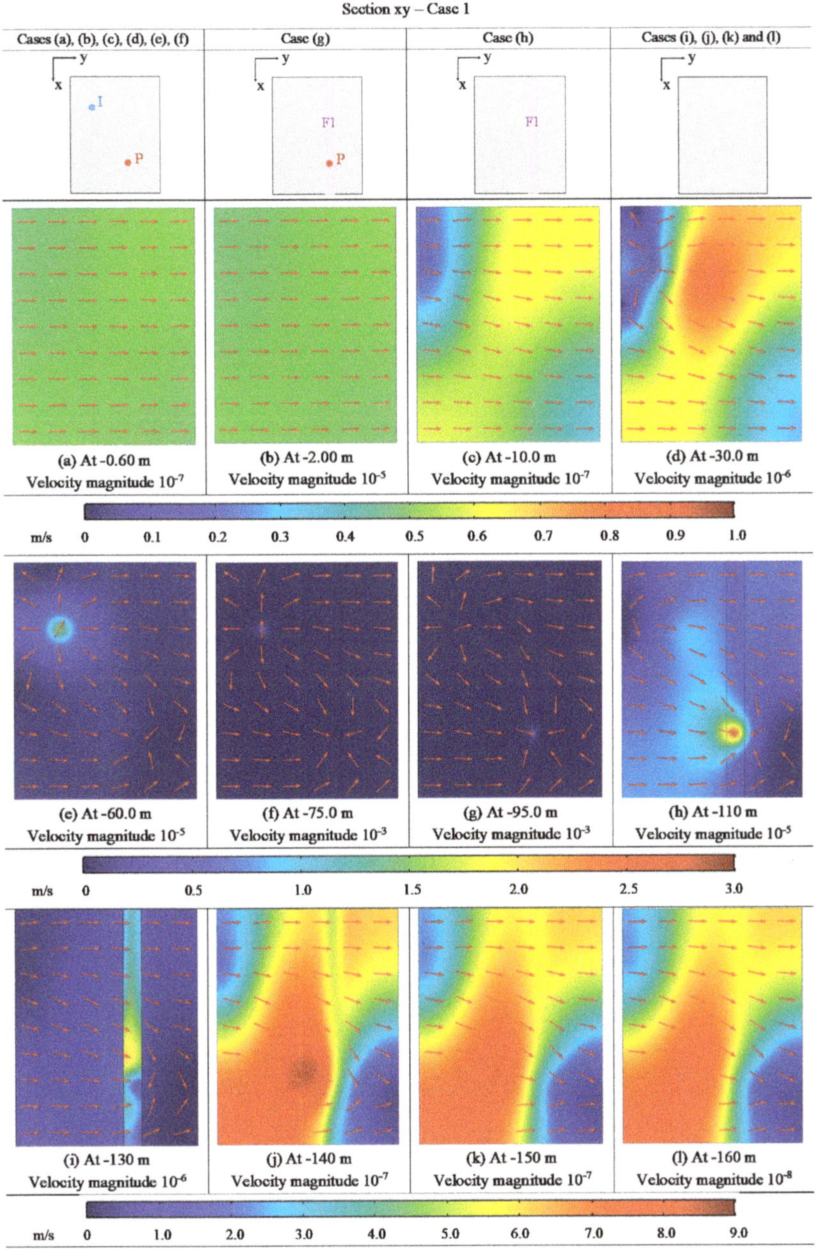

Figure 13. Darcy velocity field at different depths for Case 1 (single fault).

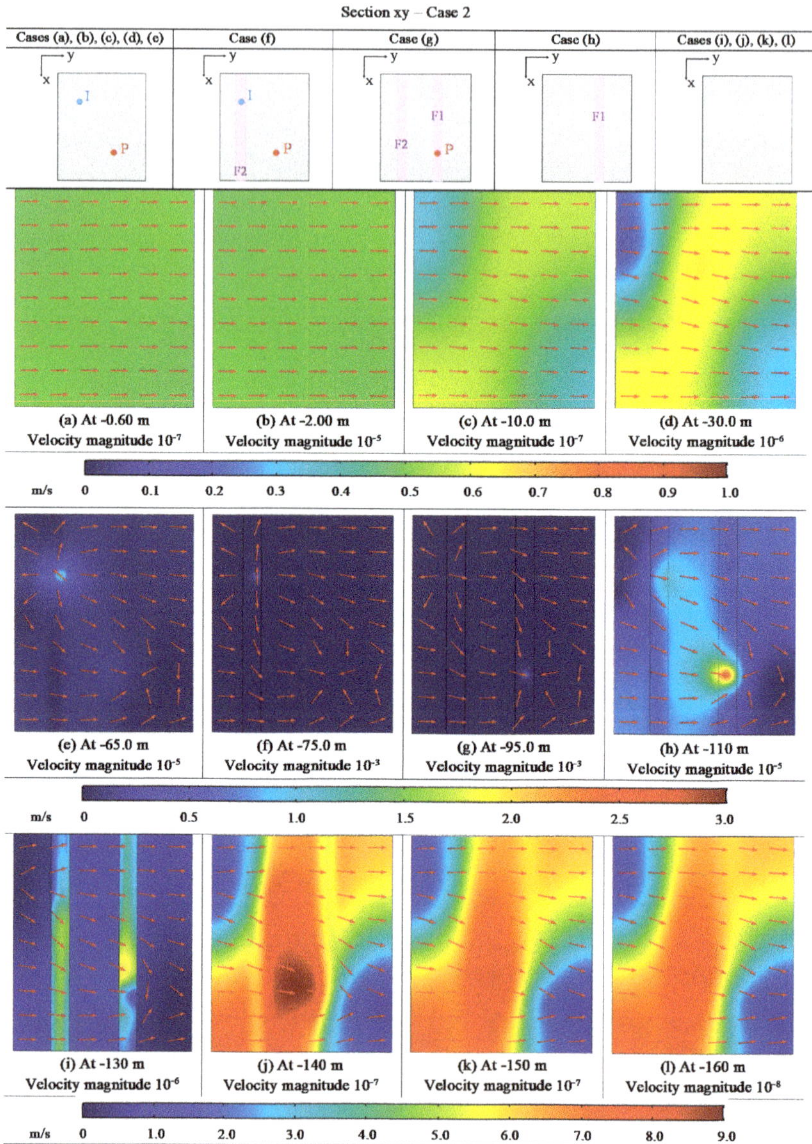

Figure 14. Darcy velocity field at different depths for Case 2 (two faults).

Finally, in Figure 15, the temperature over time is plotted at three different distances from the wells (10, 50, and 100 m) at a depth equal to half of the filtering sections of the wells, for the two cases analyzed.

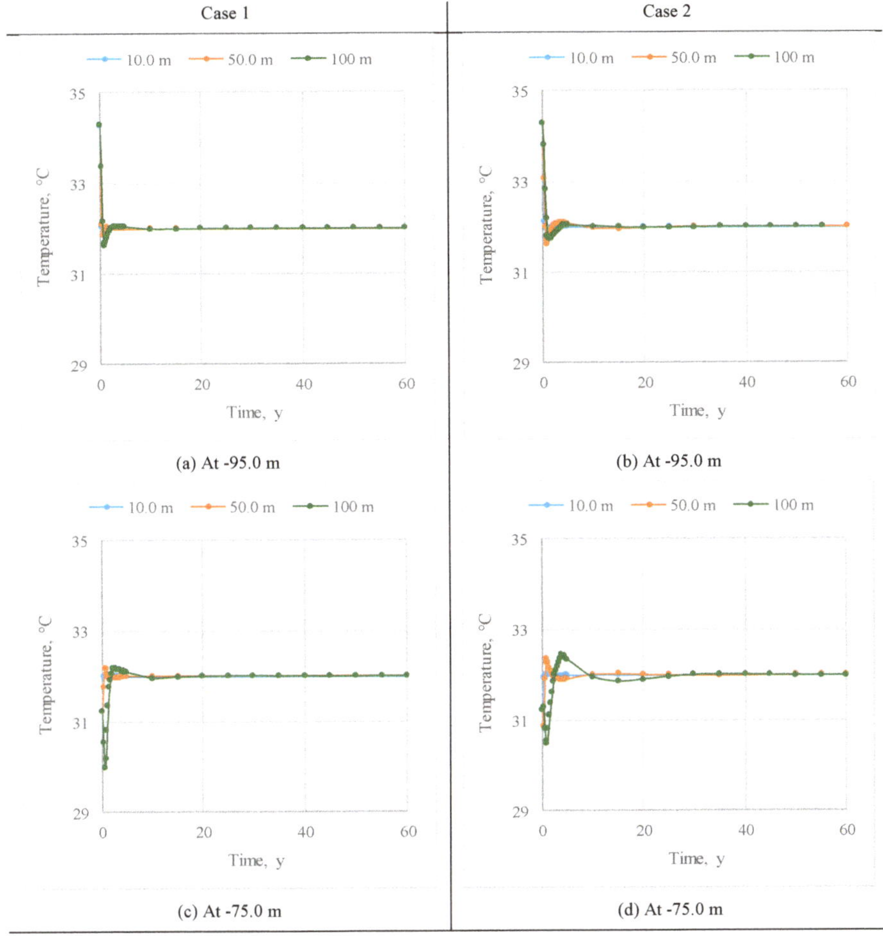

Figure 15. Temperature over time at three different distances from the wells (10.0, 50.0, and 100 m), for two different depths (−95.0 and −75.0m).

Considering the production well (see Figure 15a,b) whose filtering section goes from −90 to −100 m, the geothermal exploitation causes a temperature decrease of around 2 °C during the considered period of time. The temperature becomes constant after 10 years in Case 1 and 20 years in Case 2. It is worth noticing that in Case 2, the temperature decrease at a distance of 100 m is slightly lower than in Case 1, suggesting a positive effect of the injection in the fault. A similar conclusion can be drawn analyzing the results related to the injection well (see Figure 15c,d). The variation in temperature with respect to the initial condition is lower than 1 °C, therefore significantly lower than that observed in the region of the production well. The period of time needed to reach a constant temperature is longer than for the production well: 20 years for the Case 1 and 25 years for the Case 2. The injection in the fault has a positive effect, reducing the temperature oscillations during the first years of exploitation in Case 2. The low decrease of the basin temperature is coherent with the results found in the literature related to the geothermal exploitation of low-temperature reservoirs [2].

So far, the numerical results show that the geothermal exploitation modifies the groundwater flow in correspondence with the filtering sections of the wells, whereas only a weak effect can be observed in the rest of the domain, in terms of speed and direction. In order to account for the possible presence of a fault crossing the injection well, two case studies were analyzed. When the injection occurs in the fault the variation of the velocity field is lower, thanks to the permeability in correspondence with the fault. The injection in the fault positively affects the temperature distribution as well. The thermal disturbance does not reach the top layers of the domain due to their hydraulic and thermal properties in both cases. It is worth noticing that in the case of injection in the fault the disturbance tends to be more concentrated in the same fault region, decreasing its effect on the other layers of the domain. In addition, in this case, also the temperature oscillation over time observed at the depth of the wells is lower than in the case without the injection in the fault.

The proposed model allows estimation of the effect of geothermal exploitation not only on the reservoir but also on the operation of the technological system that uses the geothermal source for thermal energy production. The results of the simulations indicate a temperature decrease of around 2.00 °C in the extraction temperature during the considered period of time. Such a decrease will reduce the thermal energy available from geothermal exploitation. In the specific case analyzed in this work, where a heat pump was considered to supply heat to two schools, a decrease in the COP of the heat pump is expected.

6. Conclusions

The numerical results on the geothermal exploitation here reported are of global interest for the following reasons: (a) the nature of aquifer represented by carbonate rocks with strong anisotropy, due to the presence of several fault systems, with confined groundwater; (b) the environmental effects of an open-loop well-doublet plant are assessed considering the geometry of the wells (imposed by logistical constraints) which is designed in a different way from the one usually adopted, where the injection well is placed downstream of the extraction well. In fact, even if thermal feedback is present at the production well due to the position of injection well, the sustainability of the use of the geothermal resource over time (60 years here tested) is asserted by the numerical simulation. In fact, the main results show that the temperature decrease is only 2.00 °C in correspondence with the production well filtering section, and even less at greater depths.

It is likely that the sustainability is linked, considering the hydrogeological scheme, to the characteristics of groundwater flow (direction and pressure) and to the position of the filtering section of the extraction well, that is placed near the main cataclastic zone of a fault along which the permeability is higher than the surrounding rocks and the upwelling of deep hot fluids is easier.

Even assuming the presence of another fault at the injection well, the numerical simulation still indicates the full sustainability of the geothermal plant. It is worth noticing that in the case of injection in the fault the disturbance tends to be more concentrated in the same fault region, decreasing its effect on the other layers of the domain. Finally, in all the cases considered, the thermal disturbance does not reach the top of the domain due to the hydraulic and thermal properties of the rocks closest to the ground level.

The definition of the hydrogeological and stratigraphic scheme of the geothermal area together with the numerical modelling are therefore decisive either for the final choice of the well plant localization or for the evaluation over time of the environment effects of the use of the geothermal resource in faulted carbonate rocks, which are common in Southern Italy and peri-Mediterranean regions.

Author Contributions: Conceptualization, M.I., N.M., and L.V.; Data curation, M.I. and A.C. (Alfonso Corniello); Formal analysis, S.D.F.; Funding acquisition, M.I., A.C. (Alberto Carotenuto), N.M., and L.V.; Methodology, S.D.F. and N.M.; Project administration, M.I., A.C. (Alberto Carotenuto) and N.M.; Writing—original draft, S.D.F.; Writing—review and editing, M.I., A.C. (Alberto Carotenuto), A.C. (Alfonso Corniello), S.D.F., N.M., R.S., L.V., and A.M. All authors have read and agreed to the published version of the manuscript.

Funding: The authors gratefully acknowledge the financial support of VIGOR project (Valutazione del Potenziale Geotermico delle Regioni Convergenza) CUP: B72J10000060007, GEOGRID project CUP: B43D18000230007 and project M027061 funded by Italian Ministry of the Foreign Affairs.

Conflicts of Interest: The authors declare no conflict of interest. The funders had no role in the design of the study; in the collection, analyses, or interpretation of data; in the writing of the manuscript, or in the decision to publish the results.

References

1. Amoresano, A.; Angelino, A.; Anselmi, M.; Bianchi, B.; Botteghi, S.; Brandano, M.; Brilli, M.; Bruno, P.P.; Caielli, G.; Caputi, A.; et al. VIGOR: Sviluppo geotermico nella regione Campania—Studi di Fattibilità a Mondragone e Guardia Lombardi. In *Progetto VIGOR—Valutazione del Potenziale Geotermico delle Regioni della Convergenza, POI Energie Rinnovabili e Risparmio Energetico 2007-2013*; CNR-IGG: Pisa, Italy, 2014; ISBN 9788879580151.

2. Kiryukhin, A.V.; Vorozheikina, L.A.; Voronin, P.O.; Kiryukhin, P.A. Thermal and permeability structure and recharge conditions of the low temperature Paratunsky geothermal reservoirs in Kamchatka, Russia. *Geothermics* **2017**, *70*, 47–61. [CrossRef]

3. Manni, M.; Coccia, V.; Nicolini, A.; Marseglia, G.; Petrozzi, A. Towards zero energy stadiums: The case study of the Dacia arena in Udine, Italy. *Energies* **2018**, *11*, 2396. [CrossRef]

4. Bonati, A.; De Luca, G.; Fabozzi, S.; Massarotti, N.; Vanoli, L. The integration of exergy criterion in energy planning analysis for 100% renewable system. *Energy* **2019**, *174*, 749–767. [CrossRef]

5. Østergaard, P.A.; Lund, H. A renewable energy system in Frederikshavn using low-temperature geothermal energy for district heating. *Appl. Energy* **2011**, *88*, 479–487. [CrossRef]

6. Kiryukhin, A.V.; Asaulova, N.P.; Vorozheikina, L.A.; Obora, N.V.; Voronin, P.O.; Kartasheva, E.V. Recharge Conditions of the Low Temperature Paratunsky Geothermal Reservoir, Kamchatka, Russia. *Procedia Earth Planet. Sci.* **2017**, *17*, 132–135. [CrossRef]

7. Pocasangre, C.; Fujimitsu, Y.; Nishijima, J. Interpretation of gravity data to delineate the geothermal reservoir extent and assess the geothermal resource from low-temperature fluids in the Municipality of Isa, Southern Kyushu, Japan. *Geothermics* **2020**, *83*, 101735. [CrossRef]

8. Zimmermann, G.; Moeck, I.; Blöcher, G. Cyclic waterfrac stimulation to develop an Enhanced Geothermal System (EGS)—Conceptual design and experimental results. *Geothermics* **2010**, *39*, 59–69. [CrossRef]

9. Chen, Y.; Ma, G.; Wang, H.; Li, T.; Wang, Y.; Sun, Z. Optimizing heat mining strategies in a fractured geothermal reservoir considering fracture deformation effects. *Renew. Energy* **2020**, *148*, 326–337. [CrossRef]

10. Simsek, S. Hydrogeological and isotopic survey of geothermal fields in the Buyuk Menderes graben, Turkey. *Geothermics* **2003**, *32*, 669–678. [CrossRef]

11. Gouasmia, M.; Gasmi, M.; Mhamdi, A.; Bouri, S.; Dhia, H.B. Prospection géoélectrique pour l'étude de l'aquifère thermal des calcaires récifaux, Hmeïma–Boujabeur (Centre ouest de la Tunisie). *Comptes Rendus Geosci.* **2006**, *338*, 1219–1227. [CrossRef]

12. Mohammadi, Z.; Bagheri, R.; Jahanshahi, R. Hydrogeochemistry and geothermometry of Changal thermal springs, Zagros region, Iran. *Geothermics* **2010**, *39*, 242–249. [CrossRef]

13. Duan, Z.; Pang, Z.; Wang, X. Sustainability evaluation of limestone geothermal reservoirs with extended production histories in Beijing and Tianjin, China. *Geothermics* **2011**, *40*, 125–135. [CrossRef]

14. Pasquale, V.; Verdoya, M.; Chiozzi, P. Heat flow and geothermal resources in northern Italy. *Renew. Sustain. Energy Rev.* **2014**, *36*, 277–285. [CrossRef]

15. Homuth, S.; Götz, A.; Sass, I. Reservoir characterization of the Upper Jurassic geothermal target formations (Molasse Basin, Germany): Role of thermofacies as exploration tool. *Geotherm. Energy Sci.* **2015**, *3*, 41–49. [CrossRef]

16. Montanari, D.; Minissale, A.; Doveri, M.; Gola, G.; Trumpy, E.; Santilano, A.; Manzella, A. Geothermal resources within carbonate reservoirs in western Sicily (Italy): A review. *Earth-Sci. Rev.* **2017**, *169*, 180–201. [CrossRef]

17. Cataldi, R.; Mongelli, F.; Squarci, P.; Taffi, L.; Zito, G.; Calore, C. Geothermal ranking of Italian territory. *Geothermics* **1995**, *1*, 115–129. [CrossRef]

18. Minissale, A.; Duchi, V. Geothermometry on fluids circulating in a carbonate reservoir in north-central Italy. *J. Volcanol. Geotherm. Res.* **1988**, *35*, 237–252. [CrossRef]

19. Della Vedova, B.; Bellani, S.; Pellis, G.; Squarci, P. Deep temperatures and surface heat flow distribution. In *Anatomy of An Orogen: The Apennines and Adjacent Mediterranean Basins*; Springer: Berlin/Heidelberg, Germany, 2001; pp. 65–76.

20. Chiarabba, C.; Chiodini, G. Continental delamination and mantle dynamics drive topography, extension and fluid discharge in the Apennines. *Geology* **2013**, *41*, 715–718. [CrossRef]

21. Giambastiani, B.; Tinti, F.; Mendrinos, D.; Mastrocicco, M. Energy performance strategies for the large scale introduction of geothermal energy in residential and industrial buildings: The GEO. POWER project. *Energy Policy* **2014**, *65*, 315–322. [CrossRef]

22. Corniello, A.; Cardellicchio, N.; Cavuoto, G.; Cuoco, E.; Ducci, D.; Minissale, A.; Mussi, M.; Petruccione, E.; Pelosi, N.; Rizzo, E.; et al. Hydrogeological Characterization of a Geothermal system: The case of the Thermo-mineral area of Mondragone (Campania, Italy). *Int. J. Environ. Res.* **2015**, *9*, 523–534.

23. Taillefer, A.; Soliva, R.; Guillou-Frottier, L.; Le Goff, E.; Martin, G.; Seranne, M. Fault-related controls on upward hydrothermal flow: An integrated geological study of the têt fault system, eastern pyrénées (France). *Geofluids* **2017**, *2017*, 8190109. [CrossRef]

24. Chi, G.; Xue, C. An overview of hydrodynamic studies of mineralization. *Geosci. Front.* **2011**, *2*, 423–438. [CrossRef]

25. Lowell, R.P.; Farough, A.; Hoover, J.; Cummings, K. Characteristics of magma-driven hydrothermal systems at oceanic spreading centers. *Geochem. Geophys. Geosystems* **2013**, *14*, 1756–1770. [CrossRef]

26. Lowell, R.P. A fault-driven circulation model for the Lost City Hydrothermal Field. *Geophys. Res. Lett.* **2017**, *44*, 2703–2709. [CrossRef]

27. Moeck, I.S. Catalog of geothermal play types based on geologic controls. *Renew. Sustain. Energy Rev.* **2014**, *37*, 867–882. [CrossRef]

28. Blöcher, G.; Cacace, M.; Reinsch, T.; Watanabe, N. Evaluation of three exploitation concepts for a deep geothermal system in the North German Basin. *Comput. Geosci.* **2015**, *82*, 120–129. [CrossRef]

29. Calise, F.; Di Fraia, S.; Macaluso, A.; Massarotti, N.; Vanoli, L. A geothermal energy system for wastewater sludge drying and electricity production in a small island. *Energy* **2018**, *163*, 130–143. [CrossRef]

30. Di Fraia, S.; Macaluso, A.; Massarotti, N.; Vanoli, L. Energy, exergy and economic analysis of a novel geothermal energy system for wastewater and sludge treatment. *Energy Convers. Manag.* **2019**, *195*, 533–547. [CrossRef]

31. Franco, A.; Vaccaro, M. Numerical simulation of geothermal reservoirs for the sustainable design of energy plants: A review. *Renew. Sustain. Energy Rev.* **2014**, *30*, 987–1002. [CrossRef]

32. Carotenuto, A.; Massarotti, N.; Mauro, A. A new methodology for numerical simulation of geothermal down-hole heat exchangers. *Appl. Therm. Eng.* **2012**, *48*, 225–236. [CrossRef]

33. Carotenuto, A.; Ciccolella, M.; Massarotti, N.; Mauro, A. Models for thermo-fluid dynamic phenomena in low enthalpy geothermal energy systems: A review. *Renew. Sustain. Energy Rev.* **2016**, *60*, 330–355. [CrossRef]

34. Carotenuto, A.; Marotta, P.; Massarotti, N.; Mauro, A.; Normino, G. Energy piles for ground source heat pump applications: Comparison of heat transfer performance for different design and operating parameters. *Appl. Therm. Eng.* **2017**, *124*, 1492–1504. [CrossRef]

35. Sanyal, S.K.; Butler, S.J.; Swenson, D.; Hardeman, B. Review of the state-of-the-art of numerical simulation of enhanced geothermal systems. In Proceedings of the World Geothermal Congress 2000, Kyushu, Japan, 28 May–10 June 2000.

36. Burnell, J.; Clearwater, E.; Croucher, A.; Kissling, W.; O'Sullivan, J.; O'Sullivan, M.; Yeh, A. Future directions in geothermal modelling. In Proceedings of the (Electronic) 34rd New Zealand Geothermal Workshop, Auckland City, New Zealand, 19–21 November 2012.

37. White, J.A.; Foxall, W. Assessing induced seismicity risk at CO_2 storage projects: Recent progress and remaining challenges. *Int. J. Greenh. Gas Control* **2016**, *49*, 413–424. [CrossRef]

38. Moridis, G.J.; Pruess, K. T2SOLV: An enhanced package of solvers for the TOUGH2 family of reservoir simulation codes. *Geothermics* **1998**, *27*, 415–444. [CrossRef]

39. O'Sullivan, M.J.; Pruess, K.; Lippmann, M.J. State of the art of geothermal reservoir simulation. *Geothermics* **2001**, *30*, 395–429. [CrossRef]

40. O'Sullivan, M.J.; Yeh, A.; Mannington, W.I. A history of numerical modelling of the Wairakei geothermal field. *Geothermics* **2009**, *38*, 155–168. [CrossRef]

41. Bujakowski, W.; Barbacki, A.; Miecznik, M.; Pająk, L.; Skrzypczak, R.; Sowiżdżał, A. Modelling geothermal and operating parameters of EGS installations in the lower triassic sedimentary formations of the central Poland area. *Renew. Energy* **2015**, *80*, 441–453. [CrossRef]

42. Carlino, S.; Troiano, A.; Di Giuseppe, M.G.; Tramelli, A.; Troise, C.; Somma, R.; De Natale, G. Exploitation of geothermal energy in active volcanic areas: A numerical modelling applied to high temperature Mofete geothermal field, at Campi Flegrei caldera (Southern Italy). *Renew. Energy* **2016**, *87*, 54–66. [CrossRef]

43. Zyvoloski, G.; Dash, Z.; Kelkar, S. *FEHMN 1.0: Finite Element Heat and Mass Transfer Code*; Los Alamos National Lab: Los Alamos, NM, USA, 1992.

44. Zyvoloski, G. FEHM: A control volume finite element code for simulating subsurface multi-phase multi-fluid heat and mass transfer. In *Los Alamos Unclassified Report LA-UR-07-3359*; Los Alamos National Lab: Los Alamos, NM, USA, 2007.

45. Kelkar, S.; Lewis, K.; Karra, S.; Zyvoloski, G.; Rapaka, S.; Viswanathan, H.; Mishra, P.; Chu, S.; Coblentz, D.; Pawar, R. A simulator for modeling coupled thermo-hydro-mechanical processes in subsurface geological media. *Int. J. Rock Mech. Min. Sci.* **2014**, *70*, 569–580. [CrossRef]

46. Dempsey, D.; Kelkar, S.; Davatzes, N.; Hickman, S.; Moos, D. Numerical modeling of injection, stress and permeability enhancement during shear stimulation at the Desert Peak Enhanced Geothermal System. *Int. J. Rock Mech. Min. Sci.* **2015**, *78*, 190–206. [CrossRef]

47. Faust, C.R.; Mercer, J.W. Geothermal reservoir simulation: 1. Mathematical models for liquid-and vapor-dominated hydrothermal systems. *Water Resour. Res.* **1979**, *15*, 25–30. [CrossRef]

48. Higuchi, S.; Nishijima, J.; Fujimitsu, Y. Integration of geothermal exploration data and numerical simulation data using GIS in a hot spring area. *Procedia Earth Planet. Sci.* **2013**, *6*, 177–186. [CrossRef]

49. Pola, M.; Fabbri, P.; Piccinini, L.; Zampieri, D. Conceptual and numerical models of a tectonically-controlled geothermal system: A case study of the Euganean Geothermal System, Northern Italy. *Cent. Eur. Geol.* **2015**, *58*, 129–151. [CrossRef]

50. Aliyu, M.D.; Chen, H.-P. Optimum control parameters and long-term productivity of geothermal reservoirs using coupled thermo-hydraulic process modelling. *Renew. Energy* **2017**, *112*, 151–165. [CrossRef]

51. Bakhsh, K.J.; Nakagawa, M.; Arshad, M.; Dunnington, L. Modeling thermal breakthrough in sedimentary geothermal system, using COMSOL multiphysics. In Proceedings of the 41st Workshop on Geothermal Reservoir Engineering, Stanford, CA, USA, 22–24 February 2016.

52. Romano-Perez, C.; Diaz-Viera, M. A Comparison of Discrete Fracture Models for Single Phase Flow in Porous Media by COMSOL Multiphysics®Software. In Proceedings of the 2015 COMSOL Conference in Boston, Boston, MA, USA, 7–9 October 2012.

53. Pandey, S.; Chaudhuri, A.; Kelkar, S. A coupled thermo-hydro-mechanical modeling of fracture aperture alteration and reservoir deformation during heat extraction from a geothermal reservoir. *Geothermics* **2017**, *65*, 17–31. [CrossRef]

54. Pandey, S.; Chaudhuri, A.; Kelkar, S.; Sandeep, V.; Rajaram, H. Investigation of permeability alteration of fractured limestone reservoir due to geothermal heat extraction using three-dimensional thermo-hydro-chemical (THC) model. *Geothermics* **2014**, *51*, 46–62. [CrossRef]

55. Cherubini, Y.; Cacace, M.; Blöcher, G.; Scheck-Wenderoth, M. Impact of single inclined faults on the fluid flow and heat transport: Results from 3-D finite element simulations. *Environ. Earth Sci.* **2013**, *70*, 3603–3618. [CrossRef]

56. Carotenuto, A.; De Luca, G.; Fabozzi, S.; Figaj, R.D.; Iorio, M.; Massarotti, N.; Vanoli, L. Energy analysis of a small geothermal district heating system in Southern Italy. *Int. J. Heat Technol.* **2016**, *34*, S519–S527. [CrossRef]

57. Billi, A.; Bosi, V.; De Meo, A. Caratterizzazione strutturale del rilievo del M. Massico nell'ambito dell'evoluzione quaternaria delle depressioni costiere dei fiumi Garigliano e Volturno (Campania Settentrionale). *Il Quat.* **1997**, *10*, 15–26.

58. Milia, A.; Torrente, M.M. Tectono-stratigraphic signature of a rapid multistage subsiding rift basin in the Tyrrhenian-Apennine hinge zone (Italy): A possible interaction of upper plate with subducting slab. *J. Geodyn.* **2015**, *86*, 42–60. [CrossRef]

59. Luiso, P.; Paoletti, V.; Nappi, R.; La Manna, M.; Cella, F.; Gaudiosi, G.; Fedi, M.; Iorio, M. A multidisciplinary approach to characterize the geometry of active faults: The example of Mt. Massico, Southern Italy. *Geophys. J. Int.* **2018**, *213*, 1673–1681. [CrossRef]

60. Corniello, A.; Ducci, D.; Ruggieri, G.; Iorio, M. Complex groundwater flow circulation in a carbonate aquifer: Mount Massico (Campania Region, Southern Italy). Synergistic hydrogeological understanding. *J. Geochem. Explor.* **2018**, *190*, 253–264. [CrossRef]
61. Cuoco, E.; Minissale, A.; Tamburrino, S.; Iorio, M.; Tedesco, D. Fluid geochemistry of the Mondragone hydrothermal systems (southern Italy): Water and gas compositions vs. geostructural setting. *Int. J. Earth Sci.* **2017**, *106*, 2429–2444. [CrossRef]
62. Adinolfi, M.; Maiorano, R.M.S.; Mauro, A.; Massarotti, N.; Aversa, S. On the influence of thermal cycles on the yearly performance of an energy pile. *Geomech. Energy Environ.* **2018**, *16*, 32–44. [CrossRef]
63. Di Fraia, S.; Massarotti, N.; Nithiarasu, P. Modelling electro-osmotic flow in porous media: A review. *Int. J. Numer. Methods Heat Fluid Flow* **2018**, *28*, 472–497. [CrossRef]
64. Arpino, F.; Carotenuto, A.; Ciccolella, M.; Cortellessa, G.; Massarotti, N.; Mauro, A. Transient natural convection in partially porous vertical annuli. *Int. J. Heat Technol.* **2016**, *34*, S512–S518. [CrossRef]
65. Massarotti, N.; Ciccolella, M.; Cortellessa, G.; Mauro, A. New benchmark solutions for transient natural convection in partially porous annuli. *Int. J. Numer. Methods Heat Fluid Flow* **2016**, *26*, 1187–1225. [CrossRef]
66. Banks, D. Thermogeological assessment of open-loop well-doublet schemes: A review and synthesis of analytical approaches. *Hydrogeol. J.* **2009**, *17*, 1149–1155. [CrossRef]
67. McLean, K.; Zarrouk, S.J. Pressure transient analysis of geothermal wells: A framework for numerical modelling. *Renew. Energy* **2017**, *101*, 737–746. [CrossRef]

Article

A Novel Ground-Source Heat Pump with R744 and R1234ze as Refrigerants

Giuseppe Emmi [1,*], Sara Bordignon [1], Laura Carnieletto [1], Michele De Carli [1], Fabio Poletto [2], Andrea Tarabotti [2], Davide Poletto [3], Antonio Galgaro [4], Giulia Mezzasalma [5] and Adriana Bernardi [6]

[1] Department of Industrial Engineering, University of Padova, 35131 Padova, Italy; sara.bordignon.2@phd.unipd.it (S.B.); laura.carnieletto@unipd.it (L.C.); michele.decarli@unipd.it (M.D.C.)
[2] HiRef S.p.A., 35020 Tribano, Padova, Italy; fabio.poletto@hiref.it (F.P.); andrea.tarabotti.engineering@hiref.it (A.T.)
[3] UNESCO Regional Bureau for Science and Culture in Europe-Science Unit (United Nations Educational, Scientific and Cultural Organization), 30122 Venice, Italy; d.polettodlh@gmail.com
[4] Department of Geosciences, University of Padova, 35131 Padova, Italy; antonio.galgaro@unipd.it
[5] R.E.D. Srl (Research and Environmental Devices), 35129 Padova, Italy; giulia.mezzasalma@red-srl.com
[6] National Research Council, Institute of Atmospheric Sciences and Climate, 35127 Padova, Italy; a.bernardi@isac.cnr.it
* Correspondence: giuseppe.emmi@unipd.it

Received: 22 September 2020; Accepted: 26 October 2020; Published: 29 October 2020

Abstract: The energy-saving potential of heat pump technology is widely recognized in the building sector. In retrofit applications, especially in old and historic buildings, it may be difficult to replace the existing distribution and high-temperature emission systems. Often, historical buildings, especially the listed ones, cannot be thermally insulated; this leads to high temperatures of the heat carrier fluid for heating. In these cases, the main limits are related, on the one hand, to the reaching of the required temperatures, and on the other hand, to the obtaining of good performance even at high temperatures. To address these problems, a suitable solution can be a two-stage heat pump. In this work, a novel concept of a two-stage heat pump is proposed, based on a transcritical cycle that uses the natural fluid R744 (carbon dioxide) with an ejector system. The second refrigerant present in the heat pump and used for the high-temperature stage is the R1234ze, which is an HFO (hydrofluoro-olefin) fluid. This work aims to present the effective energy performance based on real data obtained in operating conditions in a monitoring campaign. The heat pump prototype used in this application is part of the H2020 Cheap-GSHP project, which was concluded in 2019.

Keywords: heat pump; R1234ze; CO_2; geothermal; renewable energy; historic building; energy saving

---

1. Introduction

The European Union (EU) set ambitious targets to fight the issue of climate change, to guarantee the security of supply and increase competitiveness in the energy sector. The building sector, responsible for about 40% of total EU energy consumption, is a great contributor to greenhouse gas (GHG) emissions [1]. When considering the household sector, the share of the European final energy consumption is around 26%, with about 80% due to heating, cooling, and domestic hot water production [2]. An important step to meet the goals of reducing GHG emissions and increasing the renewable energy sources (RES) share can be obtained by adopting technologies that do not employ fossil fuels for the heating and cooling of buildings.

Heat pumps are electrical appliances that can be used for this purpose and allow an improvement in the efficiency of the energy system.

These machines can be employed when a replacement of the energy plant is needed for the retrofit of existing buildings [3,4]. For these applications, high-temperature heat pumps (HTHPs) can be used when heat is demanded at high-temperature levels (80–150 °C) and COP (coefficient of performance) values range between about 2.4 and 5.8 with a temperature lift of 95 to 40 °C, respectively [5]. Traditionally, HTHPs are built using a single-stage thermodynamic cycle, representing a challenge for obtaining favorable COPs with high-temperature heat production. In the literature, studies investigating the use of new refrigerants can be found [6,7], highlighting the need for the adoption of fluids with low GWP (global warming potential) that show acceptable system performance at high temperature.

Other studies analyzed different machine configurations, where the thermodynamic cycle is modified to increase the efficiency and reach higher temperature levels. Dai et al. [8] investigated five configurations of dual-pressure level condensations for exploiting waste heat, obtaining an improvement in COP of more than 9%, compared to traditional cycles. One possible layout is the cascade cycle, composed of two independent single-stage cycles, the high-temperature-stage cycle and the low-temperature-stage cycle [9]. The two cycles are connected by an intermediate heat exchanger that works as the evaporator for the high-temperature cycle and as the condenser for the low-temperature cycle.

Fine et al. [10] developed a numerical model of a solar-assisted cascade heat pump using photovoltaic thermal panels to provide the heat source and electricity to the system. Xu et al. [11] presented an experimental investigation on an air-source cascade heat pump operating in cold climate conditions, demonstrating a higher efficiency in comparison to other heat solutions at low ambient temperature and in a high water-supply temperature region. Yang et al. [12] proposed a single-fluid cascade air-source heat pump that can operate in different modes, and the analysis of the monitored data showed a linear increase in the heating capacity with the lower-stage compressor speed.

Le et al. [13] analyzed an air-source cascade heat pump coupled with thermal storage through laboratory and field results and dynamic simulation models, finding that the heat pump can be used to obtain a CO_2 emission reduction (up to 57%) compared to the traditional boilers and obtain a seasonal COP of 2.12 with the direct heating of the retrofitted building. Mota-Babiloni et al. [14] presented an optimization of the intermediate temperature and the internal heat exchanger effectiveness in both stage cycles of a cascade HTHP using low-GWP refrigerants, finding the maximum COP (3.15) in the combination of pentane and butane.

A different layout can be adopted by including an ejector in the thermodynamic cycle, which allows the decrease in the compression work by reducing the throttling losses and the liquid overfeeding and lifting the compressor inlet pressure [15]. The ejector is a flow device with two intake ports and one discharge port. The primary high-pressure stream and the secondary low-pressure stream are mixed inside the ejector and discharged at some intermediate or back-pressure [16]. Brodal and Eiksund [17] investigated the heat pump performance with and without an ejector or a suction gas heat exchanger, as a modification to conventional transcritical CO_2-based heat pump systems. They found that for pure CO_2-based heat pumps, the systems with an ejector are more efficient than systems without, with an increase in the COP up to 19%. Liu and Lin [18] carried out a thermodynamic analysis of an air-source heat pump producing heat at two temperature levels and using a zeotropic mixture refrigerant (R1270/R600a) of three different configurations, while Besagni et al. [19] studied the influence of different refrigerants on ejector refrigeration systems.

A further way to enhance the efficiency of the heat pump could be to use the soil as the heat source, instead of the external air. These heat pumps are called ground-source heat pumps (GSHP) [20–23] and their advantages range from the high energy performance to environmental friendliness due to the exploitation of renewable energy sources and ease of integration with other energy systems. D'Agostino et al. [24] compared GSHPs to air-source heat pumps or a condensing boiler coupled to a chiller to analyze the energy savings of the investigated system that reaches up to 55% for one of the two locations. Christodoulides et al. [25] provided the cost analysis and comparisons between an air-source heat pump and a ground-source heat pump showing that, for the case study in Cyprus, the air source HP is highly competitive, while Li et al. [26] theoretically demonstrated the general exergy convenience of GSHPs compared to air-source heat pumps.

In this work, the monitored data of a novel partial cascade GSHP are presented. Indeed, the heat pump project is currently involved in a patenting process (Patent application number: 102020000021097, presented on 7th September 2020). The heat pump configuration includes an ejector and two gas-coolers in series in the low-temperature cycle, where CO_2 is used as the refrigerant in a transcritical cycle. The high-temperature-stage fluid is the refrigerant R1234ze. This last cycle is not properly the high-temperature one, as the analyzed layout of the heat pump is different from a common cascade cycle as described in the following section of the text.

2. Materials and Methods

The present paper provides the description and the energy analysis of a heat pump prototype, which uses low-GWP fluids as refrigerants. Moreover, the heat pump has been developed to supply the high-temperature heat carrier fluid to the terminal units used for heating a real building in Zagreb (Croatia). The first part of the paper summarizes the main information about the case study and the details of the heat pump from the mechanical and thermodynamic points of view. In the second part of the work, the results of the operating conditions of the heat pump are described and discussed. The energy analysis has been carried out using the measured data obtained from the monitoring system installed in the plant.

2.1. The Building

The ground-source heat pump has been installed in the building that hosts the Technical Museum Nikola Tesla in Zagreb. The entire complex is listed within the National Register of Cultural Property. The building area is located in the historic city center, which is protected by a special regime of conservation and heritage protection. The building complex is enlisted within the National Register of Cultural Property, by the decision of the Ministry of Culture of the Republic of Croatia in 2005, and it is located in the historic urban areas of the city of Zagreb. The national law on the Protection and Conservation of Cultural Goods forces any structural work planned for the museum to be greenlit by the Urban Office for the Protection of Cultural and Natural Monuments. A limited part of the museum was renovated, which was an exhibition room of 380 m^2 with a volume equal to 1463 m^3, as shown in Figure 1.

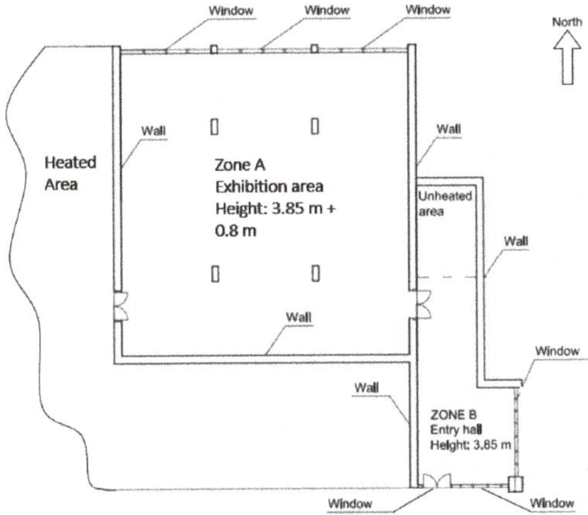

Figure 1. Plan of the exhibition room (ZONE A) of the Technical Museum Nikola Tesla in Zagreb (Croatia). Source: [27].

The original plant system consisted of only six electrical heaters originally used for heating, and no devices for cooling were present. This had a serious impact on the environmental comfort of both visitors and personnel with a possible threat to the fire safety of the structure, which is made mainly of wood components both in its interior and exterior.

The refurbishment intervention was approved by the museum and municipal authorities and included the installation of a CO_2 heat pump with a capacity of about 30 kW. The installed heat pump is a prototype device, specifically designed to provide water at high temperature to terminals.

The refurbishment also included the drilling of a borehole heat exchanger (BHE) field consisting of six 100 m-deep BHEs in the courtyard, plus the relative hydraulic connections. In order to assure heating and cooling in the exhibition room of the museum, 10 fan coils were installed [27].

Unlike the previous work, the present analysis investigates in detail the real behavior of the heat pump, using the data of the plant obtained from a monitoring campaign. The previous paper focused on a possible application of different management strategies for the air-conditioning of the plant and, in that context, a dynamic simulation tool was used for deriving the energy performance of the heat pump considering a mean seasonal coefficient of performance. The museum and the municipal authorities approved the retrofit of the plant after the involvement of the Museum in the Cheap-GSHP European Project [27].

2.2. The Thermal Power Plant

The reversible heat pump is a water-to-water machine that uses a geothermal field as the source and sink in heating and cooling operation, respectively. It consists of two cycles: The low-temperature cycle uses the natural fluid CO_2 as the refrigerant, and the high-temperature cycle uses the hydrofluoro-olefin (HFO) refrigerant R1234ze.

The design of this machine came from the necessity to develop and build a heat pump generation system suitable for a retrofit application, where the heat carrier fluid is required at high-temperature levels, to be used in existing terminal units. As the existing terminal units were electric in the analyzed case study, the use of the heat pump presented several advantages. Indeed, the heat pump reduces the electric energy demand and, at the same time, can produce chilled water to cool the exhibition room in summer, while in the previous configuration, the building was not provided with a cooling system. The solution proposed in the present work is suitable for buildings characterized by a dominant heating thermal load profile. Usually, commercial heat pumps can be used for cooling, through the use of a four-way valve in the refrigerant circuit that reverses the thermodynamic cycle (the component that was used as the condenser is used as the evaporator). On the contrary, in this study, as the heat pump has a double cycle in cascade, the switch of the operating mode is obtained by changing the hydraulic circuit side, not the refrigerant circuit.

The switch of the operation mode, heating and cooling, is allowed using the two 4-way valves and is highlighted in the two schemes of Figure 2. The figure shows the connections of the thermal power plant located outside the building, inside a dedicated room used by the Museum as a showroom for didactic visits. The plant is divided into four main parts: The heat pump, the hydronic module, the borehole heat exchanger (BHE) field, and the distribution pipelines.

The refurbishment of the thermal power plant also included the drilling of a BHE field, consisting of six 100 m-deep ground heat exchangers positioned in the garden and the relative horizontal hydraulic connections. In order to assure heating and cooling provision to the exhibition room of the museum, 10 fan coils were installed. The fluid that circulates in the BHE field is a mixture of pure water and antifreeze. For ground-source heat pump applications, the antifreeze is normally used to prevent the freezing of the horizontal connections during the stop of the thermal power plant, or to prevent the damaging of the plant and of the heat pump when the borehole field is not properly designed and the ground is affected by thermal drift.

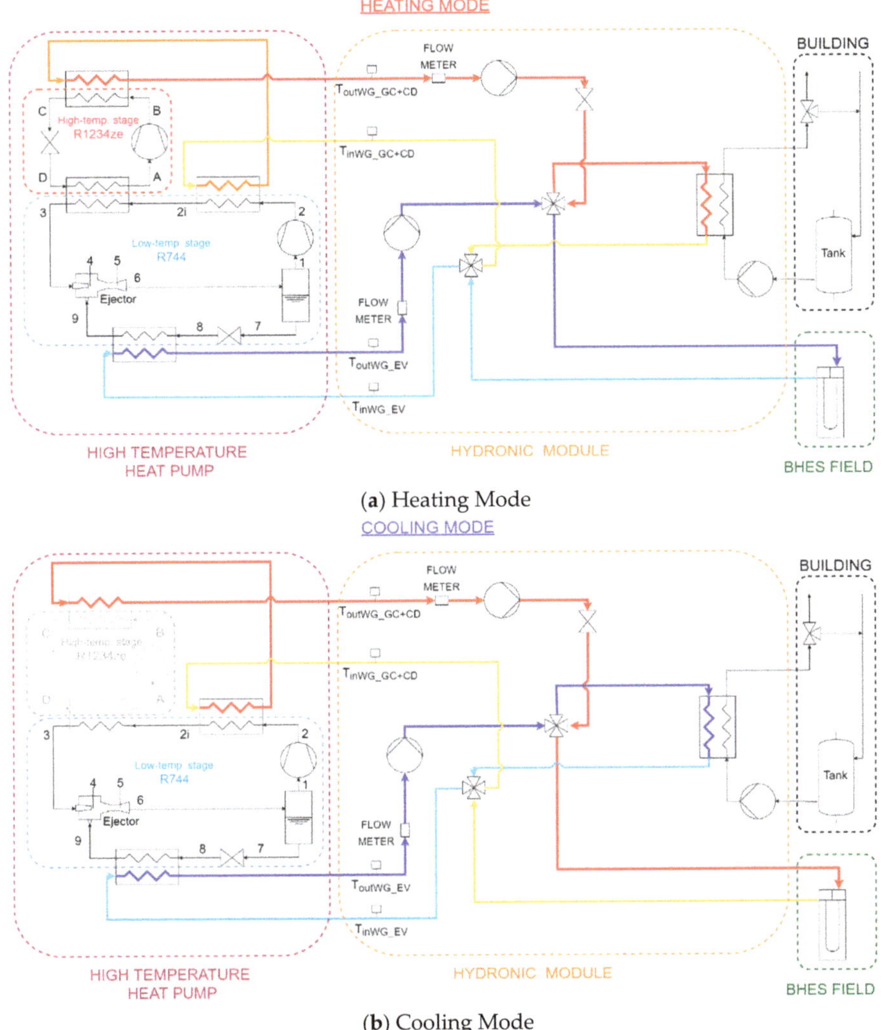

(**a**) Heating Mode

(**b**) Cooling Mode

Figure 2. Scheme of the thermal power plant of the system.

2.3. The Heat Pump

As can be seen from the scheme in Figure 2, the heat pump configuration differs from the more common double cascade thermodynamic cycles that can usually be found in the market. The heat carrier fluid exchanges heat with the two thermodynamic cycles by the use of two flat plate heat exchangers connected in series, while usually, the double cascade cycles use only one heat exchanger at the condenser side of the high-temperature cycle. For this reason, the configuration in the figure could be defined as a "partial" double cascade cycle. In the market, high-temperature transcritical heat pumps that use CO_2 as the refrigerant are quite common, but among the main limits of these machines, the high values of temperature difference required at the condenser side limit the achievement of good energy performances. Consequently, the terminal units used for heating should work with low flow rates and high temperature drops to go below the critical temperature of the CO_2. As widely known in the air-conditioning field, the only terminal units that can work below the critical temperature are radiant systems. However, in these applications, the temperature drops do not exceed 3–4 °C, as suggested by the standards, in order to guarantee a uniform temperature of the radiant surface and good thermal comfort for the users. The layout of the heat pump prototype allows this last problem that limits the use of the transcritical cycle in the field of air conditioning of buildings to be overcome.

In this case study, the heat pump exploits the advantages of the medium/high-temperature cycle, using the CO_2 refrigerant in the transcritical cycle and the heat exchanged at high temperature by the subcritical cycle, which uses the R1234ze refrigerant. The total heat flux exchanged with the heat carrier fluid is the sum of the heat flux exchanged at the gas cooler of the CO_2 transcritical cycle and the heat flux exchanged at the condenser of the high-temperature cycle working with the HFO refrigerant. As is well known, the CO_2 is not suitable for every application in refrigeration, air-conditioning, and heat pump systems, and it presents some limitations in order to guarantee good energy performance. The CO_2 has a vapor pressure much higher than other traditional refrigerant fluids and its critical temperature is around 31 °C. This last characteristic leads to the impossibility of discharging heat to the external air through condensation in a subcritical cycle, when the air temperature, or in general, the heat carrier fluid, is above the critical temperature. The main consequence is that CO_2 can be used efficiently for cooling applications when heat is discharged below the critical temperature, while at higher temperatures, when the cycle works at supercritical pressures, only gas cooling is possible. Therefore, in cooling applications, the use of the ground as the heat sink is a possible and efficient solution. On the other hand, when the heat pump operates in heating mode, the HVAC (Heating Ventilation Air Conditioning) applications could result more complex because, generally, the supply temperature for the terminal units is above 30 °C, except for radiant systems. However, CO_2 is a natural fluid, its ODP (ozone depletion potential) is null, it is not flammable and not toxic (A1), and its GWP is equal to 1. For this reason, CO_2 was also used in the past when environmental advantages and safety restrictions reasons overcame the energy drawbacks. The R1234ze refrigerant is an alternative fluid to the R134a and its environmental impact is lower if compared to the latter. The main properties of the two refrigerants are summarized in Table 1.

Table 1. Main properties of the refrigerants used in the heat pump [28].

	ODP	GWP	Boiling Point @ 1 atm [°C]	Critical Temperature [°C]	Critical Pressure [bar$_a$]	ASHRAE Safety Group
CO_2 (R744)	0	1	−57	31	74	A1
R1234ze	0	6	−19	109	36	A2

The thermodynamic cycles of the heat pump are represented in Figure 3 in a temperature–entropy chart. The diagram shows, from a qualitative point of view, the operating points of the thermodynamic cycles during the heat pump operations. The letters and the numbers in the chart refer to the points reported in Figure 2, which shows the scheme of the heat pump's high and low-temperature cycles.

The properties of the main components of the heat pump are summarized in Table 2. The heat pump is a prototype because it is not available in the market yet, even though all the devices constituting the machine are employed in other applications. For example, the ejector used in the CO_2 cycle is not commonly used in HVAC application, but rather in commercial refrigeration. One of the main objectives during the development of this prototype was to investigate the energy performance potential of this technology at the time of the project. In the last two years, the HVAC market started to introduce new devices and components properly developed for HVAC applications that use the CO_2 as the refrigerant. These new products can certainly guarantee better results than those obtained with the technology available about 3–4 years ago and used in the prototype; therefore, the results of this study can be considered the starting point of this emerging technology. The investigated heat pump was thought to be used in plug-and-play applications, as the generation module and the hydraulic module have been included in a common box. The control of the heat pump works at two levels: The high-temperature stage is controlled by the regulation system that keeps the gas cooler outlet temperature at a setpoint value of 35 °C, with a dead band of 5 °C. If the temperature drops below 30 °C, the thermal efficiency of the high stage cycle decreases under acceptable values as the evaporating temperature is too low. However, the dead band of 5 °C prevents the continuous on/off of the HFO compressor. The compressor of the CO_2 sets its working conditions by monitoring the return temperature of the heating and cooling terminal units. The volume flow rate of the BHE field circuit is modified according to the following principles: In heating mode, when the evaporator supply temperature decreases, the flow rate increases, and the control is done evaluating a set point evaporator pressure; in cooling mode, the flow rate is modified by keeping a constant temperature difference between the inlet and the outlet at the evaporator side. The volume flow rate at the user side of the heat pump is set to maintain a constant temperature difference between the supply and return temperatures of the circuit, to 5 °C and 10 °C in cooling mode and heating mode, respectively.

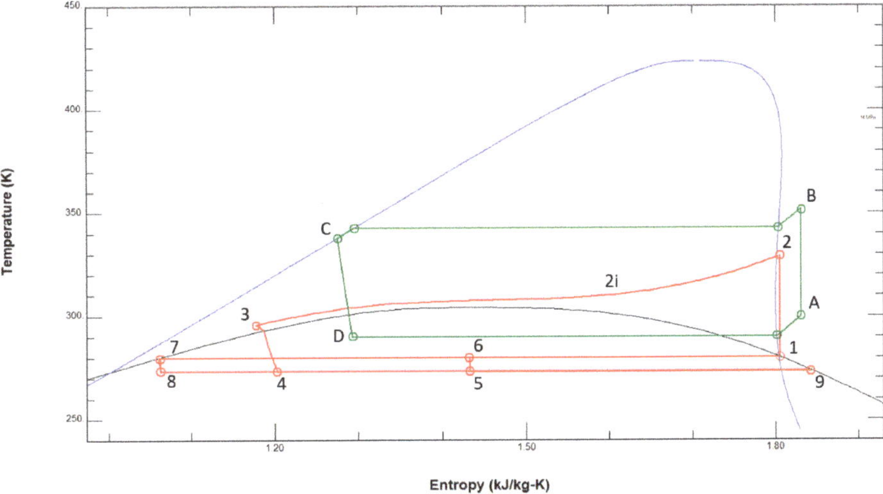

Figure 3. Thermodynamic cycles of the high-temperature heat pump.

2.4. The Monitoring System

The monitoring system consists of flow meters and temperature probes for the measurement of the heat fluxes exchanged between the heat carrier fluids and the refrigerants, at the user side (fan coil terminal units) and source side (ground heat exchangers field), respectively. Other parameters have been monitored in the thermodynamic cycle in order to evaluate the behavior of the cycle during the operation of the system. In particular, considering the CO_2 cycle, the pressures of the refrigerant inside the gas cooler (GC) and inside the evaporator (EV) have been logged by the system. The electrical consumption of the heat pump, necessary for the evaluation of the COP and EER (Energy Efficiency Ratio), has been measured using two energy meters. The first electricity meter monitored the electrical energy demanded by the compressors and the auxiliary devices of the heat pump. These consisted of the auxiliary electric resistances present in the heat pump for the safety of the components during the stop of the plant in the heating period, in the unit controller and in other electric minor devices present inside the heat pump box. The second is located in the hydronic module of the thermal power plant, which has two pumps (user and source side). The electrical energy meter measures their energy consumption and the demand of the controller, which manages the operation of this system. As for the first energy meter, a small part of the total electrical energy consumption is due to other devices located in the hydraulic box, installed in the thermal power plant of the Museum.

Table 2. Properties of the heat pump and design conditions for the compressors.

	Parameter	Value	Unit	Note
Heat Pump	Heating Capacity (60/70 °C–10/5 °C)	30.0	kW_h	Water Flow User side = 3028 L/h (30% propylene glycol/70% pure water) Water Flow Source side = 3831 L/h (30% propylene glycol/70% pure water)
	Electric Power Input	11.0	kW_{el}	
	COP	2.72	-	
	Cooling Capacity (12/7 °C–30/35 °C)	25.0	kW_c	Water Flow User side = 4759 L/h (30% propylene glycol/70% pure water) Water Flow Source side = 5847 L/h (30% propylene glycol/70% pure water)
	Electric Power Input	6.0	kW_{el}	
	EER	4.12	-	
CO_2 Compressor	Heating Capacity	24.7	kW_h	$T_{in\,gc}/T_{out\,gc}$ = 89/30 °C (100 bar) T_{evap} = 9 °C (ΔT_{sh} = 5 °C)
	Cooling Capacity	19.0	kW_c	
	Electric Power Input	5.8	kW_{el}	
	COP/EER	4.30/3.30	-/-	
R1234ze Compressor	Heating Capacity	32.0	kW_h	T_{cond} = 72 °C (ΔT_{sc} = 5 °C) T_{evap} = 25 °C (ΔT_{sh} = 10 °C)
	Cooling Capacity	24.2	kW_c	
	Electric Power Input	7.7	kW_{el}	
	COP/EER	3.14/4.14	-/-	

The main properties of the monitoring units installed in the thermal power plant are summarized in Table 3. The experimental uncertainty of the measurement chain is within ±10% for the analyzed case study. The time step of the logging data has been set to 30 s.

Table 3. Properties of the measuring devices.

Parameter	Type	Accuracy	Note
Temperature	PT500	Class B (±(0.3 + 0.005 * t))	4-wires, EN 60751
Flow meter	Electromagnetic	±0.5%	User side and source side
Data Logger	Energy meter	±(0.18 + 0.1/Δt)%	
Pressure	Piezoresistive	±4% Full Scale	Refrigerant cycle, Range 0–150 bar
Electrical Energy	Metering and monitoring device	Class 0.5S	IEC 62053-22

3. The Monitoring Data

The thermal power plant of the air conditioning system has been monitored for about one year, from the beginning of March 2018 to the end of April 2019. The energy data and the values of pressures of the thermodynamic cycle have been analyzed to evaluate the thermal and energy behavior of the system in cooling and heating operation mode, with the aim of checking the trend of the pressures during the switch-on cycle of the system and the energy performance of the heat pump. The air conditioning system operation, during the monitoring period, was scheduled according to the opening days of the Museum, when possible. For this reason, when the building was closed to visitors, the heat pump switched on only a few times during the day to keep the water inside the tank at the set point temperature of about 67.5 °C in heating and 7 °C in cooling, as can be seen in the trends of the storage temperature. In some cases, with the purpose of gathering enough data for the energy analysis of the system, the system was switched on more often and independently from the opening hours of the Museum. Two examples of operating conditions of a day are shown in Figures 4 and 5 for the heating and cooling periods, respectively. In the chart, some details about the switching-on and -off of the heat pump are reported. Indeed, the trends of the temperature of the thermal storage and of the heat carrier fluids are highlighted.

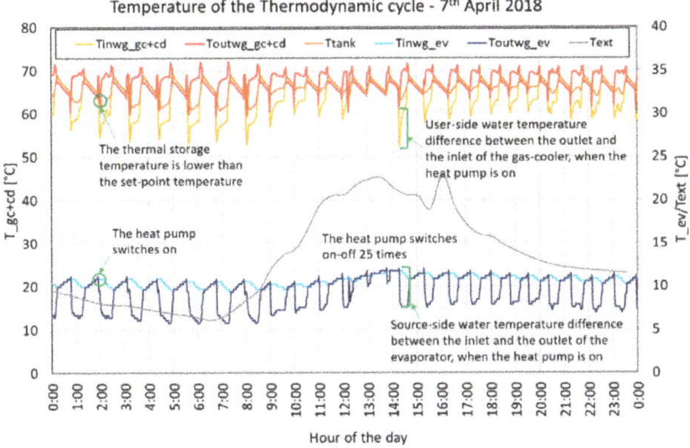

Figure 4. Example of day—Heating mode.

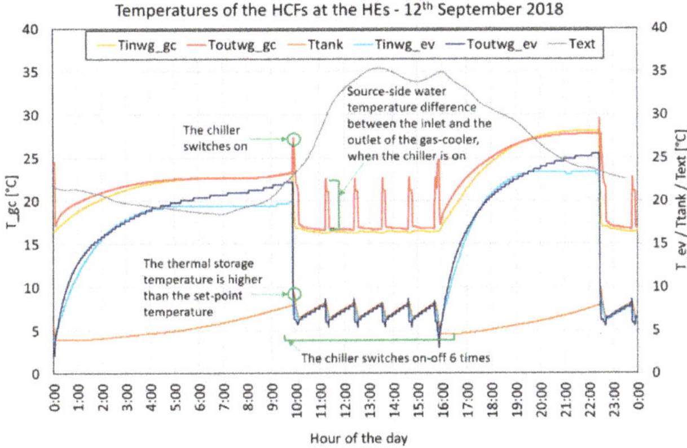

Figure 5. Example of day—Cooling mode.

The main objective of this work was the investigation of the novel reversible heat pump producing high-temperature water for heating and chilled water for cooling a historic building during the winter and summer period, respectively. This goal was one of the main fields of investigation of the European Cheap-GSHP H2020 project.

4. Results and Discussion

In this section, the data of the monitoring campaign are discussed in detail, evaluating the behavior of the system from the thermal and electrical points of view. The operating conditions for some selected days and the energy fluxes involved in the system have been investigated. In particular, the heating mode and the cooling mode of the heat pump were considered. Some days have been chosen and analyzed considering the days with continuous operation and the days with only a few starts of the heat pump during the day.

In Figure 6, the trend of the system's temperatures and R744 pressures are represented for 18 November 2018. This is a representative day of continuous operation mode in heating. In particular, the chart shows the heat carrier fluid temperatures at the inlet and outlet of the high-temperature heat exchangers and the heat carrier fluid temperatures at the inlet and outlet of the evaporator coupled with the BHE field. The temperatures of the fluids have been measured in the points of the system, as represented in the scheme of the thermal power plant in Figure 2. The switch-on of the heat pump is controlled by the value of the temperature of the fluid in the thermal storage tank (T_{tank}). The setpoint temperature of the storage tank is 67.5 °C with an upper dead band of 1.5 °C, so the heat pump is shut down when the temperature of the storage tank is higher than 69 °C. As can be seen in the chart during this day, the system works continuously, the heat pump starts for 29 times, and each on and off cycle lasts about 50 min. The pressure of the R744 transcritical cycle is shown in Figure 6b. The high pressure (HP—Point 3 of the scheme in Figure 2) of the cycle at the gas cooler varies between 95 and 80 bar while the low pressure (LP—Point 9 of the scheme in Figure 2) and suction pressure (SP—Point 6 of the scheme in Figure 2) at the compressor vary between 40 and 50 bar. A zoom of the on and off cycle of the heat pump is reported in Figure 7. In the charts, temperatures and pressures between 6:28 a.m. and 7:26 a.m. are reported. At the beginning of the on and off cycle, the high pressure rises to the maximum value, decreasing gradually until the stop of the heat pump when the setpoint temperature of the storage tank was reached. As can be seen in the chart in Figure 7a, before and during the first part of the cycle, the values of T_{inwg_gc+cd} and T_{outwg_gc+cd} were the same. This happens because the heat pump compressor was off until 6:37 a.m., but the circulator of the hydronic module started about two minutes before the compressor. Indeed, the compressor has a time delay of switch-on after the start of the circulator for safety and protection of the components of the heat pump. The details of the pressures of the R744 cycle are reported in Figure 7b. As can be seen in the chart, the SP and evaporator pressure (LP) are the same during the operation of the heat pump, as expected, while the SP is equal to the HP when the compressor is off because the two pressures are in equilibrium through the compressor. The LP is less than the other pressures because of the expansion valves present in the refrigerant loop.

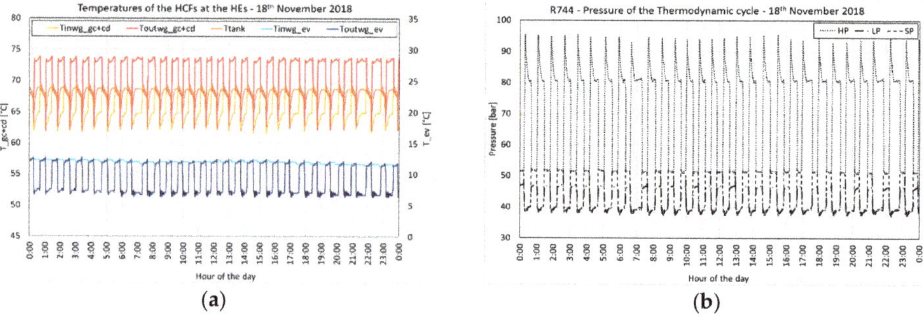

Figure 6. System's temperatures (**a**) and pressures (**b**)—18 November 2018.

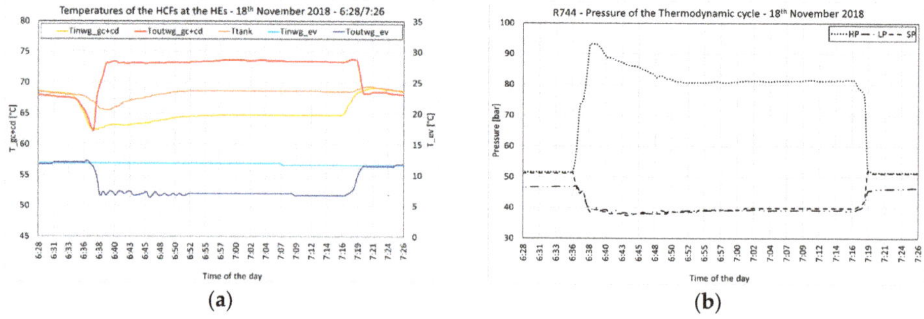

Figure 7. System's temperatures (**a**) and pressures (**b**)—18 November 2018—6:28/7:26.

In Figures 8–11, similar charts are presented for two representative days in December. As for 4 December 2018, the plant was working for 14.9 h, providing water to the storage tank at 67.8 °C on average. From the graph in Figure 8a, it can be noticed that the switching-on cycles were less frequent in the warmer hours of the day (the external air temperature is also shown in this figure). In this case, the high pressure of the cycle at the gas cooler also varied between 95 and 80 bar, while the low pressure and suction pressure at the compressor varied between 40 and 50 bar.

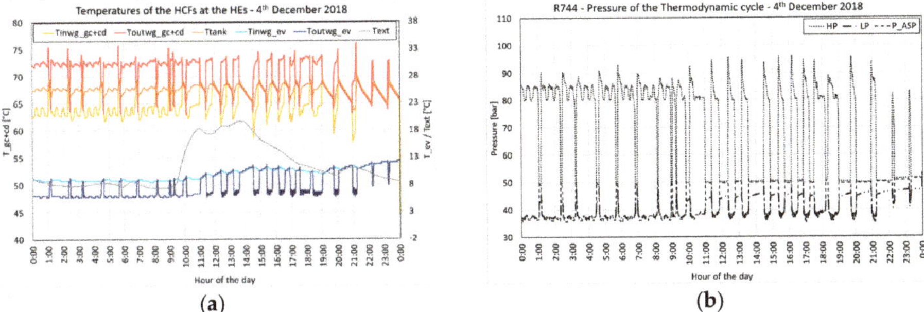

Figure 8. System's temperatures (**a**) and pressures (**b**)—4 December 2018.

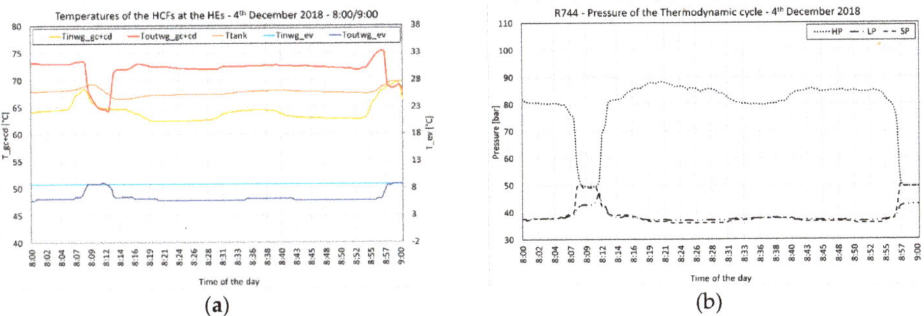

Figure 9. System's temperatures (**a**) and pressures (**b**)—4 December 2018—8:00/9:00.

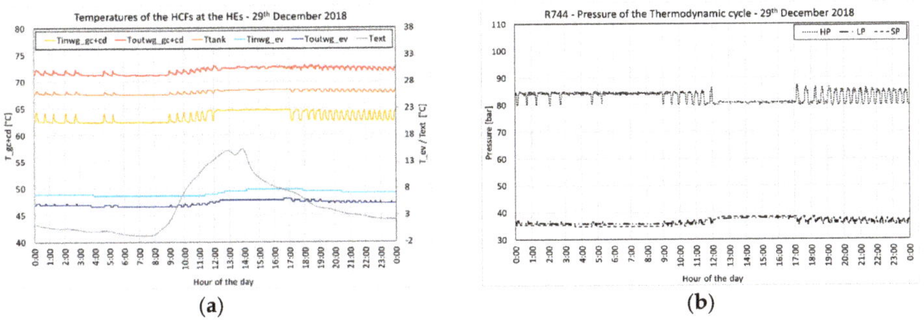

Figure 10. System's temperatures (**a**) and pressures (**b**)—29 December 2018.

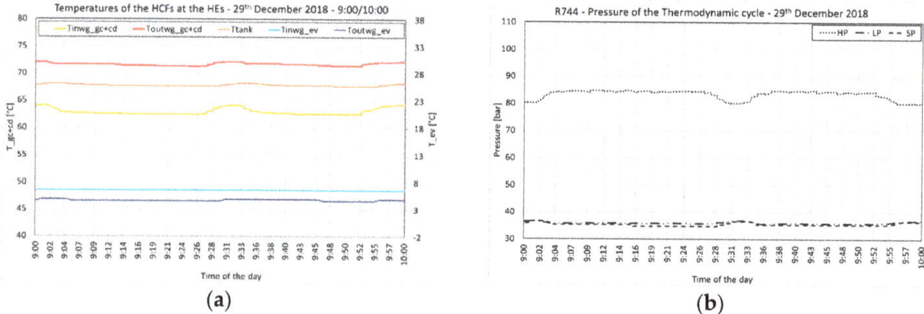

Figure 11. System's temperatures (**a**) and pressures (**b**)—29 December 2018—9:00/10:00.

Similar considerations can be done for 29 December 2018, when the heat pump results to be operating nearly without interruptions. In this case, the electrical energy demanded by the compressors of the heat pump compressors during the whole day is 291 kWh, nearly 56% more than for the day of 4 December. In the same way, the electrical energy absorbed by the auxiliaries is nearly three times with 29 against 11 kWh for the day of 4 December. In addition, the thermal energy released at the user-side of the HP and withdrawn from the ground is about 54% higher for 29 December, with an amount of 631 and 350 kWh, respectively. As shown in Figure 11b, the high pressure of the cycle at the gas cooler varies between 80 and 85 bar, while the low pressure and suction pressure at the compressor are between 35 and 40 bar. In this last case, the values of LP are lower than the other two days because of the lower temperature of the heat source that affects the evaporating temperature and the relative pressure at the evaporator side. As can be seen in the charts, the inlet temperature of the heat carrier fluid at the evaporator moves from about 11 °C to about 8 °C. On the other hand, at the gas cooler, the steady-state pressure after the first phase of the switching-on of the heat pump is about 80 bar. This is due to the control system that regulates the outlet temperature at the gas cooler at 35 °C, as previously described in the Heat Pump section of the text.

Considering the months of November and December, the temperature of the water entering the condenser at the user side of the heat pump is about 60 °C and 63 °C, respectively. The water is then heated up to 67/71 °C, maintaining an average storage tank temperature of 67.5 °C. At the source side of the heat pump, the heat carrier fluid from the BHE field loop has a monthly average temperature of 13.5 °C in November and 8.4 °C in December. The fluid is then cooled inside the evaporator to 9 °C and 6 °C in November and December, respectively.

In Table 4, the values for the heat pump COP and for the system COP, for the investigated days in November and December, and for the selected on-off cycle of the heat pump are shown.

Table 4. COP_{HP} and COP_{sys} in heating period.

Day	Period	COP_{HP}	COP_{sys}
18 November	All day	2.16	2.05
	6:28/7:26	2.15	2.06
4 December	All day	2.16	2.04
	8:09/9:00	2.18	2.07
29 December	All day	2.17	1.98
	9:00/10:00	2.17	1.96

The COP_{HP} of the heat pump is defined as the thermal energy delivered at the user-side of the heat pump (Q_1), divided by the electrical energy absorbed by the compressors of the low ($P_{comp,L}$) and high-temperature cycles ($P_{comp,H}$). On the other hand, the COP_{sys} of the system is defined as the thermal energy delivered at the user-side of the heat pump (Q_1), divided by the electrical energy absorbed by the compressors of the low and high-temperature cycles ($P_{comp,H}$ + $P_{comp,L}$) and by the auxiliaries of the hydronic system (P_{aux}).

$$COP_{HP} = \frac{Q_1}{P_{comp,L} + P_{comp,H}} \tag{1}$$

$$COP_{sys} = \frac{Q_1}{P_{comp,L} + P_{comp,H} + P_{aux}} \tag{2}$$

During these days, the source temperatures are 12 °C, 9.5 °C, and 7.1 °C for November and 4 and 29 December.

In cooling mode, the cycle working with R1234ze is off and only the CO_2 thermodynamic cycle is switched on. The hydronic module modifies its layout by changing the position of the two 4-way valves. In this configuration, the BHE field loop exchanges heat with the gas cooler heat exchanger while the user tank exchanges heat with the evaporator heat exchanger. Unlike the heating mode, the pressures of the CO_2 cycle are different because the high pressure (HP) is set by the controller of the heat pump in order to obtain the optimal value, which is a function of the outlet temperature at the gas cooler heat exchanger. Therefore, in this case, the pressure depends on the temperature of the heat carrier fluid in the BHE field loop. Similarly, as shown for the heating mode, few days have been analyzed in detail to show the behavior of the heat pump from the temperature and pressure point of views.

In particular, the temperature profile and the operating pressures of the cycle have been analyzed for one day in June, July, and August. Moreover, the one on-off cycle of the heat pump can be seen in detail for each representative day.

The temperature of the water entering the evaporator and, therefore, at the user side of the heat pump is about 7.4 °C for the months of June and July, while it is an average of 7 °C in August. On the other hand, the temperature exiting the evaporator is set to 6 °C. Considering the source side of the heat pump, the temperature of the heat carrier fluid coming from the BHE field loop is around 17.7 °C in June on a monthly average, 18.8 °C in July, and 17.3 °C in August. Respectively, inside the gas cooler, its temperature is increased to 23.2 °C, 24.3 °C, and 22.4 °C.

The average monthly temperature of the water inside the storage tank is 6.6 °C in June, and 6.9 °C in July and August.

In Figures 12–17, the trends of the system's temperatures and R744 pressures are represented for 29 June, 19 July, and 22 August 2018. The charts show the heat carrier fluid temperatures at the inlet and outlet of the evaporator at the user-side of the HP and the heat carrier fluid temperatures at the inlet and outlet of the condenser coupled with the BHE field.

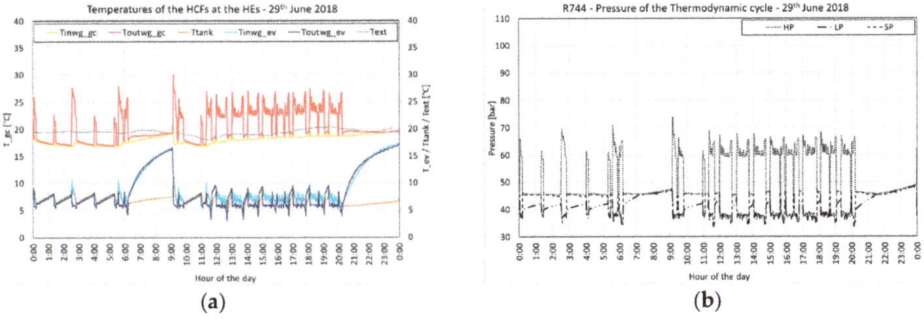

Figure 12. System's temperatures (**a**) and pressures (**b**)—29 June 2018.

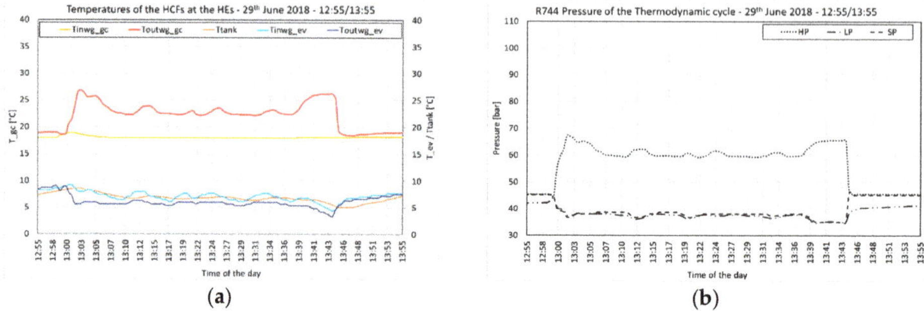

Figure 13. System's temperatures (**a**) and pressures (**b**)—29 June 2018—12:55/13:55.

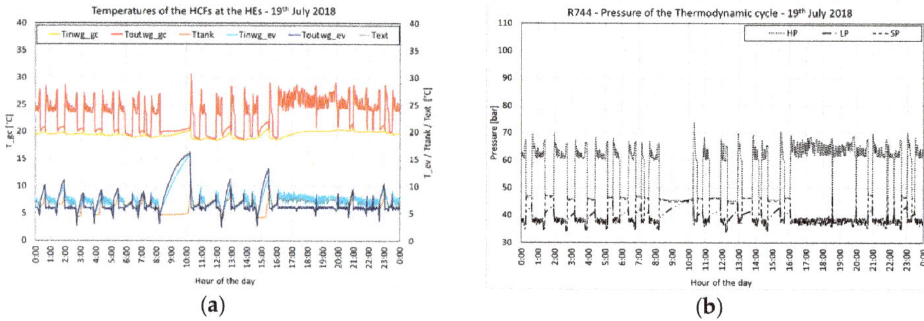

Figure 14. System's temperatures (**a**) and pressures (**b**)—19 July 2018.

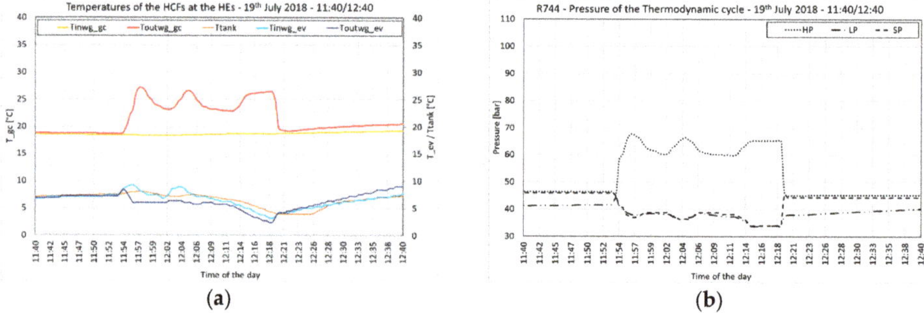

Figure 15. System's temperatures (**a**) and pressures (**b**)—19 July 2018—11:40/12:40.

The representative day for June was 29, when the heat pump was on nearly 9 h. The high pressure of the cycle at the gas cooler varied between 45 and 70 bar while the low pressure and suction pressure were in the range of 25 and 50 bar, as for the other analyzed summer days.

Table 5 summarizes the EER values for the heat pump and for the heat pump coupled with the hydronic system, for the investigated days in June, July, and August, and for the selected on-off cycle of the heat pump.

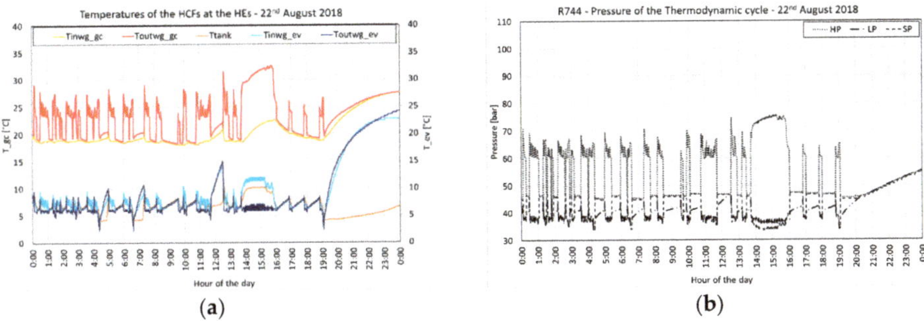

(a)   **(b)**

Figure 16. System's temperatures (**a**) and pressures (**b**)—22 August 2018.

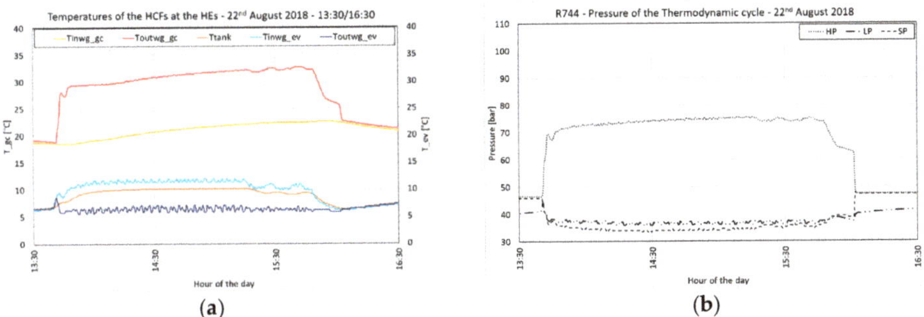

(a)   **(b)**

Figure 17. System's temperatures (**a**) and pressures (**b**) —22 June 2018—11:30/16:30.

Table 5. EER_{HP} and EER_{sys} in cooling period.

Day	Period	EER_{HP}	EER_{sys}
29 June	All day	5.89	3.79
	12:55/13:55	5.17	3.62
19 July	All day	3.50	2.76
	11:40/12:40	3.78	2.70
22 August	All day	4.13	4.09
	13:30/16:30	3.61	3.47

The EER of the heat pump is defined as the thermal energy withdrawn at the user-side of the heat pump (Q_2), divided by the electrical energy absorbed by the compressor of the CO_2 cycle ($P_{comp,L}$). On the other hand, the EER of the system is defined as the thermal energy withdrawn at the user-side of the heat pump (Q_2), divided by the electrical energy absorbed by the compressor of the low-temperature cycles ($P_{comp,L}$) and by the auxiliaries of the hydronic system (P_{aux}).

$$EER_{HP} = \frac{Q_2}{P_{comp,L}} \tag{3}$$

$$EER_{sys} = \frac{Q_2}{P_{comp,L} + P_{aux}} \tag{4}$$

During these days, the source temperatures are 18 °C, 19.4° C, and 19.6 °C for June, July, and August, respectively, and it can be seen that higher source temperatures lead to lower EER values.

From the monitoring campaign, the EER and COP of the heat pump and of the entire system were evaluated. In particular, when considering the operation during the month of July, a heat pump

EER of 3.55 and a system EER of 2.74 have been found. As for the heating operation, in December, the heat pump COP was 2.15, while the system COP was 2.01.

5. Conclusions

In the present work, the configuration and monitored data for a novel cascade GSHP developed within the H2020 Cheap-GSHP project, concluded in 2019, are presented. The heat pump, installed at the Technical Museum Nikola Tesla in Zagreb, uses CO_2 and R1234ze as the working fluids in a transcritical cycle where the common expansion valve is replaced by the ejector technology and in the high-temperature cycle, respectively. During the heating season, the reversible heat pump provides high-temperature water used for the space heating of the historical building provided with fan coil terminal units, rejecting heat to the ground through the use of six borehole heat exchangers. Data for temperatures and pressures at the heat exchangers are shown for some representative days of the warm and cold seasons, and the logic controls regulating the operations of the heat pump are explained. Even though the heat pump is a prototype built with devices and components normally used in commercial refrigeration and not for HVAC systems, the results in terms of COP and EER values are not different if compared with standard high-temperature double-stage heat pumps present in the market. Moreover, the case study presented has several positive characteristics and innovations, such as the use of low-GWP refrigerant, CO_2 (R744), and R1234ze.

Author Contributions: Conceptualization, G.E.; methodology, G.E., S.B., L.C. and M.D.C.; validation, G.E., S.B., L.C., F.P. and A.T.; formal analysis, G.E., S.B. and L.C.; investigation, G.E., S.B. and F.P.; resources, F.P.; data curation, G.E., S.B. and F.P.; writing—original draft preparation, G.E. and S.B.; writing—review and editing, M.D.C., F.P., A.G., D.P., G.M. and A.B.; visualization, G.E. and S.B.; supervision, G.E., S.B. and A.B.; project administration, A.B.; funding acquisition, A.B. All authors have read and agreed to the published version of the manuscript.

Funding: Cheap-GSHPs project has received funding from the European Union's Horizon 2020 research and innovation program under grant agreement No. 657982.

Acknowledgments: The authors thank many people that contributed, participated and supported this work. For the work performed at the Technical Museum Nikola Tesla in Zagreb, the director Franulić and Eng. Branimir Prgomet, Davor Trupoković, Assistant Minister, Ministry of Culture of Croatia, Vladimir Soldo, Faculty of Mechanical Engineering and Naval Architecture University of Zagreb, Eng. Ante Pokupčić and Arch. Željko Kovačić, designer of the technical room, heat pump installer Eng. Franjo Banić, Trojanović local REHAU staff, Veljko Mihalić, Ines Franov Beoković, Narcisa Vrdoljak and Jasna Šćavničar Ivković, Municipality and City office of Zagreb, Vesna Drasal from E-kolektor services, Miljenko Sedlar, Deputy Principal of Department for project implementation, Rut Carek, Secretary General, Croatian Commission for UNESCO.

References

1. Eurostat. *Energy Data*, 2020 ed.; Eurostat: Luxembourg, Luxembourg, 2020. [CrossRef]
2. Energy Consumption in Households–Statistics Explained. Available online: https://ec.europa.eu/eurostat/statistics-explained/index.php/Energy_consumption_in_households (accessed on 3 May 2020).
3. Piselli, C.; Romanelli, J.; Di Grazia, M.; Gavagni, A.; Moretti, E.; Nicolini, A.; Cotana, F.; Strangis, F.; Witte, H.J.L.; Pisello, A.L. An Integrated HBIM Simulation Approach for Energy Retrofit of Historical Buildings Implemented in a Case Study of a Medieval Fortress in Italy. *Energies* **2020**, *13*, 2601. [CrossRef]
4. Hirvonen, J.; Jokisalo, J.; Kosonen, R. The Effect of Deep Energy Retrofit on The Hourly Power Demand of Finnish Detached Houses. *Energies* **2020**, *13*, 1773. [CrossRef]
5. Arpagaus, C.; Bless, F.; Uhlmann, M.; Schiffmann, J.; Bertsch, S.S. High temperature heat pumps: Market overview, state of the art, research status, refrigerants, and application potentials. *Energy* **2018**, *152*, 985–1010. [CrossRef]
6. Luo, B.; Zou, P. Performance analysis of different single stage advanced vapor compression cycles and refrigerants for high temperature heat pumps. *Int. J. Refrig.* **2019**, *104*, 246–258. [CrossRef]

7. Wu, D.; Hu, B.; Wang, R.Z.; Fan, H.; Wang, R. The performance comparison of high temperature heat pump among R718 and other refrigerants. *Renew. Energy* **2020**, *154*, 715–722. [CrossRef]

8. Dai, B.; Liu, X.; Liu, S.; Zhang, Y.; Zhong, D.; Feng, Y.; Nian, V.; Hao, Y. Dual-pressure condensation high temperature heat pump system for waste heat recovery: Energetic and exergetic assessment. *Energy Convers. Manag.* **2020**, *218*, 112997. [CrossRef]

9. Pan, M.; Zhao, H.; Liang, D.; Zhu, Y.; Liang, Y.; Bao, G. A Review of the Cascade Refrigeration System. *Energies* **2020**, *13*, 2254. [CrossRef]

10. Fine, J.P.; Friedman, J.; Dworkin, S.B. Detailed modeling of a novel photovoltaic thermal cascade heat pump domestic water heating system. *Renew. Energy* **2017**, *101*, 500–513. [CrossRef]

11. Xu, L.; Li, E.; Xu, Y.; Mao, N.; Shen, X.; Wang, X. An experimental energy performance investigation and economic analysis on a cascade heat pump for high-temperature water in cold region. *Renew. Energy* **2020**, *152*, 674–683. [CrossRef]

12. Yang, Y.; Li, R.; Zhu, Y.; Sun, Z.; Zhang, Z. Experimental and simulation study of air source heat pump for residential applications in northern China. *Energy Build.* **2020**, *224*, 110278. [CrossRef]

13. Le, K.X.; Huang, M.J.; Shah, N.N.; Wilson, C.; Mac Artain, P.; Byrne, R.; Hewitt, N.J. Techno-economic assessment of cascade air-to-water heat pump retrofitted into residential buildings using experimentally validated simulations. *Appl. Energy* **2019**, *250*, 633–652. [CrossRef]

14. Mota-Babiloni, A.; Mateu-Royo, C.; Navarro-Esbrí, J.; Molés, F.; Amat-Albuixech, M.; Barragán-Cervera, Á. Optimisation of high-temperature heat pump cascades with internal heat exchangers using refrigerants with low global warming potential. *Energy* **2018**, *165*, 1248–1258. [CrossRef]

15. Zhang, Z.; Feng, X.; Tian, D.; Yang, J.; Chang, L. Progress in ejector-expansion vapor compression refrigeration and heat pump systems. *Energy Convers. Manag.* **2020**, *207*, 112529. [CrossRef]

16. Tashtoush, B.M.; Al-Nimr, M.A.; Khasawneh, M.A. A comprehensive review of ejector design, performance, and applications. *Appl. Energy* **2019**, *240*, 138–172. [CrossRef]

17. Brodal, E.; Eiksund, O. Optimization study of heat pumps using refrigerant blends–Ejector versus expansion valve systems. *Int. J. Refrig.* **2020**, *111*, 136–146. [CrossRef]

18. Liu, J.; Lin, Z. Thermodynamic analysis of a novel dual-temperature air-source heat pump combined ejector with zeotropic mixture R1270/R600a. *Energy Convers. Manag.* **2020**, *220*, 113078. [CrossRef]

19. Besagni, G.; Mereu, R.; Di Leo, G.; Inzoli, F. A study of working fluids for heat driven ejector refrigeration using lumped parameter models. *Int. J. Refrig.* **2015**, *58*, 154–171. [CrossRef]

20. Ma, Z.; Xia, L.; Gong, X.; Kokogiannakis, G.; Wang, S.; Zhou, X. Recent advances and development in optimal design and control of ground source heat pump systems. *Renew. Sustain. Energy Rev.* **2020**, *131*, 110001. [CrossRef]

21. Nouri, G.; Noorollahi, Y.; Yousefi, H. Solar assisted ground source heat pump systems—A review. *Appl. Therm. Eng.* **2019**, *163*, 114351. [CrossRef]

22. Karytsas, S.; Choropanitis, I. Barriers against and actions towards renewable energy technologies diffusion: A Principal Component Analysis for residential ground source heat pump (GSHP) systems. *Renew. Sustain. Energy Rev.* **2017**, *78*, 252–271. [CrossRef]

23. Luo, J.; Rohn, J.; Xiang, W.; Bertermann, D.; Blum, P. A review of ground investigations for ground source heat pump (GSHP) systems. *Energy Build.* **2016**, *117*, 160–175. [CrossRef]

24. D'Agostino, D.; Mele, L.; Minichiello, F.; Renno, C. The Use of Ground Source Heat Pump to Achieve a Net Zero Energy Building. *Energies* **2020**, *13*, 3450. [CrossRef]

25. Christodoulides, P.; Aresti, L.; Florides, G. Air-conditioning of a typical house in moderate climates with Ground Source Heat Pumps and cost comparison with Air Source Heat Pumps. *Appl. Therm. Eng.* **2019**, *158*, 113772. [CrossRef]

26. Li, R.; Ooka, R.; Shukuya, M. Theoretical analysis on ground source heat pump and air source heat pump systems by the concepts of cool and warm exergy. *Energy Build.* **2014**, *75*, 447–455. [CrossRef]

27. Cadelano, G.; Cicolin, F.; Emmi, G.; Mezzasalma, G.; Poletto, D.; Galgaro, A.; Bernardi, A. Improving the Energy Efficiency, Limiting Costs and Reducing CO_2 Emissions of a Museum Using Geothermal Energy and Energy Management Policies. *Energies* **2019**, *12*, 3192. [CrossRef]

28. Refrigerants. Product Data Summary. Available online: http://www.linde-gas.com/en/images/Refrigerants-Product-Data-Summary_tcm17-108590.pdf (accessed on 22 October 2020).

Publisher's Note: MDPI stays neutral with regard to jurisdictional claims in published maps and institutional affiliations.

Article

The Effect of Groundwater Flow on the Thermal Performance of a Novel Borehole Heat Exchanger for Ground Source Heat Pump Systems: Small Scale Experiments and Numerical Simulation

Ahmed A. Serageldin [1,2,*], **Ali Radwan** [1,3], **Yoshitaka Sakata** [1], **Takao Katsura** [1] and **Katsunori Nagano** [1]

[1] Environmental System Research Laboratory, Division of Human Environmental Systems, Hokkaido University, Sapporo 060-8628, Japan; ali.radwan@ejust.edu.eg (A.R.); y-sakata@eng.hokudai.ac.jp (Y.S.); katsura@eng.hokudai.ac.jp (T.K.); nagano@eng.hokudai.ac.jp (K.N.)

[2] Mechanical Power Engineering Department, Faculty of Engineering at Shoubra, Benha University, Shoubra 11629, Egypt

[3] Mechanical Power Engineering Department, Faculty of Engineering, Mansoura University, El Mansoura 35516, Egypt

* Correspondence: ahmed.serageldin@eng.hokudai.ac.jp

Received: 13 February 2020; Accepted: 16 March 2020; Published: 18 March 2020

Abstract: New small-scale experiments are carried out to study the effect of groundwater flow on the thermal performance of water ground heat exchangers for ground source heat pump systems. Four heat exchanger configurations are investigated; single U-tube with circular cross-section (SUC), single U-tube with an oval cross-section (SUO), single U-tube with circular cross-section and single spacer with circular cross-section (SUC + SSC) and single U-tube with an oval cross-section and single spacer with circular cross-section (SUO + SSC). The soil temperature distributions along the horizontal and vertical axis are measured and recorded simultaneously with measuring the electrical energy injected into the fluid, and the borehole wall temperature is measured as well; consequently, the borehole thermal resistance (R_b) is calculated. Moreover, two dimensional and steady-state CFD simulations are validated against the experimental measurements at the groundwater velocity of 1000 m/year with an average error of 3%. Under saturated conditions without groundwater flow effect; using a spacer with SUC decreases the R_b by 13% from 0.15 m·K/W to 0.13 m·K/W, also using a spacer with the SUO decreases the R_b by 9% from 0.11 m·K/W to 0.1 m·K/W. In addition, the oval cross-section with spacer SUO + SSC decreases the R_b by 33% compared with SUC. Under the effect of groundwater flow of 1000 m/year; R_b of the SUC, SUO, SUC + SSC and SUO + SSC cases decrease by 15.5%, 12.3%, 6.1% and 4%, respectively, compared with the saturated condition.

Keywords: Ground source heat pump; oval cross-section; groundwater; borehole heat exchanger; CFD; Sandtank

---

1. Introduction

The ground source heat pumps (GSHP) systems become one of the leading technologies used recently worldwide to provide heating, cooling and domestic hot water (DHW) due to its privileges as a cost-effective operation and being environmentally friendly [1]. GSHP system consists of three main parts: a) the ground heat exchanger (GHX) which is absorbing/rejecting the heat from/to the subsurface layers of the surrounding soil; b) the heat pump (HP) is conveying the heat energy in bi-directional between fluid flowing through the GHX and the operating fluid from the building; c) the fan-coil system which transmits the heat energy to the desired space. The GHX could be horizontal or vertical types. The vertical GHX is extended deeply for more than 100 m underneath the soil surface, and it is preferred in a compact area as a high densely populated area [2]. However, high drilling cost is the main issue which is an obstacle preventing the wide spreading of it. On the other hand, horizontal GHX is extended in a horizontal plan, which requires a vast plan area, so it is not preferred in urban areas [3].

The thermal performance of GHX depends on several operating parameters, including inlet velocity, temperature and operating interval, and the hydrological composition of the surrounding soil [4]. In addition, accurate estimation of the soil thermo-physical properties (soil thermal conductivity, soil volumetric heat capacity and soil temperature distribution profile) leads to the proper design of the GHX dimensions. In-situ measurements by thermal response test (TRT) are the dominant common approach to determine these values and the borehole thermal resistance. The TRT methodology is injecting/extracting a constant rate of heat energy at a constant fluid flow rate inside BHX, and the inlet and outlet fluid temperature is recorded for continuous 48 hours [5–9]. The Infinite Line Source Model (ILSM) is the typical model to calculate and predict the soil properties; the LSM in an infinite homogenous medium, borehole and grouting is not accounted for in the ILS but is added through the borehole resistance term [10].

Moreover, the presence of groundwater advection can significantly enhance heat transfer and accelerate soil recovery state possibility [11]. Therefore, the thermal convection of the groundwater flow could impact the results calculated by the LSM [12]. Hence, many studies are recommended to perform multi TRTs at different places with different groundwater flow velocities, to assure accurate calculations of the soil properties and borehole thermal resistance [13–21].

Due to the high cost and long-time experimental effort of TRT, small-scale laboratory test and numerical simulation are of great importance for designing and predicting the performance of different configurations and designs of BHX. Examples of laboratory tests are: Li W et al. [22] investigated the effects of ground stratification on soil temperature distribution by an experimental apparatus with dimensions of 6.25 m × 1.5 m × 1m. Double Copper U-tube was buried horizontally inside a well-insulated box filled with sand and clay. Wan R. et al. [23] studied the effect of inlet water temperature, Reynolds number and backfilling material on thermal performance of a single U-tube heat exchanger. The single U-tube inserted inside a circular tank with diameter and depth of 240 mm and 1100 mm, respectively. Erol S. et al. [24] evaluated the performance of various grouting materials, through thermal, hydraulic and mechanical laboratory characterizations by sandbox with dimensions of 1 m × 1 m × 1m. Li H. et al. [25] investigated the effect of groundwater flow on the performance of spiral heat exchanger. Their experimental set-up consisted of a well-insulated stainless-steel box divided into three sections that were separated by a metal mesh and filter. The middle part was the test section, which had dimensions of 0.4 × 0.5 × 0.8. A spiral heater was filled with saturated silica sand to represent the spiral heat exchanger. Beier R.A. et al. [26] introduced a reference data set from a large laboratory "sandbox" containing a borehole with a U-tube. Their experimental measurements included thermal response tests on the borehole include temperature measurements on the borehole wall and within the surrounding soil. The test data provide independent values of soil thermal conductivity and borehole thermal resistance for verifying borehole models and TRT analysis procedures.

While numerical methods present another effective alternative with an acceptable accuracy to full-scale experimental tests. Computational fluid dynamics (CFD) approach can study the impacts of the diversity of factors on the performance of BHX with various configurations and arrangements [27–36]. For example, Jahangir MH et al. [37] utilized the classical finite element method to study the thermal and moisture behavior of the heat exchanger during the heat transfer process of several U-shape straight pipes. In addition, Biglarian H. et al. [38] developed a numerical model to simulate the short-term and long-term performance of the borehole heat exchanger. Their model simulated the fluid transport through the U-tube and the temperature variation of the borehole components with depth.

From this literature, the performance of ground heat exchanger under the impact of saturated conditions with groundwater flow was extensively investigated. Either by full scale experimental analysis or numerical simulations. All of these studies used the conventional costumery U-tube heat exchanger with a circular cross-section, no one tried to use an oval cross-section U-tube. The oval shape has many advantages over the circular tube as the author has shown in former research [39]. In this paper, the authors developed a new small-scale experimental set-up to compare the thermal performance of four different heat exchangers under the effect of groundwater flow. Finally, a two-dimensional, steady-state computational fluid dynamics simulation is carried out by ANSYS FLUENT software (ANSYS 2019 R3, Free student version, ANSYS INC, USA) to simulate and predict the soil temperature distribution, borehole thermal resistance and borehole temperature under different groundwater flow velocities.

2. Experimental Set-Up

A small-scale experimental set-up investigates the effect of groundwater flow on the thermal performance of GHX in laboratory scale. Comparison of the thermal performance of oval and circular cross-sections is carried out. The apparatus consists of 4 tanks (a, b, c and d) as shown in Figure 1. A comprehensive explanation about experimental set-up and instrumentations are indicated as follows:

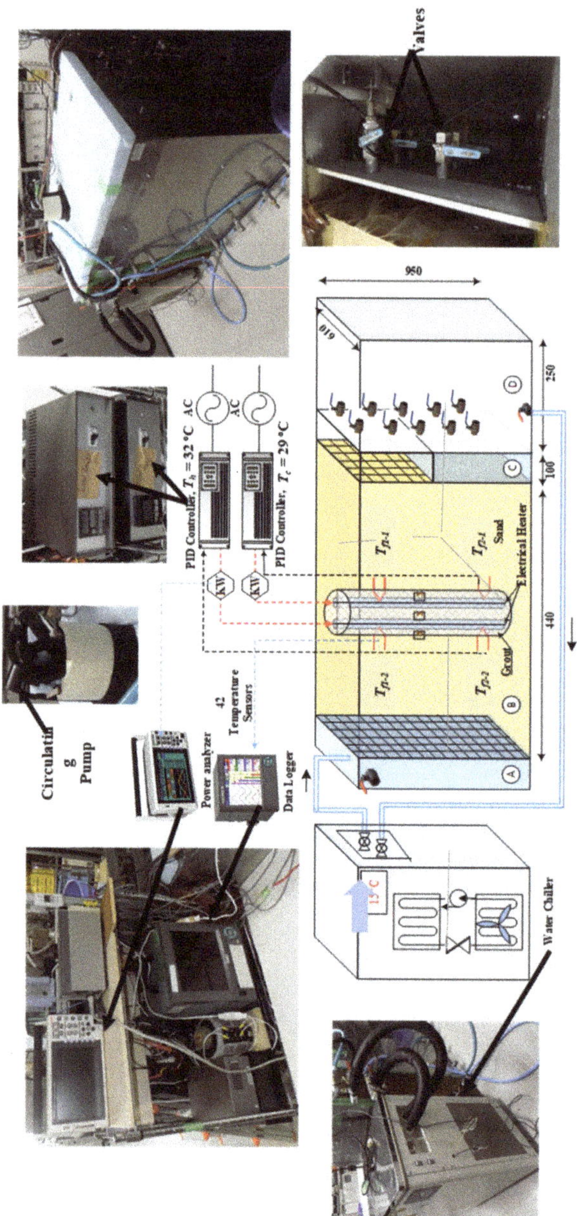

Figure 1. Schematic diagram and pictures of the experimental set-up.

2.1. SandBox

The main test section (Tank B) is consisting of a double-wall stainless steel tank with dimensions of 400 mm x 500 mm x 800 mm, (W x L x H), the front and back walls thickness is 5 cm and filled with an insulation foam layer, while left and right walls are made from steel net and a semi-permeable membrane. The membrane is separating the sand and water interfaces; it permits the water to flow and prevents the sand particle to sweep into water tanks. The test section filled with wet silica sand and sand gravel filled the borehole at the middle of the tank.

2.2. Groundwater Flow

The water flows from tank A to tank C passing through the voids and pores of the sand and grouting material in tank B. The flow rate is depending on the head difference between the water level in tank A and tank C, also the permeability of sand and gravel domains followed the Darcy flow principle. The flow rate is measured in each test by measuring the volume of water in liters that permeated through the sand and gravel in one minute. The head difference is controlled using ten ball valves fixed at the interface between tanks C and D; it is distributed every 5 cm from top to bottom. The upper surface of the four tanks is covered and insulated by high-density polyurethane foam insulation panel of 5 cm thickness.

2.3. U-Tube Heat Exchanger

The U-tube heat exchanger is simplified by using two parallel PVC pipes (the pipes are provided by Inoac Housing and Construction Material, Co., Ltd., Nagoya, Japan), it is filled with tap water and blocked from the bottom end. Two different cross-sections, oval and circular, are used with dimensions as indicated in Figure 2. Additionally, two different spacing distances are investigated. A total of four cases are studied; a single U-tube with a circular cross-section (SUO), a single U-tube with an oval cross-section (SUO), a single U-tube with a circular cross-section and a single spacer with a circular cross-section (SUC + SSC) and a single U-tube with an oval cross-section and a single spacer with a circular cross-section (SUO + SSC). The tubes are inserted inside the borehole which is filled with sand gravel with an average particle diameter of 5.5 mm as shown in Figure 3.

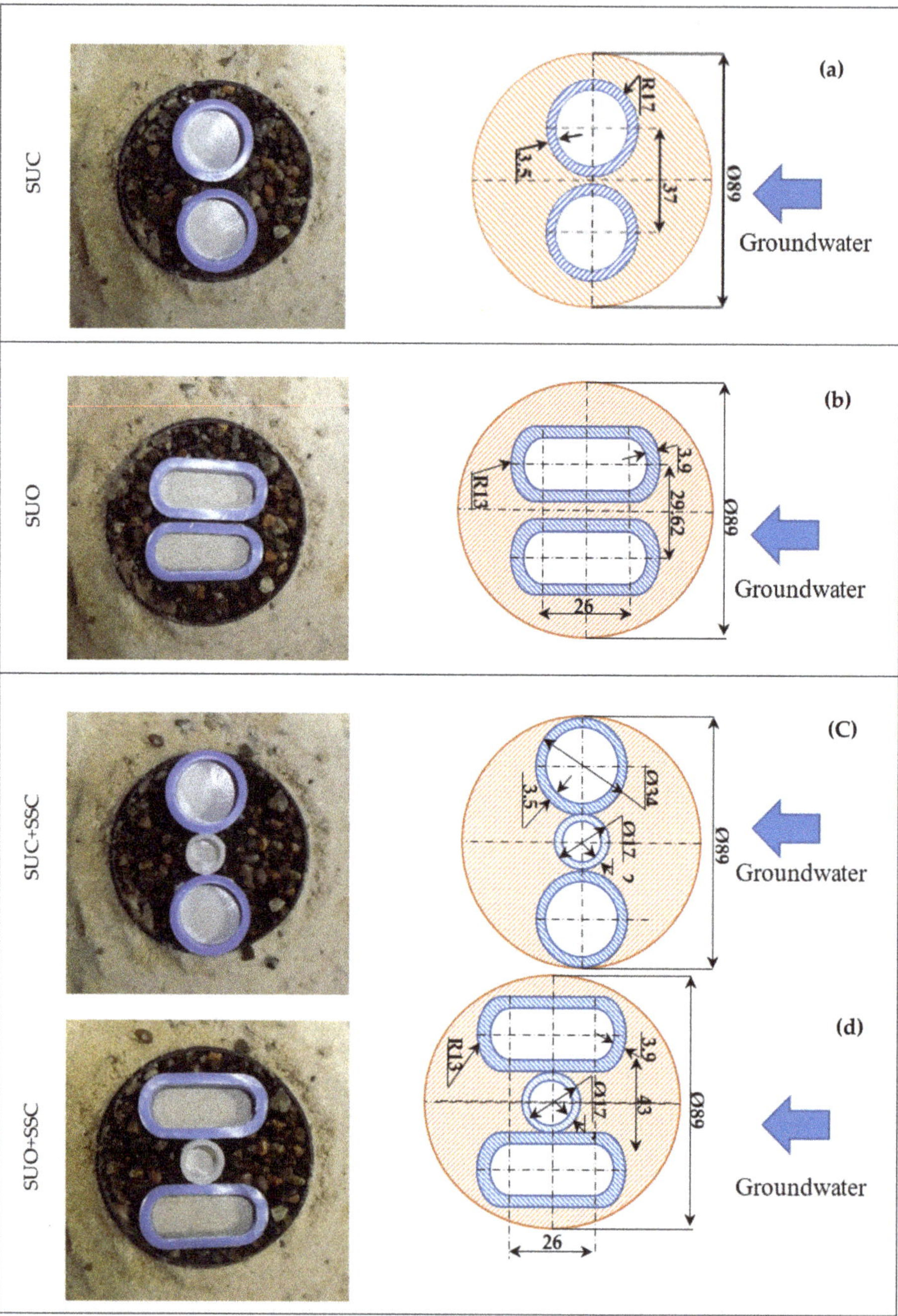

Figure 2. Photos and dimensions of each case: (**a**) circle; (**b**) oval; (**c**) circle with spacer and (**d**) oval with spacer.

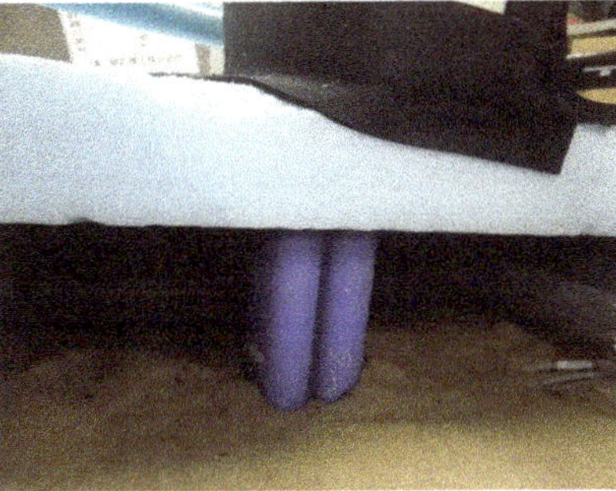

Figure 3. Photos of U-tube inside the borehole and surrounding soil.

2.4. Initial Groundwater Temperature Measurements

The initial sand temperature is maintained at 15 °C using constant temperature water chiller (ADVANTEC LV 600, ADVANTEC INC., Tokyo, Japan). The chiller pump circulates the cold water through the sand pores in a closed loop circulation as shown in Figure 1. PT100 temperature sensor is inserted inside sand to controlling the temperature feedback to the chiller cooling circuit. After the temperature reaches the setting temperature, the chiller cooling circuit is switched off. While, the circulating pump remains working to keep the water level in tank A at the determined level.

2.5. Fluid Temperature Difference

The temperature difference between fluids inside the right and left tubes is 3 °C. It is controlled using two orders made heaters with a maximum heating power of 350 W. The two heaters are consisting of the heating element in the core made from NCHW1 (Nickel Chromium Wire, 80% Nickle, 20% chromium), the wire is insulated by Magnesium oxide powder, and it is shielded by an Aluminum layer. The heating element had dimensions of 860 mm × 9.45 mm × 5.9 mm (length, width and depth).

The input electrical power is controlled and regulated by a handmade proportional-integral-derivative (PID) controller circuit. The PID controls and regulates the input electrical power with a feedback signal from temperature sensor inserted inside the tube. The PID controller circuit consists of PID controller (Omron E5CK, OMRON Corporation, Kyoto, Japan) with sampling accuracy of 100 ms, circuit breaker (MCB1, C 10A), solid-state relay (TOHO TRS 1225, TOHO INC., Tokyo, Japan) and fuses 15 A.

Two small centrifugal electrical pumps (0.4 L/min and 10 W) are used to circulate water inside each tube to guarantee a uniform fluid temperature along the pipe length.

2.6. Instrumentations

2.6.1. Soil Temperature Measurements

Forty thermocouples measure the soil temperature (Type-T, red cross x), it is divided into two groups; each group containing 20 sensors which are divided into two levels at heights of 50 cm and 80 cm from the bottom, each level is distributed as shown in Figure 4. The first level at a height at 50 cm contains sensors from Ts1-1 to Ts1-20, while the second level at a depth of 80 cm contains sensors from Ts2-1 to T2-20. The borehole temperature is measured by four PT100 (blue circle), they are fixed at the mid plan and distributed counterclockwise from Tb1 to Tb4 as shown in Figure 4. The fluid temperatures are measured by four PT100s (green circle), each tube has two sensors at upper and lower parts. Before the sensors are fixed to the experimental set-up, they had been calibrated against a standard thermometer measurements in the temperature range between 15.25 °C to 30.2 °C and a temperature step of 5 °C by means of constant temperature water bath. The errors absolute and relative values for a sample of these PT100 temperature sensors and the average error value of the thermocouples are indicated in Table 1. Finally, 48 temperature sensors are connected to a data logger (Yokogawa DX2048, YOKAGAWA Electric, Tokyo, Japan) which has a sampling interval every 1 min.

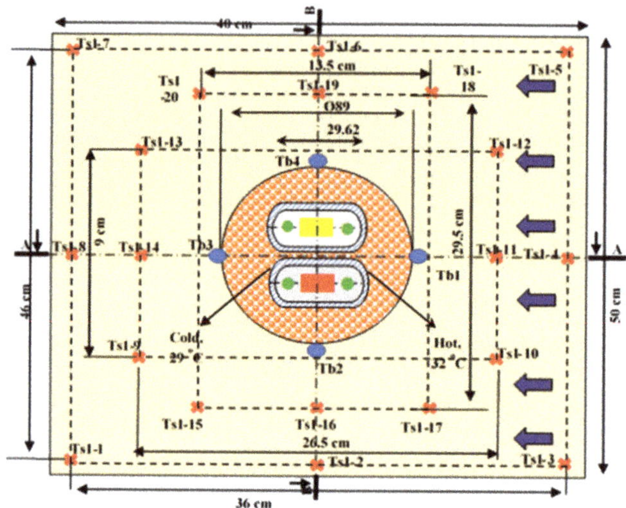

Figure 4. Temperature sensors positions and section A-A and B-B for horizontal and vertical orientation.

Table 1. Typical error of the PT100 temperature sensors and thermocouples.

PT100	@15.25 °C		@25.2 °C	
	Error, °C	Error, %	Error, °C	Error, %
Tb1	±0.15	±0.98	±0.14	±0.55
Tb2	±0.15	±0.98	±0.10	±0.40
Tb3	±0.15	±0.98	±0.11	±0.44
Tb4	±0.11	±0.72	±0.09	±0.37
Tf_h	±0.15	±0.98	±0.12	±0.46
Tf_c	±0.16	±1.04	±0.17	±0.68

Thermocouples
The average Error is ±0.51 °C

2.6.2. Heating Rate Measurements

The electrical power supplied to each electrical heater are measured by measuring the electrical current and voltage. The electrical current is measured by clamp on sensor (Hioki 9272). The electrical voltage and current measurements are recorded and analyzed by power analyzer (Hioki 3390).

2.7. Physical, Thermal and Hydraulic Properties

The thermophysical properties of both soil and gravel are measured according to the associated standard. The Particle size distribution (PSD), porosity, thermal conductivity and heat capacity are measured and used in the numerical simulation.

a) Particle-size distribution and porosity

The particle-size distribution (PSD) is measured by the mechanical method according to ISO 13503-2, the sand/gravel sample is dried and weighted. Then it is shacked carefully to pass through a series of sieves with a coarse sieve at the top and finest one at the bottom. From measuring the weight percentages of sand/gravel retained in each sieve, then plotted against the sieve size, as shown in Figure 5a,b. This figure shows that the sand particle size is in the range of 150–300 μm, while the gravel size ranged from 2.8 mm to 5.6 mm.

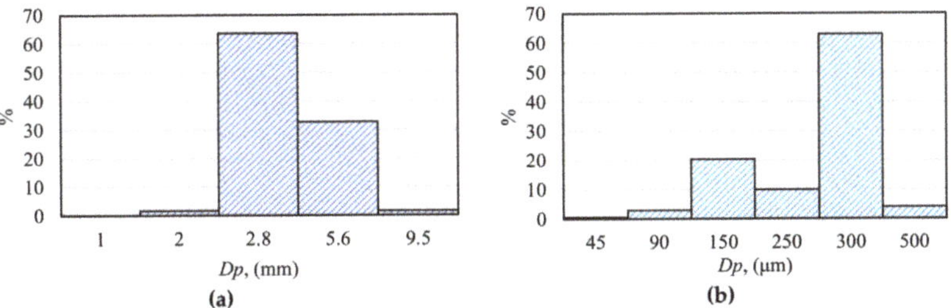

Figure 5. Particle Size Distribution of: (**a**) Gravel and (**b**) Sand.

In addition, the porosity of both sand and gravel are measured following the standard method, the porosity (*n*) are 0.4 and 0.36 for sand and gravel when tightly packed, respectively.

b) Thermal conductivity

The thermal conductivity of sand and gravel is measured by the method of thermal response test analysis under dry and saturated conditions. The thermal conductivity, including the conductivity of solid grains and the pore-filling material. The measurements rely on a one-dimensional conduction test using the needle probe method. The experiment set-up includes the packing tube, with a length of 25 cm and a diameter of 5 cm, which is packed tightly with solid grains, then a needle heater with a length of 15 cm is inserted inside the porous material. A PT-100 thermocouple is integrated with the needle heater to measure the soil temperature gradient. The electrical voltage (*V*, [volt]), current (*I*, [Ambers]) and soil temperatures are measured and recorded via data logger as shown in Figure 6. The time step is 30 sec, and each experiment lasted for 10 minutes. Then, a linear relationship between temperature and log scale of time is derived. Then heating power (*P*, [W]) and the thermal conductivity (*λ*, [W/m·K] are calculated according to Equation 1 and Equation 2:

$$P = IV \tag{1}$$

$$\lambda = \frac{P}{4\pi Hk} \tag{2}$$

where *P* is the heating rate in *W*, *H* is the tube length in m and *k* is the slop coefficient

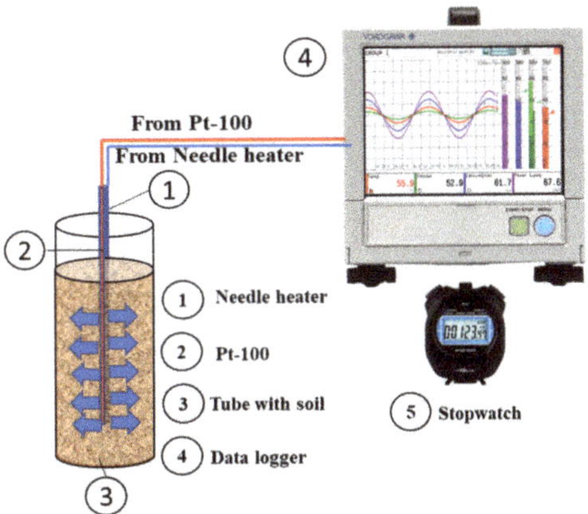

Figure 6. Thermal conductivity measurement schematic diagram.

3. Error Analysis

The borehole thermal resistance calculations are affected by the error in the measurements of primary variables. The primary variables include the hot and cold fluid temperatures, borehole temperatures, soil temperatures and electrical voltage and current values. Improperly calibrated sensors, sensor installation and connections cause errors in these measurements. Therefore, error analysis is essential to clarify the error values in the desired calculated results—heating power and borehole thermal resistance. The following equations are applied to calculate the error values and percentages.

3.1. Error in Thermal Conductivity Measurements

The error of the thermal conductivity depends on the error of temperature measurement slop coefficient, heating power measurement and the tube length measurement and how these errors are combined according to Equation (1). The error values of P and λ are calculated by Equation (3) and Equation (4) with the reference values and error values listed in Table 2.

$$\frac{\delta P}{P} = \pm \sqrt{\left(\frac{\delta I}{I}\right)^2 + \left(\frac{\delta V}{V}\right)^2} \tag{3}$$

where V and I are the voltage and current values, δV and δI are the error values in voltage and current, respectively.

$$\frac{\delta \lambda}{\lambda} = \pm \sqrt{\left(\frac{\delta P}{P}\right)^2 + \left(\frac{\delta H}{H}\right)^2 + \left(\frac{\delta k}{k}\right)^2} \tag{4}$$

where δP is the error value on heating rate in W, δH is the error value in tube length in m and δk is the error value of the slop coefficient.

Table 2. Thermal conductivity and error % of sand and gravel in W/m.K.

Variable		Sand			Gravel		
		Value	Error	Error, %	value	Error	Error, %
Dry	I, [Amber]	0.23	0.00023	0.1	0.1	0.0001	0.1
	V, [Volt]	8.2	0.0082	0.1	3.8	0.0038	0.1
	P, [W]	1.89	0.00189	0.1	0.38	0.00038	0.1
	K, [-]	1.87	0.02805	1.5	0.29	0.00435	1.5
	H, [m]	0.25	0.0025	1	0.25	0.0025	1
	λ, [W/m·K]	0.32	0.00576	1.8	0.42	0.00756	1.8
Saturated	I, [Amber]	0.23	0.00023	0.1	0.1	0.0001	0.1
	V, [Volt]	**8.2**	0.0082	**0.1**	3.8	0.0038	0.1
	P, [W]	1.89	0.00189	0.1	0.38	0.00038	0.1
	K, [-]	0.29	0.00435	1.5	0.05	0.00075	1.5
	H, [m]	0.25	0.0025	1	0.25	0.0025	1
	λ, [W/m·K]	2.1	0.0378	1.8	2.2	0.0396	**1.8**

3.2. Error in Heating Power Measurement

Two electric heaters are injecting heat to the surrounding fluid to keep the temperatures inside the two tubes at 32 °C and 29 °C respectively and maintain the temperature difference at 3 K. the current

I (Ambers) and voltage V (Volt) are measured and recorded simultaneously and the electric power P (W) of each heater is calculated, then the heating rate per unit length q' (W/m) are calculated as follows

$$q' = \frac{P}{L} \tag{5}$$

where P is the total heating power of the two heaters in W

The error % values in q' is calculated as shown in Equation (6):

$$\frac{\delta q'}{q'} = \pm \sqrt{\left(\frac{\delta P}{P}\right)^2 + \left(\frac{\delta L}{L}\right)^2} \tag{6}$$

Also, P and q' are the heating power and heating power per unit length, while δP *and* $\delta q'$ are the error values of heating power and heating power per unit length, respectively.

3.3. Error in Borehole Thermal Resistance Calculation

The borehole thermal resistance is calculated by the following equation:

$$Rb = \frac{\Delta T}{q'} = \frac{T_f - T_b}{q'} \tag{7}$$

The error value of the R_b is calculated as follows:

$$\frac{\delta R_b}{R_b} = \pm \sqrt{\left(\frac{\delta \Delta T}{\Delta T}\right)^2 + \left(\frac{\delta q'}{q'}\right)} \tag{8}$$

where T_f, T_b and q' are the fluid temperature, borehole temperature and heating power per unit length, respectively. In addition, $\delta \Delta T$, *and* $\delta q'$ are the error values of temperature difference between fluid temperature and borehole temperature and heating rate per unit length, respectively. Table 3 listed the error percentage of these variables.

Table 3. Error values of different variables and R_b.

Variable	SUO			SUO + SSC		
	Value	Error	Error, %	Value	Error	Error, %
$I1$	0.54	0.00054	0.1	0.65	0.00065	0.1
$V1$	51.6	0.0516	0.1	50.9	0.0509	0.1
$P1$	27.9	0.0279	0.1	33.1	0.0331	0.1
$I2$	0.34	0.00034	0.1	0.37	0.00037	0.1
$V2$	51.7	0.0517	0.1	51.9	0.0519	0.1
$P2$	17.6	0.0176	0.1	19.2	0.0192	0.1
L	0.8	0.008	1	0.8	0.008	1
q'	56.9	0.569	1	65.4	0.654	1
T_f	30.8	0.2464	0.8	30.9	0.2472	0.8
T_b	23.5	0.2115	0.9	23.0	0.207	0.9
$\Delta T = T_f - T_b$	7.3	0.0876	1.2	7.9	0.0948	1.2
R_b	0.128	0.002176	1.7	0.121	0.002057	1.7

4. CFD Simulation Set-Up

A two-dimensional finite volume model is developed by ANSYS FLUENT environment, the conjugated fluid flow and energy is solved simultaneously at a horizontal plane. The geometry is consisting of four domains for water filling left and right tanks (tanks 1 and 3), grout material inside borehole and sand filling the outer volume. The soil and gravel material are considered as porous media with the corresponding porosity and particle diameter. The following assumptions are applied in this study:

1. The simulation is done under steady-state condition
2. The ground water flow is considered as laminar flow through the porous voids
3. Temperature independence of the solid materials' properties (density, thermal conductivity and specific heat capacity).
4. The gravitational force effect was not considered.

    a. The fluid volume is omitted, and the tube inner wall temperature had been set at the same fluid temperature

5. The soil and borehole are considered as homogenous porous in all directions.

The governing equations controlling the heat transfer and mass transfer through the porous voids inside the grout and sand materials are shown in the following subsections:

a) Mass conservation equation

The mass conservation equation or "continuity equation" Equation (9 is written as follows:

$$\frac{\partial \rho_f}{\partial t} + \nabla \cdot \left(\rho_f \vec{v} \right) = 0 \tag{9}$$

where ρ_f is fluid density in (kg/m3), $\vec{v}$ is velocity vector in (m/s)

b) Momentum conservation equation

The general form of the momentum equation inside porous media is the standard fluid flow equation with additional momentum source term S_i

$$\frac{\partial}{\partial t}\left(\rho_f \vec{v} \right) + \nabla \cdot \left(\rho_f \vec{v}\,\vec{v} \right) = -\nabla p + \nabla \cdot \left(\overline{\overline{\tau}} \right) + \rho \vec{g} + \vec{F} + S_i \tag{10}$$

where p is the static pressure in (N/m^2), $\overline{\overline{\tau}}$ is the surface shear stress in (N/m^2), $\rho \vec{g}$ is the gravitational body force in (N/m^3) which is neglected in this study, $\vec{F}$ is the external body force in (N/m^3) and S_i is the momentum source that is calculated by Equation (11)

$$S_i = -\left(\frac{\mu}{\alpha} v_i + C_2 \frac{1}{2} \rho v v_i \right) \tag{11}$$

The first term in Equation (11) is the viscous term and the second term is the inertia loss term. Where S_i is the source term for the ith momentum equations (x, y and z momentum). α is the permeability and C_2 is the inertia resistance factor calculated by Equation (12) and Equation (13), respectively.

$$\alpha = \frac{D_p^2}{150} \frac{\epsilon^3}{(1-\epsilon)^2} \tag{12}$$

$$C_2 = \frac{3.5}{D_p} \frac{(1-\epsilon)}{\epsilon^3} \tag{13}$$

where, D_p is the particle size (m), while ϵ is the porosity.

In laminar flow, the momentum sink creates a pressure drop through the porous media, ∇p is the pressure drop is proportional to the velocity as indicated in Equation (14)

$$\nabla p = -\frac{\mu}{\alpha}\vec{v} \tag{14}$$

where μ is the fluid viscosity (kg/ms) and $\vec{v}$ is the velocity vector.

c) Energy conservation equation

The energy equation for porous media is a blending between the energy of the fluid and the solid, the blending factor is the porosity. In fluent, the standard energy transport equation is modified by include the effective conductivity and the thermal inertia of the solid region on the medium as shown in Equation (15)

$$\frac{\partial}{\partial t}\left(\epsilon\rho_f E_f + (1-\epsilon)\rho_s E_s + \nabla.\left(\vec{v}\left(\rho_f E_f + p\right)\right)\right) = \nabla.\left[k_{eff}\nabla T + \left(\bar{\bar{\tau}}\vec{v}\right)\right] + S \tag{15}$$

where k_{eff} is the effective thermal conductivity in the porous medium and it is calculated by Equation (16)

$$k_{eff} = \epsilon k_f + (1-\epsilon)k_s \tag{16}$$

where k_f and k_s are fluid phase and solid medium thermal conductivity, respectively.

A two-dimensional simplified geometry is considered as shown in Figure 7a. The left and right water tanks are considered as a fluid domain, while the soil and the borehole are assigned as a porous medium with water as the continuum medium. The dimensions are corresponding to the exact experimental dimensions.

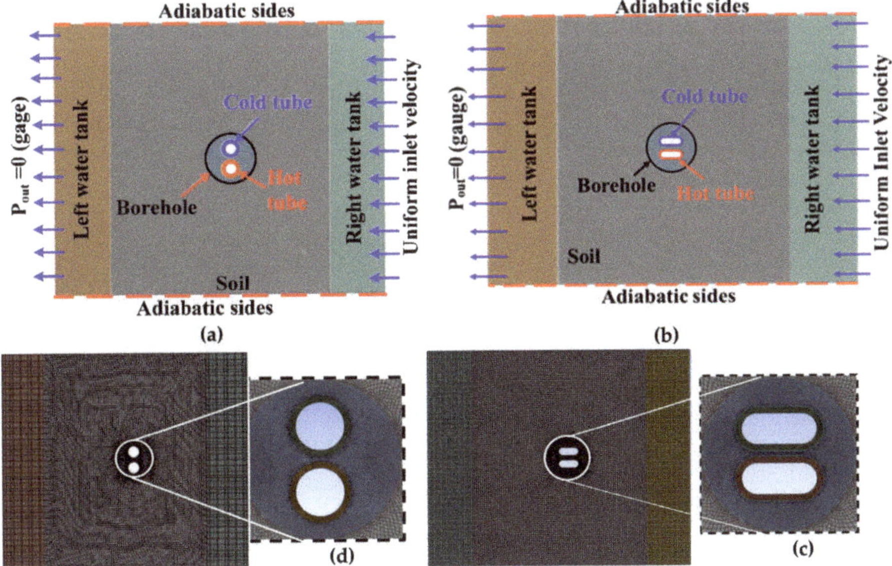

Figure 7. Geometry and boundary conditions used in this study: (**a**) Geometry of single U-tube with circular cross-section (SUC); (**b**) geometry of), single U-tube with an oval cross-section (SUO); (**c**) mesh of SUC and (**d**) mesh for SUO used in this study.

4.1. Mesh Independence Test

As seen in Figure 7b, structured mesh is discretized in the fluid domains of both inlet and outlet storage tanks. While the unstructured element is selected for the porous zones including the sand zone and the borehole zone. To confirm that the results are not dependent on the grid size, a mesh independence test was conducted. The mesh test results indicated that increasing the number of elements more than 27,000 has negligible impact on the simulation accuracy. Therefore, this number of elements were adopted through this study.

4.2. Boundary Conditions

In the present simulation, the right and left edges were defined as a velocity inlet and pressure outlet boundary condition, respectively. The velocity inlet changes from 0 (Saturation condition), 100 m/year, 300 m/year, 500 m/year and 1000 m/year. While the other outer edges are considered as adiabatic boundary as the outer surfaces of the experiment apparatus are perfectly insulated. The groundwater flows from the right water tank through the porous media to the left water tank with a constant inlet temperature of $16 \pm 0.5\ ^\circ$C. In addition, the initial temperature of $16 \pm 0.5\ ^\circ$C is patched for all domains.

4.3. Solver Schemes

In this study, the fluid flow and temperature distribution were solved by couple arithmetic, while a second order discretization scheme was used to discretize pressure, X-momentum, Y-momentum, Z-momentum and energy equation. The convergence criteria are 10^{-3} for all equations except 10^{-6} for the energy equation.

5. Results and Discussions

A new small-scale laboratory test is built to investigate the impact of groundwater flow on the thermal performance of the ground heat exchanger with a unique oval cross-section. Four different configurations are investigated under the effect of groundwater velocity of 1000 m/year. The soil temperature distribution on a horizontal plane at a depth of 0.5 m, the fluid temperatures and heating injection rate are measured and recorded instantaneously. Simultaneously, two-dimensional and steady-state CFD simulations are validated against the experimental measurements. After validating the CFD model, it is used to compare the thermal performance of oval and circle cross-sections under different groundwater velocities. Four different cases have been studied, the cases are circle and oval shapes, and another two cases with a spacer. Furthermore, the borehole thermal resistances are calculated and compared.

The experiments were conducted for one week, and each experiment continues for seven hours, followed by twelve hours without heating. This period is needed for complete recovery of the soil temperature. In addition, the initial soil temperature, the hot and cold water temperatures and groundwater flow rate are controlled as possible to maintain the same values for all cases, but unfortunately, there is a bit fluctuation in the operating conditions from one case to another as listed in Table 4.

Table 4. Experimental operating conditions and standard deviation (SD) in temperature and heating rate.

Geometry	$T_{initial}$		T_{hot}		T_{cold}		T_f		q	
	avg	SD	avg	SD	avg	SD	avg	SD	avg	SD
SUC	16.8	0.14	33.5	1.17	29.0	0.23	31.25	0.05	68.5	3.5
SUO	16.6	0.38	33.1	1.19	29.4	0.21	31.25	0.05	56.9	9.24
SUC + SSC	16.5	0.71	33.2	1.54	29.2	0.32	31.3	0.08	70.8	8.4
SUO + SSC	16.7	0.27	32.5	1.34	28.9	0.45	30.7	0.19	65.4	9.12

5.1. Experimental Results

5.1.1. Soil Temperature Contours

The soil temperature distribution contours for each configuration are drawn from the temperature measurements on a plane at a depth of 0.8 m, as shown in Figure 8. Figure 8 concluded that heat energy is advected by the effect of groundwater flow from right to left. Hence, the soil temperature at the upstream is lower than that at the downstream side. Additionally, it is clear that the impact of groundwater is changed from one case to another. The results of SUC and SUO cases are identical with ignorable differences due to the different operating conditions. While the other two cases SUC + SSC and SUO + SSC show distinct differences where the groundwater impact becomes much more apparent.

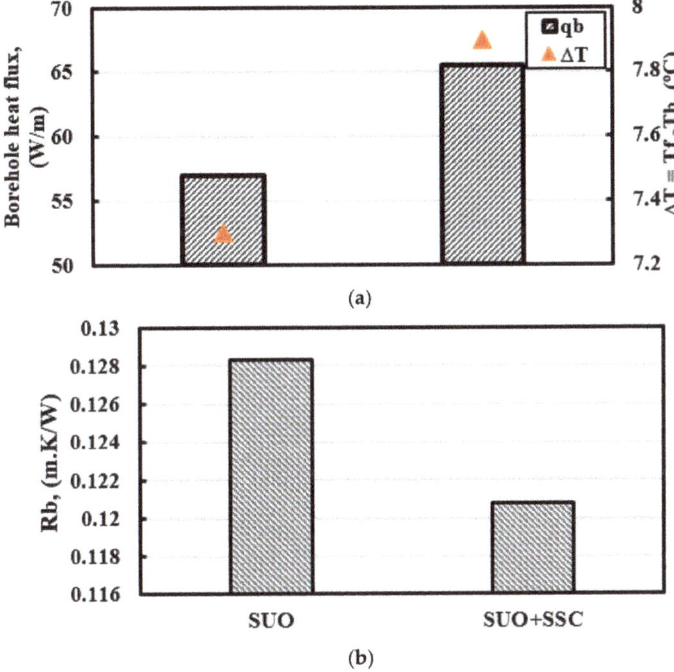

Figure 8. Experimental results for: (**a**) borehole heat flux and (**b**) borehole thermal resistance.

The comparison between the heat fluxes and average borehole temperatures for each case could approve these results, as indicated in the following subsection.

5.1.2. Borehole Thermal Resistance

The borehole thermal resistance for each case is calculated based on Equation (7) where T_b is the average temperature of the borehole outer wall, T_f is the average fluid temperature and q' is the average borehole heat flux per unit length. Figure 8a shows the heat flux per unit length (W/m) and the difference between the fluid temperature and borehole wall temperature for SUO and SUO + SSC cases. The heat flux is changed from 56.9 ± 1% W/m to 65.4 ± 1% W/m for SUO and SUO + SSC, respectively under the effect of ground water flow velocity of 1000 m/year, while the temperature differences are 7.3 ± 1.2% °C and 7.9 ± 1.2% °C. This indicates that the effect of groundwater flow is dominant in the SUO case. In addition, Figure 8b indicates that effect on the R_b values which decreases by 5% from 0.128 ± 1.7% m·K/W to 0.121 ± 1.7% m·K/W for SUO and SUO + SSC cases, respectively.

Moreover, the borehole temperature of the SUO is 22.5 °C, which is more than borehole temperature of SUC at 22.0 °C, means that the heat transfer rate inside the borehole for SUO is more than that in the case of SUC at which the groundwater flow cannot overcome this current. The reason for that is the supplied electrical power to the heater for SUC to maintain the fluid temperature at the predefined values is more than that of the SUO to compensate the heat energy advected by the groundwater flow. The results of SUC and SUC + SSC are not consistent. In this case, the difference between the SUC and SUO cannot be detected.

5.2. Validation of CFD Model

The temperature changes along the A-A section and B-B section in X and Y directions are used to validate the CFD simulation at a ground-water velocity of 1000 m/year for each configuration as shown in Figure 9. Figure 9 is divided into two columns, temperature variation along section A-A is shown in the left column, while temperature variation along section B-B is shown in the right column. The black scattered circles and the red triangles corresponding to the experimental and CFD results. The section A-A is extended from 0.2 m to the 0.36 m on the horizontal axis, where the water flows in the reverse direction. Section B-B is extended from 0.2 m to 0.48 m in the vertical direction. Figure 9 indicates that the soil temperature increases in the direction of groundwater flow, so the soil temperature at the upstream is lower than that at downstream of the borehole. In addition, the borehole temperature at an upstream is more than that of the downstream. More details and explanations are included in the following sections. These figures show good agreement between the CFD results and the experimental measurements with an average error value listed in Table 5.

Table 5. Average error values between CFD and experimental measurements.

Geometry	SUC	SUO	SUC + SSC	SUO + SSC
Error, %	3.2	3.3	2.8	3.4

After validating the CFD simulation, it was used to study the impact of changing the groundwater velocity on the heat transfer inside the borehole for each case. In these simulations, the groundwater changes from 0 m/year, which corresponds to the heat transfer through saturated soil, 100 m/year, 300 m/year, 500 m/year and 1000 m/year. To conduct a fair comparison, the operating conditions are fixed in all cases as listed in Table 6.

Table 6. Operating conditions used in these simulations.

$T_{initial}$	$T_{groundwater}$	T_{hot}	T_{cold}	T_f
16	16	32	29	30.5

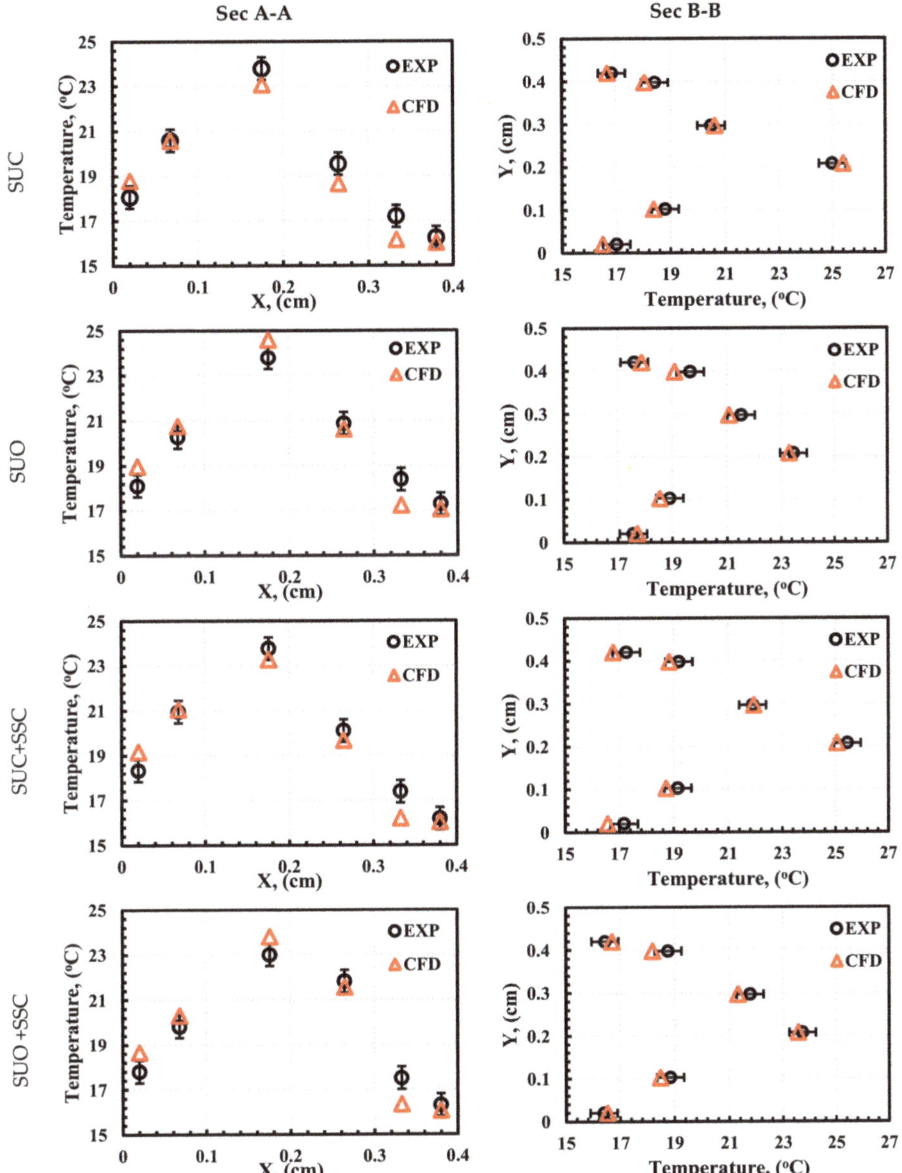

Figure 9. CFD model validation.

5.3. Simulation Results

5.3.1. Borehole Temperature Contours

As mentioned in Section 5.1.2, the differences in R_b between SUC and SUO cannot be detected due to a large error of ±10 % in the measurement of the R_b. Hence, a two-dimensional CFD model is developed to predict the values of R_b in all cases under different groundwater velocities. By CFD simulation, the borehole outer wall temperature and the borehole wall heat flux are predicted, and R_b is analyzed as well.

The borehole temperature distribution contours for each case at every groundwater velocity is summarized in Figure 10. Figure 10 demonstrates the effect of increasing groundwater flow velocity on transferring the heat by advection from the center of the borehole to the outer boundary.

For the saturated case without groundwater flow, the temperature inside the borehole is almost symmetrical around the vertical section B-B and the borehole wall temperature at T_{b1} and T_{b2} is 21.8 °C in SUC and SUC+SSC cases, while it is 22.6 °C in SUO and SUO + SSC cases. The heat is transferred mainly by conduction and convection through the fluid-filled bores between solid particles, therefore this result leads to that the heat transfer rate from the SUO to the borehole is more than that transferred from SUC, and it enhanced more when using the spacer.

Increasing the groundwater flow increases the temperature differences between T_{b1} and T_{b2}, where T_{b1} decreases and T_{b2} increases accordingly. The maximum temperature differences occurred at 1000 m/year which reached 6.1 °C, 5.7 °C, 5.3 °C and 5.3 °C for SUC, SUO, SUC + SSC and SUO + SSC, respectively. In addition, the temperature at the center of the borehole in between the two tubes is correspondingly changed from one case to another. For SUC, it changed from 26.2 °C to 28.0 °C when groundwater velocity changed from 0 m/year to 1000 m/year. However, it changes from 29 °C to 30 °C for the SUO case. For SUC + SSC and SUO + SSC, the maximum center temperature at a velocity of 1000 m/year was 26 °C and 27.5 °C, respectively as shown in Figure 11.

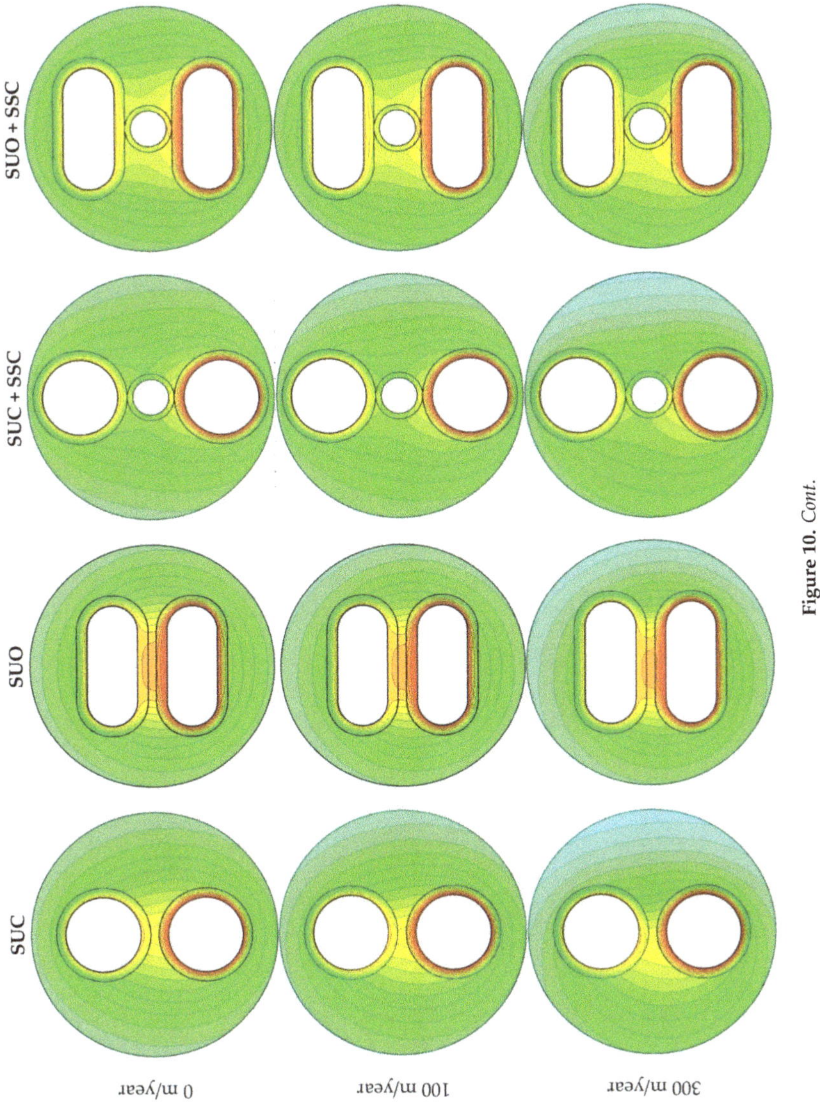

Figure 10. *Cont.*

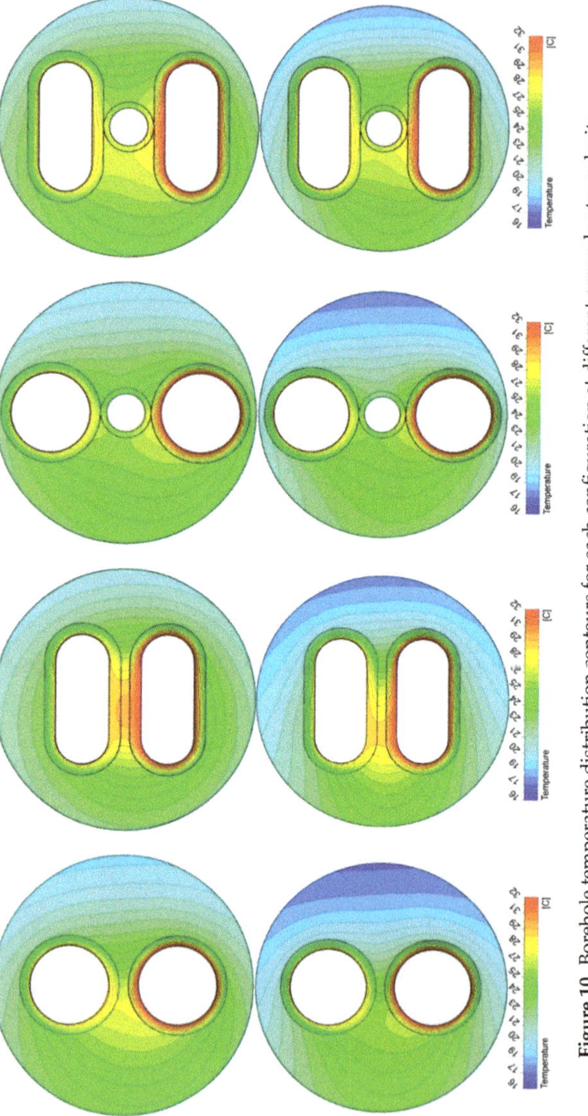

Figure 10. Borehole temperature distribution contours for each configuration at different groundwater velocity.

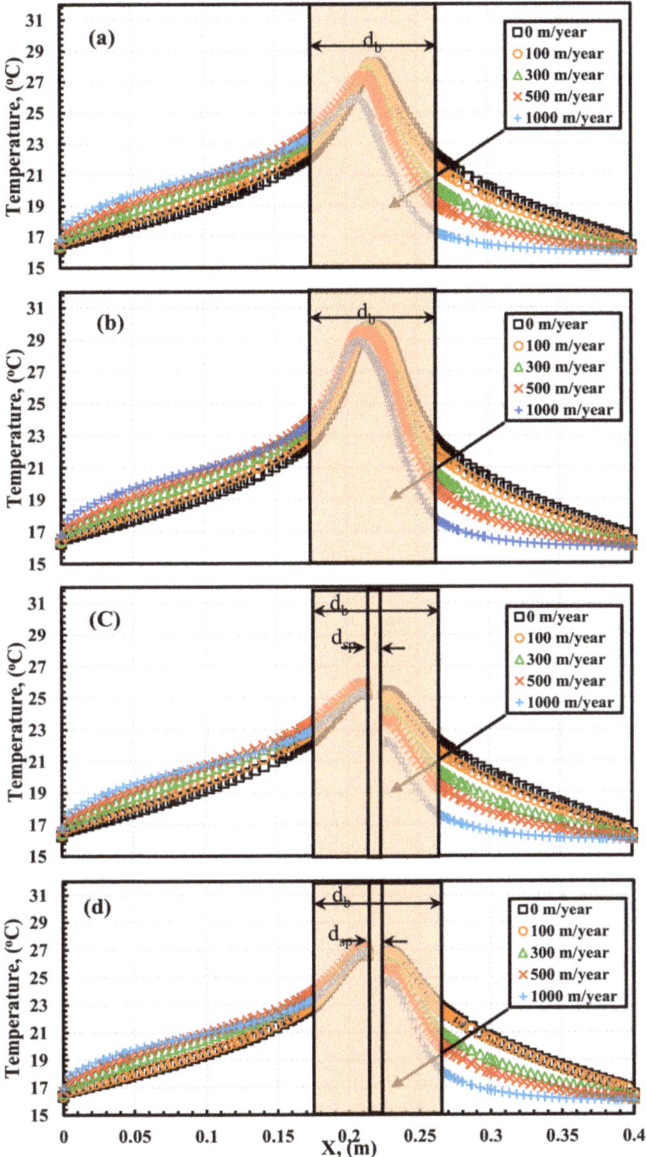

Figure 11. Soil temperature along X-X line: (**a**) SUC; (**b**) SUO, (**c**) SUC + single spacer with circular cross-section (SSC) and (**d**) SUO + SSC.

5.3.2. Borehole Thermal Resistance (R_b)

As a result of the fluid temperature being fixed in the CFD simulation at 30.5 °C, the borehole thermal resistance (Rb) depends on both the borehole wall temperature (T_b) and the borehole wall heat flux per length (q'). Figure 12 shows the changes in T_b and q' for each configuration at different groundwater velocities. At the same groundwater velocity, the SUO case shows higher T_b than SUC, and adding a spacer increases the T_b more.

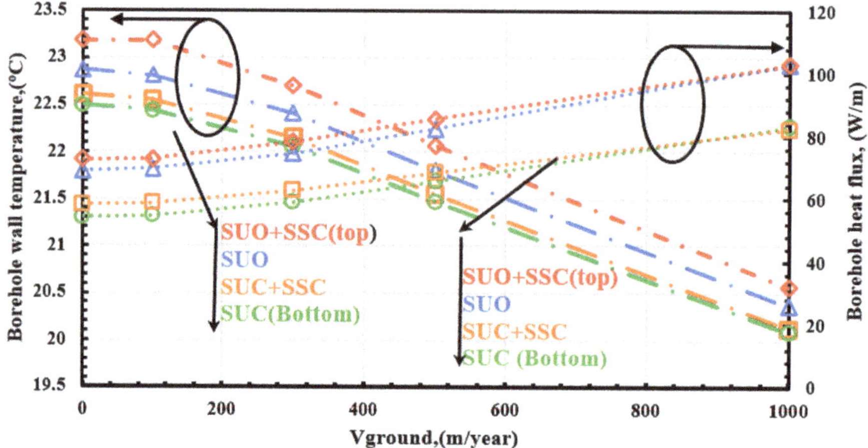

Figure 12. Borehole heat flux and borehole wall temperature.

At saturation condition, the T_b was 22.5 °C and 23.2 °C for SUC and SUO + SSC with a difference of 0.7 °C, while increasing the groundwater velocity to 1000 m/year decreases the difference to 0.5 °C. In addition, for each case, increasing the groundwater velocity decreases the borehole water temperature due to dissipating the heat energy transferred by advection from the borehole wall in the direction of groundwater flow. Additionally, it decreases the thermal interaction between the two legs of the U-tube.

In addition, the same explanation can be used to describe the difference between q' values of SUC and SUC + SSC. However, increasing the groundwater flow velocity increases the heat transfer rate from the borehole wall to the undisturbed soil. There is no significant impact of using a spacer in between the U-tube legs when the groundwater flow velocity increased. This is because the existence of the spacer blocked the path of groundwater between the U-tube legs. For SUO + SSC, the q' increased from 72.5 W/m at 0 m/year to 103.0 W/m at 1000 m/year, the heat flux increased by 42.1%, while it increased by 52.3% for SUC case.

At a groundwater velocity of 1000 m/year, there is no noticeable impact of using a spacer. The oval cross-section increases the q' by 27.3% at saturation condition and by 25.1% at the groundwater velocity of 1000 m/year.

Finally, the borehole thermal resistance (R_b) for each case at various groundwater velocities is shown in Figure 13. Under saturation condition, the R_b for SUC was 0.15 m·K/W and changed to 0.13 m·K/W with using the spacer in the SUC + SSC case. As the spacer decreases the thermal interaction between the two legs of the U-tube, it decreases the R_b by 13.3%. In addition, the R_b decreased by 9% and it changed from 0.11 m·K/W to 0.1 m·K/W with a spacer. In addition, the oval cross-section with spacer SUO + SSC decreases the R_b by 33% compared with SUC.

At the groundwater velocity of 1000 m/year, the R_b for SUC decreases to 0.12 m·K/W with a 15.5% decrement compared with saturation condition. Additionally, it decreased only by 4% for the SUO + SSC case as listed in Table 7. The impact of using a spacer decreased the R_b by only 1.5% and 4% for the SUC and SUO cases, respectively.

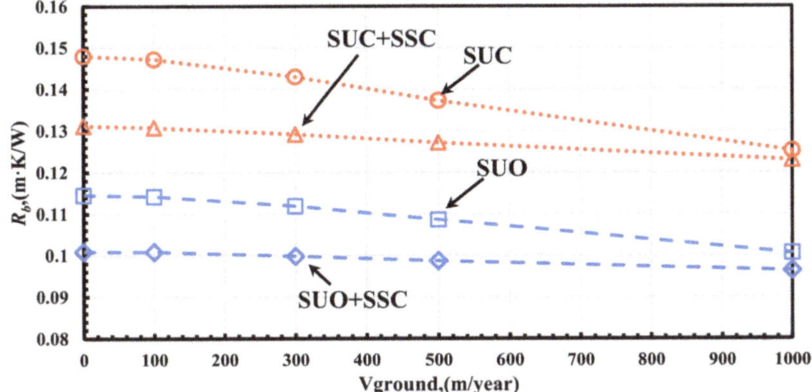

Figure 13. Borehole thermal resistance at different groundwater flow velocities.

Table 7. Decrement percentage in R_b when groundwater increased from 0 m/year to 1000 m/year.

SUC	SUO	SUC + SSC	SUO + SSC
15.5%	12.3%	6.1%	4%

6. Conclusions

The effect of groundwater flow on the thermal performance of the ground heat exchanger with a unique oval cross-section is experimentally and numerically investigated and compared with the customary circle cross-section tubes. A new small-scale laboratory experiment was built to mimic the full-scale ground heat exchanger where two separated tubes filled with water and equipped with two electrical heaters are used to replicate the U-tube heat exchanger. The two heaters keep the temperature difference between the two tubes at 3 °C using two PID controllers and temperature feedback. Four configurations are investigated; single U-tube with circular cross-section (SUC), single U-tube with an oval cross-section (SUO), single U-tube with circular cross-section and single spacer with circular cross-section (SUC + SSC) and single U-tube with an oval cross-section and single spacer with circular cross-section (SUO + SSC). The soil temperature distributions along the horizontal and vertical axis are measured and recorded simultaneously with measuring the electrical energy injected into the fluid, and the borehole wall temperature is measured as well; consequently, the borehole thermal resistance (R_b) is calculated. The experiments are carried out with a groundwater velocity of 1000 m/year. Moreover, two-dimensional and steady-state CFD simulations are developed to predict the R_b of each configuration at various groundwater flow velocities. The main conclusions are summarized as follows:

1. The differences in R_b between the SUC and SUO cannot be detected experimentally due to large error of ±10% in the measurement of the R_b;
2. The CFD model shows a good agreement with the experimental results with an average error value of 3%;
3. For the saturated case without groundwater flow:

    i. The temperature inside the borehole is almost symmetrical around the vertical section because the heat is transferred mainly by conduction and convection through the fluid-filled bores between solid particles;
    ii. The heat transfer rate from the SUO to the borehole increased by around 15% of that transferred from SUC;

    iii.   The R_b for SUC is decreased by 13%, it was 0.15 m·K/W and changed to 0.13 m·K/W with using spacer in the SUC + SSC case, and by 9% for SUO case, it changes from 0.11 m·K/W to 0.1 m·K/W;

    iv.   In addition, the oval cross-section with spacer SUO + SSC decreases the R_b by 33% compared with SUC;

4.   At the groundwater velocity of 1000 m/year:

    i.   The R_b for SUC decreases to 0.12 m·K/W with a 15.5% decrement compared with saturation condition, while it decreased by 4% for the SUO + SSC case;

    ii.   in addition, using the spacer decreases the R_b by 1.5% and 4% for the SUC and SUO cases.

Accordingly, oval shape shows a better thermal performance under the impact of groundwater flow than the customary U-tube. It boosts the COP and cut down the operating cost of the GSHP systems. Therefore, the author recommends to use the U-tube with an oval cross-section in regions with a potential of fast groundwater flow. In addition, the SUO U-tube can fit easily inside the tight casing and borehole which is practically more cost-effective for construction, environmentally friendly as it saves the surrounding environment and it is preferable in a high-density population urban area.

7. Future Recommendations

As thermally proved through this work, using the oval cross-section with spacer attained the best performance compared to the conventional cylindrical cross-section. However, a cost analysis, payback period and long-term simulation for both designs are worthy to be investigated in the future.

Author Contributions: Conceptualization, A.A.S. and Y.S.; methodology, T.K.; software, A.A.S.; validation, A.R.; formal analysis, A.A.S.; investigation, A.A.S.; writing—original draft preparation, A.A.S.; writing—review and editing, A.A.S., A.R.; visualization, A.A.S.; supervision, K.N. All authors have read and agreed to the published version of the manuscript.

Funding: This research received no external funding

Acknowledgments: In this section you can acknowledge any support given which is not covered by the author contribution or funding sections. This may include administrative and technical support, or donations in kind (e.g., materials used for experiments).

Conflicts of Interest: The authors declare no conflict of interest.

References

1.   Self, S.J.; Reddy, B.V.; Rosen, M.A. Geothermal heat pump systems: Status review and comparison with other heating options. *Appl. Energy* **2013**, *101*, 341–348. [CrossRef]

2.   Sarbu, I.; Sebarchievici, C. General review of ground-source heat pump systems for heating and cooling of buildings. *Energy Build.* **2014**, *70*, 441–454. [CrossRef]

3.   Serageldin, A.A.; Abdelrahman, A.K.; Ookawara, S. Earth-Air Heat Exchanger thermal performance in Egyptian conditions: Experimental results, mathematical model, and Computational Fluid Dynamics simulation. *Energy Convers. Manag.* **2016**, *122*, 25–38. [CrossRef]

4.   Zhu, L.; Chen, S.; Yang, Y.; Sun, Y. Transient heat transfer performance of a vertical double U-tube borehole heat exchanger under different operation conditions. *Renew. Energy* **2019**, *131*, 494–505. [CrossRef]

5.   Nagano, K. *Standard Procedure of Standard Thermal Response Test, IEA ECES, ANNEX21*; IEA report; IEA: Paris, France, 2011; Volume 21.

6.   Han, Z.; Zhang, S.; Li, B.; Ma, C.; Liu, J.; Ma, X.; Ju, X. Study on the influence of the identification model on the accuracy of the thermal response test. *Geothermics* **2018**, *72*, 316–322. [CrossRef]

7.   Ma, L.; Gao, Z.; Wang, Y.; Sun, Y.; Zhao, J.; Feng, N. Numerical Simulation of Soil Thermal Response Test with Thermal-dissipation Corrected Model. *Energy Procedia* **2017**, *143*, 512–518. [CrossRef]

8.   Zhang, Y.; Hao, S.; Yu, Z.; Fang, J.; Zhang, J.; Yu, X. Comparison of test methods for shallow layered rock thermal conductivity between in situ distributed thermal response tests and laboratory test based on drilling in northeast China. *Energy Build.* **2018**, *173*, 634–648. [CrossRef]

9. Sakata, Y.; Katsura, T.; Nagano, K. Multilayer-concept thermal response test: Measurement and analysis methodologies with a case study. *Geothermics* **2018**, *71*, 178–186. [CrossRef]

10. Bandos, T.V.; Montero, A.; Fernández, E.; Santander, J.L.; Isidro, J.M.; Pérez, J.; de Córdoba, P.J.; Urchueguía, J.F. Finite line-source model for borehole heat exchangers: Effect of vertical temperature variations. *Geothermics* **2009**, *38*, 263–270. [CrossRef]

11. Li, B.; Han, Z.; Hu, H.; Bai, C. Study on the effect of groundwater flow on the identification of thermal properties of soils. *Renew. Energy* **2018**, 6–13. [CrossRef]

12. Luo, J.; Tuo, J.; Huang, W.; Zhu, Y.; Jiao, Y.; Xiang, W.; Rohn, J. Influence of groundwater levels on effective thermal conductivity of the ground and heat transfer rate of borehole heat exchangers. *Appl. Therm. Eng.* **2018**, *128*, 508–516. [CrossRef]

13. Samson, M.; Dallaire, J.; Gosselin, L. Influence of groundwater flow on cost minimization of ground coupled heat pump systems. *Geothermics* **2018**, *73*, 100–110. [CrossRef]

14. Smith, D.C.; Elmore, A.C. The observed effects of changes in groundwater flow on a borehole heat exchanger of a large scale ground coupled heat pump system. *Geothermics* **2018**, *74*, 240–246. [CrossRef]

15. Akhmetov, B.; Georgiev, A.; Popov, R.; Turtayeva, Z.; Kaltayev, A.; Ding, Y. A novel hybrid approach for in-situ determining the thermal properties of subsurface layers around borehole heat exchanger. *Int. J. Heat Mass Transf.* **2018**, *126*, 1138–1149. [CrossRef]

16. Hu, J. An improved analytical model for vertical borehole ground heat exchanger with multiple-layer substrates and groundwater flow. *Appl. Energy* **2017**, *202*, 537–549. [CrossRef]

17. Zhang, L.; Zhang, Q.; Huang, G.; Ma, X. Transient Ground and Grout Parameters Estimation Method for a Ground-Coupled Heat Pump System with Sandbox TRT Reference Data. *Procedia Eng.* **2017**, *205*, 2662–2669. [CrossRef]

18. Choi, W.; Ooka, R. Effect of natural convection on thermal response test conducted in saturated porous formation: Comparison of gravel-backfilled and cement-grouted borehole heat exchangers. *Renew. Energy* **2016**, *96*, 891–903. [CrossRef]

19. Zheng, T.; Shao, H.; Schelenz, S.; Hein, P.; Vienken, T.; Pang, Z.; Kolditz, O.; Nagel, T. Efficiency and economic analysis of utilizing latent heat from groundwater freezing in the context of borehole heat exchanger coupled ground source heat pump systems. *Appl. Therm. Eng.* **2016**, *105*, 314–326. [CrossRef]

20. Coleman, T.I.; Parker, B.L.; Maldaner, C.H.; Mondanos, M.J. Groundwater flow characterization in a fractured bedrock aquifer using active DTS tests in sealed boreholes. *J. Hydrol.* **2015**, *528*, 449–462. [CrossRef]

21. Li, H.; Nagano, K.; Lai, Y. A new model and solutions for a spiral heat exchanger and its experimental validation. *Int. J. Heat Mass Transf.* **2012**, *55*, 4404–4414. [CrossRef]

22. Li, W.; Li, X.; Peng, Y.; Wang, Y.; Tu, J. Experimental and numerical investigations on heat transfer in stratified subsurface materials. *Appl. Therm. Eng.* **2018**, *135*, 228–237. [CrossRef]

23. Wan, R.; Chen, M.; Huang, Y.; Zhou, T.; Liang, B.; Luo, H. Evaluation on the heat transfer performance of a vertical ground U-shaped tube heat exchanger buried in soil–polyacrylamide. *Exp. Heat Transf.* **2017**, *30*, 427–440. [CrossRef]

24. Erol, S.; François, B. Efficiency of various grouting materials for borehole heat exchangers. *Appl. Therm. Eng.* **2014**, *70*, 788–799. [CrossRef]

25. Li, H.; Nagano, K.; Lai, Y. Heat transfer of a horizontal spiral heat exchanger under groundwater advection. *Int. J. Heat. Mass Transf.* **2012**, *55*, 6819–6831. [CrossRef]

26. Beier, R.A.; Smith, M.D.; Spitler, J.D. Reference data sets for vertical borehole ground heat exchanger models and thermal response test analysis. *Geothermics* **2011**, *40*, 79–85. [CrossRef]

27. Li, W.; Dong, J.; Wang, Y.; Tu, J. Numerical Modeling of a Simplified Ground Heat Exchanger Coupled with Sandbox. *Energy Procedia* **2017**, *110*, 365–370. [CrossRef]

28. Zhao, T.; Yu, M.; Rang, H.; Zhang, K.; Fang, Z. The influence of ground heat exchangers operation modes on the ground thermal accumulation. *Procedia Eng.* **2017**, *205*, 3909–3915. [CrossRef]

29. Morrone, B.; Coppola, G.; Raucci, V. Energy and economic savings using geothermal heat pumps in different climates. *Energy Convers. Manag.* **2014**, *88*, 189–198. [CrossRef]

30. Eslami-nejad, P.; Bernier, M. Freezing of geothermal borehole surroundings: A numerical and experimental assessment with applications. *Appl. Energy* **2012**, *98*, 333–345. [CrossRef]

31. Lee, C.K.; Lam, H.N. A modified multi-ground-layer model for borehole ground heat exchangers with an inhomogeneous groundwater flow. *Energy* **2012**, *47*, 378–387. [CrossRef]

32. Gustafsson, A.M.; Westerlund, L.; Hellström, G. CFD-modelling of natural convection in a groundwater-filled borehole heat exchanger. *Appl. Therm. Eng.* **2010**, *30*, 683–691. [CrossRef]

33. Fan, R.; Jiang, Y.; Yao, Y.; Shiming, D.; Ma, Z. A study on the performance of a geothermal heat exchanger under coupled heat conduction and groundwater advection. *Energy* **2007**, *32*, 2199–2209. [CrossRef]

34. Diao, N.; Li, Q.; Fang, Z. Heat transfer in ground heat exchangers with groundwater advection. *Int. J. Therm. Sci.* **2004**, *43*, 1203–1211. [CrossRef]

35. Sutton, M.G.; Nutter, D.W.; Couvillion, R.J. A Ground Resistance for Vertical Bore Heat Exchangers With Groundwater Flow. *J. Energy Resour. Technol.* **2003**, *125*, 183. [CrossRef]

36. Gu, Y. Effect of Backfill on the Performance of a Vertical U-tube Ground-Coupled Heat Pump. Ph.D. Thesis, Texas A&M University, College Station, TX, USA, 1995.

37. Jahangir, M.H.; Sarrafha, H.; Kasaeian, A. Numerical modeling of energy transfer in underground borehole heat exchanger within unsaturated soil. *Appl. Therm. Eng.* **2018**, *132*, 697–707. [CrossRef]

38. Biglarian, H.; Abbaspour, M.; Saidi, M.H. A numerical model for transient simulation of borehole heat exchangers. *Renew. Energy* **2017**, *104*, 224–237. [CrossRef]

39. Serageldin, A.A.; Sakata, Y.; Katsura, T.; Nagano, K. Thermo-hydraulic performance of the U-tube borehole heat exchanger with a novel oval cross-section: Numerical approach. *Energy Convers. Manag.* **2018**, *177*, 406–415. [CrossRef]

Article

The Optimization of the Thermal Performances of an Earth to Air Heat Exchanger for an Air Conditioning System: A Numerical Study

Adriana Greco [1] and Claudia Masselli [2,*]

[1] Department of Industrial Engineering, University of Naples Federico II, P.le Tecchio 80, 80125 Napoli, Italy; adriana.greco@unina.it
[2] Department of Industrial Engineering, University of Salerno, Via Giovanni Paolo II 132, 84084 Fisciano (SA), Italy
* Correspondence: cmasselli@unisa.it

Received: 2 November 2020; Accepted: 2 December 2020; Published: 4 December 2020

Abstract: The aim of this paper is to research the parameters that optimize the thermal performances of a horizontal single-duct Earth to Air Heat eXchanger (EAHX). In this analysis, the EAHX is intended to be installed in the city of Naples (Italy). The study is conducted by varying the most crucial parameters influencing the heat exchange between the air flowing in the duct and the ground. The effect of the geometrical characteristics of the duct (pipe length, diameter, burial depth), and the thermal and flow parameter of humid air (inlet temperature and velocity) has been studied in order to optimize the operation of this geothermal system. The results reveal that the thermal performance increases with length until the saturation distance is reached. Moreover, if the pipe is designed with smaller diameters and slower air flows, if other conditions remain equal, the outlet temperatures come closer to the ground temperature. The combination that optimizes the performance of the system, carried out by forcing the EAHX with the design conditions for cooling and heating, is: $D = 0.1$ m s^{-1}; $v = 1.5$ m s^{-1}; $L = 50$ m. This solution could also be extended to horizontal multi-tube EAHX systems.

Keywords: geothermal energy; earth-to-air; horizontal pipe; 2D model; air conditioning; renewable energy sources; thermal performances; energy efficiency; parametric study

---

1. Introduction

1.1. General Concepts and Context

Heating Ventilation and Air Conditioning (HVAC) systems contribute 10–20% to worldwide energy consumption [1]. A relevant amount (20–40%) of the energy consumption attributed to HVAC is required for building air conditioning. Thus, in addition to the increase in the use of renewable energy sources, energy saving solutions are also being adopted [2–6]. The basic imperative recommended by energy policies is to consider energetically improved solutions that often could not be satisfied only through the vapor compression technology, but which have limits linked to the use of refrigerants with high Global Warming Potential [7–10]. A viable path is a Not-In-Kind cooling technology [11–13] where the refrigerants are solid-state materials showing caloric effects, i.e., their Global Warming Potential is zero [14]. Although they are a promising solution, the bottleneck is the limited range of use of these technologies due to the limited caloric effects of the materials. This can constitute some difficulties in the development of environmental conditioning systems that operate on a wide range. Another valid solution is the utilization of renewable energy sources that must increasingly become a shared responsibility. Currently, only 14% of the global energy demand is satisfied by means of renewable energy [15], and many renewable energy sources are employable for these purposes.

Among them, geothermal energy has been revealed to be very promising for employment as an answer to the energy demand attributable to HVAC systems. The recently emerging geothermal hybrid solutions with solar collectors of PV panels are also very promising [16–18] for building conditioning. Geothermal energy is very promising because of the properties of the soil to be, from a certain depth onward, be a constant temperature throughout the year. Specifically, during winter the temperature is higher than the temperature of the external air, and lower during summer [19]. For this reason, the soil assumes the double role of heat sink/source despite it functioning in cooling/heating operation mode.

There are three geothermal systems that exploit this property and could be used for air conditioning [20]: earth homes (the idea of using a partially buried building to minimize the heating and cooling loads), ground-source heat pumps (heat pumps where the water is the secondary fluid that flows underground to heat transfer with the soil), and Earth to Air Heat eXchangers (EAHX). The latter are geothermal systems formed by a certain number of pipes buried to a depth that allows the exploitation of the undisturbed temperature properties of the ground. The external air flows within the pipes and transfers heat from/to the soil.

An EAHX could be arranged through horizontal or vertical banks of pipe. These could be placed in parallel with each other or following other configurations. The specificity of the design depends on the project and the space available for the installation. However, they must be designed to optimize the performance. Although valid research exists [21] on numerical optimization of EAHX with vertical ducts, in most cases, EAHX is formed by horizontal ducts [22].

A well-designed EAHX should satisfy the requirement to bring the temperature of the external air as close as possible to the undisturbed temperature of the ground. Furthermore, efficiency and economic aspects should also be optimized [23].

1.2. State of the Art

There are several parameters that influence the performances of the EAHX: they are both geometric factors of the system and the thermo-physical parameters of the soil and air.

The geometric configuration of the tube of a heat exchanger is a key factor both because its performance depends on it, and because an adequate sizing optimizes both the excavation costs and the area occupied by the plant. Selamat et al. [24] researched, through numerical modelling, the best configuration for an EAHX during winter in the city of Dublin (Ireland). A sensitivity analysis was carried out by varying the key parameters influencing the thermal performance of the EAHX. The analysis clearly shows that an increase in the length of the tube (from 30 to 70 m), a decrease in the diameter of the tube (from 150 to 100 mm) and a decrease in the air velocity (from 15 to 5 m s^{-1}) provide an improvement of the system heating capacity. The study was also focused on the thermal performance of 4 parallel tubes: 1.5 m distance, 30 m length and 125 mm diameter, with an air velocity of 8 m s^{-1}. The simulations clearly show that with the multiple-duct configuration, due to heat loss from adjacent pipes, there was a 0.6 °C as decrement of the air temperature.

Mathur et al. [25] introduced an experimental study where an earth to air spiral heat exchanger was designed with a 60-m pipe length, a diameter of 0.1 m, and a 1-m distance among the spirals, and it was buried at a depth of 3 m. The results obtained were compared to straight tube EAHX-system results: they noted that the coefficient of performance of the latter configuration is 5.94 in the cooling phase and 1.92 in the heating phase, whereas for a spiral configuration these values are 6.24 and 2.11, respectively. As Benrachi et al. reported [26], the use of the spiral configuration can be an interesting alternative to straight tube exchangers if the space available for excavation is limited. This advantage in terms of occupied area tends to disappear when dealing with the installation of several pipes in parallel.

Many studies focus on parametric research into the influence of pipe diameter on system performance. Sodha et al. [27] made a comparison between single-tube and multiple-tube exchangers. It has been observed that, with the same air flow, the heating and cooling potential increases as the number of tubes of smaller diameter increase, because of the consequent augmentation of the heat exchange area. With the increase in pipe diameter, the absolute value of temperature variation reduces, reducing the heating/cooling capacity of the system.

Wu et al. [28] estimated the daily cooling capacity of EAHX with a 0.4 m and 0.6 m pipe diameter. The results obtained (respectively 43.2 kWh and 74.6 kWh) underline the advantage that can be gained from using a smaller diameter. Mihalakakou et al. [29,30] evidenced that an increase in the diameter of the pipe corresponds to a decrease in the convective heat transfer coefficients which leads to a higher temperature at the outlet of the EAHX in summer and lower temperature in winter. Typical diameter values range from 0.1 m to 0.3 m, up to 1 m for commercial applications. Goswami and Biseli [31] suggested that the ideal performance for single-tube EAHX is achieved with a diameter of 0.3 m; for systems with parallel pipes, the ideal diameter varies between 0.2 m and 0.25 m. Sehli et al. [32], considering an EAHX placed in the Algerian arid climate and operating in cooling mode, estimated the variation of air temperature at the pipe outlet as a function of the tube form factor (the ratio between the length and the diameter of the pipe). They present the effect of both the form factor and the Reynolds number on the outlet air temperature, for an ambient temperature of 45 °C and a tube depth set at 4 m. As the form factor increases, the outlet air temperature decreases because a longer tube provides a longer path to the air, favoring heat exchange. As the Reynolds number increases, the outlet air temperature also increases, since the air remains in the tube for a shorter time with respect to the cases with smaller Reynolds numbers. The study therefore concludes that the optimal value of the form factor σ was 250.

Hanby et al. [33] reported that, for a fixed length of the duct, at fixed air flow, there is always an optimal diameter of the pipe that minimizes the energy consumption of the system. Niu et al. [34] found that, for an EAHX working in cooling mode, the advantage of a smaller diameter is not only linked to a faster lowering of the temperature in the duct, but also to a lower presence of humidity in air passing through the heat exchanger. With small pipe diameters, the air temperature drops faster than when using larger diameters. For this reason, the air temperature recorded at the outlet of the tube with a pipe diameter of 0.7 m is 17 °C, whereas for a diameter of 0.3 m the temperature is 13 °C.

The length of the tube is the main geometric factor affecting the performance of the earth to air heat exchanger. It is implicit that it must be chosen appropriately to reduce the initial costs of construction and installation of the system without compromising its performance. As the length of the tube increases, the temperature difference between the inlet and outlet ends of the air increases. Derbel and Kanoun [35] found out that the energy load of the heat exchanger increases with the length of the buried pipe; this is mainly due to the growth in the crossing time of the air flow. Furthermore, Benhammou and Draoui [36] asserted that the efficiency of heat transmission does not increase from a certain tube length onward: this is called the saturation length, and increases with increasing air flow rate. Lee et al. [37] noted that, with respect to their experimental test study focused on four locations of USA with different climates, there were no further advantages in using pipes longer than 70 m. Ahmed et al. [38] established that the pipe length is the greatest influencing factor on thermal performances of an EAHX. They chose four different pipe lengths (7.5 m, 15.0 m, 30.0 m and 60.0 m), achieving better performance with maximum length, while the system was operating in cooling mode.

The study of the temperature profile in the soil and its variation with depth is essential for the optimization of an EAHX system. The temperature distribution in the soil is mainly influenced by its physical properties and its structure. The soil temperature varies with time (daily and seasonal temperature) and with depth. The amplitude of this variation decreases with increasing soil depth until a certain value reaching the so-called Undisturbed Ground Temperature (UGT). This value is typically reached at a depth of 2–4 m, but in some latitudes, it is reached at greater depths. Hanby et al. [33] showed that there is an ideal diameter at each depth that allows the system to save energy. They also show that the energy saving increases by increasing the tube depth (from 2 to 4 m). Popiel et al. [39] measured the soil temperature at different depths and found that short-term variations occurred down to a depth of 1 m. Sanusi et al. [40] investigated on an EAHX buried at different depths: 0.5 m, 1.0 m and 1.5 m in a humid and warm climate like that of Malaysia. The maximum soil temperatures at 0.5 m, 1.0 m and 1.5 m are respectively: 28.3 °C, 28.5 °C and 28.6 °C in the wet season and 30.3 °C, 30.1 °C and 30.1 °C in the dry season. The optimum depth for burying the EAHX considering both economic and thermal performance terms is 1 m. Mihalakakou et al. [41] analysed the influence of the depth of tube of the EAHX on the cooling performances considering the values of 1.2 m, 2.0 m, and 3.0 m and they observed that the best results are obtained for the deepest buried tube. Ahmed et al. [38] buried the duct at 0.6 m, 2.0 m, 4.0 m, and 8.0 m, finding that the best cooling performances are associated with a 8-m burieb pipe. Wu et al. [28] evaluated the thermal performances of the EAHX at two depths (1.6 m and 3.2 m) detecting that for air cooling in summer, the temperature at the EAHX outlet varies between 27.2 °C and 31.7 °C at 1.6 m and between 25.7 °C and 30.7 °C at 3.2 m. Generally, as the depth of the pipe increases, the potential for heating and cooling increases, but beyond a certain depth there is no noticeable increase in performance. Badescu [42] noticed that for depths greater than 4 m, the performance of the system remains unchanged, so a depth value of 2 m is indicated as a good compromise between the excavation costs (which increase with increasing depths) and the annual variation of the temperature (which decreases with increasing depths). Hermes et al. [43] conducted a research on the soil of Rio Grande, Brazil and underlined how problematic the installation of a heat exchanger can be in a coastal city, where the aquifer can be very close to the ground surface. They estimated that below the depth of 2 m the temperature remains constant both in the warm months and in the cold ones.

Lee et al. [37] studied the effect of the air velocity inside the tube on the outlet air temperature in four different locations (Key West, Peoria, Phoenix and Spokane) in both heating and cooling mode. They found that as the air flow velocity increases (from 2 to 14 m s^{-1}) the air temperature at the outlet of the pipe in cooling mode also increases, as the air spends less time in the pipe and less time in contact with the ground. Likewise, the range and rate of this increase was a function of the different locations depending on the soil conditions.

Niu et al. [34] analyzed the effect of air velocity on the thermal performances of an EAHX in cooling mode. The study was conducted by varying the air velocity from 0.5 to 2.5 m s^{-1}. The smaller the air velocity was, the faster the air temperature decreasing rate was, and the lower the outlet air temperature was. Bansal et al. [44] conducted an experimental study considering air velocity as 2.0, 3.0, 4.0, and 5.0 m s^{-1}, in the case of operation of the EAHX in heating mode. They observed that 2 m s^{-1} is the velocity at which the most noticeable temperature increase was registered. In the case of cooling [45], the most pronounced decrease in temperature was always recorded for an air speed of 2 m s^{-1}. For the cooling operation, Benhammou and Draoui [36] observed the increase in the outlet temperature of the exchanger as the velocity of the inlet air increased. As a consequence of increasing the speed from 1 to 3 m s^{-1}, the outlet temperature increases by 5.6 °C. This phenomenon is attributed to the reduction of heat transfer from the air to the ground. While the air velocity rose from 1 to 3 m s^{-1}, the average daily efficiency decreased by 31.6%, together with a significant increase in the COP which can be attributed to the increasing of the pressure losses.

1.3. Aim of the Paper

Beyond the generally accepted results on qualitative trends following the variation of the key parameters for the EAHX, the above introduced state of the art also presents peculiarities that in some cases may seem conflicting. Most simulations focus on only very limited configurations, most of them based on simple one-dimensional models that can hardly predict the real behavior of an EAHX. In addition, most research ignored the latent heat transfer related to the water vapor condensation during humid air cooling in the summer season.

The aim of this paper is to develop an extensive analysis of the parameters that would optimize the thermal performances of a horizontal single-duct system to give clear indications for the design of a real application for an EAHX. The behavior of an EAHX is influenced by the geographical and climatic conditions where the system is installed. There are national or global directives that suggest the climatic design parameters in projecting a ground to air heat exchanger system. The ASHRAE identifies [46] the reference climate design parameters such as intensity of the solar irradiation, temperature and relative humidity of the external air, both in winter and in summer operation modes for many cities. In this paper, the EAHX is intended to be installed in the city of Naples (Italy), which has a typical Mediterranean climate. The study is conducted by means of a numerical two-dimensional model based on finite element method and experimentally validated [47]. The step of the investigation concerning the optimization of geometrical factors of the duct and air velocity is conducted by imposing the design parameters for cooling and heating operation modes prescribed by ASHRAE. Subsequently, once the optimal parameters have been identified, the effect of the external air temperature on the thermal performances of the geothermal system is analyzed.

2. Methods

2.1. The Design and the Assumptions Made

The numerical tool employed in the analysis is a two-dimensional numerical model, based on a finite element method, of a horizontal single-tube earth to air heat exchanger surrounded by the ground domain. A schematic of the model design is visible in Figure 1. The horizontal disposition was adopted since a vertical EAHX is generally connected with much more expansive installation and maintenance costs, whereas the reason for the single-tube choice lies in the desire to preserve as much genericity as possible, to allow the optimization. To this purpose, the parameters of the tube such as burial depth (z), diameter (D), length (L), and inlet air velocity (v), are held as variables. Once the optimal combination of design parameters has been identified, the number of parallel tubes can be increased to supply larger demands of volumetric flow rate.

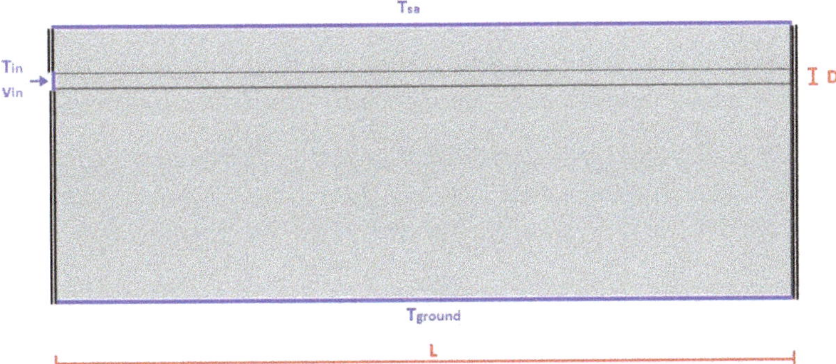

Figure 1. A schematic of the design of the numerical model of the horizontal Earth to Air Heat eXchanger (EAHX).

The mathematical model is forced on the earth to air heat exchanger considering the following assumptions:

- the computational domain has two-dimensional geometry (formed by the ground and the duct) modelled in a longitudinal section for symmetry;
- the thermal resistance of the pipe is not considered (the reason for this is explained in the following subsection);
- a time-dependent analysis is conducted on the model until the stationarity is observed;
- 20 m is the depth where temperature of the ground is considered undisturbed;
- the properties of the soil in the whole domain are assumed to be constant;
- the soil is considered as an isotropic medium.

Humid air is the fluid crossing the pipe, whose thermodynamic properties (such dry bulb temperature; relative and specific humidity) are punctually evaluated (timely and spatially) in the numerical simulation of the model. Furthermore, the tool allows us to identify and quantify the amount of water condensation of the humid flow.

2.2. The Thermal Resistance Approach in the EAHX System

According to the design of Figure 1 and following the approach of thermal resistance [48], the model could be schematized as the series of three resistances:

$$R_{eq} = R_{ground} + R_{pipe} + R_c, \tag{1}$$

Both the conductive thermal resistances of the surrounding ground and the pipe are evaluated as:

$$R_{ground} = \frac{\ln\left(\frac{r_{ground}}{r_{pipe,ext}}\right)}{2\pi k_{ground}L}, \tag{2}$$

$$R_{pipe} = \frac{\ln\left(\frac{r_{pipe,ext}}{r_{pipe,int}}\right)}{2\pi k_{ground}L}, \tag{3}$$

The thermal resistance due to the convective heat exchange between the air flowing in the tube and the pipe is:

$$R_c = \frac{1}{2\pi r_{pipe,int}LU}, \tag{4}$$

Considering a PVC pipe ($k = 0.16$ W m^{-1} K^{-1}; $\rho = 1380$ kg m^{-3}; $c = 900$ J kg^{-1} K^{-1}), the orders of magnitude for R_c and R_{ground} are $10^{-2} \div 10^{-3}$ K W^{-1} whereas 10^{-4} K W^{-1} is the dimension proper of R_{pipe}.

For this reason, in our model, the thickness of the tube is not considered since the thermal resistance associated with the conduction in the thickness of the tube is negligible with respect to the two others. As a matter of fact, for modeling needs with a software CFD, such a small thickness, compared to the orders of magnitude of the dimensions of the other domains of the model (ground and air-occupied area in the tube), would have required a much smaller thickening of the FEM grid. Thus, the dimensions of the elements of the mesh should have assumed values small enough to make it really difficult to converge the model in finite or acceptable times.

The problem is very common in the literature and this approximation (calculation of the temperature field using CFD basing of FEM, therefore not through a resistive approach) is usually accepted for approaches similar to ours [45,49,50].

2.3. The Mathematical System

The fluid domain is described by the following differential equations:

- the mass conservation of the humid air:

$$\frac{\partial \rho}{\partial t} + \nabla\left(\rho\vec{v}\right) = \dot{S}_{mass}, \tag{5}$$

where $\dot{S}_m$ is a negative term that represents the mass of the condensed water;
- the conservation momentum is guaranteed by the Navier-Stokes equations for turbulent air flow:

$$\rho\frac{\partial\vec{v}}{\partial t} + \rho\left(\vec{v}\cdot\nabla\right)\vec{v} = \nabla\cdot\left[-p\vec{I} + (\mu + \mu_T)[\nabla\vec{v} + \left(\nabla\vec{v}\right)^T]\right], \tag{6}$$

where μ_T is the turbulent viscosity defined as:

$$\mu_T = \rho C_\mu \frac{K}{\varepsilon},\tag{7}$$

with C_μ, that is one of the constants of the K-ε model for turbulent flow [47].

- the energy equation for the air flow:

$$\frac{\partial(\rho E)}{\partial t} + \nabla\cdot\left[\vec{v}(\rho E + p)\right] = \nabla\cdot\left[k_{eff}\nabla T - \sum_j h_j \vec{J}_j + \left(\tau_{\text{eff}}\cdot\vec{v}\right)\right],\tag{8}$$

where k_{eff} is the effective conductivity defined as the sum of the conventional thermal conductivity of the fluid (k_f) and the thermal conductivity of the turbulent flow (k_T) and thus modeled as:

$$k_{eff} = k_f + k_T\tag{9}$$

- Using the K-ε model for turbulent flow, the turbulence kinetic energy equation is:

$$\frac{\partial(\rho K)}{\partial t} + \rho\vec{v}\cdot\nabla K = \nabla\cdot\left(\left(\mu + \frac{\mu_T}{\sigma_K}\right)\nabla K\right) + P_K - \rho\varepsilon,\tag{10}$$

where P_K can be evaluated as:

$$P_K = \mu_T\left(\nabla\vec{v} : \left(\nabla\vec{v} + \left(\nabla\vec{v}\right)^T\right) - \frac{2}{3}\left(\nabla\cdot\vec{v}\right)^2\right) - \frac{2}{3}\rho K\nabla\cdot\vec{v}\tag{11}$$

The specific dissipation rate equation is:

$$\frac{\partial(\rho\varepsilon)}{\partial t} + \rho\vec{v}\cdot\nabla\varepsilon = \nabla\cdot\left(\left(\mu + \frac{\mu_T}{\sigma_\varepsilon}\right)\nabla\varepsilon\right) + C_{\varepsilon 1}\frac{\varepsilon}{K}P_K - C_{\varepsilon 2}\rho\frac{\varepsilon^2}{K}\tag{12}$$

Table 1 reports the experimental constants of the K-ε model.

Table 1. The K-ε model constants.

Constant	Value
C_μ	0.09
$C_{\varepsilon 1}$	1.44
$C_{\varepsilon 2}$	1.92
σ_k	1.0
σ_ε	1.3

Where the fluid flows in laminar motion, the turbulent flow parameters k_T and μ_T become zero and Equations (2) and (4) are incorporated respectively in:

$$\rho\frac{\partial\vec{v}}{\partial t} + \rho\left(\vec{v}\cdot\nabla\right)\vec{v} = \nabla\cdot\left[-p\vec{I} + \mu[\nabla\vec{v} + \left(\nabla\vec{v}\right)^T]\right],\tag{13}$$

$$\frac{\partial(\rho E)}{\partial t} + \nabla\cdot\left[\vec{v}(\rho E + p)\right] = \nabla\cdot\left[k\nabla T - \sum_j h_j\vec{J}_j + \left(\tau\cdot\vec{v}\right)\right],\tag{14}$$

The differential equation of heat transfer in ground domain is the energy equation for solid medium:

$$\frac{\partial(\rho_{soil}c_{soil}T_{soil})}{\partial t} = \nabla\cdot(k_{soil}\nabla T_{soil}),\tag{15}$$

The soil humidity is considered balancing water and solid properties according to the porosity (ψ) with the following equation:

$$z_{soil} = \psi z_{liquid} + (1 - \psi) z_{solid},$$ (16)

The above introduced mathematical model must be coupled with a number of boundary conditions forced on the domain. Specifically, the adopted ones are reported below:

- to the lateral sides of the ground domain there is symmetry with the behavior of the ground beyond the sides; thus, adiabatic (2nd type) conditions are imposed;
- a 1st type condition is considered at the bottom of the ground domain (as shown in Figure 1), since 20 m is a depth where the ground is certainly undisturbed by the weather conditions. For this reason, the enforced condition is the undisturbed temperature calculated though the Kusuda [51] equation:

$$T_{ground}(D, t) = T_m - A$$
$$\cdot \exp\left[-Depth \cdot \sqrt{\tfrac{\pi}{365 \cdot \alpha}} \right] \cdot \cos\left[\tfrac{2\pi}{365} \cdot \left(t - t_{min} - \tfrac{Depth}{2} \cdot \sqrt{\tfrac{365}{\pi \cdot \alpha}} \right) \right]$$ (17)

- the *sun-air temperature* model is a 1st type boundary condition imposed at the top of the ground domain, as visible in Figure 1. This temperature takes into account the influence of both the incident solar radiation on the ground surface and the convective heat exchange with the external air, according to the following equation:

$$T_{sa}(x, 0, t) = T_{air,ext}(t) + \frac{\alpha G(t)}{U_c},$$ (18)

- at the inlet of the pipe, the temperature and relative humidity of the external air are imposed, whereas the inlet velocity of the air is a variable to be optimized through the investigation introduced in this paper (Figure 1).

The model is solved through the finite element method applied to the domain meshed in free triangular elements; the dimension of the mesh has been chosen according to a grid independence study that has been the object of a previous investigation [47].

The model has been experimentally validated in three case studies of horizontal ducts placed in three different countries (Algeria, Morocco, Egypt) and the maximum relative error between the experimental and the numerical data is 2.55%. Further details are reported in [47,52].

2.4. Parameters of the Optimization and Operative Conditions

Using the tool introduced in the previous section, a parametric analysis is carried out on the earth to-air heat exchanger, verifying the main design parameters such as: velocity of the air flow, length, diameter, and burial depth of the pipe affect the thermal performance of the EAHX itself, both in summer (cooling) and winter (heating) operation modes. In particular, we consider the influence of the three parameters on:

- the temperature of the outlet air;
- absolute value of the temperature difference between the inlet and outlet sections of the tube;
- efficiency of the EAHX, that is the ratio between the EAHX temperature span and the ideal temperature difference and it is defined as:

$$\varepsilon = \frac{T_{out} - T_{in}}{T_{ground} - T_{in}},$$ (19)

For this purpose, the Italian location of Naples (Lat. 40° 51′ 22.72″ N; Long. 14° 14′ 47.08″ E) has been considered. According to Köppen climate classification [53], which divides the areas of the globe into five main climatic zones mainly based on temperature and vegetation criteria, Naples belongs to a *Csa* climatic zone (hot-summer Mediterranean climate). The thermodynamical properties of the soil of the city of Naples are the ones typical of an Italian locality (k = 1.63 W m^{-1} K^{-1}; ρ = 1700 kg m^{-3}; c = 1600 J kg^{-1} K^{-1}) with 37% as porosity. Through the Kusuda relation [51] and using the data coming from the ASHRAE climate database, in a previous investigation [54], the undisturbed temperature of the ground for the city of Naples was calculated (17.0 °C). Furthermore, always referring to ASHRAE climate data [46], the reference climate design parameters such as intensity of the solar irradiation, temperature and relative humidity of the external air, both in winter and in summer operation modes, were identified and listed in Table 2.

Table 2. The reference climate design parameters for the city of Naples.

Climatic Zone—Locality	Geographic Coordinates	Winter Design Parameters			Summer Design Parameters		
		T [°C]	Φ [%]	G [Wm^{-2}]	T [°C]	Φ [%]	G [Wm^{-2}]
Csa—Naples	Lat. 40°51′22″ N Long. 14°14′47″ W	1.90	52.00	808	31.90	48.60	825

The investigation has been conducted for different values of length, diameter, and burial depth of the tube. The speed of the air flow entering the pipe was also varied. Specifically, below are reported the values under which the EAHX has been tested:

$$z = [1.5; 3.5; 5.5; 7.5; 9.5]\ \text{m}, \tag{20}$$

$$L = [20; 50; 60; 80; 100; 120; 140; 160]\ \text{m}, \tag{21}$$

$$D = [0.1; 0.2; 0.3; 0.4]\ \text{m}, \tag{22}$$

$$v = [0.5; 1.0; 1.5; 2.0; 2.5]\ \text{m s}^{-1}, \tag{23}$$

The regime of motion that is going to develop into the pipe, in dependence with the choice of different values of the parameters D and v is well ascertainable through the estimation of the corresponding Reynolds number, whose mathematical formulation is given by:

$$Re = \frac{vD}{v}, \tag{24}$$

The establishment of the fully turbulent regime inside a duct is guaranteed by Reynolds numbers higher than 10^4. Values lower than 2300 ensure the laminar one. In the range of numbers delimited by these two extremes, a transition zone takes place in which the fluid gradually evolves from laminar to turbulent, or it is in a metastable state. Table 3 reports the proper Reynolds number, evaluated at 20 °C, for each diameter and inlet air velocity couple.

Table 3. Reynolds numbers evaluated at 20 °C for each diameter and inlet velocity couple. In bold the couples for which the motion in not fully turbulent.

D [m]	u [m s^{-1}]	Re [10^4]
0.1	**0.5**	**3312**
0.1	**1**	**6624**
0.1	1.5	9936
0.1	2	13,248
0.1	2.5	16,560
0.2	**0.5**	**6624**
0.2	1	13,248
0.2	1.5	19,873
0.2	2	26,497
0.2	2.5	33,121
0.3	0.5	9936
0.3	1	19,873
0.3	1.5	29,809
0.3	2	39,745
0.3	2.5	49,681
0.4	0.5	13,248
0.4	1	26,497
0.4	1.5	39,745
0.4	2	52,993
0.4	2.5	66,242

For almost all the couples of (D,v) tested, the motion is fully turbulent since the coupled Re are greater than 10^4, except the following ones (bold marked in Table 3):

- $(D = 0.1$ m, $v = 0.5$ m s^{-1}), where the fluid motion is associated to Re $= 3312$;
- $(D = 0.1$ m, $v = 1.0$ m s^{-1}) and $(D = 0.2$ m, $v = 0.5$ m s^{-1}) where Re $= 6624$;

which fall in the transition zone between laminar and fully developed turbulent flow. In any case, there are no cases where the flow is laminar. For this reason, according to a study in the literature [55,56], the effect of natural convection was not considered since it becomes relevant and comparable with a forced one, for Reynolds up to 600.

3. Results

Numerous series of numerical tests were carried out to develop maps of performances of the modelled earth to air heat exchanger with the final purpose of optimizing the thermal performances with respect to one working parameter at time. Specifically, to accurately evaluate the incidence of the different parameters on the thermal performances of the EAHX, a single parameter (burial depth, pipe length, diameter and air velocity) was varied at a time while the other ones remain constant.

3.1. Effect of the Burial Depth

The first parameter to be analyzed was the burial depth of the tube. The burial depth z was varied and the air temperature at the outlet of the pipe was evaluated while flowing, according to different velocities, in tubes with different diameters and lengths. The resulting data in all the tests show that if the pipe is buried from a depth of 1.5 m onwards, the outlet air temperature is not affected by the variation of z, whereas the diameter, length, and air velocity are. The reasons for such behavior can be found with the help of Figure 2.

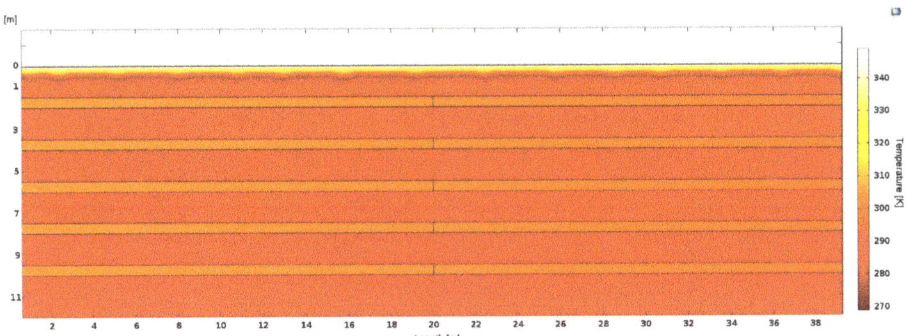

Figure 2. Thermal penetration depth generated by the earth surface temperature.

Figure 2 shows a zoomed-in image of the investigation domain. From the figure it is clearly visible that the boundary condition (14) forced on the top of the domain, and representing both the influence of the solar radiation incident on the surface and the convective heat exchange, generates a thermal disturbance that has an influence up to 1.0 m of depth. For this reason, the tubes placed at 1.5 m and 9.5 m are not disturbed by the presence of the surface of the ground. Anyhow, as we detected and revised in previous studies published in the literature [22,38,57], the undisturbed ground temperature is generally observed at a depth of 2–4 m, but at some latitudes, depending on radiation and soil properties, it could be more than 4 m. Correspondingly, in many studies [58–60] it has been noticed that that diurnal variation (observation period of 24 h) of earth surface temperature does not penetrate more than 0.5 m, whereas, for annual variation (observation period of 365 days) it is no more than 4.0 m. The trends detected in this analysis are in accordance with the studies [58–61] because our numerical tests are run with fixed design conditions of solar radiation and external air temperature and relative humidity; thus, the simulation period is the time needed to the domain to reach thermal steady-state (about 3000–5000 s seconds), which is much less than diurnal variation.

3.2. Effect of the Pipe Diameter

The effect of the pipe diameter on the thermal performances of the EAHX is analysed in this subsection. In Figure 3 one can appreciate the trends of the air temperature at the pipe outlet at different tube diameters while the air flow enters the tube with: 0.5 m s^{-1} (Figure 3a), 1.0 m s^{-1} (Figure 3b), 1.5 m s^{-1} (Figure 3c), 2.0 m s^{-1} (Figure 3d), and 2.5 m s^{-1} (Figure 3e). In the upper part of the figures, we show the temperature profile during summer, and, in the lower part, winter. The temperature profiles are reported for each diameter as a function of the tube length. For all the figures an analogous tendency can be noted: the air flow temperature increases/decreases through the tube length (in winter/summer). This increment/decrement is faster for the initial length of EAHX and thereafter it becomes moderate.

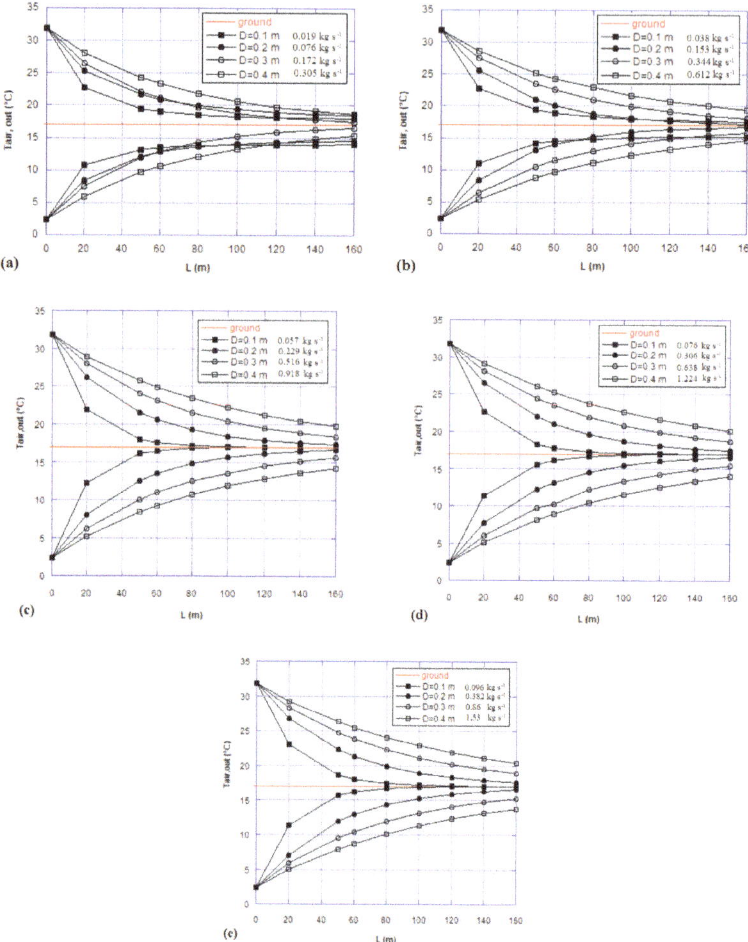

Figure 3. Outlet air temperature vs. length parametrized for diameter while the inlet air velocity is: (**a**) 0.5 m s^{-1}; (**b**) 1.0 m s^{-1}; (**c**) 1.5 m s^{-1}; (**d**) 2.0 m s^{-1}; (**e**) 2.5 m s^{-1}.

Figure 3 reveals that, for a fixed length, the smaller the diameter is, the closer the outlet temperature is to the ground temperature (which thermodynamically constitutes an upper limit). As a consequence, at fixed length and fluid velocity, the air temperature during the summer always reaches lower values for smaller tube diameters, and the opposite trend is seen during the winter. The result is that decreasing the tube diameter at fixed air velocity increases the convective heat transfer coefficient, enhancing the heat transfer between the air flow and the tube wall. The difference is more marked corresponding to smaller tube length. As an example, at a fluid velocity of 0.5 m s^{-1} the outlet temperature at a tube length of 60 m and diameter of 0.1 m is 17% lower/higher than that corresponding to a diameter of 0.4 m; at a length of 160 m the absolute value of the difference reduces to 3.7. This trend can be explained because, by increasing the tube length, the increment/decrement of air temperature becomes moderate because of the driving force of the heat exchange process, which is the difference between the temperature of the air and that of the undisturbed ground. This decreases, making the convective heat exchange less effective. Hence the influence of the decrease in the heat transfer coefficient is less sensitive. Observing Figure 3, one can note that the temperature profiles varying the tube diameter are closer for low air velocities: at a speed of 0.5 m s^{-1}, the temperature profiles corresponding to the different diameters tend to overlap at a tube length greater than 100 m. As an example, at a fluid velocity of 2.5 m s^{-1} the outlet temperature at a tube length of 60 m and diameter of 0.1 m is 29% lower/higher than that corresponding to a diameter of 0.4 m; at a length of 160 m the absolute value of the difference reduces to 17%. Comparing these values with those found for the speed of 0.5 m s^{-1} it is evident that the difference is more marked. Therefore, at higher air flow velocities the influence of the tube diameter is more marked. The results shown in Figure 3 are in agreement with the inherent literature. Mihalakakou et al. in their investigation [29], found that a reduction of the pipe diameter from 0.5 m to 0.25 m resulted in an improvement of the outlet air temperature drop by 1.5–2.5 °C during cooling mode operation. Moreover, they detected [30] that the reduction the diameter from 0.3 m to 0.2 m, during heating mode operation, reflected in an increase of outlet air temperature by 0.9–1.8 °C. They concluded that augmenting pipe diameter lowers convective heat transfer coefficients [62,63].

The effect of the diameter also reflects on the efficiency of the EAHX. Figure 4 shows the efficiency vs. length parametrized for diameter for different air velocities at inlet of the pipe in cooling mode. Similar trends have been observed in heating mode. In the cases plotted in the Figure 4c–e, the air flow always follows turbulent motion; here one can see that, for a fixed length, the lower the diameter, the higher the efficiency. For a tube length of L = 100 m, the efficiency is 98.6% if D = 0.1 m, with respect to 60.1% if D = 0.4 m; i.e., at this length the reduction of the diameter from 0.4 m to 0.1 m carries an increment of the efficiency of +32%, whereas the medium increment is +38.4%. The situation in Figure 4a,b is different. In this case, for D = 0.1 m, v = 0.5 m s^{-1} and 1.0 m s^{-1} an for D = 0.2 m and v = 0.5 m s^{-1}, the flowing of the fluid according to a metastable state (transition between laminar and turbulent motion) results in a degradation of the efficiency for large lengths with respect to the closer turbulent points.

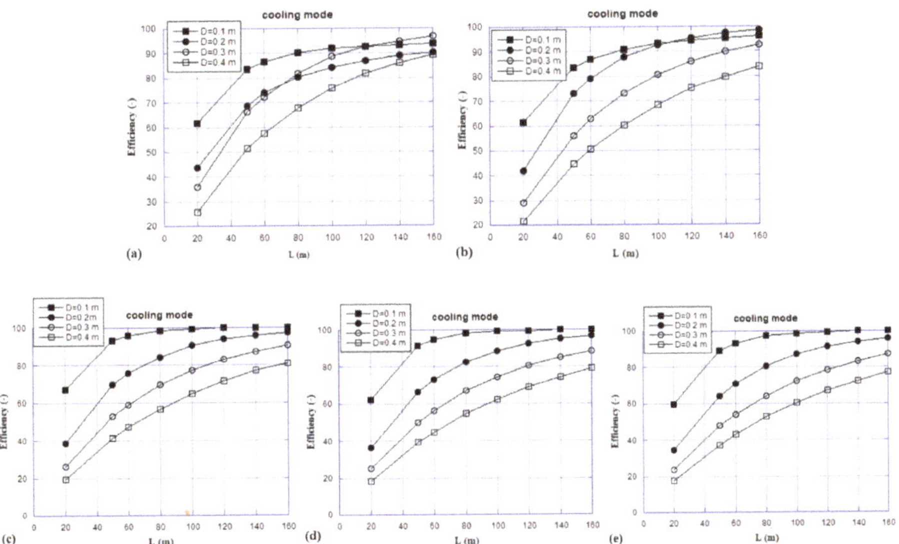

Figure 4. Efficiency of the EAHX vs. pipe length parametrized for diameter for different air velocities at inlet of the pipe: (a) 0.5 m s^{-1}; (b) 1.0 m s^{-1}; (c) 1.5 m s^{-1}; (d) 2.0 m s^{-1}; (e) 2.5 m s^{-1}.

From this section an important conclusion can be drawn: to increase the volumetric flow rate, it is better to project the system with a certain number of parallel tubes with smaller diameters rather than working with only one bigger duct.

In any case, one must ensure to set v and D opportunely in order that the turbulent motion is guaranteed.

3.3. Effect of the Pipe Length

The length of the tube is a crucial parameter that influences the thermal performances of an earth to air heat exchanger. In this subsection, all the aspects inherent to this variable are analyzed.

Figure 5 reports the inlet-outlet temperature span of the EAHX operating in cooling as a function of the air velocity for different tube lengths.

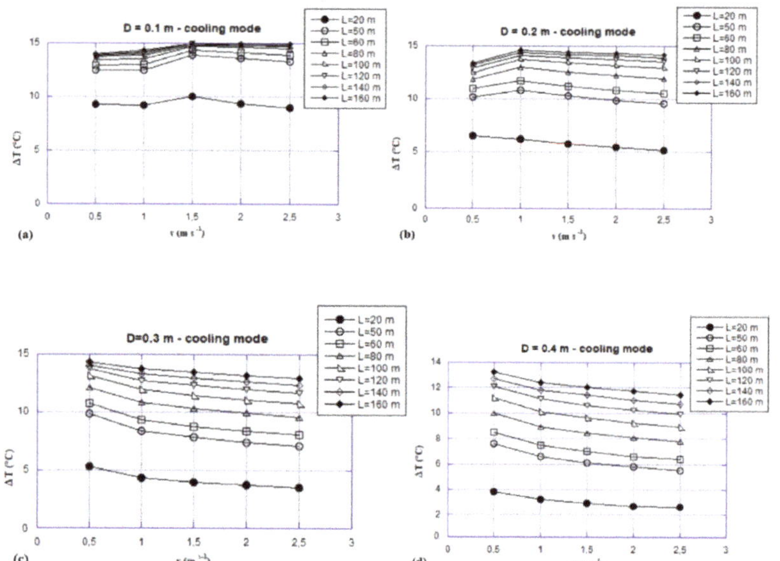

Figure 5. Inlet-outlet temperature span of the EAHX operating in cooling mode vs. air velocity, parametrized along the length for: (**a**) D = 0.1 m; (**b**) D = 0.2 m; (**c**) D = 0.3 m; (**d**) D = 0.4 m.

Specifically, in Figure 5a–d, this effect is studied, respectively, for the following diameters: 0.1 m; 0.2 m; 0.3 m; 0.4 m. At a diameter of 0.1 m, increasing the length, one can observe how the curves approach each other and from a certain length onwards they almost overlap. At higher diameters, the curves approach but they did not overlap. There is a limit in which a further increase in length no longer leads to an improvement in thermal performances of the EAHX: this limit is called saturation length. The saturation length constitutes a knee point and it is defined as the length of the tube at which more than 90% of the global increase or decrease of the air temperature has been obtained. More generally, in correspondence with the saturation length, the thermal performances of the EAHX, such as outlet temperature, temperature span and efficiency, reach 90% of their upper limits. The aspect was plainly detected in many studies [22,41,64] where it was detected that, after a limit, further increments in pipe length do not lead to further improvements of thermal performances. The physical explanation lies in the fact that the air temperature can at most reach that of the ground; therefore, the convective heat exchange is more noticeable the greater the ΔT is. As the air flows in the pipe (and therefore as the length increases) its temperature gets closer and closer to that of the ground. This is the reason why, when the air temperature approaches the ground temperature, the heating and cooling powers associated with the heat exchanged by convection become very small. As a consequence, during the projecting of the earth to air heat exchanger the pipe lengths of the tubes should be optimized to balance the trade-off between the investment cost for the construction and the expected thermal and energy performances.

To better understand this theme, in Figure 6 the saturation lengths with respect to inlet air velocity, parametrized for the diameter are plotted while the EAHX works in cooling mode. As a general trend, the Figures clearly show that for smaller diameters the saturation of the thermal performances is reached in correspondence of smaller lengths. Moreover, the figure shows that the incidence of the regime of fluid motion evidently emerges: if the fluid flows in the transition zone between laminar and turbulent motions, on equal diameter, the saturation lengths are higher than if the turbulent flow is fully developed. Specifically, if we collocate in turbulent motion, we can observe that, on equal velocity, the saturation length for pipes with D = 0.4 m is four times (+200%) the one established at D = 0.1 m. This is due to the above-mentioned influence of the diameter of the pipe on the thermal performances of the EAHX. Therefore, with a tube diameter of 0.1 m, a tube length lower than 80 m can be used. In contrast, with a tube diameter of 0.4 m, the tube length should always exceed 150 m.

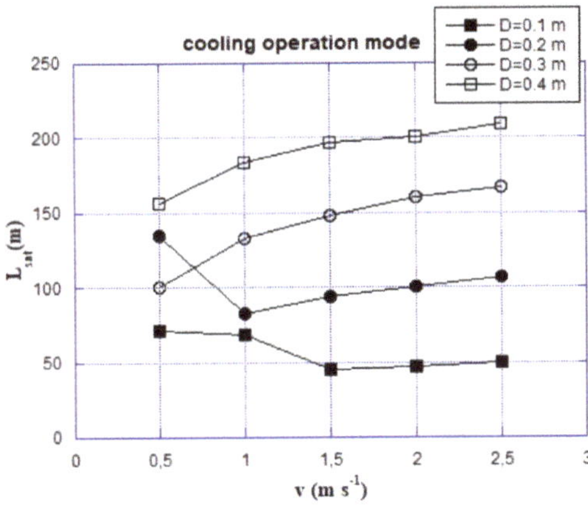

Figure 6. Saturation lengths vs. inlet air velocity parametrized for the diameter while the EAHX works in cooling mode.

The pipe form factor (i.e., the ratio between the tube length and diameter) takes into account both the effect of tube length and diameter on the thermal performances of the EAHX. Figure 7 presents the effect of the pipe form factor on the air temperature variation in the EAHX during summer for different Reynolds numbers. For each Reynolds number, as the form factor increases the temperature variation also increases. This increase is faster for low form factors and then (for form factor greater than 500) becomes moderate: by increasing the form factor the length of the tube increases and the diameter decreases. This leads to a longer path over which heat transfer between the air flow and the ground can take place together with an increase of the heat transfer coefficient. At fixed form factor, the Reynolds number increases with the air velocity. For L/D less than 500, by increasing the Reynolds number the temperature difference decreases because air spends less time in the tube and thus in contact with soil. For L/D greater than 1000, all the curves merge together and therefore the influence of the Reynolds number on the thermal performance of the EAHX is negligible. This behavior can be found only for Reynolds numbers that follow in a turbulent regime. In all the explored range of L/D, the worst values of the temperature difference are those corresponding to Reynolds number that follow in the transition regime. Therefore, to optimize the thermal performance of an EAHX, a form factor L/D greater than 500 is recommended with a Reynolds number that follows in the turbulent regime. This result agrees with the results of the investigation by Sehli et al. [32].

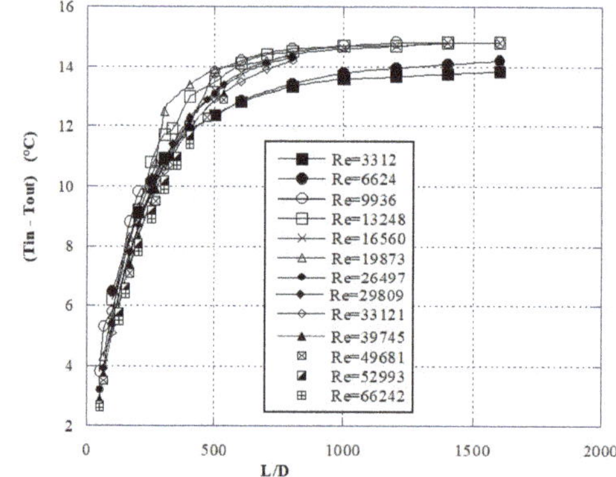

Figure 7. Temperature variation in the EAHX during summer as a function of the tube form factor for different Reynolds numbers.

3.4. Effect of the Air Flow Velocity

Another design parameter playing a key role in the performance of the earth to air heat exchanger is the velocity of the air flow. As shown in Figure 8, the outlet temperatures are reported to be parametrized for velocity variation when the diameters are: (a) 0.1 m; (b) 0.2 m; (c) 0.3 m; (d) 0.4 m.

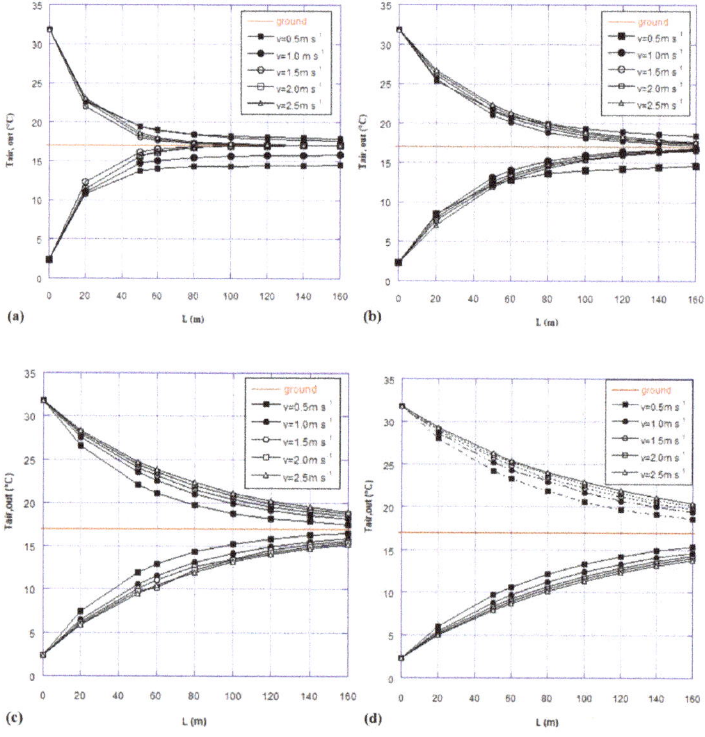

Figure 8. Outlet air temperature vs length parametrized for air velocity while the diameter is: (**a**) 0.1 m; (**b**) 0.2 m; (**c**) 0.3 m; (**d**) 0.4 m.

Figure 8 reveals that the outlet temperature is closer to the ground temperature (both for winter and summer) the air flow is slower. The reason of such behavior is that air flows with lower velocities result in a longer time at which the air is in contact with the tube wall: consequently, the heat transfer increases with reducing air velocity. For a fixed diameter, comparing the data carried out with the smaller and the larger velocities, a medium decrement of +2.5 °C in the temperature difference between the air outlet temperature and the ground temperature, is observed in correspondence with the smaller velocity. We can also note that the outlet temperature of the air approaches the undisturbed ground temperature more quickly if the velocity is smaller. For a fixed diameter, if the motion is turbulent, the most promising temperature drop is observed for $v = 0.5$ m s^{-1}, as Figure 8c,d show. In Figure 8a,b the situation is slightly different where the fluid is not fully developed in turbulent: here is clearly noticeable that the best performances are detected: for $v = 1.5$ m s^{-1}, if $D = 0.1$ m (Figure 8a); $v = 1.0$ m s^{-1}, if $D = 0.2$ m (Figure 8b). In other words, there is a lower limit to the increment of heat transfer with velocity reduction: the latter mentioned points ($D = 0.1$ m, $v = 1.5$ m s^{-1}; $D = 0.2$ m, $v = 1.0$ m s^{-1}) represent these limits (one for each diameter) since, for a fixed diameter, they are the smaller (according to velocity growing) where the fluid is considered turbulent. The same behavior is seen for both working modalities of the EAHX.

The trend is also in agreement with a comparison to the literature [29,30,61,64,65] where a decrement of the temperature span was shown to lead to an increment of the flow speed. Furthermore, Niu et al. found the highest temperature drop to be at 0.5 m s^{-1} because a slow flow is more in contact with the pipe walls, leading to a more efficient heat exchange [34].

Figure 9 reports the efficiency of the EAHX operating in cooling mode vs. length, parametrized for air velocity while: (a) $D = 0.1$ m; (b) $D = 0.2$ m; (c) $D = 0.3$ m; (d) $D = 0.4$ m. In Figure 9a,b it is clearly noticeable that reduced efficiency (9.5% medium reduction) is registered for the D and v couples that do not ensure the fully turbulent motion. In fully turbulent motion, the effect of the air velocity on the thermal performances of the EAHX reflects in an improvement of the efficiency for slower speed for a fixed diameter. The larger the diameter, the more remarkable the improvement is: for $D = 0.4$ m, a medium (with respect to the length) increment of +18% is registered if the velocity falls from 2.5 m s^{-1} to 0.5 m s^{-1}, whereas for smaller diameters the increment is progressively reduced.

The trends in Figure 10 are in agreement with the data of Figures 8 and 9 since, at fixed diameter and velocity (i.e., at fixed air flow rate), the heating power grows with the length since as the air passes through the tube its temperature get closer to the temperature of the ground and therefore inlet and outlet temperature span increases. At fixed D and L, $\dot{Q}_h$ augments with the air flow velocity, since the air flow increases. Similarly, $\dot{m}_{air}$ also grows with the diameter, at fixed v and L, i.e., the heating power follows the same trend.

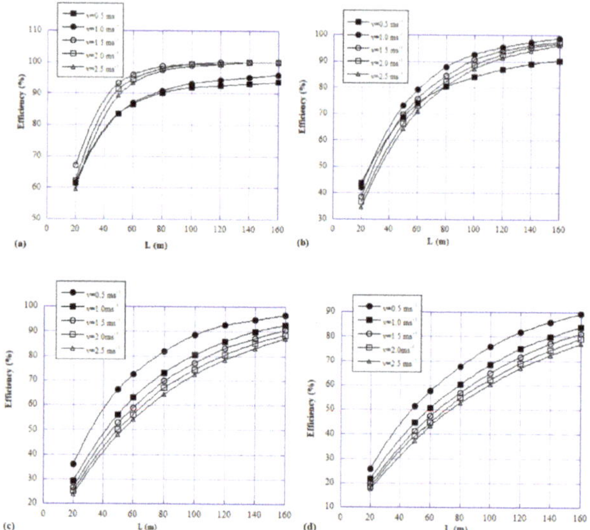

Figure 9. Efficiency of the EAHX operating in cooling mode vs length, parametrized for air velocity with: (**a**) D = 0.1 m; (**b**) D = 0.2 m; (**c**) D = 0.3 m; (**d**) D = 0.4 m.

In Figure 10, the heating powers evaluated while the EAHX works in heating mode are reported as a function of the pipe length whereas the diameter is: (a) 0.1 m; (b) 0.2 m; (c) 0.3 m; (d) 0.4 m. The heating power has been evaluated according to the following relation:

$$\dot{Q}_h = \dot{m}_{air} c_p (T_{out} - T_{in}),$$

(25)

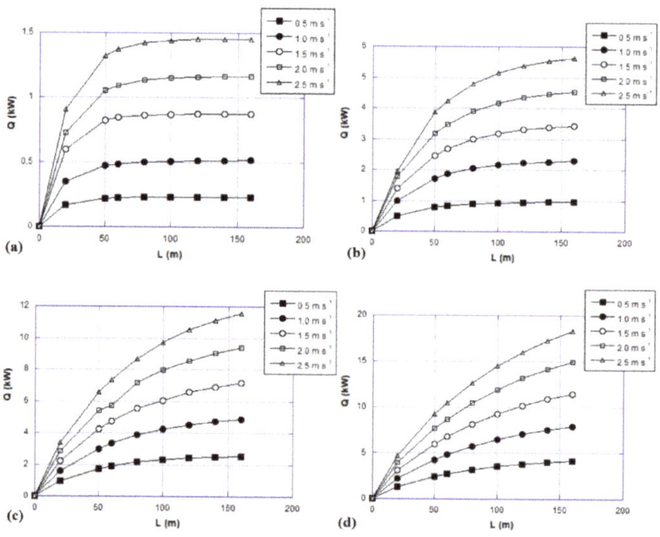

Figure 10. Power of the EAHX operating in heating mode vs length, parametrized for air velocity with: (**a**) D = 0.1 m; (**b**) D = 0.2 m; (**c**) D = 0.3 m; (**d**) D = 0.4 m.

3.5. Effect of the Air Temperature

The evaluation of the effects of the burial depth, the pipe lengths, the pipe diameter and the air velocity at pipe inlet, illustrated in the previous sections, allows us to identify, with respect to our case-study, the combinations of such parameters that enhance the thermal performances of the EAHX. These results can be generalized to other cities with a Mediterranean climate. In order to optimize a horizontal single tube EAHX system placed in the city of Naples, its thermal performances must be formed by tubes with D = 0.1 m where the air flows at an inlet velocity of v = 1.5 m s^{-1}, buried at z = 1.5 m. For the above chosen combination (*D,v,z*) the length that would satisfy the trade-off between thermal performances and installation and excavation costs is 50 m.

As next step of the investigation introduced in this paper, we studied the effect of the air temperature on the EAHX formed by one horizontal single tube and characterized by (*D,v,z*) selected above. Furthermore, we decided to leave a degree of freedom on *L* (not fixing it) to better study the effects of the thermal transient on the length.

Specifically, fixing the relative humidity rate to 50%, the EAHX was tested while the temperature at the inlet of the air assumes the following values:

$$T_{air,in} = [1.9; 5; 10; 15; 20; 25; 31.9; 35] \;°C, \tag{26}$$

Figure 11 plots the outlet temperature (Figure 10a) and the specific humidity (Figure 10b) of the air flowing with 1.5 m s^{-1} in the tube with D = 0.1 m as the diameter for different inlet temperatures.

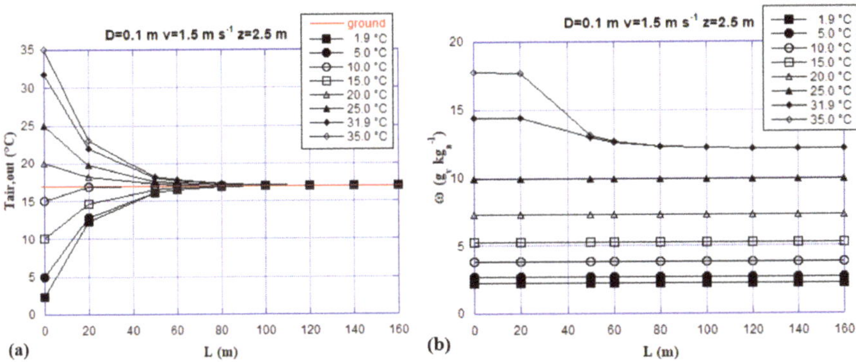

Figure 11. (a) Outlet temperature and (b) specific humidity, parameterized with respect to inlet temperature of the air flowing with 1.5 m s^{-1} in the tube with D = 0.1 m as diameter.

From Figure 11a one can observe that the highest temperature gradients are in the part of the tube defined as initial with respect to the fluid flow. In this area, for a fixed length, the gradients are larger for greater inlets ΔT between the air and the ground. Consequently, in the initial part of the tube the heat transfer is much more pronounced. Such trends correspond with the work of Niu et al. [34].

For L = 50 m, the maximum deviation detected between the air temperature and the ground temperature is −5.23% during heating operation mode (for air entering the tube with 2.3 °C) and +7.05% during cooling operation mode (for air entering the tube with 35 °C). This data confirms the choice of L = 50 m as optimal length of the tube both for cooling and heating operation modes, if the EAHX works with optimized (*D,v,z*). Therefore, if the tube length exceeds the saturation value, the inlet air temperature has a negligible influence on the thermal performance of the EAHX.

Figure 11b allows us to detect the length where the air achieves the saturation point (Φ= 100%) and consequently the water begins the condensation process. As long as the air does not saturate, the humidity ratio remains constant, then it begins to condense and the amplitude of the negative gradient is a function of the temperature, since higher temperature corresponds with larger specific humidity. These circumstances occur only during cooling operation mode and while the inlet air temperature is 31.9 °C and 35 °C. The condensation lengths in the tube are: 32.5 m if $T_{air,in} = 31.8$ °C; 19.9 m if $T_{air,in} = 35$ °C.

The amounts of condensed water are reported in Figure 12 with respect to the length. According to the considerations made for Figure 11, for a fixed tube length, the larger the inlet temperature is, the higher the specific humidity, and the greater the amount of condensed water. Furthermore, from the knee length of the EAHX onward, amount of the condensed water at the end of the condensation process remains constant.

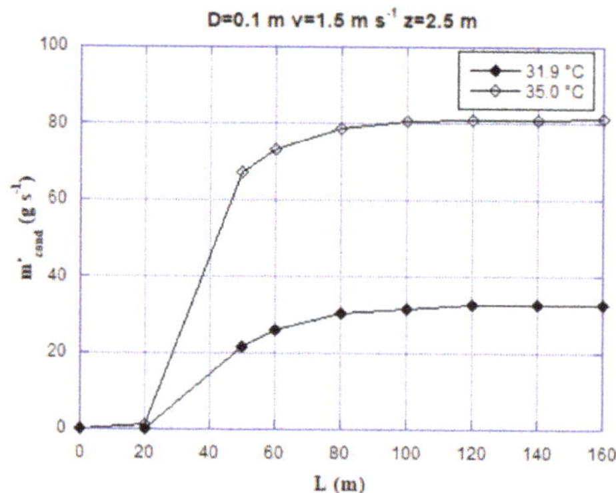

Figure 12. Amount of condensed water with length parameterized for inlet temperature of the air.

4. Discussion and Conclusions

The analysis introduced in this paper aims to explore the thermal performances of a horizontal single-tube earth to air heat exchanger to be placed in the city of Naples (Italy), by varying the most crucial parameters influencing the system. Both geometrical (pipe length, diameter, burial depth) and physical (velocity and temperature of the air flow at pipe inlet) properties are varied in an appropriate range, with the final aim to find the combinations that would optimize the operation of this geothermal system. The following conclusion can be drawn:

- the burial depth does not affect the thermal performances from 1.5 m onwards because of the typology of the study conducted (forcing design conditions to observe the steady-state answer of the EAHX).
- The thermal performances increase with the length of the pipe up to a certain limit, called saturation length over which no longer enhancements are registered. The smaller the diameter, the lower the saturation length;
- with the same length, the smaller the diameter the closer the outlet temperature to the ground temperature both for cooling and heating operation modes;
- a form factor greater than 500 is recommended with a Reynolds number that follows in the turbulent regime in order to optimize the thermal performances of the EAHX.
- Supposing that the fluid motion is fully turbulent, the slower the air flow, the closer the outlet temperature is to the ground temperature.
- The combination that optimizes the performances of the EAHX system that works under the design conditions for cooling and heating is $D = 0.1$ m s^{-1}; $v = 1.5$ m s^{-1}; $L = 50$ m.
- The effect of the inlet temperature of the air, if $D = 0.1$ m and $v = 1.5$ m s^{-1}, reveals that the heat transfer is more pronounced at the beginning of the tube and that the temperature gradients along the pipe are greater for greater values of ΔT between the inlet air and the ground.

Nevertheless, if the tube length exceeds the saturation value, the influence of this parameter is negligible.

- If the EAHX works in cooling mode, the air entering at higher temperatures could during the flowing process, generate water condensation along the tube. For fixed length and relative humidity, the entity of such a phenomenon grows with the inlet temperature.

In conclusion, the investigation has identified the optimal combination of parameters that optimize the thermal performances of a single-tube horizontal EAHX system; consequently, an optimal air flow rate also emerges. In any case, for applications needing higher air flow rates, the optimal parameters of this optimization could be applied for each duct composing a horizontal multi-tube geothermal system. These results can be generalized to all cities with a Mediterranean climate.

Author Contributions: Conceptualization, A.G. and C.M.; methodology, A.G. and C.M.; formal analysis, A.G. and C.M.; investigation, A.G. and C.M.; data curation, A.G. and C.M.; writing—original draft preparation, A.G. and C.M.; writing—review and editing, A.G. and C.M.; visualization, A.G. and C.M. All authors have read and agreed to the published version of the manuscript.

Funding: This research received no external funding.

Conflicts of Interest: The authors declare no conflict of interest.

Nomenclature

Roman symbols

A	amplitude of the temperature variation, °C
C	constant of K-ε model
c	specific heat, J kg^{-1} K^{-1}
D	diameter of the pipe, m
E	energy, J
EAHX	Earth-to-Air-Heat-eXchanger
G	incident radiation, W m^{-2}
h	specific enthalpy, kJ kg^{-1}
HVAC	Heating, Ventilation & Air Conditioning
$\vec{I}$	identity vector, -
$\vec{J}_j$	component of diffusion flux, kg m^{-3} s^{-1}
K	turbulent kinetic energy, J
k	thermal conductivity, W m^{-1} K^{-1}
L	length of the pipe, m
$\dot{m}$	flow rate, kg s^{-1}
$\dot{Q}$	power, kW
p	pressure, Pa
R	thermal resistance, W m^{-1} K^{-1}
Re	Reynolds number, -
$\dot{S}$	source term, kg m^{-3} s^{-1}
T	temperature, °C
t	time, s
U	convective heat transfer coefficient, W m^{-2} K^{-1}
$\vec{v}$	fluid velocity vector, m s^{-1}
z	generic property, m

Greek symbols

α	absorbance of the surface
Δ	finite difference
∂	partial derivative
ε	efficiency, %

$\hat{\varepsilon}$	turbulent cinematic viscosity, $m^2\ s^{-1}$
μ	dynamic viscosity, $Pa\ s^{-1}$
ν	cinematic viscosity, $m^2\ s^{-1}$
ρ	density, $kg\ m^{-3}$
σ	constant of K-$\hat{\varepsilon}$ model
τ	tangential stress, $Pa\ m^{-1}$
Φ	relative humidity, %
ψ	porosity, %
ω	humidity ratio, $g_v\ kg_a^{-1}$

Subscripts

0	phase constant of the lowest average/mean soil surface temperature since the beginning of the year
air	air
c	convective
cond	condensed
eff	effective
eq	equivalent
ext	external
$\hat{\varepsilon}$	related to turbulent cinematic viscosity
f	fluid
ground	ground
h	heating
in	inlet of the EAHX
int	internal
j	species
liquid	liquid
m	annual mean soil
mass	mass
min	minimum
out	outlet of the EAHX
μ	related to the evaluation of turbulent dynamic viscosity
sa	sun-air
soil	soil
solid	solid
T	turbulent

References

1. Perez-Lombard, L.; Ortiz, J.; Maestre, I.R. The map of energy flow in HVAC systems. *Appl. Energy* **2011**, *88*, 5020–5031. [CrossRef]
2. Lu, Y.; Wang, S.; Zhao, Y.; Yan, C. Renewable energy system optimization of low/zero energy buildings using single-objective and multi-objective optimization methods. *Energy Build.* **2015**, *89*, 61–75. [CrossRef]
3. Buonomano, A.; Guarino, F. The impact of thermophysical properties and hysteresis effects on the energy performance simulation of PCM wallboards: Experimental studies, modelling, and validation. *Renew. Sustain. Energy Rev.* **2020**, *126*, 109807. [CrossRef]
4. Aprea, C.; Greco, A.; Maiorino, A.; Masselli, C. The employment of caloric-effect materials for solid-state heat pumping. *Int. J. Refrig.* **2020**, *109*, 1–11. [CrossRef]
5. Buonomano, A. Building to Vehicle to Building concept: A comprehensive parametric and sensitivity analysis for decision making aims. *Appl. Energy* **2020**, *261*, 114077. [CrossRef]
6. Aprea, C.; Greco, A.; Maiorino, A. The application of a desiccant wheel to increase the energetic performances of a transcritical cycle. *Energy Convers. Manag.* **2015**, *89*, 222–230. [CrossRef]
7. Aprea, C.; Greco, A.; Maiorino, A.; Masselli, C. The drop-in of HFC134a with HFO1234ze in a household refrigerator. *Int. J. Therm. Sci.* **2018**, *127*, 117–125. [CrossRef]

8. McLinden, M.O.; Domanski, P.A.; Kazakov, A.; Heo, J.; Brown, J.S. *Possibilities, Limits, and Tradeoffs for Refrigerants in the Vapor Compression Cycle*; ASHRAE Transactions: Gaithersburg, MD, USA, 2012.

9. Greco, A.; Vanoli, G.P. Experimental two-phase pressure gradients during evaporation of pure and mixed refrigerants in a smooth horizontal tube. Comparison with correlations. *Heat Mass Transf.* **2006**, *42*, 709–725. [CrossRef]

10. Greco, A.; Vanoli, G.P. Flow boiling heat transfer with HFC mixtures in a smooth horizontal tube. Part II: Assessment of predictive methods. *Exp. Therm. Fluid Sci.* **2005**, *29*, 199–208. [CrossRef]

11. Aprea, C.; Greco, A.; Maiorino, A.; Masselli, C. Analyzing the energetic performances of AMR regenerator working with different magnetocaloric materials: Investigations and viewpoints. *Int. J. Heat Technol.* **2017**, *35*, S383–S390. [CrossRef]

12. Aprea, C.; Greco, A.; Maiorino, A.; Masselli, C. The use of barocaloric effect for energy saving in a domestic refrigerator with ethylene-glycol based nanofluids: A numerical analysis and a comparison with a vapor compression cooler. *Energy* **2020**, *190*, 116404. [CrossRef]

13. Aprea, C.; Greco, A.; Maiorino, A.; Masselli, C. Solid-state refrigeration: A comparison of the energy performances of caloric materials operating in an active caloric regenerator. *Energy* **2018**, *165*, 439–455. [CrossRef]

14. Aprea, C.; Greco, A.; Maiorino, A.; Masselli, C. The environmental impact of solid-state materials working in an active caloric refrigerator compared to a vapor compression cooler. *Int. J. Heat Technol.* **2018**, *36*, 1155–1162. [CrossRef]

15. Coulomb, D. Air conditioning environmental challenges. *REHVA J.* **2015**, *4*, 30–34.

16. Li, Z.X.; Shahsavar, A.; Al-Rashed, A.A.; Kalbasi, R.; Afrand, M.; Talebizadehsardari, P. Multi-objective energy and exergy optimization of different configurations of hybrid earth-air heat exchanger and building integrated photovoltaic/thermal system. *Energy Convers. Manag.* **2019**, *195*, 1098–1110. [CrossRef]

17. Akhtari, M.R.; Shayegh, I.; Karimi, N. Techno-economic assessment and optimization of a hybrid renewable earth—Air heat exchanger coupled with electric boiler, hydrogen, wind and PV configurations. *Renew. Energy* **2020**, *148*, 839–851. [CrossRef]

18. Mahdavi, S.; Sarhaddi, F.; Hedayatizadeh, M. Energy/exergy based-evaluation of heating/cooling potential of PV/T and earth-air heat exchanger integration into a solar greenhouse. *Appl. Therm. Eng.* **2019**, *149*, 996–1007. [CrossRef]

19. Kurevija, T.; Vulin, D.; Krapec, V. Influence of undisturbed ground temperature and geothermal gradient on the sizing of borehole heat exchangers. In Proceedings of the World Renewable Energy Congress, Linkoping, Sweden, 8–11 May 2011; pp. 8–13.

20. Samuel, D.L.; Nagendra, S.S.; Maiya, M.P. Passive alternatives to mechanical air conditioning of building: A review. *Build. Environ.* **2013**, *66*, 54–64. [CrossRef]

21. Liu, Z.; Yu, Z.J.; Yang, T.; Roccamena, L.; Sun, P.; Li, S.; Zhang, G.; El Mankibi, M. Numerical modeling and parametric study of a vertical earth-to-air heat exchanger system. *Energy* **2019**, *172*, 220–231. [CrossRef]

22. Agrawal, K.K.; Agrawal, G.D.; Misra, R.; Bhardwaj, M.; Jamuwa, D.K. A review on effect of geometrical, flow and soil properties on the performance of Earth air tunnel heat exchanger. *Energy Build.* **2018**, *176*, 120–138. [CrossRef]

23. Agrawal, K.K.; Misra, R.; Agrawal, G.D.; Bhardwaj, M.; Jamuwa, D.K. The state of art on the applications, technology integration, and latest research trends of earth-air-heat exchanger system. *Geothermics* **2019**, *82*, 34–50. [CrossRef]

24. Selamat, S.; Miyara, A.; Kariya, K. Numerical study of horizontal ground heat exchangers for design optimization. *Renew. Energy* **2016**, *95*, 561–573. [CrossRef]

25. Mathur, A.; Mathur, S.; Agrawal, G.D.; Mathur, J. Comparative study of straight and spiral earth air tunnel heat exchanger system operated in cooling and heating modes. *Renew. Energy* **2017**, *108*, 474–487. [CrossRef]

26. Benrachi, N.; Ouzzane, M.; Smaili, A.; Lamarche, L.; Badache, M.; Maref, W. Numerical parametric study of a new earth-air heat exchanger configuration designed for hot and arid climates. *Int. J. Green Energy* **2020**, *17*, 115–126. [CrossRef]

27. Sodha, M.S.; Mahajan, U.; Sawhney, R.L. Thermal performance of a parallel earth air-pipes system. *Int. J. Energy Res.* **1994**, *18*, 437–447. [CrossRef]

28. Wu, H.; Wang, S.; Zhu, D. Modelling and evaluation of cooling capacity of earth–air–pipe systems. *Energy Convers. Manag.* **2007**, *48*, 1462–1471. [CrossRef]

29. Mihalakakou, G.; Santamouris, M.; Asimakopoulos, D. On the cooling potential of earth to air heat exchangers. *Energy Convers. Manag.* **1994**, *35*, 395–402. [CrossRef]

30. Mihalakakou, G.; Lewis, J.O.; Santamouris, M. On the heating potential of buried pipes techniques—Application in Ireland. *Energy Build.* **1996**, *24*, 19–25. [CrossRef]

31. Goswami, D.Y.; Biseli, K.M. Use of underground air tunnels for heating and cooling agricultural and residential buildings. *Fact Sheet EES* **1993**, *78*, 1–4.

32. Sehli, A.; Hasni, A.; Tamali, M. The Potential of Earth-air Heat Exchangers for Low Energy Cooling of Buildings in South Algeria. *Energy Procedia* **2012**, *18*, 496–506. [CrossRef]

33. Hanby, V.I.; Loveday, D.L.; Al-Ajmi, F. The optimal design for a ground cooling tube in a hot, arid climate. *Build. Serv. Eng. Res. Technol.* **2005**, *26*, 1–10. [CrossRef]

34. Niu, F.; Yu, Y.; Yu, D.; Li, H. Heat and mass transfer performance analysis and cooling capacity prediction of earth to air heat exchanger. *Appl. Energy* **2015**, *137*, 211–221. [CrossRef]

35. Derbel, H.B.J.; Kanoun, O. Investigation of the ground thermal potential in tunisia focused towards heating and cooling applications. *Appl. Therm. Eng.* **2010**, *30*, 1091–1100. [CrossRef]

36. Benhammou, M.; Draoui, B. Parametric study on thermal performance of earth-to-air heat exchanger used for cooling of buildings. *Renew. Sustain. Energy Rev.* **2015**, *44*, 348–355. [CrossRef]

37. Lee, K.H.; Strand, R.K. The cooling and heating potential of an earth tube system in buildings. *Energy Build.* **2008**, *40*, 486–494. [CrossRef]

38. Ahmed, S.F.; Amanullah, M.T.O.; Khan, M.M.K.; Rasul, M.G.; Hassan, N.M.S. Parametric study on thermal performance of horizontal earth pipe cooling system in summer. *Energy Convers. Manag.* **2016**, *114*, 324–337. [CrossRef]

39. Popiel, C.O.; Wojtkowiak, J.; Biernacka, B. Measurements of temperature distribution in ground. *Exp. Therm. Fluid Sci.* **2001**, *25*, 301–309. [CrossRef]

40. Sanusi, A.N.; Shao, L.; Ibrahim, N. Passive ground cooling system for low energy buildings in Malaysia (hot and humid climates). *Renew. Energy* **2013**, *49*, 193–196. [CrossRef]

41. Mihalakakou, G.; Santamouris, M.; Asimakopoulos, D.; Papanikolaou, N. Impact of ground cover on the efficiencies of earth-to-air heat exchangers. *Appl. Energy* **1994**, *48*, 19–32. [CrossRef]

42. Badescu, V. Simple and accurate model for the ground heat exchanger of a passive house. *Renew. Energy* **2007**, *32*, 845–855. [CrossRef]

43. Hermes, V.F.; Ramalho, J.V.A.; Rocha, L.A.O.; dos Santos, E.D.; Marques, W.C.; Costi, J.; Rodrigues, M.; Isoldi, L.A. Further realistic annual simulations of earth-air heat exchangers installations in a coastal city. *Sustain. Energy Technol. Assess.* **2020**, *37*, 100603. [CrossRef]

44. Bansal, V.; Misra, R.; Agrawal, G.D.; Mathur, J. Performance analysis of earth–pipe–air heat exchanger for winter heating. *Energy Build.* **2009**, *41*, 1151–1154. [CrossRef]

45. Bansal, V.; Misra, R.; Agrawal, G.D.; Mathur, J. Performance analysis of earth–pipe–air heat exchanger for summer cooling. *Energy Build.* **2010**, *42*, 645–648. [CrossRef]

46. American Society of Heating, Refrigerating and Air-Conditioning Engineers (ASHRAE). *ASHRAE Handbook Fundamentals*; SI Ed.; ASHRAE: Atlanta, GA, USA, 2017.

47. D'Agostino, D.; Greco, A.; Masselli, C.; Minichiello, F. The employment of an earth-to-air heat exchanger as pre-treating unit of an air conditioning system for energy saving: A comparison among different worldwide climatic zones. *Energy Build.* **2020**, *229*, 110517. [CrossRef]

48. Misra, R.; Jakhar, S.; Agrawal, K.K.; Sharma, S.; Jamuwa, D.K.; Soni, M.S.; Agrawal, G.D. Field investigations to determine the thermal performance of earth air tunnel heat exchanger with dry and wet soil: Energy and exergetic analysis. *Energy Build.* **2018**, *171*, 107–115. [CrossRef]

49. Hatraf, N.; Chabane, F.; Brima, A.; Moummi, N.; Moummi, A. Parametric Study of to Design an Earth to Air Heat Exchanger with Experimental Validation. *Eng. J.* **2014**, *18*, 41–54. [CrossRef]

50. Hasan, M.I.; Noori, S.W.; Shkarah, A.J. Parametric study on the performance of the earth-to-air heat exchanger for cooling and heating applications. *Heat Transf.-Asian Res.* **2019**, *48*, 1805–1829. [CrossRef]

51. Kusuda, T.; Achenbach, P.R. *Earth Temperature and Thermal Diffusivity at Selected Stations in the United States (No. NBS-8972)*; ASHRAE Transactions: Gaithersburg, MD, USA, 1965; Volume 71.

52. D'Agostino, D.; Esposito, F.; Greco, A.; Masselli, C.; Minichiello, F. The Energy Performances of a Ground-to-Air Heat Exchanger: A Comparison Among Köppen Climatic Areas. *Energies* **2020**, *13*, 2895. [CrossRef]

53. Chen, D.; Chen, H.W. Using the Köppen classification to quantify climate variation and change: An example for 1901–2010. *Environ. Dev.* **2013**, *6*, 69–79. [CrossRef]

54. D'Agostino, D.; Esposito, F.; Greco, A.; Masselli, C.; Minichiello, F. Parametric Analysis on an Earth-to-Air Heat Exchanger Employed in an Air Conditioning System. *Energies* **2020**, *13*, 2925. [CrossRef]

55. Everts, M.; Bhattacharyya, S.; Bashir, A.I.; Meyer, J.P. Heat transfer characteristics of assisting and opposing laminar flow through a vertical circular tube at low Reynolds numbers. *Appl. Therm. Eng.* **2020**, *179*, 115696. [CrossRef]

56. Bashir, A.I.; Everts, M.; Bennacer, R.; Meyer, J.P. Single-phase forced convection heat transfer and pressure drop in circular tubes in the laminar and transitional flow regimes. *Exp. Therm. Fluid Sci.* **2019**, *109*, 109891. [CrossRef]

57. Agrawal, K.K.; Misra, R.; Agrawal, G.D.; Bhardwaj, M.; Jamuwa, D.K. Effect of different design aspects of pipe for earth air tunnel heat exchanger system: A state of art. *Int. J. Green Energy* **2019**, *16*, 598–614. [CrossRef]

58. Bharadwaj, S.S.; Bansal, N.K. Temperature distribution inside ground for various surface conditions. *Build. Environ.* **1981**, *16*, 183–192. [CrossRef]

59. Bansal, N.K.; Sodha, M.S.; Bharadwaj, S.S. Performance of earth air tunnels. *Int. J. Energy Res.* **1983**, *7*, 333–345. [CrossRef]

60. Bansal, N.K.; Sodha, M.S. An earth-air tunnel system for cooling buildings. *Tunn. Undergr. Space Technol.* **1986**, *1*, 177–182. [CrossRef]

61. Wang, H.; Qi, C.; Wang, E.; Zhao, J. A case study of underground thermal storage in a solar-ground coupled heat pump system for residential buildings. *Renew. Energy* **2009**, *34*, 307–314. [CrossRef]

62. Mihalakakou, G.; Santamouris, M.; Asimakopoulos, D. Modelling the thermal performance of earth-to-air heat exchangers. *Sol. Energy* **1994**, *53*, 301–305. [CrossRef]

63. Mihalakakou, G.; Lewis, J.O.; Santamouris, M. The influence of different ground covers on the heating potential of earth-to-air heat exchangers. *Renew. Energy* **1996**, *7*, 33–46. [CrossRef]

64. Ghosal, M.K.; Tiwari, G.N. Modeling and parametric studies for thermal performance of an earth to air heat exchanger integrated with a greenhouse. *Energy Convers. Manag.* **2006**, *47*, 1779–1798. [CrossRef]

65. Kabashnikov, V.P.; Danilevskii, L.N.; Nekrasov, V.P.; Vityaz, I.P. Analytical and numerical investigation of the characteristics of a soil heat exchanger for ventilation systems. *Int. J. Heat Mass Transf.* **2002**, *45*, 2407–2418. [CrossRef]

Publisher's Note: MDPI stays neutral with regard to jurisdictional claims in published maps and institutional affiliations.

Article

Dynamic Simulation and Energy Economic Analysis of a Household Hybrid Ground-Solar-Wind System Using TRNSYS Software

Rafał Figaj, Maciej Żołądek * and Wojciech Goryl

Department of Sustainable Energy Development, Faculty of Energy and Fuels, AGH University of Science and Technology, 30059 Krakow, Poland; figaj@agh.edu.pl (R.F.); wgoryl@agh.edu.pl (W.G.)
* Correspondence: mzoladek@agh.edu.pl

Received: 18 June 2020; Accepted: 5 July 2020; Published: 8 July 2020

Abstract: The adoption of micro-scale renewable energy systems in the residential sector has started to be increasingly diffused in recent years. Among the possible systems, ground heat exchangers coupled with reversible heat pumps are an interesting solution for providing space heating and cooling to households. In this context, a possible hybridization of this technology with other renewable sources may lead to significant benefits in terms of energy performance and reduction of the dependency on conventional energy sources. However, the investigation of hybrid systems is not frequently addressed in the literature. The present paper presents a technical, energy, and economic analysis of a hybrid ground-solar-wind system, proving space heating/cooling, domestic hot water, and electrical energy for a household. The system includes vertical ground heat exchangers, a water–water reversible heat pump, photovoltaic/thermal collectors, and a wind turbine. The system with the building is modeled and dynamically simulated in the Transient System Simulation (TRNSYS) software. Daily dynamic operation of the system and the monthly and yearly results are analyzed. In addition, a parametric analysis is performed varying the solar field area and wind turbine power. The yearly results point out that the hybrid system, compared to a conventional system with natural gas boiler and electrical chiller, allows one to reduce the consumption of primary energy of 66.6%, and the production of electrical energy matches 68.6% of the user demand on a yearly basis. On the other hand, the economic results show that that system is not competitive with the conventional solution, because the simple pay back period is 21.6 years, due to the cost of the system components.

Keywords: household hybrid system; dynamic simulation; economic analysis; TRNSYS software

---

1. Introduction

Over the last few years, the main goal of worldwide energy policy is decreasing the use of fossil fuels, and, consequently, greenhouse gas emissions. Many changes are taking place to make possible the achievement of such a goal, especially in the form of policy of incentives for investments in renewable energy, scientific grants for actions in developing innovative energy systems, or requirements for energy-efficient buildings [1]. In this context, positive effects in the energy sectors may be achieved. In fact, due to policy, research, and incentive strategies, European total energy consumption in the last decade decreased by about 10%, to 1105 Mtoe in 2017. In general, all the energy-consumption sectors, such as the transport, industry and residential, are involved in the developed strategies. In the framework of energy use, about 25% of European final energy demand, 283 Mtoe, is used in the residential sector, especially for heating and cooling purposes [1].

Further decreasing demand in this sector is possible due to the development of energy-efficient buildings, especially nearly zero-energy buildings (NZEB) [2–4]. The most common energy sources for such buildings are photovoltaic (PV) and photovoltaic-thermal collector (PVT) systems [5–7], wind turbines [8,9] and heat pumps [10,11]. Nevertheless, efforts in reducing energy consumption lead also to the development of highly efficient hybrid and polygeneration systems [12].

Geothermal energy technologies, involving earth-air heat exchangers or electric heat pumps (EHP), can be coupled with other renewable energy devices to improve the efficiency of the whole system [11]. Li et al. [13] presented a parametric study of a standalone installation with a 7 kW wind turbine, a set of batteries, and a heat pump for a single-family house in Sweden. Dynamic simulations of such a system proved that wind energy, due to its intermittent characteristics, cannot fully satisfy the needs of the heat pump.

Roselli et al. [14] provided an analysis of ground source heat pump supported by a 5 kW wind turbine with battery storage. The model was prepared for 200 m^2 office and analyzed for two locations in Italy—Cagliari and Naples. The dynamic model developed in the Transient System Simulation (TRNSYS) software showed that the coefficient of performance (COP) and energy efficiency ratio (EER) for Naples were consequently 3.97 and 4.59. The primary reduction per kWh of final energy demand was 0.8 for Naples and 1.24 for Cagliari. The use of batteries allowed to lower the value of energy exported to the grid in the range of 27% to 63%, depending on the capacity. The fraction of the energy demand met by renewables was about 25% for Naples and 48% for Cagliari.

PV-based systems also are unsuitable to fully cover the electrical energy needs of heat pumps. Kemmler [15] provided a simulation of 4 reference buildings in Germany with a PV-EHP system, controlled by an algorithm that was maximizing the self-consumption of PV energy. The authors proved that 25.3–41.0% of the electricity demand may be covered with a PV system. This value increases in the case of adding batteries, but such a solution increases the payback time.

Psimopoulos et al. [16] evaluated the impact of advanced control strategies on energy and economic performance of a residential heat pump system with a photovoltaic field and electrical energy storage operating under a wide range of climate conditions by means of TRNSYS software. Presented results show that forecast-based control systems lead to greater final energy savings. The presented strategy allowed to reduce final energy use up to 842 kWh per year, which leads to about 175 EUR savings per year in Spain and Germany and about half of this value for Italy, France, and the United Kingdom.

Another trend in the hybridization of EHP systems consists in connecting them with phase-changing material thermal storage. The study reported in ref. [17] showed that in Italy, due to the use of PCM storage it was possible to reduce electric energy consumption by 18%, thanks to improved COP in terms of heating (increase from 3.5 to 4.13) and cooling (increase from 4.0 to 5.9).

Another interesting concept that may be used with EHP installation is the installation of PVT collectors. Such systems provide electricity for operating the heat pump and also generating heat with high efficiency, allowing to lower the electrical energy consumption of ECH needed to match the thermal demand. Research presented in reference [18] showed that the installation of PVT collectors in Italy leads to a reduction of the non-renewable primary energy requirements of buildings, nonetheless, it is connected with higher investment costs.

Most of the literature studies describe EHP driven by renewable energy provided only by one source, like wind turbines or PV fields. However, those energy resources are mostly available in different seasons—solar energy abounds in summer months, and wind energy may be available during the whole year or seasonally, depending on local wind resource conditions. Connecting those technologies in one hybrid installation may provide benefits in terms of matching a significant part of electrical energy demand with renewables.

Ozgener provided an experimental study of a solar-assisted geothermal heat pump connected with a 1.5 kW wind turbine to meet the heating demand of a greenhouse in Izmir, Turkey. Only about 3% of energy consumption was provided by the turbine. The author concluded that such a system could be economically viable in areas with higher wind resources, and the results of the presented study could be extended to the residential sector [19].

Vanhoudt et al. [20] described an installation based on EHP for a residential building located in Belgium. The system consisted of 7.7 kW of PV panels and a 5.8 kW wind turbine. Authors proved that a dynamic control strategy allows us to decrease the one percent peak power demand by up to 5%, and the average power demand about 17%. The self-consumption of renewable energy increased by about 20%. Due to frequent switching, energy consumption increased by 8–12%.

Stanek et al. [21] proposed an energy and thermo-ecological analysis to evaluate the impact of a 3.5 kW PV field and small (3 kW) wind turbine on the operation of EHP in a 163 m^2 residential building in Katowice, Poland. Authors considered a combination of operation of both systems, a wind-turbine (WT) system only and three configurations of PV fields. The average exergy efficiency of wind-turbine and PV panels was equal to 25.9% and 12.94% respectively, while the value for system power plants was equal to 33.67%. Average thermo-ecological costs for EHP installation supported by a wind turbine and grid, PV field and grid and wind turbine, PV field, and grid were, consequently, 2.14, 2.16, 1.24.

Li et al. [22] presented an analysis of a residential building HP system with a 5 kW wind turbine and flat-plate solar collectors. Collectors were used for heating of domestic hot water and increasing heat pump evaporation temperature for room heating. Wind power was used for meeting the heat pump power demand. Wind turbines provided about 7.5% of the yearly power demand of heat pump to satisfy thermal load of 198 m^2 residential building located in Beijing.

Rivoire et al. [23] analyzed the application of a ground-coupled heat pump (GCHP) in different buildings and climates. The highest profitability of such systems was found for poorly insulated buildings in cold climates, because of its high demands (8.6–9.9 years). However, a balance between heating and cooling needs was found in temperate climate zones for highly insulated buildings. This parameter is important for a ground heat pump (GHP), because of reducing the thermal imbalance of the ground. Authors proved that HP with a capacity of 60% of peak load meets 82–96% of the annual demand. The reduction of primary energy consumption was 33–75% and CO_2 emission was reduced by 27–56%.

A numerical study taken for the tropical environment in Thailand [24] proved, that using a ground-source heating pump (GSHP) only for cooling purposes allows us to achieve about 40% energy savings. Analysis was performed for three types of administration buildings (about 9 working hours per day, 200, 40, and 25 m^2). Electricity consumption for air conditioning in such buildings was 29,835, 4874, 3372 kWh/year respectively. The authors concluded, that GSHP may be feasible in Thailand's condition, however, there are problems with installing the required amount of borehole heat exchangers. As a solution, the authors proposed installing them in car parks.

Another study for Thailand described the installation of GSHP with horizontal heat exchangers [25]. The authors compared experimental data from two months of operation for the GSHP and air-source heat pump (ASHP). It was found that the GSHP consumed about 19% less electricity than the ASHP system. Comparing those two systems it was proved that the reduction of CO_2 emission in the case of GSHP was about 3000 kg. Because of the high initial cost, such installation was found to be unprofitable, however, the growth of electricity costs may significantly increase this parameter.

The literature analysis shows that existing works are mainly dedicated to the description and investigation of systems with one renewable energy source connected to electric ground source heat pumps. Applications with hybridization of energy sources in GHP systems are not so numerous. This paper presents the numerical analysis of a hybrid HP system driven by renewable electrical energy provided by PVT collectors and a wind turbine. To the authors' best knowledge, this is the first time in literature that a system like the one considered in this paper is comprehensively investigated. In fact, there are no papers available in literature dealing with the proposed system configuration, detailed simulation of the system operation and adoption of realistic user demand. The presented control strategy leads to a maximum reduction of non-renewable primary energy requirements. Analysis was provided for a year's operation of the system, considering the energy and economic aspects of the presented system. Furthermore, a parametric analysis of the system is performed, where solar field extension and wind turbine power are varied to investigate the effect of these parameters on the system performance.

2. Methodology

The proposed hybrid geothermal-solar-wind system was modelled and dynamically simulated adopting TRNSYS software [26]. The selected tool is widely used in commercial and scientific applications to perform advanced multipurpose analyses of complex, hybrid and novel energy systems based on close to the reality simulations [27,28]. The software comprises a vast built-in library of components, spacing between different categories of equipment (heating cooling and ventilation, solar devices, etc.), models of which are validated experimentally and/or are intrinsically validated because they are based on manufacturer data [26]. Therefore, the simulation of systems developed within TRNSYS software carries out reliable results.

In the software used, selected components and connections among them were used to develop the system structure in a user-friendly graphical environment. In particular, the modelling and the simulation of the system were performed using both built-in library components (e.g., ground heat exchanger, photovoltaic thermal collectors, wind turbine, etc.) and user-defined components (control system, energy and economic model). As regards the building, Type 56 was used to simulate its thermal behavior, while the SketchUp TRNSYS3d plug-in [29] was used to implement the geometric structure of the building (thermal model).

The list of the built-in components used to develop the model of the system is shown in Table 1. In this section, only a brief introduction of the main system components models was provided, since the detailed presentation of the models is available in the software reference [26]. The model of the fan-coil operating in heating model was taken from the literature [30], as for the heat pump, manufacturer data (Aermec WRL 026/161 [31]) was used based on the approach reported in [32].

Table 1. List of the built-in components used to develop the hybrid system.

Component	Type	Component	Type
weather data reader	15–2, 109	domestic hot water profile	14h
photovoltaic/thermal collector	50b	inverter controller	48b
ground heat exchanger	557a	virtual storage	47a
storage tanks	534, 340, 60f	building	56
natural gas boiler	751	schedulers	519a, 517, 516, 515
pipes	31	psychrometric calculator	33
pumps	110, 3b	winter fan coil	508a
diverters	11f, 649	on-off hysteresis controller	2b
mixers	11d, 11h, 647	proportional controller	1669
tempering valve	11b	integrators	24
reversible heat pump	927	plotters	65d, 65c
wind turbine	90	printers	25c

2.1. Layout and Operation Strategy of the System

The proposed system includes vertical ground heat exchangers coupled with a reversible heat pump for space heating and cooling purposes, and a photovoltaic/thermal solar collector field assisting the ground heat exchanger in the heating operation during winter and preheating the domestic hot water all year long. The solar field and the micro wind turbine are used for the generation of electrical energy, matching in part the demand of the user.

The basic layout of the system, including all the main components and loops, is shown in Figure 1.

The loops of the system were designed to manage adequately the thermal energy flows in order to match the thermal energy demand of the user. The loops consist of:

-   solar fluid (SF) consisting of a 40% propylene glycol-water mixture used to cool the photovoltaic/thermal collector and supply thermal energy to assist the heating of the ground heat exchanger tank and to preheat the domestic hot water;

- ground heat exchanger fluid (GF) consisting of the same glycol-water mixture of SF and used to extract or dissipate the heat form or to ground, store the thermal energy and supply the source side of the water-to-water heat pump;
- heating and cooling water (HCW) used as working fluid at the load side of the heat pump and supplied to the user fan-coil system;
- gas boiler water (GBW) used to transfer the heat from the natural gas boiler to the domestic hot water tank;
- Aqueduct water (AW), fresh water used for sanitary purposes;
- domestic hot water (DHW), water heated by the solar energy and natural gas boiler operation supplied to the user;
- electrical energy (EE) produced by the photovoltaic cells and wind turbine in order to match a part of the user demand.

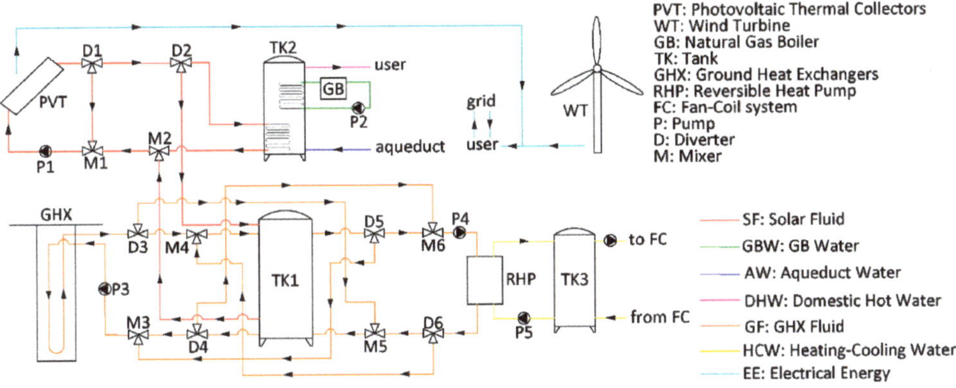

Figure 1. Layout of the system with main loops and components.

The devices that were integrated in the system layout are the following:

- ground heat exchangers (GHX) consisting of 2 vertical U-tube shape polyethylene heat exchanges arranged in parallel, used to extract and dissipate the thermal energy during the winter and summer operation of the heat pump;
- photovoltaic/thermal collectors (PVT) consisting of one glazed flat plate units integrating monocrystalline photovoltaic cells;
- wind turbine (WT), a micro-scale wind turbine
- reversible heat pump (RHP), a water-to-water unit single-stage unit allowing to modulate the thermal power;
- thermal storage tank (TK1), consisting of a thermally stratified unit allowing to store the thermal energy produced by GHX and PVT during winter used to supply RHP, and to buffer the heat rejected by the heat pump during summer before the dissipation by GHX;
- domestic hot water tank (TK2), consisting of a tank with two internal heat exchangers used to heat the water by the solar loop (HX1) and by the natural gas boiler (HX2);
- natural gas boiler (GB) used as an auxiliary device producing DHW by means of TK2 in case of scarce availability of solar energy;
- hydronic system buffer tank (TK3), consisting of a double inlet-output storage used as inertial tank for fan-coil the heating and cooling system;
- fan-coil system (FC), consisting a water to air units providing space heating and cooling for the rooms of the building, supplied by heated or cooled water produced by RHP.

The system layout included also several devices in order to manage properly the fluid loops, as diverters (D), mixers (M), and single (P2, P3, P4, and P5) and variable speed (P1) pumps. It is worth noting that the layout presented in Figure 1, is a simplified version of the system implemented in the simulation environment, since this one includes also other components, like controllers, weather database and components generating the results of the simulation as plotters, variable integrators and printers. The controller components were coupled together in order to develop an efficient control strategy of the system, described as follows.

In the heating season, the ground heat exchanger heated the working fluid provided by the bottom of the storage tank. The heated fluid was supplied to TK1 by means of a stratification supply system that allows one to supply the fluid to the tank to the node with the temperature closest to the entering fluid. In this way, the fluid with a higher temperature was supplied to the top part of the tank while the fluid with a lower temperature was supplied to the bottom part, ensuring thermal stratification of the storage. From top of TK1, the mixture of glycol and water was supplied to the source side of the reversible heat pump, which operated in order to keep the temperature of the buffer tank TK3 within a fixed dead band. RHP was activated when the temperature inside the tank drops to 35 °C while it was turned off when the temperature increased to 38 °C, in order to avoid an unnecessarily heating of TK2. Moreover, during the heating operation, the heating power of RHP was modulated by means of a proportional controller as a function of the tank temperature. The control strategy was developed to set the power of RHP to 100% when a temperature of 35 °C inside the TK3 tank was reached, while it was set to 20% for a temperature of 37 °C. It is worth noting that the activation of the pump dedicated to GHX was activated only when RHP was turned on in order to avoid an unnecessarily electrical energy consumption, and this control strategy was performed during both heating and cooling periods.

RHP ensured a proper operation of TK3 in terms of temperature required by the fan-coil system for the space heating, with the last one activated in order to ensure a temperature of the air inside the building rooms between 20 and 22 °C.

During the cooling season, the RHP was activated in a chilling mode in order the keep the TK3 temperature between 10 and 7 °C, and similarly in the winter period the unit cooling power was modulated proportionally the TK3 temperature. In fact, the load factor of RHP was set to 100% when a temperature of 10 °C inside the TK3 tank was reached, while it was set to 20% for a temperature of 8 °C. During the operation of RHP in cooling mode, the heat rejected by the condenser of the unit was supplied by means of GF to the top of TK1, and from there was dissipated by GHX.

The operation of the loops at the load and source side of TK1 were managed by a diverter and mixing valve system, consisting of D3–D6 diverters and M3–M6 mixers. In particular, the valves were used to supply GHX with the fluid stored in the bottom part of TK1 (lowest temperature) and to supply RHP with the fluid stored in the top part of TK1 (highest temperature) during winter. The same system was used in summer in order to supply from the bottom part of TK1 tank the fluid entering the condenser of RHP and to supply GHX with the fluid stored in the top part. The adoption of this valve system allowed one to operate both GHX and RHP with in a proper way from the level of temperature point of view.

The photovoltaic/thermal collectors were used all year long to heat the water stored in the bottom part of TK3, in order to produce DHW, while during winter they were used also to heat TK1 aiding the space heating operation performed by RHP. The pump of the solar collectors was activated when the solar radiation increases above 100 W/m^2 in order to avoid a possible cooling of the fluid flowing within the collectors. Moreover, in order to avoid a possible cooling of TK1, during winter, and TK2, all year long, a by-pass consisting of D1 and M1 valves was adopted. On the basis of which the tank was supplied by solar collector, the control strategy checked continuously that the fluid temperature exiting the collectors was at least 2 °C higher than the one inside the tanks before supplying it to the storage. If such a condition was not met, SF was recirculated within the collectors and M1 and D1 valves. In detail, in this control strategy, the solar collectors' outlet temperature was compared to the top temperature of TK1 and to the water temperature inside TK2 at the level of the internal heat exchanger supplied by the solar loop. The setpoint temperature of the solar collectors was set by means of the operation of the variable speed pump and depended on the tank to be loaded: 30 °C for TK1 and 60 °C for TK2.

The heating operation of TK2 by the solar system was set to be performed until the water near the bottom internal heat exchanger of the tank reached the setpoint value of 55 °C. In case the DHW demand is not met by solar thermal energy, GB is activated. In particular, then the TK2 top temperature drops to 50 °C GB is turned on in order to heat the water up to 55 °C. The auxiliary operation of GB allows one to match the user DHW independently from the availability of solar energy.

The solar loop operation during winter was designed to heat primarily TK3 and secondarily TK1. The heating of TK1 was allowed only when the temperature of the fluid exiting the solar collectors was not enough high to heat the water near the lower internal exchanger of TK2 (SF temperature 2 °C higher than that of the water). Once the temperature of the fluid exiting PVT increases above the threshold relative to the temperature of TK2, the heating of the DHW is allowed.

Finally, the electrical energy produced by the PVT and WT is supplied to the user or to the grid depending on the demand. The grid allows one to store virtually the energy produced in excess and use it part when the demand overcomes the production.

2.2. Ground Heat Exchanger

The component has been implemented using the Type557 model, simulating a vertical heat exchanger that transfers heat with the ground. The adopted model allowed one to simulate both U-tube ground heat exchanger or a concentric tube ground heat exchanger by means of a subroutine. Nevertheless, for the purposes of this study, the U-tube sub-model was adopted.

The component simulated both heating and cooling operation of the heat exchanger, since the working fluid may absorb heat from or reject it to the ground as a function of the temperatures of the fluid and the ground surrounding the heat exchanger.

The adopted model was based on the following assumptions:

- uniform placement of the boreholes within a cylindrical storage volume of ground;
- conductive heat transfer between the heat exchanger and the storage volume, and convective heat transfer between the fluid and the pipes;
- adoption of global temperature, local solution, and steady-flux solution methods for the calculation of the ground temperature, where an explicit finite difference method was used for the global and local problems, while an analytical approach was used for the calculation of the steady flux solution. The approaches were coupled together by means of a superposition method in order to calculate the temperature within the ground storage.

The detailed description of the model is here omitted fake of brevity. Its comprehensive presentation is available in reference [33].

2.3. Phototoltaic/Thermal Collectors

The solar field is equipped with flat-plate photovoltaic/thermal solar collectors, consisting of a conventional solar thermal device with an absorber covered by a PV cell layer encapsulated within a transparent protective layer, typical of PV panels. The model of the PVT collector was based on the flat-plate thermal collector mathematical model modified in order to take into account the presence of the PV cell panel. In detail, the TRNSYS model Type50 was used for the simulation of the solar field, where constant values for the overall energy loss coefficient, the glass transmittance, and the absorbance of the absorber were assumed. The model was based on several equations, and the most important ones have been presented below.

The modified Hottel–Willier–Bliss method [34] and an energy balance was adopted in order to calculate the PV cell temperature (T_{cell}). The overall thermal loss coefficient of the collector per unit area (U_l^*), overall collector heat removal efficiency factor (FR) and the PVT thermal power output (Q_{PVT}) were calculated with the following equations:

$$U_l^* = U_l - \tau_g I_{tot} \eta_{PVT}(T_{cell} - T_{amb})$$ (1)

$$FR = m_f c_f \frac{1 - e^{\frac{f_p U_l^* A_{PVT}}{m_f c_f}}}{U_l^* A_{PVT}}$$ (2)

$$Q_{PVT} = A_{PVT} I_{tot} FR \tau_g \alpha_{PVT} U_l^*(T_{f,in} - T_{amb})$$ (3)

where I_{tot} is the solar radiation, τ_g is the glass transmission coefficient, f_p is the efficiency factor, m_f is the mass flow rate, c_f is the fluid specific heat, A_{PVT} is the area of PVT, α_{PVT} is the PVT absorption coefficient $T_{f,in}$ is the fluid inlet temperature and T_{amb} is the ambient temperature. The PVT electrical efficiency (η_{PVT}) was calculated on the basis of the temperature of the photovoltaic cell:

$$\eta_{PVT} = \eta_{PVT,ref}\left[1 - \beta_{PVT}(T_{cell} - T_{ref})\right]$$ (4)

where $\eta_{PVT,ref}$ is the reference PVT electrical efficiency, β_{PVT} is the efficiency temperature coefficient, T_{cell} is the photovoltaic cell temperature and T_{ref} is the reference temperature. In addition, energy balances of the collector were used to calculate the electric power and the outlet temperature of the fluid.

2.4. Wind Turbine

The built-in library TRNSYS component Type90 was used to simulate the micro-wind turbine. The adopted model calculated the power produced (P_{WT}) as a function of air density (ρ), power coefficient (c_p), rotor area (A_r) and wind speed (v), as follows:

$$P_{WT} = \frac{1}{2}\rho_{air} c_p A_r v^3$$ (5)

where c_p is dependent on the axial induction factor (a) according to the following equation:

$$c_p = 4a(1 - a)^2$$ (6)

When a reaches the value of 1/3, c_p reaches the maximum theoretical value of 0.593, according to the Betz limit [35].

The effect of the variation of the height above the ground of the site of WT installation was taken into account by the model in terms of the air density variation and the increase of wind speed. The density of air as a function of the height (z) was calculated with the ideal gas law where temperature (T_z) and pressure (p_z) are considered:

$$\rho_z = \frac{p_z}{RT_z}$$ (7)

T_z is determined on the basis of the vertical gradient of temperature ("lapse rate" [36]), as follows:

$$T_z = T_0 - Bz$$ (8)

where T_0 was assumed constant to 15 °C and B was equal to 6.5 °C/km [37] Moreover, the pressure p_z was calculated as follows:

$$p_z = p_0\left(1 - \frac{Bz}{T_0}\right)^{5.26}$$ (9)

where the reference pressure (p_0) is provided on the basis of dynamic local conditions. The wind speed (v) variation with the elevation was evaluated on the basis of Von Karman analysis [38]:

$$\frac{v_1}{v_2} = \left(\frac{z_1}{z_2}\right)^\alpha$$ (10)

where α is the wind shear exponent, dependent on the site location and its dynamic conditions like atmospheric conditions, presence of objects obstructing the airflow, roughness of the ground surface, etc. In this study, the shear exponent was fixed to 0.14, a value that characterized ideal boundary layer conditions. During the simulation, Equation (10) was used to calculate the wind speed at a certain elevation taking into account time-dependent wind velocities provided at ground level by the Meteonorm weather database included in TRNSYS.

Besides the mathematical formulation of the model, the component in order to calculate the power produced by WT required an external file, where the geometry of the wind turbine as well as the characteristic curve of power as a function of the wind speed were provided. In order to implement such manufacturer data, the data from the datasheet of an ENAIR 30PRO unit [39] were considered. In order to scale the wind-power characteristic, the manufacturer data was normalized with respect to the nominal power.

2.5. Energy and Economic Model

In order to assess the global energy and economic performance of the hybrid system (HS), a comparison was performed assuming a conventional system (CS) accordingly to the methodology reported in reference [27,28]. This approach was adopted since without the adoption of a reference, the energy and economic performance of the hybrid could only have been evaluated in absolute terms, with are not exhaustive for a comprehensive analysis of the system. Using this approach, it was assumed that both HS and CS must supply the same amount of final energy to the user, in terms of space heating and cooling, domestic hot water and electrical energy.

In detail, the reference system consists of:

- a natural gas boiler for the production of heat for DHW and space heating, with a seasonal efficiency ($\eta_{GB,ref}$) of 0.85 [40];
- an electrical chiller providing space cooling, with a seasonal coefficient of performance ($COP_{cool,ref}$) of 3.0 [41];
- the electric grid providing electrical energy, with a primary energy efficiency ($\eta_{el,ref}$) of 0.33 [40].

HS energy performance was evaluated on the basis of a primary energy (PE) consumption comparison on a yearly basis with respect to CS. PE consumption of CS was calculated as follows:

$$PE_{CS} = \frac{E_{th,heating} + E_{th,DHW}}{\eta_{GB,ref}} + \frac{E_{th,cooling}}{COP_{cool,ref}\,\eta_{el,ref}} + \frac{E_{el,user}}{\eta_{el,ref}} \tag{11}$$

where $E_{th,heating}$ and $E_{th,cooling}$ was the yearly thermal energy provided for space heating and cooling, respectively, and $E_{el,user}$ was the yearly electrical energy demand of the user.

The equation used to calculate the PE consumption of HS was:

$$PE_{HS} = \frac{E_{th,GB,DHW}}{\eta_{GB,ref}} + \frac{E_{el,grid,net} - E_{el,unrecovered,grid}}{\eta_{el,ref}} \tag{12}$$

where $E_{thGB,DHW}$ is the auxiliary thermal energy supplied by GB for the production of DHW and $E_{el,grid,net}$ is the energy electrical energy provided by the grid, excluding the electrical energy supplied to and recovered from the grid (net metering) and the $E_{el,unrecovered,grid}$ is the electrical energy supplied to the grid and not recovered by the user. On the basis of primary energy consumption, the primary energy saving ratio ($PESr$) was evaluated as follows:

$$PESr = \frac{PE_{CS} - PE_{HS}}{PE_{CS}} \tag{13}$$

The economic performance of HS was evaluated taking into account its investment costs and the operating costs of both HS and CS. In particular, the literature and market cost data were used to determine the capital cost of proposed system components, in accordance with the procedure adopted in reference [28]. In particular, the following cost function were adopted:

$$C_{GHX} = 20 \cdot L_{GHX} \tag{14}$$

$$C_{PVT} = 400 \cdot A_{PVT} \tag{15}$$

$$C_{WT} = c_{WT}P_{WT} = 4298.75P_{WT}^{-0.141} \cdot P_{WT} \tag{16}$$

$$C_{RHP} = 4.7108Q_{nom,heat,RHP}^2 + 139.69Q_{nom,heat,RHP} + 3845.7 \tag{17}$$

$$C_{TK1} = 494.9 + 808V_{TK1} \tag{18}$$

where L_{GHX} is the length of the vertical ground heat exchangers, A_{PVT} is the area of the PVT collector field, P_{WT} is the nominal power of WT, $Q_{nom,heat,RHP}$ is the nominal thermal power of RHP and V_{TK1} is the volume of TK1. It is worth noting that the costs of the other tanks (TK2 and TK3) were not taken into account since the devices were assumed to be already present in CS.

The total cost of HS was evaluated summing the cost evaluated by Equations (13)–(17) and adding 10% in order to take into account for the balance of plant components like valves, pumps, etc.

The operation costs for both HS and CS were calculated assuming constant electrical energy tariff (c_{el}) of 0.133 €/kWh and natural gas price (C_{NG}) of 0.50 €/m^3 characterized by a reference low heating value (LHV_{NG}) of 9.59 kWh/m^3 [40,42].

For HS, a net metering system, thus the possibility to store and utilize on demand the electrical energy produced in excess, was implemented by means of a bidirectional connection with the grid. In particular, it is assumed that in case of a system with nominal power up to 10 kW it is possible to recover freely up to 80% of the energy supplied to the grid, while the eventual quote exceeding 80% is paid with a mean price for electrical energy. In the case of a nominal power between 10 and 40 kW, the limit is lowered to 70%. This net metering method is adopted in Poland in order to increase the share of small-scale renewable systems in the electrical energy market [43].

Therefore, the operation cost of HS ($C_{op,HS}$) and CS ($C_{op,CS}$) and the savings ($\Delta C_{op,HS}$) were calculated as follows:

$$C_{op,HS} = \frac{E_{th,GB,DHW}}{\eta_{GB,ref}LHV_{NG}}c_{NG} + \left(E_{el,grid,net} + E_{el,recovered,limit,grid}\right)c_{el} \tag{19}$$

$$C_{op,CS} = \frac{E_{th,heating} + E_{th,DHW}}{\eta_{GB,ref}LHV_{NG}}c_{NG} + \left(\frac{E_{th,cooling}}{COP_{cool,ref}} + E_{el,user}\right)c_{el} \tag{20}$$

$$\Delta C_{op,HS} = C_{op,CS} - C_{op,HS} \tag{21}$$

where $E_{el,recovered,grid}$ was the electrical energy produced by the system and recovered from the grid above the limit established by the net metering prosumer contract (70 or 80% of the energy sent to the grid)

Finally, the simple pay back (SPB) index was introduced in the economic model in order to evaluate in a simple way the performance of the system in terms of profitability.

3. Case Study

The proposed system was investigated under a case study consisting of a single-family household with a ground floor, an attic and a sloped roof. This building as case study was used in a previous paper of the authors [28]. However, here it was adopted under different climatic conditions from those used in the previous paper, and with a dynamic electrical energy demand of the user and the simulation of the heating load. The building structure is reported in Figure 2.

The ground floor area was 100 m^2 with a floor height of 2.70 m, while the attic had the same height as the ground floor and a useful floor area of 75 m^2. The roof slope was 30°. The ground floor consisted of two rooms with an area of 25 m^2 and a room of 50 m^2. The components of the building envelope, like walls, roof, and floor, were implemented assuming several series of layers with a global transmittance reported in Table 2. The detailed structure of the envelope elements has not been reported here for reasons of brevity.

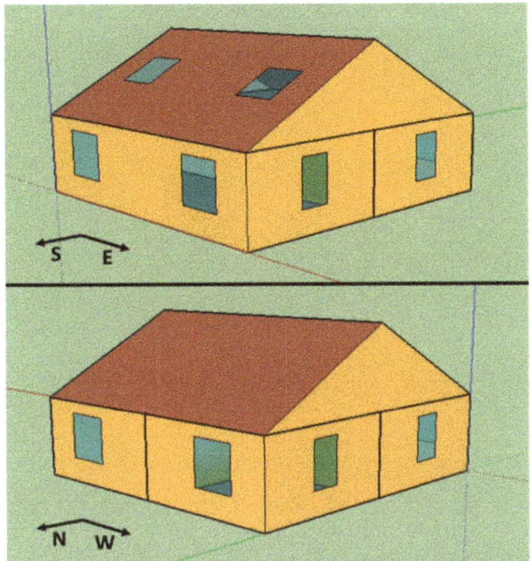

Figure 2. Structure of the case study building.

Table 2. Building elements reference transmittances [44].

Building Envelop Element	$W/(m^2K)$
External window	1.10
External wall	0.40
Adjacent wall	2.20
Ceiling	1.58
Roof	0.32
Ground floor	0.37

The climatic conditions of Gdansk, Northern Poland, were selected in order to simulate both the system and building operation. For this purpose, the Meteonorm weather database was used, implementing in the simulation a TMY2 file where the mean yearly weather conditions determined on the basis of a 10-year long period were stored. The mean hourly air temperature, horizontal solar radiation, and wind speed at ground level are shown in Figure 3.

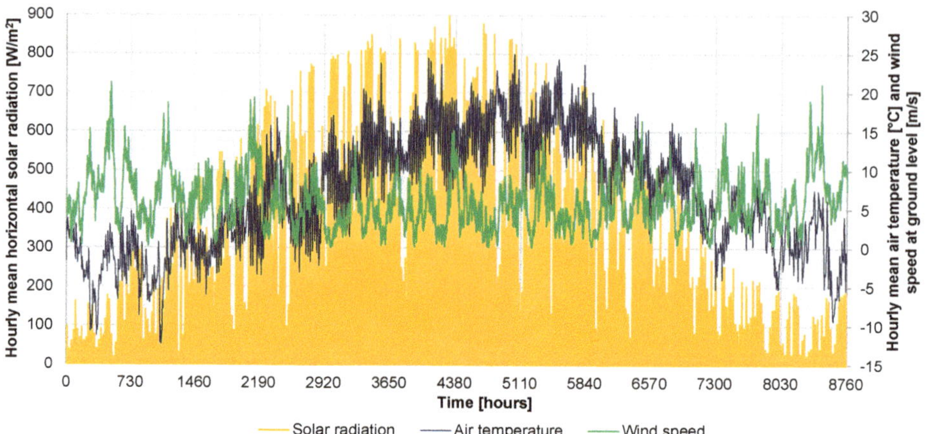

Figure 3. The mean hourly air temperature, horizontal solar radiation, and wind speed at ground level for Gdansk locality.

The air-conditioning system of the building integrates an independent fan coil unit for each zone of the house providing space heating and cooling. The space heating period was set from 15 November to 31 March, and the cooling one from 1 May to 15 October. The heating operation was allowed 24 h per day in order to keep the room temperature between 20 and 22 °C, while space cooling was performed between 8:00 am to 10:00 pm in order to maintain the air temperature between 24 and 26 °C.

The model of the building was completed by adding typical loads of a residential house, in order to simulate reliable thermal gains and losses. In particular, the following loads were considered: human activity (5 house inhabitants—sensible heat 75 W, latent heat 75 W), equipment (3.3 W/m^2), lights (5.0 W/m^2), fresh air infiltration (0.25 Vol/h). In addition, a dynamic DHW demand profile was implemented assuming that each person consumes 60 dm^3 of hot water consumption a day.

The electrical energy demand of the user was normalized with respect to the daily demand, and it is reported in Figure 4 for different days and periods of the year. The profiles were implemented for different days, workdays, Saturdays, Sundays/Holidays, for two periods in the year, period 1 and period 2 assumed from November to March and from April to October, respectively. In particular, the profiles were developed on the basis of standard electrical energy consumption data available for users as the one here investigated [45] performing a normalization of the hourly data on the basis of daily energy consumption (sum of the hourly consumption). Therefore, the profiles developed were characterized by an integral over 24 h equal to 1.0, thus, in this way it was possible to scale the normalized profiles in order to achieve a fixed annual electrical energy consumption for the user. The yearly electrical energy consumption of the building was set to 5.0 MWh, according to common energy demand levels for this kind of utility [45].

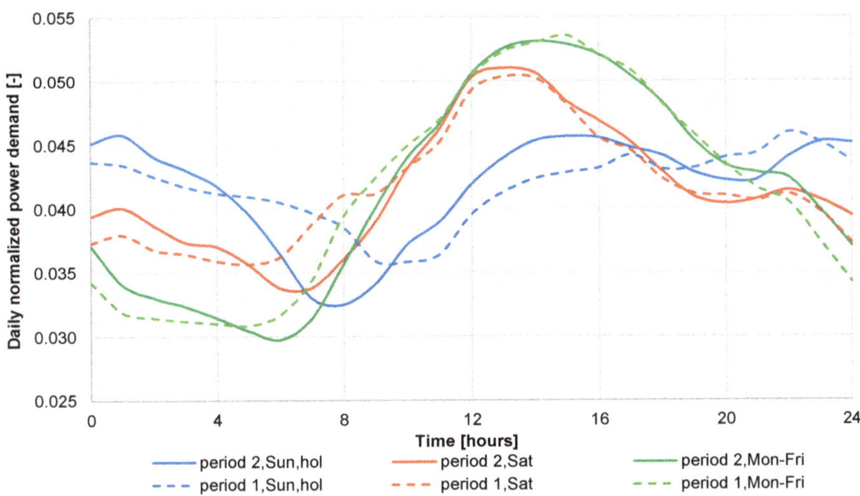

Figure 4. Normalized daily electrical energy demand of the user for different year periods and types of day.

The main design and operating parameters of the proposed system components are reported in Table 3. Such parameters were selected on the basis of manufacturer data in order to allows the system a proper operation in terms of the capability of energy transfer, number of activation-deactivation cycles of the components, and to match the user thermal demand.

Table 3. Parameters of the main system components.

Component	Parameter	Value	Unit
PVT	Area	11.07	m^2
	Slope	30	$^\circ$
	Reference electrical efficiency	0.19	-
	Efficiency of the inverter regulation system	0.90	-
	Electrical efficiency temperature coefficient	0.0035	1/K
	PV packaging factor	0.95	-
	Loss coefficient for bottom and edge	4.4	$W/(m^2{\cdot}K)$
	P1 mass flow rate per unit of PVT area	50	$kg/(h{\cdot}m^2)$
GHX	Number of boreholes in parallel	2	-
	Length	75	m
	Borehole radius	0.2	m
	Storage thermal conductivity	2.0	$W/(m{\cdot}K)$
	Storage heat capacity	2400	$kJ/(m^3{\cdot}K)$
	Storage volume	4329.5	m^3
	Outer radius of U-tube pipe	0.035	m
	Inner radius of U-tube pipe	0.030	m
	Center-to-center half distance	0.050	m
	Pipe thermal conductivity	4.0	$W/(m{\cdot}K)$
	P3 mass flow rate	2190	kg/h

Table 3. *Cont.*

Component	Parameter	Value	Unit
WT	Wind data collection height	5.0	m
	Hub height	8.0	m
	Number of wind turbines	1	-
	Site shear exponent	0.14	-
	Nominal power	1.0	kW
	Wind turbine rated wind speed	11	m/s
	Cut in and cut off wind speed	3 and 21	m/s
	Rotor diameter	2.4	m
TK1	Volume	1.0	m^3
	Height	1.5	m
	Heat loss coefficient	0.83	$W/(m^2 \cdot K)$
	Fluid specific heat	3.65	$kJ/(kg \cdot K)$
	Fluid density	1063	kg/m^3
TK2	Volume	0.3	m^3
	Height	1.2	m
	Heat loss coefficient	0.83	$W/(m^2 \cdot K)$
GB	Nominal capacity	30	kW
	Outlet set-point temperature	70	°C
	P2 mass flow rate	2400	kg/h
RHP	Nominal heating capacity	5.0	kW
	Nominal COP in heating mode	3.86	-
	Nominal cooling capacity	4.34	kW
	Nominal COP in cooling mode	4.30	-
	P4 mass flow rate	1100	kg/h
	P5 mass flow rate	870	kg/h

4. Results and Discussion

The dynamic simulation of the system was carried out for a one-year basis, from 0 to 8760 h. with a 0.04 h time step, calculating temperature and power trends and integrated variables. Under these simulation parameters, a large amount of dynamic data was generated by the simulation, thus for reasons of brevity, only the most important results were reported in this paper. In detail, the dynamic operation of the system is shown for a typical winter operation day in terms of temperature and power trends, while the behavior of the system over the yearly operation was presented on monthly basis. Finally, the yearly results are presented showing the global energy and economic performance. The system is also investigated changing the area of the PVT collector and the nominal power of the wind turbine, remaining the other parameters unchanged. In fact, the dimensioning of the ground heat pump system was performed on the basis of space heating and cooling demand of the user, which is constant under the adopted assumptions.

4.1. Daily Operation of the System in Winter

The dynamic operation of the system was analyzed for all the days of year, nonetheless due to the high amount of data this analysis cannot be presented here; thus, only the results for a winter operation day have been reported. The selected day was 18 February, falling between 1176th and 1200th hour of the year, which presented the full operation characteristic of the system operation from the point of view devices activation.

The most important temperatures of the system have been shown in Figures 5 and 6. During the firsts hours of the day, both inlet and outlet temperature of GHX (GHX, in and GHX, out) slowly decreases because the heat pump is in operation and this determines a decrease of the temperature at the bottom of TK1, which supplies GHX. The decrease is also due to the fact that thermal energy produced by the PVT during the previous day slightly increased the temperature of the GHX loop.

From 0:00 am to about 8:00 am the heat pump is supplied with GF at a temperature about 3 °C (TK1, out), while the temperature exiting the evaporator of RHP (RHP, source, out) oscillates slightly above 0 °C, due to the supplied heat to run the heat pump. After 8:00 am, PVT solar collectors start to supply heat to TK1 and the heat pump starts to reduce the heat output, determining an increase of its top temperature and as a consequence an increase of the temperatures in the GHX-RHP loop. The temperature at the outlet reaches a temperature of about 9 °C just after 12:00 am, and after then starts to decrease because TK2 starts to be heated with solar thermal energy. However, when this operation stops, the heating of TK1 is again performed but the temperature keeps to decreasing due to the operation of RHP.

It is worth noting that during the central hours of the day the temperature at the outlet temperature of RHP condenser slightly decreases while the heated water (HCW) at the outlet of TK3 (TK3, out) increases. This situation is achieved since the heat pump thermal output decreases as a function of a reduction of the space heating load of the building.

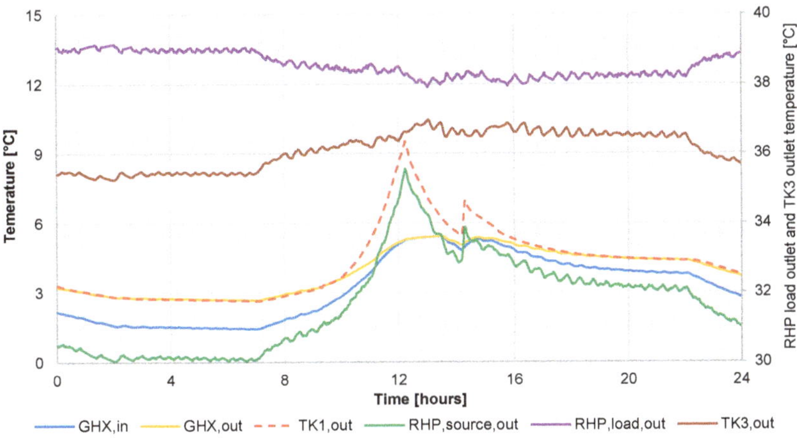

Figure 5. Main temperatures of GHX, TK1, RHP and TK3, daily analysis, winter day.

As concerns the PVT and TK2 loos, the results point out a decrease of the inlet temperature of PVT (PVT, in) in the night hours due to the heat losses on the pipes connected to the solar field. At the same time, the water temperature at the top of TK2 (TK2, out) decreases being the heat transferred to the lower parts of the tank where the temperature is slightly lower. In fact, the temperature of the water inside the tank near the solar heat exchanger (TK2, HX1) increases during the night by about 2 °C. After 8:00, am the temperature at the outlet of PVT collectors (PVT, out) increases above the inlet one allowing to start the heating of TK1. However, this operation is performed only when the outlet temperature is 2 °C higher than the top of TK1, according to the developed control strategy. The TK1 heating operation ends after 12:00 am when the PVT outlet temperature overcomes by 2 °C the temperature of the water inside TK2 near HX1, which is the condition that triggers the heating of TK2. The control system sets the PVT setpoint temperature to 60 °C from the previous value of 30 °C, thus the outlet temperature increases due to the reduction of the mass flow rate of SF operated by P1 and the proportional controller. In the result, the solar heat increases the water temperature at the bottom of TK2 to about 28 °C, allowing the preheating of DHW, and this temperature level is maintained until occurs the next DHW request near 7:00 pm. During the selected day the heating operation of DHW is completed by GB, which maintains the temperature at the top of TK2 (TK2, out) between 48 and 55 °C.

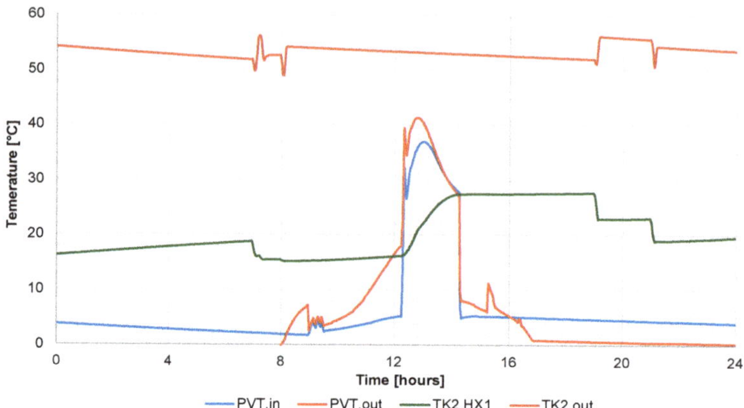

Figure 6. Main temperatures of PVT and TK2, daily analysis, winter day.

The thermal and electrical powers for the selected day are reported in Figures 7 and 8, respectively. The thermal power of the ground heat exchanger (GHX) varies as a function of the temperature of the tank (see Figure 5) and the thermal power supplied by the heat pump to the user (RHP, load). The thermal demand for space heating is higher, and the heat extracted from the ground is higher. During the day, the reduction of the heating demand along with the heating of TK1 by the solar field implies a decrease of the thermal extracted by GHX to zero just after 12:00 am, since the increase of the temperature of the working fluid of GHX is not possible due to the thermal conditions of the ground coupled with a relatively high bottom temperature of TK1. The solar field thermal power (PVT) increases up to 4 kW at noon, then the heat starts to be transferred to TK2 with lower power (TK2, HX1). This is due to a higher water temperature inside TK2 with respect to that present at the bottom of TK1, which limits the solar heat transfer rate to the tank. The heating of TK2 by renewable thermal energy is not adequate to match the DHW demand, thus the activation of the GB and the heating of the top part of TK2 by HX2 is mandatory (TK2, HX2).

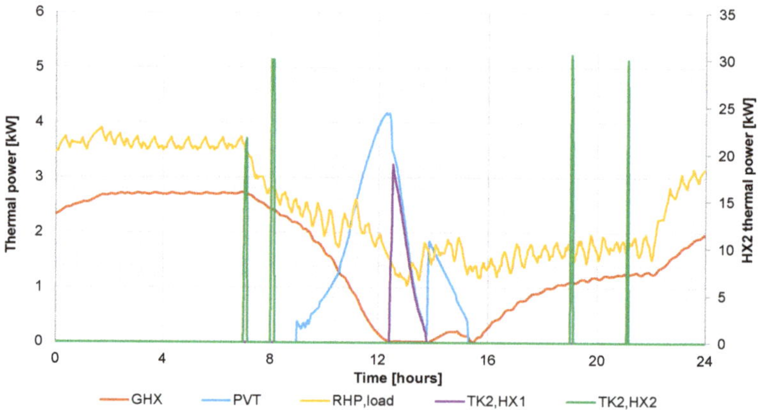

Figure 7. Main thermal powers of the system, daily analysis, winter day.

The electrical power trends highlight that during the night hours the output produced by the wind turbine (WT) is lower than the total power demand of the user, consisting of the building equipment, heat pump, and system auxiliaries operation. Therefore, a significant part (about 1/3) of the user demand (user) must be matched by the energy supplied by the grid. This mainly consists of the grid energy that is not taken into account by the net-metering system (net grid) since the contribution of the energy recovered by the grid (from the grid) is marginal. Only for about the first hour of the day, the user demand is matched by the energy that was produced during the previous day and supplied to the grid.

The system operates in the central hour of the day denotes that a certain amount of energy produced by the system is supplied to the grid (to grid), due to the output of the PVT field. The energy virtually stored by the grid is immediately supplied again to the user once the power generated by the system (PVT + WT) decreases below the demand just before 3:00 pm.

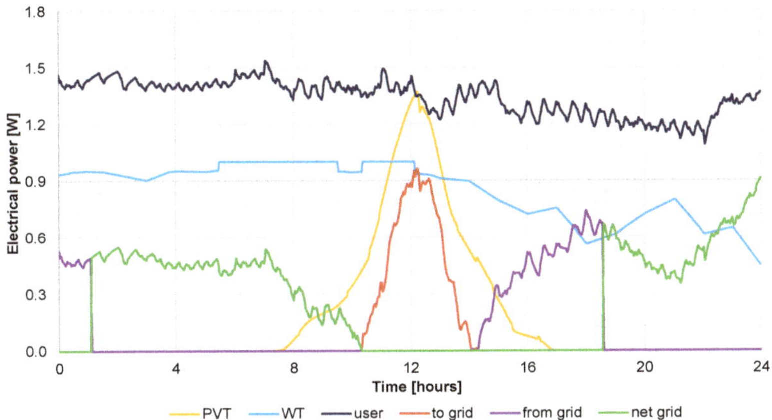

Figure 8. Main electrical powers of the system, daily analysis, winter day.

4.2. Monthly Results

The results were analysed by performing an integration of the powers on a monthly basis, thus the oscillations typical of the dynamic system operation were agglomerated in the integral results. The main thermal energy flows in the system are shown in Figure 9. Here, the effect of the user space conditioning demand on the system behavior is clearly shown, since the thermal energies flows inside the system are significantly higher during the winter heating season compared to the cooling one.

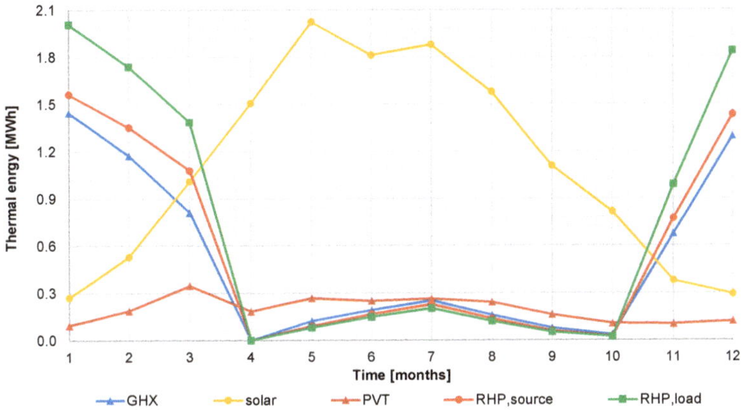

Figure 9. Main thermal energies of the system, monthly analysis.

The energy extracted by the ground heat exchanger (GHX) rages between 0.78 MWh to 1.41 MWh assuming a trend dependent on the thermal energy supplied by the load side of RHP (RHP, load). It is worth noting that as expected the major part of the thermal demand of the RHP evaporator is provided by GHX (at least 75.3% for all the months), while a relatively small part is provided by PVT (between 6.5% and 26.5%). From the point of view of summer operation, the heat rejected by GHX is from 11 to 37% higher than the one transferred by the condenser of the heat pump (RHP, source), and this is due to the fact that the pump of GHX (P3) operates independently from that of RHP (P4).

Moreover, it is important to note that the monthly thermal energy produced by PVT (PVT) outside the heating season is relatively small compared to the solar energy available (solar). This occurs because the solar field area is oversized with respect to DHW demand present during summer, and this determines a decrease in the thermal performance of the solar collector.

The monthly electrical energy flows are shown in Figure 10. The solar field electrical energy production (PVT) following the trend of the solar energy availability reported in Figure 9 ranges between 0.048 and 0.30 MWh. The energy production of the PVT field is lower than the one achieved for the wind turbine (WT) during winter, while the opposite occurs during summer. In particular, the system receives from WT between 0.11 and 0.40 MWh of electrical energy which is used to match in part the user demand.

The results also point out that the operation of the heat pump for space heating significantly affects the user's electrical energy demand during winter, since in summer the thermal energy removed by RHP from the user in the form of space cooling is relatively small, as outlined in Figure 9. Under these conditions, the production of electrical energy during the year by both PVT and WT determines the that during winter the amount of energy withdrawn from the grid excluding the net-metering (net grid) is relatively high, being between 41.1% and 58.8% of the demand (user). Conversely, it is null during the summer due to favorable conditions of demand and energy production. In addition to this, it noticeable that the energy produced in excess by the system (to grid) is always above zero, and increases significantly during the central months of the year, as well as the energy recovered from the grid in the frame of the net-metering contract. However, is it interesting to observe that the electrical energy stored virtually in the grid during April and May can be recovered in October and November when the PVT energy production drops.

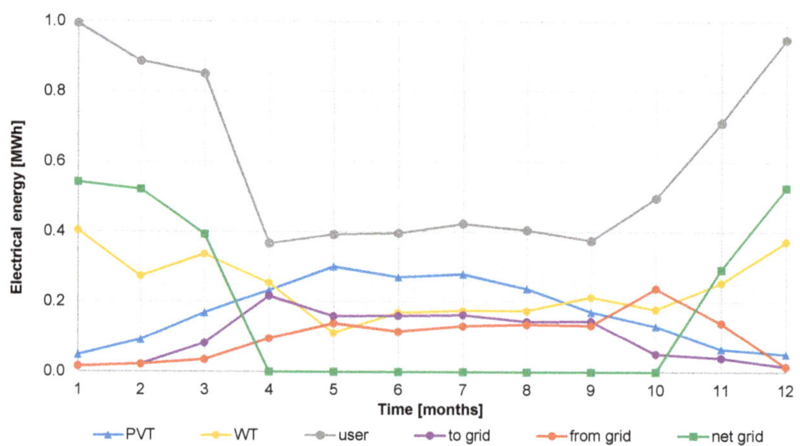

Figure 10. Main electrical energies of the system, monthly analysis.

The main efficiency parameters of the hybrid system are reported in Figure 11. As preliminarily mentioned, the thermal efficiency of the PVT field (η, *th*, *PVT*) in winter is relatively higher than the one achieved in summer, since the temperature of the solar loop increases when only DHW is produced.

In fact, the aperture area of the collector allows one to match the major part of the DHW demand in the summer months (more than 70% of the demand). It is worth noting that it is not possible to achieve a higher value due to the time of the DHW demand, occurring in the morning and late evening hours when the solar energy availability is relatively lower.

The decrease of the PVT thermal performance also negatively affects the electrical efficiency (η, *el*, *PVT*), being the PV cell affected by the operation temperature of the solar collectors. The efficiency decreases of about 15.9% with respect to the maximum value of 0.177 achieved in January and December. The wind turbine performance is presented in Figure 11 as a function of the normalized equivalent number of operation hours (H, WT) defined as the ratio between the equivalent hours of the wind turbine and the total number of hours in the considered time interval. As expected, the performance of WT decreases in summer, operating in terms of equivalent hours less than 30% of the time due to a lower availability of the wind energy source characteristic of the selected location.

Finally, the performance of the heat pump in terms of COP for space heating (COP, heat) and cooling (COP, cool) highlights that the device operates under stable conditions in terms of source and load temperatures. The COP in winter varies less than 1.5% compared to the maximum value of 4.57, while during summer the variation is 2.8% with respect to the maximum value of 8.17.

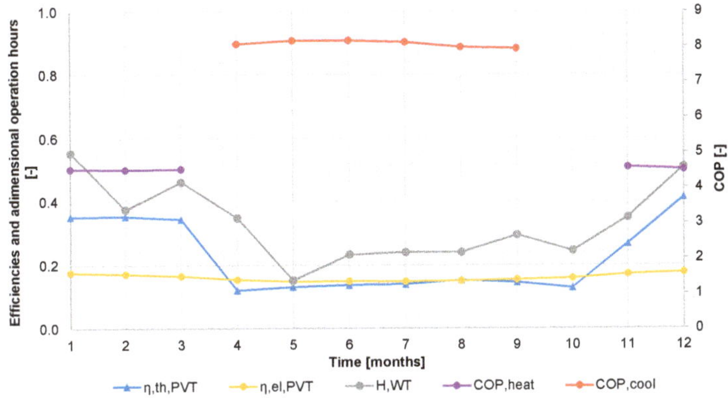

Figure 11. Energy efficiency parameters of the system, monthly analysis.

4.3. Global Energy and Economic Results

As done for the monthly analysis, the yearly results are analysed performing the integration of the variables from 0 to 8760 h. The main thermal and electrical energies of the system in the yearly time scale are reported in Table 4. Analysis of the results shows that despite a significant availability of solar energy ($E_{sol,PVT}$) the production of thermal energy by the PVT field ($E_{th,PVT}$) is less more than 50% of the thermal energy produced by GHX in total ($E_{th,GHX,heat}$ and $E_{th,GHX,cool}$). As shown by the monthly analysis, the ground heat exchanger is significantly more exploited in winter compared to summer due to different space conditioning demands, as outlined by the fact that the winter energy extracted 6.4 times higher than the heat rejected in summer.

It is interesting to note that the yearly DHW demand ($E_{th,DHW}$) is matched by solar energy only in 38.7%, and this occurs because the contribution of the solar energy to the production of DHW ($E_{th,HX1}$) during winter is almost null, while during summer it not exceeds 75.3%.

From the point of view of electrical energies, among the generation devices, WT produces more energy. In fact, the yield of PVT ($E_{el,PVT}$) is 0.87 MWh less than the one achieved by WT ($E_{el,WT}$), due to the comparatively better availability of wind energy compared to the solar one for the selected locality. The results also show that the incidence of the system auxiliary devices ($E_{el,AUX}$) on the system electrical energy balance is negligible, the energy demand of such equipment being only 5.6% of the total amount ($E_{el,demand}$), while the yearly energy demand of the heat pump is significant, since its operation determines 25.4% of the annual energy bill.

Moreover, the analysis of the energy flows among the system and the grid points out that the electrical energy supplied to the grid ($E_{el,to\ grid}$) and recovered ($E_{el,from\ grid}$) is the same, and thus the full bidirectional net-metering balance is achieved. In particular, the user sends to the grid 24.5% of the electrical energy produced, while the grid supplies the user with net-metering-free energy ($E_{el,net\ grid}$) in order to cover the demand in 31.4%.

Table 4. Main thermal and electrical energies of the system in MWh, yearly results.

Parameter	Value	Parameter	Value
$E_{sol,PVT}$	13.21	$E_{el,RHP,heat}$	1.76
$E_{th,PVT}$	2.32	$E_{th,RHP,source,cool}$	0.71
$E_{el,PVT}$	2.05	$E_{th,RHP,load,cool}$	0.63
$E_{th,GHX,heat}$	5.40	$E_{el,RHP,cool}$	0.08
$E_{th,GHX,cool}$	0.85	$E_{th,user,heat}$	7.58
$E_{el,WT}$	2.92	$E_{th,user,cool}$	0.33
$E_{th,HX1}$	1.50	$E_{el,aux}$	0.41
$E_{th,HX2}$	2.35	$E_{el,demand}$	7.24
$E_{th,DHW}$	3.88	$E_{el,to\ grid}$	1.22
$E_{th,RHP,source,heat}$	6.20	$E_{el,from\ grid}$	1.22
$E_{th,RHP,load,heat}$	7.96	$E_{el,net\ grid}$	2.28

The energy and economic indexes of the system are presented in Table 5. The savings of HS in terms of primary energy with respect to CS (*PESr*) are remarkable (66.6%), nonetheless, some primary energy must be consumed by HS due to the grid intervention to meet the electrical energy demand and to the operation of GB, especially during winter.

The scarce thermal efficiency of the PVT field affects the total efficiency of the collector units (33.1%), while the wind conditions of the selected locality allows one to achieve a good performance of WT, the normalized number of equivalent operation hour being equal to 0.333. This value underlines that the amount of electrical energy produced by WT is one-third of the maximum achievable for such a unit under ideal operational wind conditions (wind speed always above 11 m/s).

The yearly performance of the heat pump for heating and cooling overcomes the nominal conditions, indeed both COPs are higher than the values reported in Table 3. Such a result is achieved because: (i) during the winter the operation of GHX and PVT allows the evaporator of RHW to operate at a relatively higher temperature, and the outlet condenser temperature is lower than the nominal one (45 °C), (ii) in summer GHX allows one decrease the condenser temperature below the nominal operation conditions (30 °C).

On the basis of the comparison in terms of operational cost between CS and HS, the proposed innovative system allows one to save 0.900 k€/year which is 65.4% of the operational cost of the conventional system. Nevertheless, taking into account the fact that the investment for the whole hybrid system is relatively high (almost 20 k€), the economic performance of the investigated solution is scare. SPB is slightly over 21 years, which implies a feasibility of the system outside reliable criteria for economic investments. However, assuming that both PVT and WT are financed by 70% from incentive policies, SPB value decreases to 11.6 years, which is a value that starts to be economically viable. It is worth noting, that such incentive strategies, based on capital incentives, are increasingly common across Europe, and thus its applicability is reasonable. Moreover, it is worth noting that the proposed economic analysis of the system is performed under the worst scenario, where that total cost of the investment is considered. In case of the necessity of the installation of a new system providing DHW, heating and cooling, or in case of the substitution of the existing one, the cost that should be considered in the investigation would be only the difference of cost between CS and HS. Under such conditions the profitability of the system would be better compared to the present analysis. An interesting possibility of incentive for the proposed system could be that based on the consumption of produced renewable thermal and electrical energy, thus an additional saving may be generated by the hybrid system. In the invested case study, if an incentive of 0.05 €/kWh is adopted for both thermal (heating and cooling) and electrical energy produced and consumed by the user, SBP decreases to 13.1 years.

Table 5. Main energy and economic parameters of the system, yearly results.

Parameter	Value	Unit	Parameter	Value	Unit
PE_{HS}	9.7	MWh	$C_{op,CS}$	1.376	k€/year
PE_{CS}	29.0	MWh	C_{PVT}	4.43	k€
$PESr$	0.666	-	C_{GHX}	3.00	k€
$\eta_{th,PVT}$	0.176	-	C_{WT}	4.30	k€
$\eta_{el,PVT}$	0.155	-	C_{RHP}	4.66	k€
$h_{,WT}$	0.333	-	C_{TK1}	1.30	k€
COP_{heat}	4.52	-	C_{tot}	19.46	k€
COP_{cool}	8.12	-	SPB	21.6	years
$C_{op,HS}$	0.476	k€/year			

4.4. Parametric Analysis

The energy and economic performance of the system was also investigated changing the extension of the PVT field and the nominal power of WT. This analysis was performed with the aim of determining the effect of such design parameters on the global results when all the other ones are assumed to be constant. The PVT area was varied from 3.0 to 15.0 m^2, with a step to 3 m^2, whereas WT power was increased from 0.3 to 1.5 kW, with a step of 0.3 kW.

The results of the parametric analysis as a function of the PVT field area is presented from the point of view of energy in Figure 12, and from that of the efficiency and economical viewpoint in Figure 13.

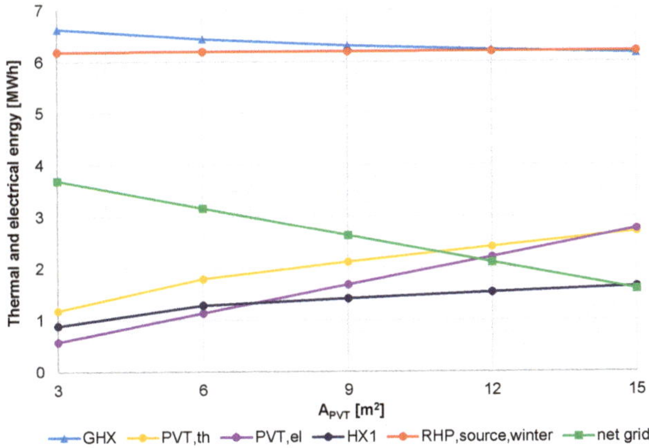

Figure 12. Main thermal and electrical energies of the system vs. variation of the PVT area, parametric analysis.

The variation of PVT area affects almost linearly the electrical energy producible by the solar field, thus the negative effect of a mean higher solar loop temperature achievable for a higher collector extension is limited. As a consequence, the electrical energy supplied by the grid and the net of the net-metering contract decreases linearly. The opposite occurs for the thermal energy produced by PVT collectors, increasing less when the area is gradually augmented. In particular, the increase of the thermal energy produced by the field configuration of 3 to 6 m^2 is 0.606 MWh, whereas from 12 to 15 m^2 it is only 0.288 MWh. The increase of the production of the thermal energy by PVT implies a decrease of the thermal energy extracted by GHX, due to a higher mean temperature of TK1, and an increase of the heat supplied to TK2 by means of HX1. Nevertheless, the yield of GHX is scarcely affected by such variation, because the additional thermal input of the PVT field to TK1 achieved in case of a higher area is marginal, and a similar situation is achieved for HX1. In this case, the solar field reaches the almost the maximum capability of providing heat to TK2, due to the DHW demand and dynamics of solar thermal energy production during the day, for and PVT area above 6 m^2. Additionally, it can be observed that the increase of the collector area does not affect the thermal energy used by the load side of RHP (evaporator) to match the used space heating demand, the latter being constant. In addition to this, also the COP of RHP in heating model remains practically constant, showing that the contribution of the solar field to the heating of TK1 is negligible.

The thermal efficiency of PVT decreases more than two times when the area increases from 3 to 15 m^2, due to the previously discussed reasons, highlighting that the increase of the PVT field extension significantly affects the operation of the solar system in terms of thermal efficiency. On the other hand, a limited decrease of the electrical efficiency of PVT is achieved along the range of investigated area values (only 5.1%), because the increase electrical losses of PVT due to a higher PV cells temperature is marginal.

The savings of primary energy achievable by HS increase as a function of the PVT area on the basis of a higher thermal and electrical energy produced and used in the system. However, the main reason for the increase of *PESr* is due to the production of electrical energy rather than the production of heat, being the effect of the last one on the energy balance of the system relatively marginal. Therefore, in terms of primary energy savings, it is convenient to increase the PVT area, and the same condition is achieved from the economic viewpoint, since *SPB* decreases for a higher area. However, the decrease is not sufficient to achieve the system economic profitability because the value remains over 21 years.

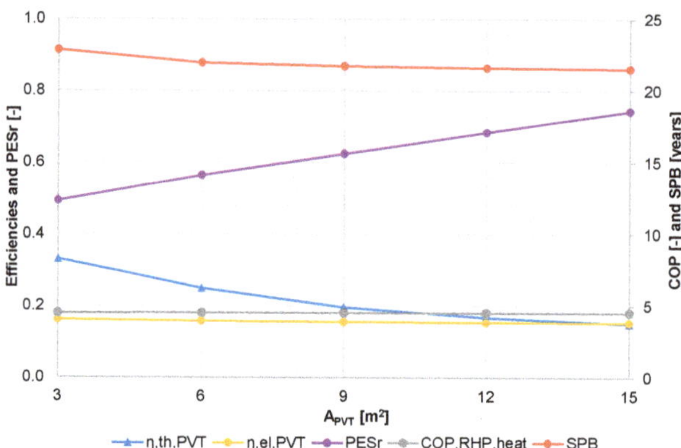

Figure 13. Efficiency, primary energy and economic parameters of the system vs. variation of the PVT area, parametric analysis.

The effect of WT power on the system performance is shown in Figures 14 and 15.

The capacity of WT affects only the electrical energy flows in the system as the thermal and electrical parts of the system are decoupled. The increase of WT nominal power, keeping constant the hub height and the wind conditions, determines a proportional increase of the generated electrical energy, which is equal to 2.92 MWh per kW. In view of this trend, the electrical energy supplied to the grid and withdrawn from the grid increases as well for a larger wind turbine, although the increase, in this case, is not linear due to the characteristic curve of the device. In fact, the increase of power from 0.3 to 0.6 kW implies a variation of energy supplied to the grid of 0.196 MWh, although the increase of 0.544 MWh for the variation from 1.2 to 1.5 kW. Moreover, it is important to note that beyond 1.2 kW of WT power, it is not possible to recover all the electrical energy supplied to the grid, since due to the intrinsic characteristic of the user demand and dynamic production of electrical energy by PVT and WT. Therefore, even when a turbine with a larger diameter is installed, the electrical energy needed from the grid is not null for the WT capacities considered. In practice, the increase of the wind turbine yield per unit of power determines the same reduction of the electrical energy produced within the grid and required by the user is needed to match the demand.

Finally, the analysis of primary energy savings shows that the trend of energy production achieved increasing the WT capacity directly affects *PESr*, and this is achieved because a higher production of electrical energy means a higher saving of non-renewable energy provided by the grid to the user. In the investigated case, the numerical data shows that increasing WT power 5 times implies an increase of *PESr* of 82.3% compared to the value achieved for a 0.3 kW unit.

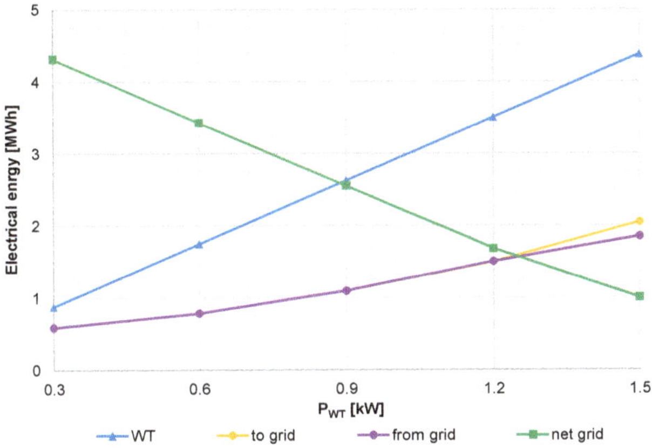

Figure 14. Main electrical energies of the system vs. variation of the WT power, parametric analysis.

As concerns the economic savings of HS with respect to CS, they increase of 67.0% passing from a wind turbine with 0.3 kW to a one with 1.5 kW, and on the other side in the same range of power, the total cost of the system (*C, tot*) increases of 30.6%. Therefore, coupling these two economic conditions, it is clear why the trend of SPB is decreasing as a function of WT power. In particular, *SPB* trend passed from 25.4 years for a 0.3 kW unit to 19.8 years for a power 5 times higher, consequently, the effect of higher savings overcomes the effect of the system cost increase when larger wind turbines are considered.

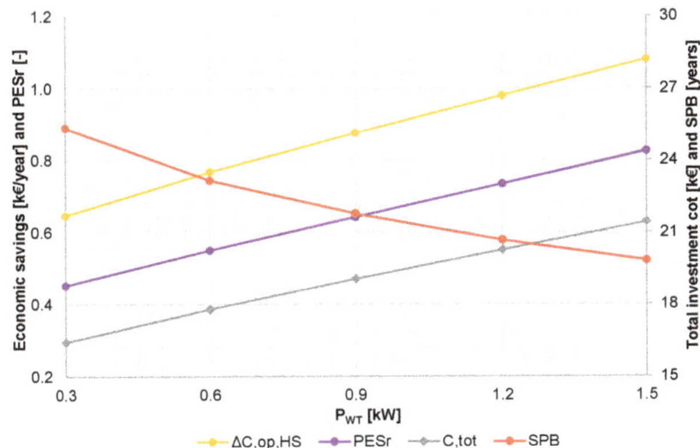

Figure 15. Primary Energy and economic parameters of the system vs. variation of the WT power, parametric analysis.

5. Conclusions

In the paper, a novel micro-scale hybrid geothermal-solar-wind system for a residential user has been investigated through dynamic simulation performed using a complex model developed in TRNSYS software [26]. The layout of the system included vertical ground heat exchangers, photovoltaic/thermal solar collectors, wind turbine, net-metered bidirectional connection with the grid, and a water-to-water reversible heat pump. The system was used to match a part of the user electrical energy demand, and to provide space heating and cooling along with DHW. The dynamic operation of the system and the building was simulated simultaneously in order to take into account their mutual interaction. In order to investigate the system, a case study of a single-family house under Polish northern climatic conditions, and a conventional reference system for the user, were considered.

The operation of the system was investigated from the viewpoints of temperature and thermal and electrical power trends for a selected winter day. Monthly and yearly analyses were carried out in order to assess the energy and economic system performance during the year and in terms of global performance. The study was concluded performing a parametric analysis of the system, where the solar collector field area and micro wind turbine power were varied.

The main results of the study were the following:

- during the firsts hours of winter operation day, both inlet and outlet temperature of the ground heat exchanger slowly decreased due to operation of the heat pump, determining a decrease of the temperature at the bottom of the tank storing the working fluid of the heat exchanger;
- the morning hours of a winter day, the hybrid solar collectors start to supply heat to the ground heat exchanger storage tank and the heat pump starts to reduce the heat output, determining an increase of the tank temperature and as a consequence an increase of the temperatures in the loop of the ground heat exchanger and source side of heat pump;
- the energy extracted by the ground heat exchanger ranges between 0.78 MWh to 1.41 MWh per winter month assuming a trend dependent on the thermal energy supplied by the load side of the reversible heat pump. The major part of the thermal demand of the heat pump evaporator is provided by the ground (at least 75.3% for all the months), while a relatively small part is provided by the solar field (between 6.5 to 26.5%);
- the production of electrical energy along the year and the user demand determine that during winter the amount of energy withdrawn from the grid excluding the net-metering is relatively high, being between 41.1% and 58.8% of the demand. Conversely, it is null during the summer due to favorable conditions of demand and energy production;
- the ground heat exchanger is significantly more exploited in winter compared to the summer due to different space conditioning demand, as outlined by the fact that the winter energy extracted was 6.4 times higher than the heat rejected in summer;
- the savings of the hybrid in terms of primary energy with respect to the reference system are 66.6%, nonetheless, some primary energy must be consumed by the proposed system due to the grid intervention to meet the electrical energy demand and to the operation of the gas boiler in order to match the domestic hot water demand, especially during winter;
- the simple pay back period is slightly over 21 years, implying a scarce economic feasibility of the system. However, assuming that both solar field and wind turbine are financed by 70% from incentive policies, the payback time decreases to 11.6 years;
- the increase of the thermal energy produced by the field configuration of 3 to 6 m^2 is 0.606 MWh, whereas from 12 to 15 m^2 it is only 0.288 MWh. The increase of the production of the thermal energy by the solar field implies a negligible decrease of the thermal energy extracted by the ground heat exchanger, due to a higher mean temperature of the storage tank;
- the increase of wind turbine nominal power, assuming constant the hub height and the wind conditions, determines a proportional increase of the generated electrical energy, which is equal to 2.92 MWh per kW.

The present study outlined that the hybrid system under consideration achieves a satisfactory performance in terms of energy, whereas its economic profitability is not adequate to permit its application as a valid alternative compared to an existing reference system based on a natural gas boiler from domestic hot water production and heating, an electrical chiller for space cooling and the grid for matching user demand. However, considering the installation of a new system matching the user demand, or in the case of the substitution of the existing one, only the extra cost for the hybrid system with respect to the reference one must to be considered. Thus, the profitability of the hybrid system will increase and the gap between the reference system in terms of economic competitiveness will be significantly reduced.

The investigations of the proposed system will be expanded by performing analyses regarding the modification of the system layout, comprehensive parametric analysis of the system, and optimizations aiming at determining the characteristics of the system depending on the design parameters and user type and behavior.

Author Contributions: Conceptualization, R.F. and M.Ż.; methodology, R.F.; software, R.F.; formal analysis, M.Ż.; investigation, R.F.; data curation, R.F.; writing—original draft preparation, R.F. and M.Ż; writing—review and editing, R.F. and W.G.; visualization, R.F.; supervision, R.F. and W.G. All authors have read and agreed to the published version of the manuscript.

Funding: Part of this work was funded by the Polish Ministry of Higher Education on the basis of the decision number 0086/DIA/2019/48.

Acknowledgments: This work was carried out under Subvention and Subvention for Young Scientists, Faculty of Energy and Fuels, AGH University of Science and Technology, Krakow, Poland. The authors acknowledge for the use of the infrastructure of the Center of Energy, AGH UST in Krakow.

Conflicts of Interest: The authors declare no conflict of interest.

Nomenclature

AW	Aqueduct Water
COP	Coefficient of Performance
CS	Conventional System
D	Diverter
DHW	Domestic Hot Water
EE	Electrical Energy
FC	Fan-Coil System
GB	Natural Gas Boiler
GBW	Gas Boiler Water
GF	Ground Heat Exchange Fluid
GHX	Ground Heat Exchangers
HCW	Heating and Cooling Water
HS	Hybrid System
M	Mixer
P	Pump
PESr	Primary Energy Saving ratio
PVT	Photovoltaic Thermal Collectors
RHP	Reversible Heat Pump
SF	Solar Fluid
SPB	Simple Pay Back
TK1	Thermal Storage Tank
TK2	Domestic Hot Water Tank
TK3	Hydronic System Buffer Tank
WT	Wind Turbine

References

1. IEA—International Energy Agency. Available online: https://www.iea.org/ (accessed on 3 May 2020).
2. Fedorczak-Cisak, M.; Furtak, M.; Hayduk, G.; Kwasnowski, P. Energy Analysis and Cost Efficiency of External Partitions In Low Energy Buildings. *IOP Conf. Ser. Mater. Sci. Eng.* **2019**, *471*, 112095. [CrossRef]
3. Fedorczak-Cisak, M.; Kotowicz, A.; Radziszewska-Zielina, E.; Sroka, B.; Tatara, T.; Barnaś, K. Multi-Criteria Optimisation of an Experimental Complex of Single-Family Nearly Zero-Energy Buildings. *Energies* **2020**, *13*, 1541. [CrossRef]
4. Simo-Tagne, M.; Ndukwu, M.; Rogaume, Y. Modelling and numerical simulation of hygrothermal transfer trough a building wall for locations subjected to outdoor conditions in Sub-Saharan Africa. *J. Build. Eng.* **2019**, *26*, 100901. [CrossRef]
5. Buonomano, A.; Calise, F.; Vicidomini, M. Design, Simulation and Experimental Investigation of a Solar System Based on PV Panels and PVT Collectors. *Energies* **2016**, *9*, 497. [CrossRef]
6. Żołądek, M.; Sornek, K.; Papis, K.; Figaj, R.; Filipowicz, M. Experimental and Numerical Analysis of Photovoltaics System Improvements in Urban Area. *Civ. Environ. Eng. Rep.* **2018**, *28*, 13–24. [CrossRef]
7. Calise, F.; Figaj, R.; Vanoli, L. Experimental and Numerical Analyses of a Flat Plate Photovoltaic/Thermal Solar Collector. *Energies* **2017**, *10*, 491. [CrossRef]
8. Filipowicz, M.; Żołądek, M.; Goryl, W.; Sornek, K. Urban ecological energy generation on the example of elevation wind turbines located at Center of Energy AGH. *E3S Web Conf.* **2018**, *49*, 00023. [CrossRef]
9. Filipowicz, M.; Goryl, W.; Żołądek, M. Study of building integrated wind turbines operation on the example of Center of Energy AGH. *IOP Conf. Ser. Earth Environ. Sci.* **2019**, *214*, 012122. [CrossRef]
10. Romanska-Zapala, A.; Bomberg, M.; Dechnik, M.; Fedorczak-Cisak, M.; Furtak, M. On Preheating of the Outdoor Ventilation Air. *Energies* **2019**, *13*, 15. [CrossRef]
11. Akhtari, M.R.; Shayegh, I.; Karimi, N. Techno-economic assessment and optimization of a hybrid renewable earth—Air heat exchanger coupled with electric boiler, hydrogen, wind and PV configurations. *Renew. Energy* **2020**, *148*, 839–851. [CrossRef]
12. Calise, F.; Dentice D'Accadia, M. Simulation of Polygeneration Systems. *Energies* **2016**, *9*, 925. [CrossRef]
13. Li, H.; Campana, P.E.; Tan, Y.; Yan, J. Feasibility study about using a stand-alone wind power driven heat pump for space heating. *Appl. Energy* **2018**, *228*, 1486–1498. [CrossRef]
14. Roselli, C.; Sasso, M.; Tariello, F. A Wind Electric-Driven Combined Heating, Cooling, and Electricity System for an Office Building in Two Italian Cities. *Energies* **2020**, *13*, 895. [CrossRef]
15. Kemmler, T.; Thomas, B. Design of Heat-Pump Systems for Single- and Multi-Family Houses using a Heuristic Scheduling for the Optimization of PV Self-Consumption. *Energies* **2020**, *13*, 1118. [CrossRef]
16. Psimopoulos, E.; Johari, F.; Bales, C.; Widén, J. Impact of Boundary Conditions on the Performance Enhancement of Advanced Control Strategies for a Residential Building with a Heat Pump and PV System with Energy Storage. *Energies* **2020**, *13*, 1413. [CrossRef]
17. Bonamente, E.; Aquino, A. Environmental Performance of Innovative Ground-Source Heat Pumps with PCM Energy Storage. *Energies* **2019**, *13*, 117. [CrossRef]
18. Conti, P.; Schito, E.; Testi, D. Cost-Benefit Analysis of Hybrid Photovoltaic/Thermal Collectors in a Nearly Zero-Energy Building. *Energies* **2019**, *12*, 1582. [CrossRef]
19. Ozgener, O. Use of solar assisted geothermal heat pump and small wind turbine systems for heating agricultural and residential buildings. *Energy* **2010**, *35*, 262–268. [CrossRef]
20. Vanhoudt, D.; Geysen, D.; Claessens, B.; Leemans, F.; Jespers, L.; van Bael, J. An actively controlled residential heat pump: Potential on peak shaving and maximization of self-consumption of renewable energy. *Renew. Energy* **2014**, *63*, 531–543. [CrossRef]
21. Stanek, W.; Simla, T.; Gazda, W. Exergetic and thermo-ecological assessment of heat pump supported by electricity from renewable sources. *Renew. Energy* **2019**, *131*, 404–412. [CrossRef]
22. Li, Q.Y.; Chen, Q.; Zhang, X. Performance analysis of a rooftop wind solar hybrid heat pump system for buildings. *Energy Build.* **2013**, *65*, 75–83. [CrossRef]
23. Rivoire, M.; Casasso, A.; Piga, B.; Sethi, R. Assessment of Energetic, Economic and Environmental Performance of Ground-Coupled Heat Pumps. *Energies* **2018**, *11*, 1941. [CrossRef]

24. Shimada, Y.; Uchida, Y.; Takashima, I.; Chotpantarat, S.; Widiatmojo, A.; Chokchai, S.; Charusiri, P.; Kurishima, H.; Tokimatsu, K. A Study on the Operational Condition of a Ground Source Heat Pump in Bangkok Based on a Field Experiment and Simulation. *Energies* **2020**, *13*, 274. [CrossRef]

25. Widiatmojo, A.; Chokchai, S.; Takashima, I.; Uchida, Y.; Yasukawa, K.; Chotpantarat, S.; Charusiri, P. Ground-Source Heat Pumps with Horizontal Heat Exchangers for Space Cooling in the Hot Tropical Climate of Thailand. *Energies* **2019**, *12*, 1274. [CrossRef]

26. Klein, S.A.; Beckman, W.A. *TRNSYS 18: A Transient System Simulation Program*; Solar Energy Laboratory, University of Wisconsin: Madison, WI, USA, 2017.

27. Calise, F.; Figaj, R.; Vanoli, L. Energy and Economic Analysis of Energy Savings Measures in a Swimming Pool Centre by Means of Dynamic Simulations. *Energies* **2018**, *11*, 2182. [CrossRef]

28. Figaj, R.; Szubel, M.; Przenzak, E.; Filipowicz, M. Feasibility of a small-scale hybrid dish/flat-plate solar collector system as a heat source for an absorption cooling unit. *Appl. Therm. Eng.* **2019**, *163*, 114399. [CrossRef]

29. Murray, M.C.; Finlayson, N.; Kummert, M.; Macbeth, J. Live Energy Trnsys: Trnsys Simulation within Google Sketchup. In Proceedings of the Eleventh International IBPSA Conference, Glasgow, UK, 27–30 July 2009; pp. 1389–1396.

30. Calise, F.; Figaj, R.D.; Massarotti, N.; Mauro, A.; Vanoli, L. Polygeneration system based on PEMFC, CPVT and electrolyzer: Dynamic simulation and energetic and economic analysis. *Appl. Energy* **2017**, *192*, 530–542. [CrossRef]

31. International—Aermec. Available online: http://global.aermec.com/ (accessed on 18 June 2020).

32. Calise, F.; Figaj, D.; Vanoli, L. A novel polygeneration system integrating photovoltaic/thermal collectors, solar assisted heat pump, adsorption chiller and electrical energy storage: Dynamic and energy-economic analysis. *Energy Convers. Manag.* **2017**, *149*, 798–814. [CrossRef]

33. HellstrÄom, G. Duct Ground Heat Storage Model. Manual for Computer Code 1996. Available online: http://repository.supsi.ch/3041/1/28-Pahud-1996-DSTP.pdf (accessed on 7 July 2020).

34. Tiwari, G.; Tiwari, A. *Hanbook of Solar Energy*; Springer: Berlin/Heidelberg, Germany, 2006.

35. West, J.R.; Lele, S.K. Wind Turbine Performance in Very Large Wind Farms: Betz Analysis Revisited. *Energies* **2020**, *13*, 1078. [CrossRef]

36. North, G.R.; Erukhimova, T.L. *Atmospheric Thermodynamics: Elementary Physics and Chemistry*; Cambridge University Press: Cambridge, UK, 2009; Volume 9780521899635.

37. White, F. *Fluid Mechanics*; Mc Graw Hill: New York, NY, USA, 1994.

38. Burton, T.; Jenkins, N.; Sharpe, D.; Bossanyi, E. *Wind Energy Handbook*, 2nd ed.; John Wiley and Sons: Chichester, UK, 2011.

39. Small Wind Turbines Enair, Efficient and Evolved Small Wind Turbines. Available online: https://www.enair.es/en/ (accessed on 18 June 2020).

40. ABB Poland—Energy Efficiency Report. Available online: https://new.abb.com/pl/efektywnosc-energetyczna/raporty-krajowe (accessed on 7 July 2020).

41. Calise, F.; Dentice d'Accadia, M.; Figaj, R.D.; Vanoli, L. Thermoeconomic optimization of a solar-assisted heat pump based on transient simulations and computer Design of Experiments. *Energy Convers. Manag.* **2016**, *125*, 166–184. [CrossRef]

42. TAURON Website. Available online: https://www.tauron-dystrybucja.pl/-/media/offer-documents/dystrybucja/uslugi-dystrybucyjne/iriesd/iriesd-teskt/2018-07-16-iriesd_tauron-dystrybucja-tekst-jednolity.ash (accessed on 18 June 2020).

43. National Fund for Environmental Protection and Water Management. Available online: https://www.nfosigw.gov.pl/en/ (accessed on 18 June 2020).

44. Polish Normative Dz.U. 2013 poz. 926. Available online: Isap.sejm.gov.pl/isap.nsf/download.xsp/WDU20130000926/O/D20130926.pdf (accessed on 2 July 2020).

45. Statistics Poland. Available online: Stat.gov.pl/en/topics/environment-energy/energy/energy-consumption-in-households-in-2018,2,5.html (accessed on 18 June 2020).

Article

Geothermal Power Production from Abandoned Oil Reservoirs Using In Situ Combustion Technology

Yuhao Zhu [1], Kewen Li [1,2,3,*], Changwei Liu [4,*] and Mahlalela Bhekumuzi Mgijimi [1]

[1] School of Energy Resources, China University of Geosciences, Beijing 100083, China; yuhao_zhu@cugb.edu.cn (Y.Z.); mahlalelab@cugb.edu.cn (M.B.M.)
[2] Key Laboratory of Strategy Evaluation for Shale Gas, Ministry of Land and Resources, Beijing 100083, China
[3] Department of Energy Resources Engineering, Stanford University, Stanford, CA 94305, USA
[4] School of Earth Sciences and Resources, China University of Geosciences, Beijing 100083, China
* Correspondence: likewen@cugb.edu.cn (K.L.); cwliu@cugb.edu.cn (C.L.)

Received: 11 October 2019; Accepted: 20 November 2019; Published: 24 November 2019

Abstract: Development of geothermal resources on abandoned oil reservoirs is considered environmentally friendly. This method could reduce the rate of energy consumption from oil fields. In this study, the feasibility of geothermal energy recovery based on a deep borehole heat exchanger modified from abandoned oil reservoirs using in situ combustion technology is investigated. This system could produce a large amount of heat compensated by in situ combustion in oil reservoir without directly contacting the formation fluid and affecting the oil production. A coupling strategy between the heat exchange system and the oil reservoir was developed to help avoid the high computational cost while ensuring computational accuracy. Several computational scenarios were performed, and results were obtained and analyzed. The computational results showed that an optimal water injection velocity of 0.06 m/s provides a highest outlet temperature of (165.8 °C) and the greatest power output of (164.6 kW) for a single well in all the performed scenarios. Based on the findings of this study, a geothermal energy production system associated with in situ combustion is proposed, specifically for economic reasons, because it can rapidly shorten the payback period of the upfront costs. Modeling was also performed, and based on the modeling data, the proposed technology has a very short payback period of about 4.5 years and a final cumulative net cash flow of about $4.94 million. In conclusion, the present study demonstrates that utilizing geothermal resources or thermal energy in oilfields by adopting in situ combustion technology for enhanced oil recovery is of great significance and has great economic benefits.

Keywords: geothermal power production; abandoned oil reservoirs; in situ combustion

1. Introduction

Generating geothermal power from abandoned oil reservoirs could be the answer to the energy crisis facing today's world. However, the upfront cost in a geothermal project is a barrier. Increasing the heat production and shortening the payback period of the upfront costs are two main objectives in a geothermal project. High outlet temperature and larger amount of geothermal fluid extraction could help to enhance the total heat production, whereas low drilling costs are conducive to reduce the payback period of the geothermal project. In this regard, most oilfields possess great potential for producing geothermal energy, because they are associated with production of geothermal fluid from oil reservoirs. Furthermore, abandoned oil reservoirs are good potential areas for harvesting and producing geothermal energy, and are most favorable because they are easily retrofitted into geothermal wells. However, the outlet temperature of geothermal wells in oil fields is usually lower, thus resulting in low heat production and long payback period.

Over recent years, the petroleum industry has seen and experienced rapid development due to the high demand of energy worldwide. In the midst of this rapid development, a larger number of oil and gas wells has been left abandoned due to either technical problems or economic benefit limitations, particularly heavy oil reservoirs. To enhance heavy oil recovery, thermal recovery technology such as in situ combustion method is often used. According to the latest update on drilled oil wells data, there exist about 20–30 million abandoned oil wells around the world [1]. Abandoned oil reservoirs may not hold oil productivity after a long-term recovery process, but such wells can produce brine fluids with high temperature. Thus, geothermal energy from such high-temperature fluids could be directly or indirectly recovered back to the surface for further utilization. Li et al. [2] and Zhang et al. [3] proposed a geothermal system combined with power generation which extracts geothermal heat from hot water in oilfields. Zheng et al. [4] proposed a concept to produce geothermal energy from abandoned oil and gas reservoirs by oxidizing the residual oil with injected air. This system, however, lacks a detailed heat exchanger model describing the heat transfer from reservoir to geothermal well, which is the essential part of evaluating the efficiency of a geothermal system. Zheng et al. [4] proposed a concept which lacks a detailed heat exchanger model, Davis and Michaelides [5] investigated the feasibility of the heat exchanger using abandoned wells for power generation by ORC (Organic Rankine Cycle). Their proposed geothermal system can provide an output power of 3.4 MW when at optimal condition.

Wight and Bennett [6] pointed out that the closed loop system (i.e., heat exchanger) retrofitted from abandoned oil well has the advantage of protecting both the equipment and the environment from performance or safety risks caused by minerals and contaminants commonly present in naturally occurring geothermal fluid. This is because the system can produce geothermal energy without extracting geothermal fluid from the formation. Another advantage of utilizing abandoned oil reservoirs is that there is little or no drilling cost at all, which is generally the main upfront investment in any geothermal project [7]. Moreover, the cost of retrofitting abandoned oil well into geothermal well using work-over techniques is equivalent to one tenth the cost of drilling a new geothermal well, and the maximum expense is even less than one seventh of the cost of drilling a new well [8]. Therefore, with regard to the above, it is clear that, if abandoned wells could be re-developed and utilized, large amount of drilling cost could be cut down, and more economic gains could be achieved.

Numerous mathematical models have been developed to both calculate the outlet temperature and evaluate the performance of the heat exchanger. Bu et al. [8] considered a two-dimensional model for numerically simulating the heat transfer between the surrounding rocks and the heat exchanger. This model neglects the variation of temperature in the tube wall of the injection annulus and the extraction well, which decreases the accuracy of the temperature distribution and overestimates the heat extracted from the surrounding rocks. Furthermore, Nian and Cheng [9] pointed out that it is unsound to use the assumption of vanishing well radius as a boundary condition in the model, because this could overestimate the heat flux from the surrounding rocks to working fluid in the heat exchanger.

Noorollahiet et al. [10] simulated a geothermal system which extracted geothermal energy by using two abandoned oil wells in Ahwaz oil field in Southern Iran. They developed a 3D numerical model using finite element method software ANSYS. They found that the sensitive parameter analysis indicates that casing geometry will strongly affect the heat transfer between the geothermal well and the surrounding rocks. Mokhtari et al. [11] found that when evaluating the pressure drop and thermal efficiency of the heat exchanger, the diameter ratio of the inner to outer pipe is very important. Their results showed that the optimal diameter ratios for pressure drop minimization and thermal efficiency maximization are 0.675 and 0.353, respectively.

In general, there are three grades of temperatures for extractive geothermal energy recognized by the industry namely: high temperature (>180 °C), intermediate temperature (100 to 180 °C) and low temperature (30 to 100 °C), respectively [12]. High-temperature geothermal resources can be used for power generation, while the low and medium temperature resources could be/are normally used directly for district heating and geothermal source heat pump (GSHP) [13,14]. However, a large number of geothermal reservoirs have low outlet temperatures. Low-temperature geothermal resources are economically not viable for commercial plants because they cannot generate sufficient heat. Thus, they cannot recover the upfront investment for retrofitting the abandoned wells and the construction of the power plants. To extract more heat from subsurface reservoir, low boiling point organic fluids with high heat-to-electricity efficiency are often used as working fluid mostly in ORC plants. Mokhtari et al. [11] investigated four organic fluids (R123, R134a, R245fa, R22) often used as working fluids in geothermal power plants. They found that R123 has the most desirable characteristics in comparison with other working fluids. Noorollahi et al. [15] further investigated two other types of working fluid (isobutene and ammonia) for power generation. A binary power cycle model was applied for power generation in their study to calculate the total electric power output. The two fluids (ammonia and isobutene) were chosen as secondary fluids in their power cycle, and their power outputs were compared. They found that for similar flow rates, the total net power for isobutene is greater than that of ammonia.

Another method for enhancing geothermal system is extracting geothermal energy from oil reservoirs with in situ combustion. Zheng et al. [4] proposed a geothermal system enhanced by oil reservoir using in situ combustion technology. They used a retrofitted abandoned well for their geothermal system. Similarly, Cinar [16] proposed a geothermal model using a perforated well to extract heated fluid from the oil reservoir where wet in situ combustion was applied. However, as pointed out in his study, wet in situ combustion may extinguish the combustion process, which causes reduction in the mass production of heated water resulting into less sufficient heated water to sustain a commercial scale power plant. Additionally, this may also result in the large depletion of the formation fluid. Cheng et al. [17] investigated the enhancement effect of geothermal power generation from abandoned oil reservoirs with thermal reservoirs. They found that a geothermal well with thermal reservoirs could produce a heat and electric power output of about four times the amount that can be produced by a geothermal well without thermal reservoirs. Although the cost of drilling can be reduced or even eliminated, the cost of the power plant installation accounts for the major portion of the total investment of the geothermal project, which further prolongs the payback period. Therefore, in this regard, more heat extraction from the subsurface is needed to sustain a commercial-sized geothermal project/plant.

In this study, a new model for recovering and re-utilizing oil for geothermal energy from abandoned wells is proposed. For this model, an in situ combustion technique is applied in the reservoir to recover heavy oil by injecting air into the formations. The air injection helps to oxidize the oil and heat up the reservoir. A schematic representation of the newly proposed model can be seen in Figure 1.

Figure 1 is a schematic representative summary of the development and utilization of the resources in abandoned oil reservoirs by in situ combustion. As can be seen from the figure, during the recovery process, a large amount of heat is generated from the in situ combustion. The abandoned well becomes a heat exchanger (i.e., geothermal well). A schematic representation of the heat exchange system is shown in Figure 2.

As can be seen from Figure 2, water is injected through the injection annulus and extracted through the extraction well. The oil reservoir with in situ combustion provides large amounts of heat at the bottom of the geothermal system. The system mainly includes heat transfer in the geothermal well (i.e., extraction well and injection annulus), oil reservoir and the surrounding rocks (i.e., strata). In general, the geothermal fluid extraction process may cause some problems in the formation, including groundwater recession, corrosion and scaling problem, the high cost of geothermal, and drilling of the re-injection well [18]. However, the system shown in Figure 2 has no direct contact in the formation fluid and no effect on oil production. It only extracts the heat from the formation and the oil reservoir by recycling the working fluid in a closed loop concentric tube, which is combined with in situ combustion by air injection in the oil reservoir. The detailed methodology adopted in this study is shown in Figure 3.

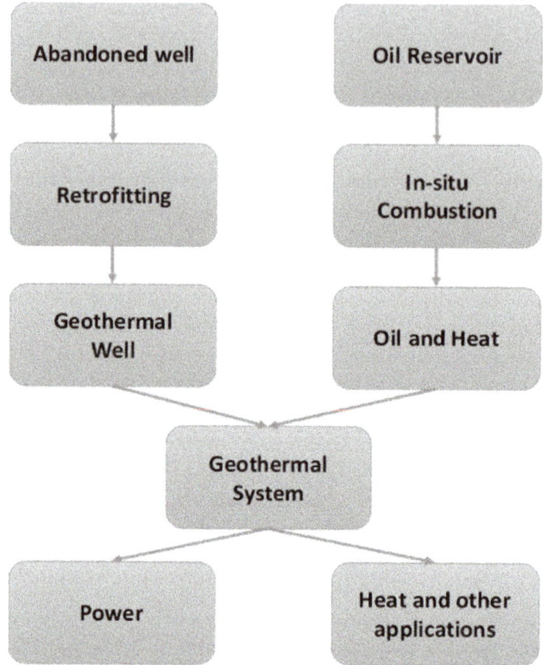

Figure 1. A new method of geothermal production utilizing abandoned wells and in situ combustion.

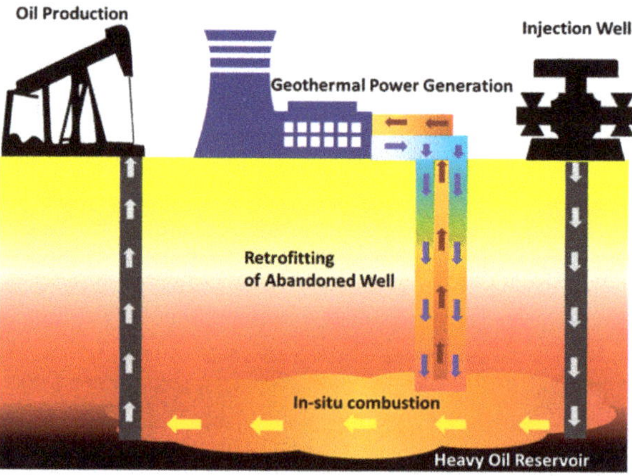

Figure 2. Schematic of heat exchange system.

In this study, a numerical model describing retrofitting abandoned oil wells into useful geothermal wells with in situ combustion for the purpose of recovering geothermal resources was developed. This model provides a new coupling strategy between the heat exchange system and the oil reservoir. The main idea is that the reservoir temperature with in situ combustion is expressed as a function of time, and then imported into the heat exchange system as a boundary condition. This model takes advantage of the commercial reservoir simulator CMG for simulating in situ combustion and helps to avoid large amounts of computational cost when calculating the heat transfer from the surrounding rocks and oil reservoir to the heat exchanger. Several parameters known to affect the system performance were analyzed using the proposed model in this study. Moreover, it should be noted that turbulent flow, which increases the computational efficiency without losing much computational accuracy is not considered in this model. For the findings of this study, the computational results indicate that the outlet temperature increases remarkably after the combustion front reaches the area near the heat exchange wellbore. Furthermore, a geothermal system using advanced in situ combustion was proposed to extract more heat from the reservoir right from the beginning, thereby significantly shortening the payback period of the upfront costs.

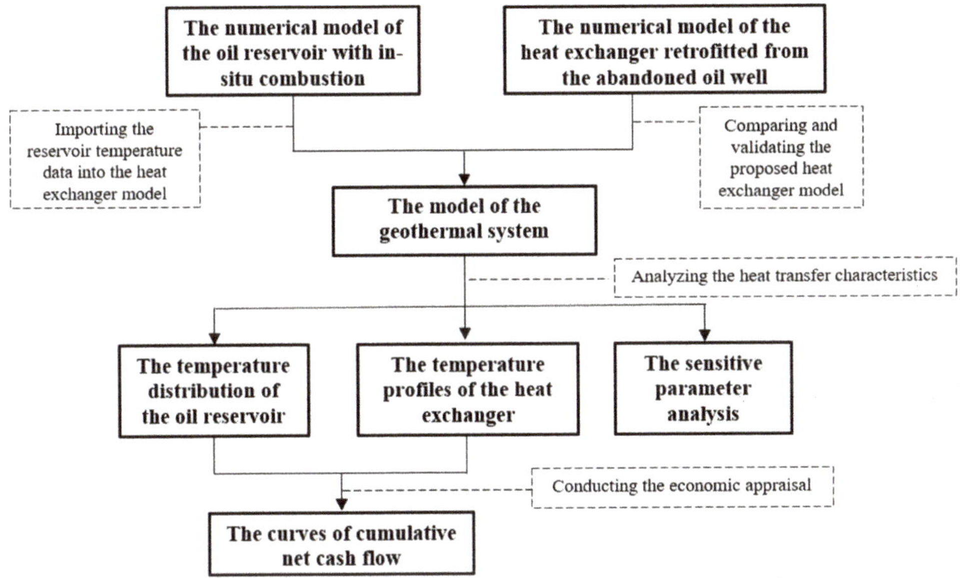

Figure 3. Flowchart followed when investigating the proposed geothermal system.

2. Model Description of In Situ Combustion

The process of in situ combustion is basically the injection of oxidized gas or oxygen-enriched air to generate heat by burning a portion of residual oil, where most of the oil is driven toward the producers by a combination of gas-drive, steam or water-drive. The main purpose of applying in situ combustion is to generate a large amount of heat for the geothermal system. For this study, the modeling was performed using STARS by CMG. The in situ combustion model assumes that the reservoir has: uniform porosity, isotropic permeability, and closed upper and lower layers. The model utilizes a five spots pattern, as shown in Figure 4.

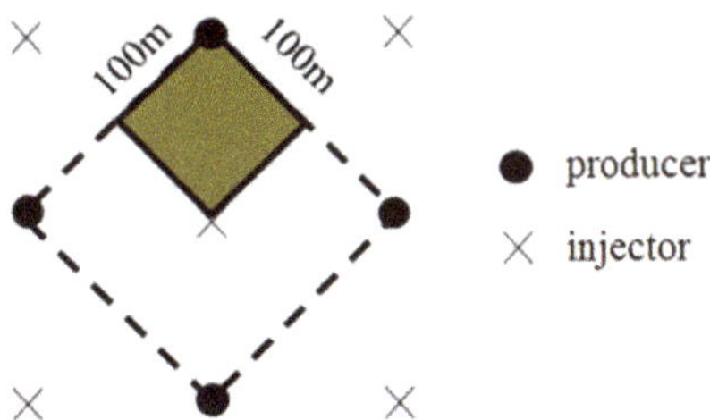

Figure 4. Five spots pattern applied in oil reservoir model.

The numerical model domain, which is the dark area shown in Figure 4, takes advantage of the symmetric flow, with injection in one quarter and a production well in two corners. To simulate the process of in situ combustion in a general form, a direct conversion cracking kinetics scheme for three components of oil was chosen in the simulation, as it does not depend upon the stoichiometry of the products, and thus reduces the degree of uncertainty in the simulation results, as the number of unknowns is reduced [19]. Here we applied the chemical reaction data of the template in STARS of CMG [20] to obtain a typical in situ combustion scenario. Seven components were considered in the process of in situ combustion, and these were: water (H_2O), heavy oil component (HC), light oil component (LC), inert gas (IG), oxygen (O_2), carbon dioxide (CO_2), and coke, respectively. Three chemical reactions were introduced to describe the components change in reservoirs and the heat generated in the process. The enthalpy and activation energy of these three reactions were as shown in Table 1.

Table 1. Combustion reactions and their respective kinetic parameters.

Combustion Reaction	Activation Energy (J/gmole)	Enthalpy (J/gmole)
$HC \rightarrow 10LC + 20coke$	2.463×105	-6.86×106
$HC + 16O_2 \rightarrow 12.5H_2O + 5LC + 9.5CO_2 + 1.277IG + 15coke$	8.41×104	6.29×106
$Coke + 1.225O_2 \rightarrow 0.5H_2O + 0.95CO_2 + 0.2068IG$	5.478×104	5.58×105

Their relative permeability curves and the reservoir properties are shown in Table 2 and Figure 5.

Table 2. STARS input parameters for in situ combustion model.

Parameters	Values
Dimension	$100 \text{ m} \times 100 \text{ m} \times 10 \text{ m}$
Grid (i, j and k)	$20 \times 20 \times 5$
Permeability	10 mD in i and j direction, 5 mD in k direction
Porosity	0.27
Thermal conductivity of reservoir rock	6×10^5 J/(m·day·K)
Overburden and under-burden volumetric heat capacity	2.350×10^6 J/(m^3·K)
Overburden and under-burden thermal conductivity	1.496×10^5 J/(m·day·K)
Oil saturation	0.4
Initial temperature	135 °C at 4 km, geothermal gradient 3 °C/km
Initial pressure	40 MPa at 4 km, pressure gradient 10 MPa/km
Well pattern	1/4 injectorlocatedat0, 0mand 1/4 producer located at 100, 100 m; both wells fully penetrate the reservoir
Air injection and oil production	5000 m^3/day and 5 m^3/day for both 1/4 wells

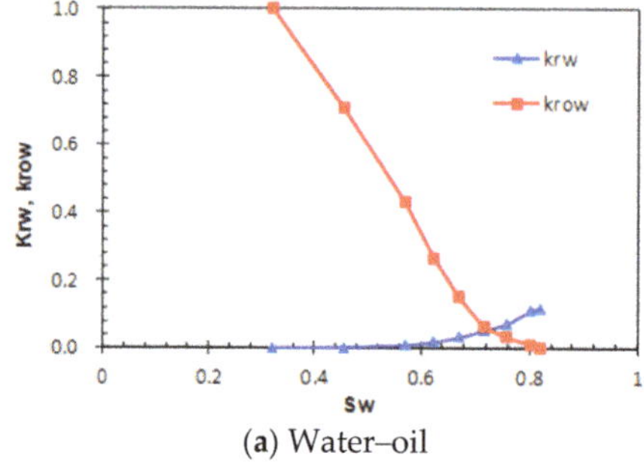

(**a**) Water–oil

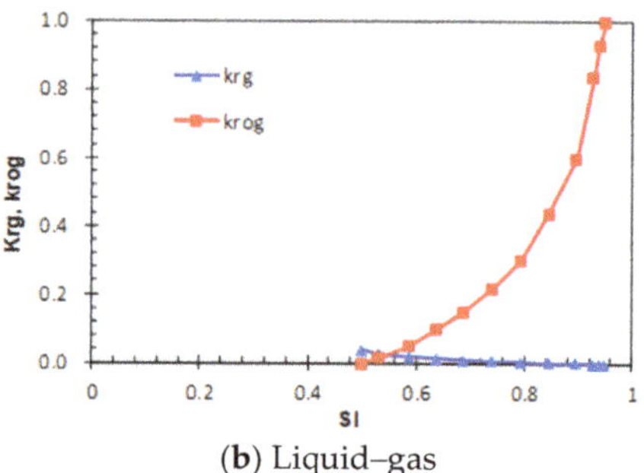

(**b**) Liquid–gas

Figure 5. Relative permeability curves used in model simulation.

3. Model Description of Geothermal Well

As seen in Figure 2, an abandoned well can be retrofitted into geothermal well by coating the inner tubing (extraction well) with insulation and sealing the bottom of the well. Water is then used as working fluid, and is injected through the annulus space and extracted from the inner tubing to the surface [21]. When the water flows downward along the annulus, it is heated by the surrounding rocks and by the reservoir, where there is a large amount of heat generated by in situ combustion. The heated water is then extracted back to the surface. This is a concentric tube heat exchanger, and the working fluid is not in direct contact with the surrounding rocks. Based on the data of the main oil fields in China, the geothermal gradient is generally about 0.03 K/m [22–26]. The model developed in this study utilized a concentric tube heat exchanger that was designed to retrofit an abandoned well with a typical casing outer diameter of 19.6 cm and an inner diameter of 15 cm. The inner diameter of the extraction well was 4 cm, the thickness of insulation was 2 cm, the well depth was 4000 m, and the thickness of the reservoir was 10 m. The whole reservoir was fully penetrated by the well. This system was modeled with the finite element modeling software COMSOL Multiphysics.

3.1. Governing Equations

A two-dimensional axi-symmetric cylindrical model was applied to describe the whole system. Heat exchange takes place between the working fluid and the rocks simultaneously [21,27], see Figure 6.

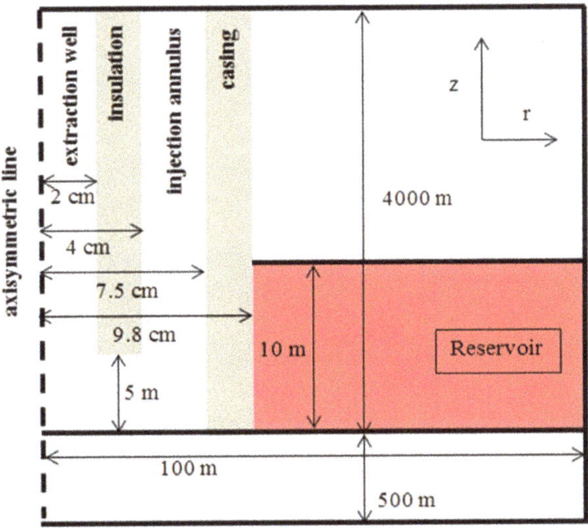

Figure 6. Dimensions of concentric pipe heat exchange system.

The dashed vertical line on the left side of Figure 6 shows the axis of symmetry, which is parallel to the direction of the z axis. The total depth of the heat exchange system is 4000 m from the surface. There is a buffer zone of 5 m at the bottom of the heat exchange system. The thickness of the oil reservoir is 10 m. The time-dependent governing equation corresponds to the convection-diffusion equation, which contains additional contributions of heat flux and no other heat source [28]. The heat flux describes the heat transfer from the rock to the injection annulus and from the injection annulus to the extraction well. Therefore, the expressional equation is

$$\rho C_p \frac{\partial T}{\partial t} + \rho C_p \vec{u} \cdot \nabla T + \nabla \vec{q} = 0 \tag{1}$$

where ρ is the density (kg/m^3), C_p is the specific heat capacity (J/(kg·K)), T is the absolute temperature (K), $\vec{u}$ is the velocity vector (m/s) and $\vec{q}$ is the heat flux by conduction (W/m^2), which can be described using the Fourier's three-dimensional diffusion law

$$\vec{q} = -k\nabla T \tag{2}$$

where k is thermal conductivity (W/(m·K)). Some parameters used in the numerical simulation are listed in Table 3. This heat transfer equation is already built into COMSOL, and can be called up in the software directly.

Table 3. The parameters used in the numerical simulation of the concentric tube heat exchanger.

Properties	Heat Capacity (J/kg·K)	Thermal Conductivity (W/(m·K))	Density (kg/m^3)
Casing	450	60	7850
Insulation	1010	0.025	1.225
Rock	1000	2	2200

For the water, the values of heat capacity, density and thermal conductivity depend on temperature and are already built into the COMSOL materials database, being expressed by the following formulas:

$$C_{p\_water} = 12010.1 - 80.4 \times T + 0.3 \times T^2 - 5.4 \times 10^{-4} \times T^3 + 3.6 \times 10^{-7} \times T^{-4} \tag{3}$$

$$\rho_{\_water} = 838.5 + 1.4 \times T - 0.003 \times T^2 + 3.7 \times 10^{-7} \times T^3 \tag{4}$$

$$k_{\_water} = -0.9 + 0.009 \times T - 1.610^{-5} \times T^2 + 8.010^{-9} \times T^3 \tag{5}$$

3.2. Initial Conditions

The velocity and the initial temperature of the injected water are both uniform along the well, and their values are 0.03 m/s and 30 °C, respectively. It should be noted that the velocity of water in the extraction well should be a corresponding value in order to retain a constant mass flow rate in the numerical model. The surface temperature is 15 °C. The initial temperature of the rocks is given by the following equation

$$T_{R,0}(z) = T_{srf} + G \cdot z \tag{6}$$

where $T_{R,0}$ is the initial rock temperature (K), T_{srf} is the temperature at surface (K), G is the geothermal gradient (K/m), and z is the depth of rock (m).

3.3. Boundary Conditions

The radius of influence may be at a modest distance from the well over the time frame modeled in this study. Therefore, the approximation that the temperature is constant at a 100-m radius is justified. The chosen distance of 100 m is explained in the following section. The boundary condition of the whole system, including the rock, is given by

$$T_{R,b}\Big|_{\substack{r=R \\ z=Z}} = T_{R,0}(z) \tag{7}$$

where $T_{R,b}$ is the rock temperature at a constant temperature boundary (K), R and Z are distances at constant temperature boundary in r and z direction (m).

However, if in situ combustion is applied in the reservoir, the boundary condition at the depth of the reservoir corresponds to the temperature influenced by the in situ combustion. Hence, the boundary condition at the depth of reservoir is given by

$$T_{R,b}\big|_{r=R} = T_{ic}(z,t) \tag{8}$$

where T_{ic} is the temperature of reservoir with in situ combustion and it is the function of depth z and time t. Please note that the distance of the constant temperature boundary is still R. This is reasonable, given that the influence of the in situ combustion on the heat exchange system is dominant considering the whole region of the oil field, while the reduction of the reservoir temperature and the heat extraction caused by geothermal well is limited.

4. Results and Discussion

A two-dimensional axi-symmetric cylindrical geothermic model was applied to simulate the temperature distribution of the geothermal well system as well as the surrounding rock system. To obtain more accurate temperature data, the two systems were coupled by using COMSOL Multiphysics simulator.

4.1. Mesh Independence Study

Before applying the heat exchanger model to calculate the outlet temperature, a mesh independence study was conducted to obtain stable temperature data. Triangular meshing was applied to the numerical model of the heat exchanger. The outlet temperatures after 5 years of operation are shown in Figure 7. When the number of triangular elements is small, the outlet temperature data oscillate and then begin to converge at certain values as the number of triangular elements becomes larger than 60,000. The numerical model of the heat exchanger in this study had 64,426 triangular elements, which guaranteed the mesh independence while using a relatively small computational time.

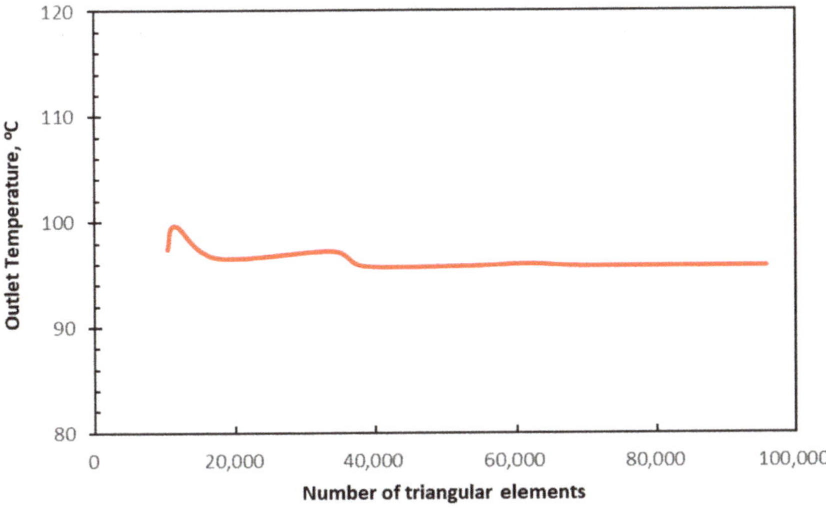

Figure 7. The outlet temperature at different numbers of triangular elements in the numerical model.

4.2. Validation and Comparison of the Model

The proposed model was compared and validated with the results from Bu et al. [8] and Templeton et al. [29]. The assumptions made by Bu et al. [8] include neglecting the variation of the tube wall temperature of the injection annulus and the extraction well, and also the Dittus-Boelter relation in the convection model. These assumptions decrease the accuracy of the temperature distribution and will overestimate the heat extracted from the surrounding rocks in their system. Furthermore, the temperature transfer is very sensitive to the properties of the extraction well and the insulation. Again, the Dittus-Boelter relation is not suitable for parts with larger temperature differences, especially near the top of well, and is also not suitable for the annular injection well [30]. For the model used by Templeton et al. [29], the whole simulation domain does not consider the rock below the heat exchanger. It should be noted that both studies neglect the changes in the thermal properties of water, which are a function of temperature.

To achieve fair comparison with the newly proposed model in this study, we applied parameters similar to those applied in the two studies done by Bu et al. [8] and Templeton et al. [29]. To be more precise, the inner diameter of the extraction and the injection wells are 10 cm and 30 cm; the thickness of the insulation is 1 cm; the thickness of casing is 2 cm; the thermal conductivity of the surrounding rocks and the insulation are 2.1 W/(m·K) and 0.027 W/(m·K); the density and the heat capacity of the surrounding rocks are 2730 kg/m^3 and 1098 J/kg/K; and the velocity of the injected water and the geothermal gradient are 0.03 m/s and 0.045 K/m, respectively. Figure 8 shows the outlet temperatures calculated by these three models over ten years.

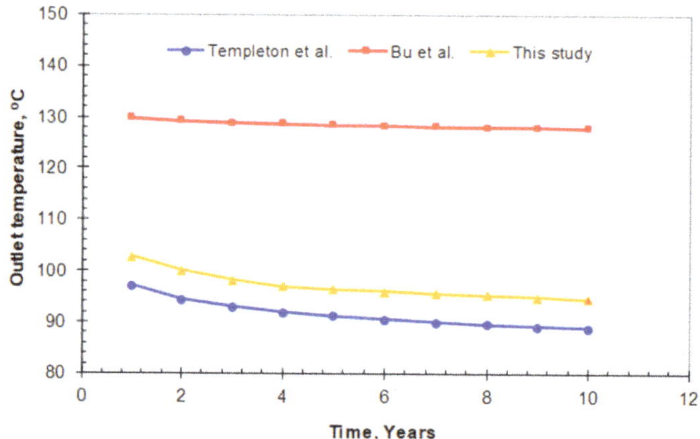

Figure 8. Verification and comparison of the results calculated by Bu et al. and Templeton et al. with the results of the proposed model under similar conditions.

As seen from Figure 8, the temperature result calculated by Bu et al. [8] model is about 26% to 44% higher than the results of both the model of Templeton et al. [29] and the proposed model in this study under similar conditions. The overestimation of the model proposed by Bu et al. [8] is because of the assumptions applied in the model, as mentioned in the previous section. The results of both the model of Templeton et al. [29] and the proposed model in this study have a similar shape of trend throughout the operational time/period. However, the negligence of the surrounding rocks below the heat exchange system causes the temperature result of the model of Templeton et al. [29] to be slightly smaller than the result of the proposed model in this study. Figure 9 shows the temperature profile of the injection annulus and the extraction well with 0.03 m/s injection velocity after a period of two months in operation.

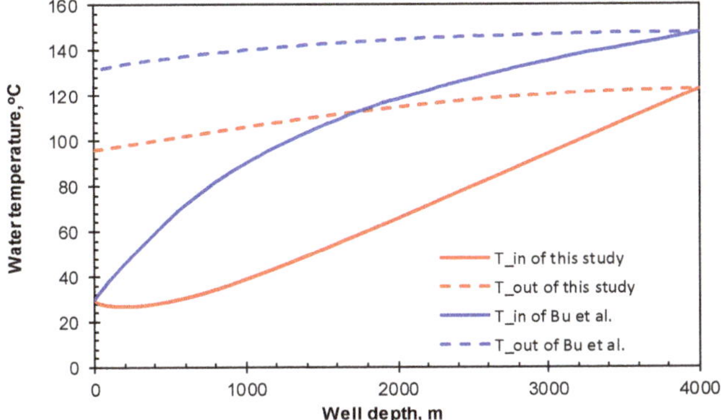

Figure 9. Comparison of the temperature profile between the proposed model of this study and the model of Bu et al.

As seen from Figure 9, the solid lines and the dashed lines show the temperature profiles of the injection annulus and extraction well, respectively. The highest temperature is found at the bottom of the well at about 4000 m. The variation of the fluid temperature curves of this study (red line) show that there is a slight decrease in the temperature near the top of the injection annulus. The reason for this is that the temperature of the injected water is 30 °C, and this is higher than that of the surface temperature, which is 15 °C. This causes the injected water to lose some heat in some regions near the wellhead. However, we cannot find any such phenomenon in the results from the model of Bu et al. [8]; thus, this overestimation reduces the accuracy of the simulation results of the model of Bu et al. [8].

4.3. Analysis of Heat Transfer Characteristics

The parameters applied in the calculation are listed in Tables 2 and 3, and Figures 5 and 6. The temperature distribution in the surrounding rocks and the heat exchange system with in situ combustion are shown in Figure 10.

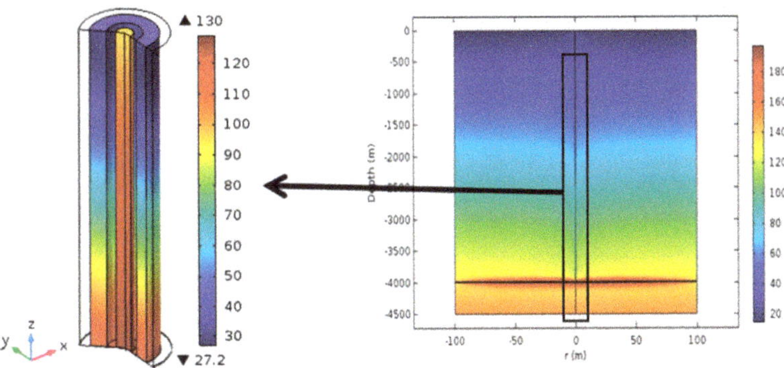

Figure 10. The temperature distribution of the heat exchange system after one year of operation (left figure), and the temperature distribution of the surrounding rocks after 50 years of operation (right figure).

As seen in Figure 10, there is a temperature drawdown near the heat exchange system. The high temperature zone is strongly marked at the depth of the in situ combustion reservoir at about 4000 m. The isothermal contours show that the initial geothermal temperature is not affected by the heat exchange system when the radius is greater than 80 m, even after a period of 50 years of operation.

This could help to explain why the 100 m distance (the radius of influence in boundary conditions) from the heat exchange system to the rock in the *r*-direction, and the 500 m distance in the *z*-direction, were chosen as the boundary of the surrounding rocks for the model proposed in this study.

4.4. Heat Production Analysis

Considering the high upfront investment required for the initial development of the geothermal power plant and retrofitting of the abandoned well, the outlet temperatures of the conventional geothermal well project may not be high enough to generate enough electric energy and recover the upfront costs. Therefore, in order to obtain more heat from the reservoir, in situ combustion was applied to supply the heat extraction from the heat exchange system. A three-dimensional model of the reservoir with in situ combustion was simulated by using the STARS simulator in CMG. The modeling grid numbers are: 20, 20 and 5 in the i, j and k directions, respectively. The temperature distributions in the oil reservoir during the 6th year and 11th year of the total operation period of 50 years are shown in Figure 11. The measurement unit of the temperatures of the reservoir is degrees Celsius.

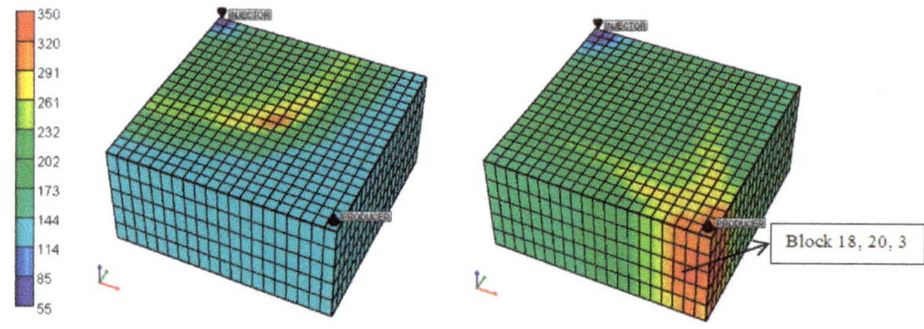

Figure 11. The temperature distribution of the oil reservoir at the 6th year (left figure) and the 11th year (right figure).

In Figure 11, the air is injected from the upper left injector. The lower right producer will be retrofitted into the heat exchange system. The extreme high temperature zone in this figure is where the combustion front is located. The temperature near the wellbore (block 18, 20, 3) and the average temperature of the whole reservoir are shown in Figure 12.

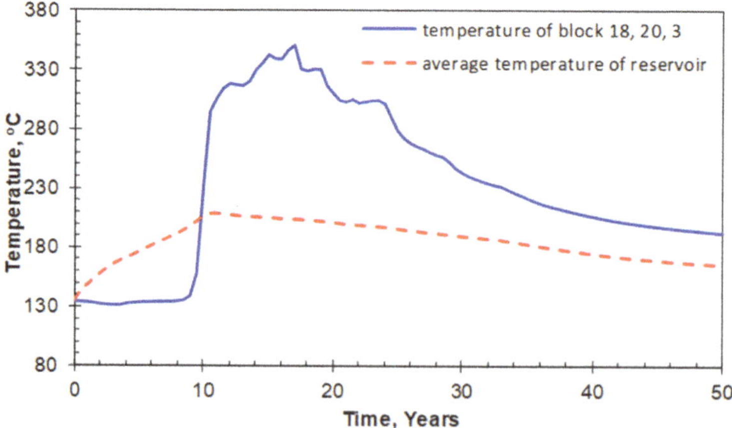

Figure 12. The average reservoir temperature (red dashed line) and the temperature near the heat exchange system wellbore (blue solid line).

As seen in Figure 12, when the combustion front approaches the wellbore (heat exchange system), the temperature around the wellbore increases sharply until reaching a peak point at about 350 °C. With continuous injection of air, the temperature decreases, and this is followed by a long stage of declination. This indicates that a large amount of heat is generated in the reservoir when in situ combustion is applied. Considerable amounts of heat can be extracted by the heat exchanger system from both the surrounding rocks and the combustion reservoir, especially when the peak point of combustion front has been reached.

The coupling between the oil reservoir model and the heat exchanger model is at the depth of the oil reservoir. If the average reservoir temperature is applied (red dashed line in Figure 12) as the boundary condition of the heat exchanger model, this may strongly underestimate the heat production of the heat exchanger. Therefore, in this study, we applied the temperature profile data of the block (18, 20, 3) into the heat exchanger model as a time-dependent boundary condition at the depth of the oil reservoir. This method reduces the computing cost while also preserving the computational validity of the result. This is reasonable, because as can be seen from Figure 10, the affected region in the surrounding rocks is limited to the place very close to the heat exchange wellbore. After applying the temperature data of the block (18, 20, 3) into the proposed heat exchanger model, the outlet temperature was strongly enhanced by the combustion reservoir. The temperature curves are as shown in Figure 13.

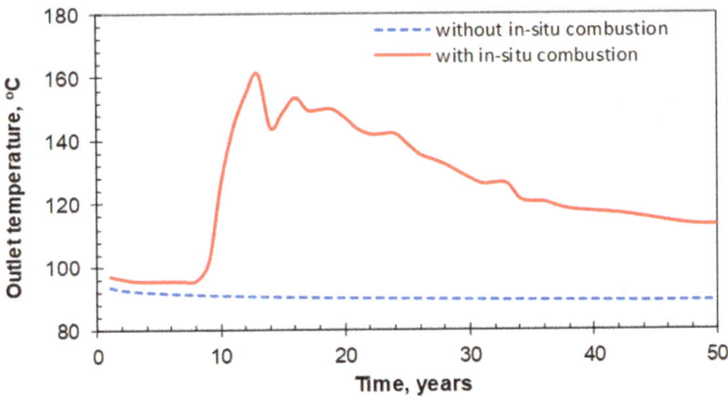

Figure 13. The effect of in situ combustion on performance.

It should be noted that in Figure 13, there is a bump up of the outlet temperature when the combustion front reaches the wellbore. The highest outlet temperature of the extraction well is about 150 °C. An enhancement effect is strongly marked when the geothermal heat is compensated by the in situ combustion technique. The corresponding temperature profiles of the injection annulus and the extraction well are shown in Figure 14.

In Figure 14, the solid lines and dashed lines represent the temperature profiles of the injection annulus and the extraction well, respectively. The lines with different colors show the temperature profiles of the heat exchange system at different operational times. The temperature bump up is also marked in the temperature profiles at the depth of reservoir (4000 m). As seen in Figure 14, before the combustion front reaches the wellbore (in less than 10 years), the temperature of both the injection annulus and the extraction well at the depth of (4000 m) are almost the same, or at least very similar. However, after the combustion front reaches the wellbore (after 10 years), the temperature of the extraction well is significantly higher than that of the injection annulus. This indicates that the in situ combustion has greatly enhanced the temperature of the water in the extraction well.

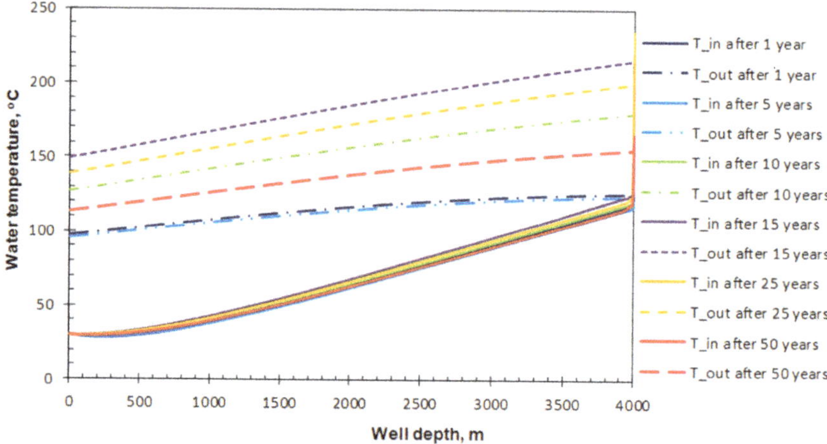

Figure 14. The temperature profiles of different operational time with in situ combustion.

According to Cheng et al. [31], the power provided by the heat exchange system can be simply given by the following equation:

$$P = M(T_{out} - T_{in})C_p\eta_{ri}\eta_m\eta_g / 1000 \tag{9}$$

where P is the actual generated power (kW), M is the mass flow rate (kg/s), T_{out} is the outlet temperature of extraction well (K), T_{in} is the inlet temperature of injection annulus (K), C_p is the specific heat capacity of water (J/(kg·K)), η_{ri} is the relative internal efficiency of steam turbine (0.8), η_m is the mechanical efficiency of steam turbine (0.97), η_g is the generator efficiency (0.98) [31]. For this study, the mass flow rate M is calculated by multiplying the injection velocity 0.03 m/s and the cross-sectional area of the injection well (see Figure 6).

Figure 15 demonstrates the variation of the outlet temperature and the actual gained power generated by the heat exchange system as functions of thermal conductivity of the insulation, inlet temperature and the velocity of the injected water in 30 years with in situ combustion.

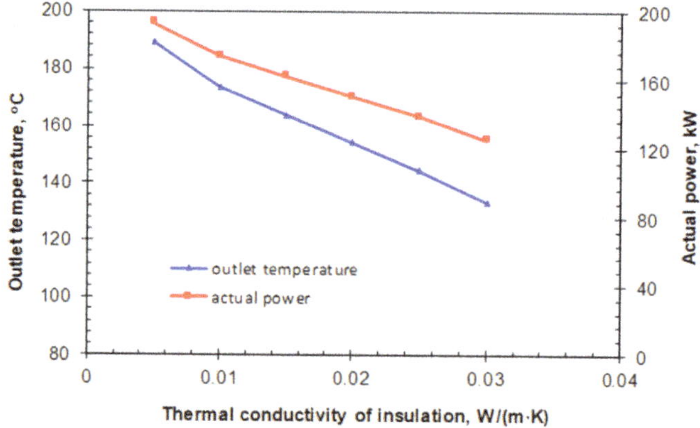

(**a**) Thermal conductivity of the insulation.

Figure 15. *Cont.*

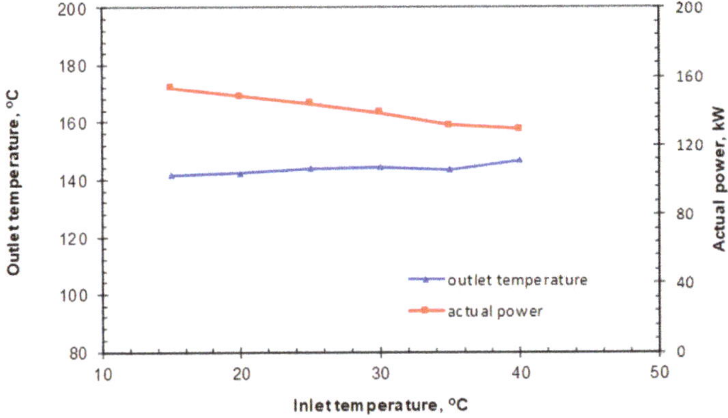

(**b**) Inlet temperature.

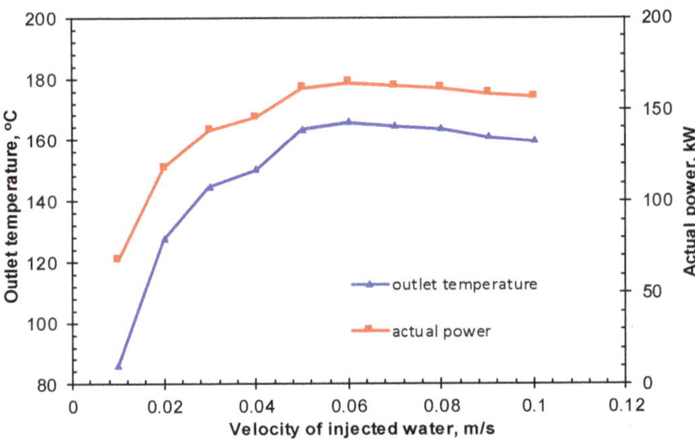

(**c**) Velocity of the injected water.

Figure 15. The variation of the outlet temperature and the actual power of the heat exchange system on different parameters.

It can be seen from Figure 15a,b that when the thermal conductivity of the insulation decreases and the inlet temperature increases, the outlet temperature increases. The outlet temperature is much more sensitive to the value of the thermal conductivity of the insulation. This indicates that insulation with better thermal resistance properties is essential to reducing the heat loss from the extraction well to the injection annulus, and to obtaining higher outlet temperature. This confirms that it helps to extract more heat from the formation. Although there is a higher outlet temperature due to higher inlet temperature, this does not guarantee a higher actual power. The reason for this is that a high inlet temperature does not provide a large temperature difference between the heat exchange system and the surrounding rocks. This reduces the efficiency of the heat transfer from the surrounding rocks to the heat exchange system. In Figure 15c, because of the increase in the velocity of the injected water, both the outlet temperature and the actual power increased first, and then decreased. This is because the water injection velocity has an optimal value of 0.06 m/s. This explains that there is a large amount of heat loss from the extraction well to the injection annulus when the water injection velocity is small.

For situations in which the water injection velocity is larger, the efficiency of the heat extraction from the surrounding rocks to heat exchange system will be very low.

With respect to the temperature near the wellbore of the heat exchange system in the early stage (about 10 years in Figures 13 and 14), it is still not significantly enhanced, even after applying in situ combustion in the reservoir. This means that in situ combustion needs to be applied in advance to avoid the low-temperature stage in order to recover the upfront costs. This is called the advanced in situ combustion method, and comes from a similar concept of advanced water injection technology. In other words, the operation of retrofitting the abandoned well will be executed only when the temperature of the bottom hole increases drastically (i.e., when the combustion front approaches the wellbore).

As can be seen from Figure 16, the outlet temperature of the scenarios without in situ combustion, with in situ combustion, and with advanced in situ combustion are compared. It can be clearly seen that the outlet temperatures of the scenarios without in situ combustion (blue dashed line) and with in situ combustion (red solid line) are relatively low at the beginning. If the heat exchange system starts to operate when the combustion front approaches the area near the wellbore, the outlet temperature will be enhanced significantly at a very early stage, and after that, decrease slowly (green dashed line). To find out how much energy could be extracted by the heat exchange system, the corresponding electricity data were calculated, and are shown in Figure 17.

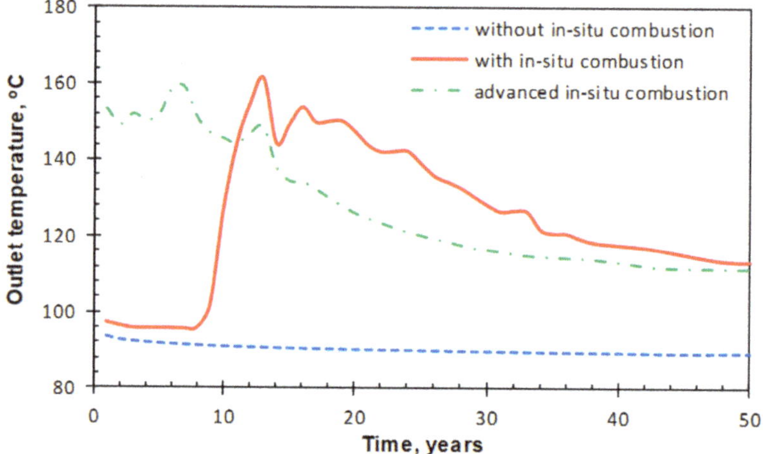

Figure 16. The effect of advanced in situ combustion on performance.

The curves demonstrate that the total electricity generated by the geothermal power plant with compensation for in situ combustion (green dashed line and red solid line) is much more than the electricity generated without in situ combustion (blue dashed line). The cumulative electric energy of the scenario with in situ combustion (red solid line) after a period of 50 years' operation is 50.3×10^6 kW·h. The scenario with advanced in situ combustion (green dashed line) has a higher cumulative electric energy of 51.4×10^6 kW·h.

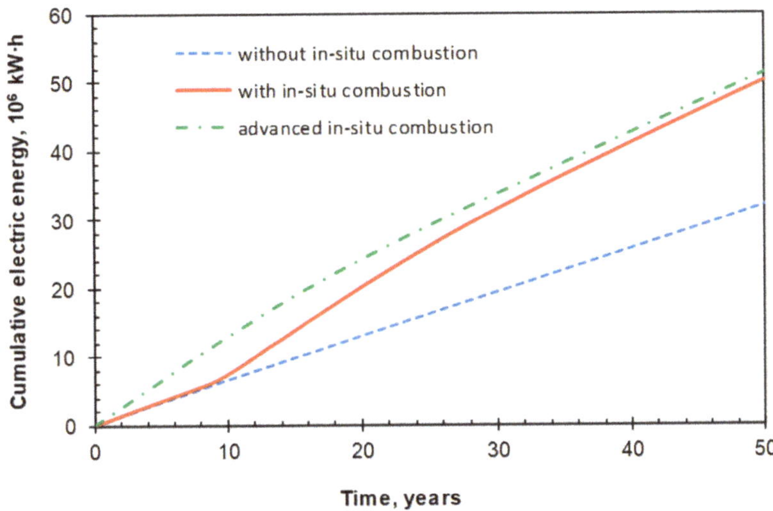

Figure 17. Cumulative electricity generated by the proposed geothermal power system.

4.5. Economic Appraisal

To determine the duration of the payback period of the upfront investment and the cumulative net cash flow after a period of 50 years' operation, an economic appraisal was performed by considering the following information: the geothermal development engineering and management engineering; the requirements of economic appraisal methods and parameters promulgated by China National Development and Reform Commission; the current fiscal and taxation system and the pricing system in China; and the current status of geothermal energy development.

A typical cost for a vertical geothermal well is about $2.3 million, based on the data of the geothermal project in Menengai, Kenya [32]. As stated previously, the cost of work over techniques in such wells is equivalent to about one tenth of the total cost of drilling a new geothermal well [33]. Hence, we assumed that the cost of retrofitting an abandoned well would be $0.23 million for a single well. For the investment of a power plant [34], the cost per installed kW comes to about $1500/kW (power peak is 159.2 kW during the 13th year in these scenarios). Based on the data given by Yambajan geothermal power generation [35], the electricity sale price is $0.14 (¥0.93) per kW·h. The management expense is 1% of the annual sales. The corporate income tax rate is 25%. Therefore, for a period of 50 years in operation, the cumulative net cash flow (NCF) curves are shown in Figure 18.

As shown in Figure 18, the scenario with advanced in situ combustion (green dashed line) has the shortest payback period (about 4.5 years) and the largest final cumulative NCF ($4.94 million). The final cumulative NCF of the two scenarios with in situ combustion (green dashed line and red solid line) are both higher than the scenario without in situ combustion (blue line). Although the scenario with simultaneous in situ combustion (red solid line) has relatively high final cumulative NCF, its payback period is very close to the scenario without in situ combustion (blue dashed line) and it is longer comarable to the scenario with advanced in situ combustion (green dashed line).

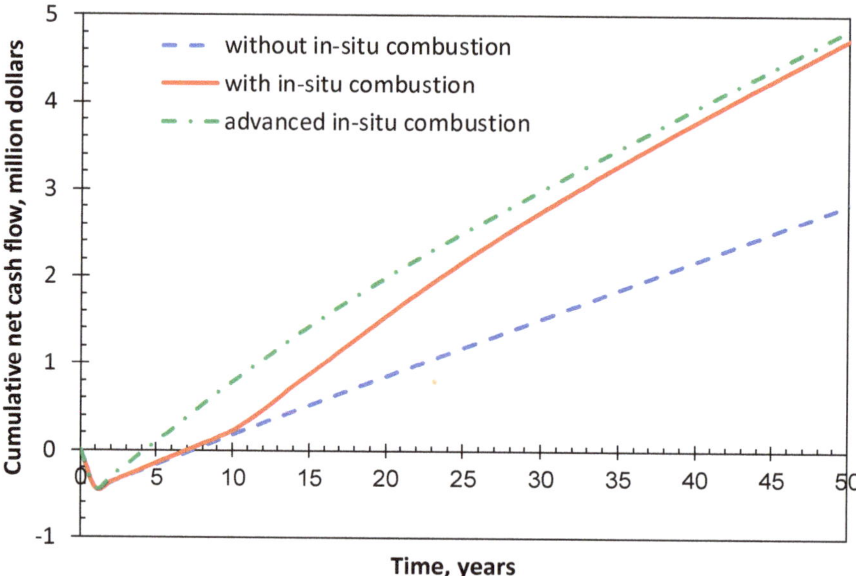

Figure 18. Cumulative net cash flow curves for the proposed geothermal power system.

In regard to the information presented above, a geothermal project with advanced in situ combustion is recommended. Please note that when evaluating geothermal projects, upfront investment, power output, fiscal policy, etc., should be fully considered.

4.6. Discussion

Note that this discussion leads to the conclusions made in this study, and was based upon a correlation of the study model with other previous models. The main theme of this paper is the application of in situ combustion for geothermal development with feasible economic gains. Although Bu et al. [8] and Templeton et al. [29] used different assumptions for their models, their purposes were similar to that of this study: the development of an economically friendly project. What makes the model of this study unique is the introduction of in situ combustion, and the fact that this model considers all of the parameters neglected by the other models. For example, the model of Bu et al. [8] neglects temperature variation in the tube wall of the injection annulus and the extraction well, and further neglects the Dittus-Boelter relation in convection model. The model of Templeton et al. [29] does not consider the rock below the heat exchanger. In short, both studies neglect the changes in the thermal properties of water, which are functions of temperature. Such negligence decreases the accuracy of the temperature distribution and may overestimate the heat extracted from the surrounding rocks in the system. However, since temperature is a major key in geothermal projects, the model proposed in this study considered all these aspects, especially temperature transfer in the extraction well and insulation. The results of this model are fair and reliable, because even with different assumptions from the previous models, the parameters applied are similar. These include the inner diameter of the extraction and the injection wells, the thickness of insulation and casing, the thermal conductivity of the surrounding rocks, the velocity of injected water, geothermal gradient, etc.

Furthermore, during the geothermal fluid extraction process, most models may cause some problems in the formation, such as groundwater recession, corrosion and scaling problems, the high cost of geothermal, and the drilling of the re-injection well, but the newly proposed model has no direct contact with the formation fluid and no effect on oil production; thus, it only extracts the heat from the formation and the oil reservoir by recycling the working fluid in a closed-loop concentric tube, which is combined with in situ combustion by air injection in the oil reservoir. The coupling between the oil reservoir model and the heat exchanger model is done at the depth of the oil reservoir. This depth may vary for each reservoir, so as not to underestimate the heat production of the heat exchanger. Again, this phenomenon is neglected in the results of the model of Bu et al. [8], and this reduces the accuracy of the simulation results. Note that the experimental data used in this study are similar to those of the two models described above, but the final results, such as temperature, are all different, which is due to the differences in the parameters that are assumed and neglected between the models. However, the results of this study model have more economic benefits than the other two models. A final point to note is that more attention should be paid to the figures, as more detail is apparent in the figures.

5. Conclusions

Based on the calculation and analysis conducted in this study, the following conclusions can be drawn:

(1)　An efficient numerical model describing the retrofitting of abandoned wells into geothermal wells with in situ combustion for the purpose of recovering geothermal energy is proposed. The reliability of the model was verified and compared with two other numerical models proposed by Bu et al. [8] and Templeton et al. [29]. The current coupling strategy of the geothermal model is a simple approach for coupling the in situ combustion model and the heat exchanger model.

(2)　Several parameters known to affect the system performance were modeled and analyzed using the proposed model in this study. Under specific conditions, the injection velocity has an optimal value of 0.06 m/s. Extreme values (either big or small) of the injection velocity will decrease the efficiency of the heat transfer from the surrounding rocks to the heat exchange system.

(3)　The scenarios considered in this study demonstrated that with the help of in situ combustion, the outlet temperature increases remarkably after the combustion front reaches the area near the heat exchange wellbore, which is about 150 °C. The cumulative electricity of the scenarios with in situ combustion and advanced in situ combustion after a period of 50 years of operation are 50.3×10^6 kW·h and 51.4×10^6 kW·h, respectively.

(4)　A geothermal system using advanced in situ combustion is proposed to extract more heat from the formation right from the beginning, thereby significantly shortening the payback period of the upfront costs. For this study, the system provides the shortest payback period of 4.5 years and a final cumulative NCF of $4.94 million.

Author Contributions: Conceptualization, C.L. and K.L.; Data curation, C.L.; Formal analysis, Y.Z., C.L. and K.L. Investigation, C.L. Methodology, Y.Z. Software, Y.Z.; Supervision, K.L.; Validation, C.L.; Writing—original draft, Y.Z.; Writing—review & editing, C.L. and K.L.; M.B.M. revised and polished the language of the final version.

Funding: This research received no external funding.

Conflicts of Interest: The authors declare no conflict of interest.

Nomenclature

C_p	specific heat capacity of water, J/(kg·K)
C_p	specific heat capacity of working fluid, J/(kg·K)
G	geothermal gradient, K/m
k	thermal conductivity, W/(m·K)
k_{og}	relative permeability of liquid phase
kr_g	relative permeability of gas phase
kr_{ow}	relative permeability of oil
kr_w	relative permeability of water
M	mass flow rate, kg/s

P	actual generated power, kW
$\vec{q}$	heat flux by conduction, W/m^2
r	radial distance from the central axis, m
R	distance at constant temperature boundary in r direction, m
t	time, s
T	absolute temperature, K
T_{ic}	temperature of reservoir with in situ combustion, K
T_{in}	inlet temperature of injection annulus, K
T_{out}	outlet temperature of extraction well, K
$T_{R,0}$	initial rock temperature, K
$T_{R,b}$	rock temperature at constant temperature boundary, K
T_{srf}	temperature at surface, K
$\vec{u}$	velocity vector, m/s
z	vertical distance from surface, m
Z	distance at constant temperature boundary in z direction, m
η_g	generator efficiency
η_m	mechanical efficiency of steam turbine
η_{ri}	relative internal efficiency of steam turbine,
ρ	density of working fluid, kg/m^3

References

1. Kotler, S. Abandoned Oil and gas Wells Are Leaking. Available online: https://zcomm.org/zmagazine/abandoned-oil-and-gas-wells-are-leaking-by-steven-kotler (accessed on 5 June 2019).
2. Li, T.; Zhu, J.; Xin, S.; Zhang, W. A novel geothermal system combined power generation, gathering heat tracing, heating/domestic hot water and oil recovery in an oilfield. *Geothermics* **2014**, *51*, 388–396. [CrossRef]
3. Zhang, Y.J.; Li, Z.W.; Guo, L.L.; Gao, P.; Jin, X.P.; Xu, T.F. Electricity generation from enhanced geothermal systems by oilfield produced water circulating through reservoir stimulated by staged fracturing technology for horizontal wells: A case study in Xujiaweizi area in Daqing Oilfield, China. *Energy* **2014**, *78*, 788–805. [CrossRef]
4. Lingyu, Z.; Jianguo, Y.; Hongbin, L.; Kewen, L. Energy from Abandoned Oil and Gas Reservoirs. In Proceedings of the SPE Asia Pacific Oil and Gas Conference and Exhibition, Perth, Australia, 20–22 October 2008.
5. Davis, A.P.; Michaelides, E.E. Geothermal power production from abandoned oil wells. *Energy* **2009**, *37*, 866–872. [CrossRef]
6. Wight, N.M.; Bennett, N.S. Geothermal energy from abandoned oil and gas wells using water in combination with a closed wellbore. *Appl. Therm. Eng.* **2015**, *89*, 908–915. [CrossRef]
7. Barbier, E. Geothermal energy technology and current status: An overview. *Renew. Sustain. Energy Rev.* **2002**, *6*, 3–65. [CrossRef]
8. Bu, X.; Ma, W.; Li, H. Geothermal energy production utilizing abandoned oil and gas wells. *Renew. Energy* **2012**, *41*, 80–85. [CrossRef]
9. Nian, Y.L.; Cheng, W.L. Insights into heat transport for thermal oil recovery. *J. Pet. Sci. Eng.* **2017**, *151*, 507–521. [CrossRef]
10. Noorollahi, Y.; Pourarshad, M.; Jalilinasrabady, S.; Yousefi, H. Numerical simulation of power production from abandoned oil wells in Ahwaz oil field in southern Iran. *Geothermics* **2015**, *55*, 16–23. [CrossRef]
11. Mokhtari, H.; Hadiannasab, H.; Mostafavi, M.; Ahmadibeni, A.; Shahriari, B. Determination of optimum geothermal Rankine cycle parameters utilizing coaxial heat exchanger. *Energy* **2016**, *102*, 260–275. [CrossRef]
12. World Energy Council. *World Energy Resources: Geothermal 2016*; World Energy Council: London, UK, 2016.
13. Østergaard, P.A.; Lund, H. A renewable energy system in Frederikshavn using low-temperature geothermal energy for district heating. *Appl. Energy* **2011**, *88*, 479–487. [CrossRef]
14. Self, S.J.; Reddy, B.V.; Rosen, M.A. Geothermal heat pump systems: Status review and comparison with other heating options. *Appl. Energy* **2013**, *101*, 341–348. [CrossRef]

15. Noorollahi, Y.; Mohammadzadeh Bina, S.; Yousefi, H. Simulation of Power Production from Dry Geothermal Well Using Down-hole Heat Exchanger in Sabalan Field, Northwest Iran. *Nat. Resour. Res.* **2016**, *25*, 227–239. [CrossRef]

16. Cinar, M. Creating Enhanced Geothermal Systems in Depleted Oil Reservoirs via In Situ Combustion. In Proceedings of the Thirty-Eighth Workshop on Geothermal Reservoir Engineering, Stanford, CA, USA, 11–13 February 2013.

17. Cheng, W.L.; Liu, J.; Nian, Y.L.; Wang, C.L. Enhancing geothermal power generation from abandoned oil wells with thermal reservoirs. *Energy* **2016**, *109*, 537–545. [CrossRef]

18. Nian, Y.L.; Cheng, W.L. Insights into geothermal utilization of abandoned oil and gas wells. *Renew. Sustain. Energy Rev.* **2018**, *87*, 44–60. [CrossRef]

19. Rabiu Ado, M.; Greaves, M.; Rigby, S.P. Dynamic Simulation of the Toe-to-Heel Air Injection Heavy Oil Recovery Process. *Energy Fuels* **2017**, *31*, 1276–1284. [CrossRef]

20. Hallam, R.J.; Donnelly, J.K. Pressure-up blowdown combustion: A channeled reservoir recovery process. *SPE Adv. Technol. Ser.* **1993**, *1*, 153–158. [CrossRef]

21. Kujawa, T.; Nowak, W.; Stachel, A.A. Utilization of existing deep geological wells for acquisitions of geothermal energy. *Energy* **2006**, *31*, 650–664. [CrossRef]

22. Wang, L.; Li, C.; Liu, S.; Li, H.; Xu, M.; Wang, Q.; Ge, R.; Jia, C.; Wei, G. Geotemperature gradient distribution of Kuqa foreland basin, north of Tarim, China. *Chin. J. Geophys.* **2003**, *46*, 403–407. [CrossRef]

23. Yuan, Y.S.; Ma, Y.S.; Hu, S.B.; Guo, T.L.; Fu, X.Y. Present-day geothermal characteristics in South China. *Chin. J. Geophys.* **2006**, *49*, 1005–1014. [CrossRef]

24. Fu-you, Z. Geothermal Gradient and HeatFlow Characteristics of Nanyang Basin. *Ground Water.* **2016**, *38*, 121–122.

25. Yusong, Y.; Lijun, M.; Gongcheng, Z.; Herong, Z. Some Remarks about Geothermal Gradient of Sedimentary Basins. *Geol. Rev.* **2009**, *55*, 531–535.

26. Liang-Ping, X.; Ju-Ming, Z. Relationship between geothermal gradient and the relief of basement rock in north China plain. *Chin. J. Geophys.* **1988**, *31*, 146–155.

27. Wenquan, T. *Numerical Heat Transfer*; Xi'an Jiaotong University Press: Xi'an, China, 1988.

28. Bird, R.B.; Stewart, W.E.; Lightfoot, E.N. *Transport Phenomena*, 2nd ed.; John Wiley Sons, Inc.: Hoboken, NJ, USA, 2006.

29. Templeton, J.D.; Ghoreishi-Madiseh, S.A.; Hassani, F.; Al-Khawaja, M.J. Abandoned petroleum wells as sustainable sources of geothermal energy. *Energy* **2014**, *70*, 366–373. [CrossRef]

30. Dittus, F.W.; Boelter, L.M.K. Heat transfer in automobile radiators of the tubular type. *Int. Commun. Heat Mass Transf.* **1985**, *12*, 3–22. [CrossRef]

31. Cheng, W.L.; Li, T.T.; Nian, Y.L.; Xie, K. Evaluation of working fluids for geothermal power generation from abandoned oil wells. *Appl. Energy* **2014**, *118*, 238–245. [CrossRef]

32. Kivure, W. *Geothermal Well Drilling Costing—A Case Study of Menengai Geothermal Field*; United Nations University Press: Tokyo, Japan, 2016.

33. Qihua, Z. The Abandoned Well can be Recycled. *China Petrochem.* **2011**, *16*, 26–27.

34. Tocci, L.; Pal, T.; Pesmazoglou, I.; Franchetti, B. Small scale Organic Rankine Cycle (ORC): A techno-economic review. *Energies* **2017**, *10*, 413. [CrossRef]

35. Although Geothermal Generation Enjoys Mature Technology, the Difficulty of Field Application Is Because of Low Feed-in Tariff. Available online: https://www.xianjichina.com/news/details_61970.html (accessed on 21 October 2018).

Article

Entropy, Entransy and Exergy Analysis of a Dual-Loop Organic Rankine Cycle (DORC) Using Mixture Working Fluids for Engine Waste Heat Recovery

Shuang Wang, Wei Zhang, Yong-Qiang Feng *, Xin Wang, Qian Wang, Yu-Zhuang Liu, Yu Wang and Lin Yao

School of Energy and Power Engineering, Jiangsu University, Zhenjiang 212013, China;
alexjuven@ujs.edu.cn (S.W.); 18852866772@163.com (W.Z.); strxinwang@163.com (X.W.);
qwang@ujs.edu.cn (Q.W.); A13844628038@163.com (Y.-Z.L.); wangyu_ujs@sina.cn (Y.W.);
YL806023480@163.com (L.Y.)
* Correspondence: yqfeng@ujs.edu.cn

Received: 12 February 2020; Accepted: 6 March 2020; Published: 11 March 2020

Abstract: The exergy, entropy, and entransy analysis for a dual-loop organic Rankine cycle (DORC) using a mixture of working fluids have been investigated in this study. A high-temperature (HT) loop was used to recover waste heat from internal combustion engine in 350 °C, and a low-temperature loop (LT) was used to absorb residual heat of engine exhaust gas and HT loop working fluids. Hexane/toluene, cyclopentane/toluene, and R123/toluene were selected as working fluid mixtures for HT loop, while R245fa/pentane was chosen for LT loop. Results indicated that the variation of entropy generation rate, entransy loss, entransy efficiency, and exergy loss are insensitive to the working fluids. The entransy loss rate and system net power output present the same variation trends, whereas a reverse trend for entropy generation rate and entransy efficiency, while the exergy analysis proved to be only utilized under fixed stream conditions. The results also showed that hexane/toluene is the preferred mixture fluid for DORC.

Keywords: dual-loop ORC; mixture working fluids; entropy generation; entransy loss; exergy loss

1. Introduction

With the depletion of coal, oil, and other non-renewable energy and the increasingly serious environmental problems, the effective use of energy has received great attention. However, the utilization rate of energy in China is only 30%, especially the recovery of low-grade waste heat energy. The recovery technology has become an important research topic. The organic Rankine cycle (ORC) has been proved an environmentally friendly and efficient solution for waste heat recovery. Dai et al. [1] choose exergy efficiency as objective function, and found that ORCs performed better cycles using water. Bombarda et al. [2] compared thermodynamic performance between Kalina cycle and ORC. They found that the environmental parameters for Kalina were too excessive to be satisfied, while ORC is more suitable. The study of ORC could provide not only necessary technical support but also create certain economic benefits for waste heat utilization.

To better improve the ORC performance, numerous researches devoted main efforts to develop new ORC configurations, such as regenerative ORC (RORC), two-staged ORC (TSORC), and DORC. Feng et al. [3] conducted thermoeconomic optimization for RORC, and found that RORC had better exergy efficiency than basic ORC by 8.1%. Mago et al. [4] selected dry organic fluids to compare the regenerative ORC with basic ORC (BORC). They found that RORC could obtain higher efficiency than BORC. Gang et al. [5] designed an regenerative ORC for solar power generation. Results showed that regenerative ORC had negative impacts on collector efficiency but benefited improving the system efficiency. The system efficiency with regenerator could improve 3.7% higher than the one without regenerator. Roy et al. [6] researched the parametric optimization of RORC using R123 and R134a. Results proved that R123 was better for converting low grade heat, and the inlet pressure of 2.70 MPa gave the maximum system efficiency. Franco et al. [7] analyzed the RORC recycling the geothermal resources and found that RORC could obtain significant reduction in cooling surface area in comparison to the basic ORC. Li et al. [8] analyzed the effects of organic fluids on ORC system performance. The experiments showed that the RORC efficiency could reach 7.98% and had an improvement of 1.83% than BORC. Li et al. [9] proposed a two-stage ORC (TSORC) for geothermal power generation. Results showed that TSORC could reduce the irreversible loss, while it was essential to improve power generation of low-pressure stage. Sun et al. [10] discussed the TSORC combining absorption refrigeration (TSORC-AR) and selected the residual heat of absorption refrigeration as heat source for low-pressure stage. Results indicated that TSORC-AR could generate more power, while its thermal efficiency was low. Yuan et al. [11] proposed a new deep super-cooling TSORC using R245fa and MM as working fluids. They found that the new TSORC could perform better than the basic ORC for high temperature waste heat recovery. The two-stage ORC and the regenerative ORC are combined by Xi et al. [12]. Results indicated that the two-stage regenerative ORC always had best thermal efficiency than TSORC or RORC.

Internal combustion engines (ICE) using fossil fuel were extensively applied to industrial machines and vehicles. It is indispensable to improve the engine efficiency to economize fuel source and reduce CO_2 emission. The ORC technology adapted to the internal combustion engine waste heat recovery (ICE-WHR) has been promising to get a considerable improvement in engine total efficiency. Traditional ORC systems using low-temperature working fluids were not applicable to ICE because of the large temperature gradient and the high temperature of engine waste heat. Accordingly, many studies were focusing on the modified ORC for better performance from two main aspects.

The first solution was adding extra loops to the basic ORC to further recover the residual heat, such as dual-loop organic Rankine cycle (DORC). Shu et al. [13] proposed the DORC system that used the fluids in low-temperature side to absorb residual heat of high-temperature fluid, waste gas and coolant. They obtained the maximum output power of 36.77 kW and exergy efficiency of 55.05%. Choi et al. [14] used DORC to recover heat of container ship engines. Water and R1234yf were selected as fluids of HT and LT loop, respectively. The final exhaust was reused to the propulsion, resulting in the reduction of fuel loss and CO_2 emissions by about 6.06%. The DORC was also proposed for the engine exhaust recovery of long-distance transportation vehicles [15]. Botaghchi et al. [16] investigated a cascade ORC with a refrigeration loop and obtained better operation parameters by optimization and decision-making. Sun et al. [17] found that higher high-level and low-level evaporation pressures in DORC caused higher investment cost, while DORC had better thermodynamic performance than BORC.

The second solution was to select high-temperature fluids that could break the limitation of thermal decomposition temperature or adopt the mixture working fluids to have a better temperature matching in thermal transfer processes. The available energy loss could decrease with the decrease of irreversible entropy increase because of the temperature slip during phase transition. The research showed that more serious the temperature slip, the more matching the temperatures between refrigerants and cold or heat source. Oyewunmi et al. [18] researched the thermodynamic and economic performance of basic ORC using different working fluid mixtures. They found that 0.5 pentane/0.5 hexane and 0.6 R245fa/0.4 R227ea has better thermodynamic performance, while the pentane and R245fa could obtain better economic performance. Wang et al. [19] designed a control strategy for ORC to adjust the composition of zeotropic mixtures according to the environmental parameters. Results showed that ORC using R245fa/R134a can have a higher thermal efficiency than Kalina cycle by 20–30%. Collings et al. [20] analyzed the dynamic ORC performance using R245fa/R134a as the working fluid and found that the energy generation could be improved by 20% under continental climates. Oyewunmi et al. [21] investigated an ORC system with CHP using working fluid mixtures. Results demonstrated that using 0.7 octane/0.3 pentane could obtain highest exergy efficiency. Another investigation by Oyewunmi et al. [22] showed that using more volatile fluids could obtain higher efficiency and lower cost than mixtures. Lecompte et al. [23] made specific exergy analysis on the ORC using working fluid mixtures, expressing that the mixtures had 7–14% higher exergy efficiency than pure refrigerants. Feng et al. [24] analyzed the entransy and entropy dissipation of basic ORC. They found that the optimal entropy generation was obtained by 0.6 R245fa/0.4 R152a, while the entransy dissipation is independent of mixture component. Zhang et al. [25] compared two groups of fluids: hydrocarbon (toluene, cyclohexane, hexane, isohexane, pentane, benzene) and common fluids (R123, R245fa, R245ca, R134). Results indicated that the using hydrocarbons had higher energy and exergy performance. Another simulation carried out by Baldi et al. [26] demonstrated that cyclohexane and benzene could best save the ship engine fuel, followed by toluene, while R245fa and R236ea performed worst. The alkanes are good environment-friendly working substances because they do not contain chlorine and their greenhouse effect is low, but most of them are much flammable and explosive. The hydrofluorocarbons also do not contain chlorine atoms, while their global warming potential value is high. The mixture of hydrofluorocarbons and alkanes complement each other.

As mentioned above, the analyses of ORCs were mainly around energy and exergy performance. Limited studies considered the entropy, exergy, and entransy analysis simultaneously. Meanwhile, the researches using mixture working fluids for DORC were not well studied. Therefore, this study conducted a detailed parametric analysis of DORC using mixture working fluids, including the entropy, exergy, and entransy analysis.

2. Modeling

2.1. System Description of DORC

The flow diagram of DORC is presented in Figure 1, including two loops with different temperature ranges, which are connected with a shared heat exchanger.

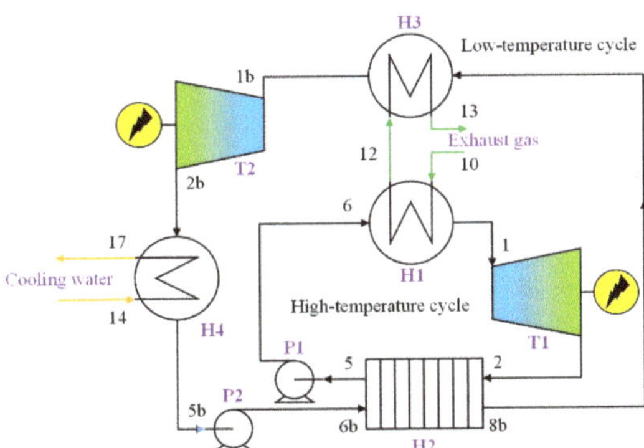

Figure 1. Flow diagram of dual-loop organic Rankine cycle (DORC) system.

The temperature-entropy diagram of DORC system is shown in Figure 2 based on hexane/toluene. In HT loop, the organic fluid from the shared heat exchanger (H2) is first converted to high-pressure state in pump (P1), then the fluid absorbs the exhaust gas heat in evaporator (H1) and is converted to high-temperature, high-pressure state. Then the overheated fluid starts expanding in expander (T1) to produce power. Afterwards, the low-temperature vapor is cooled in shared heat exchanger and flows into the pump for next cycle.

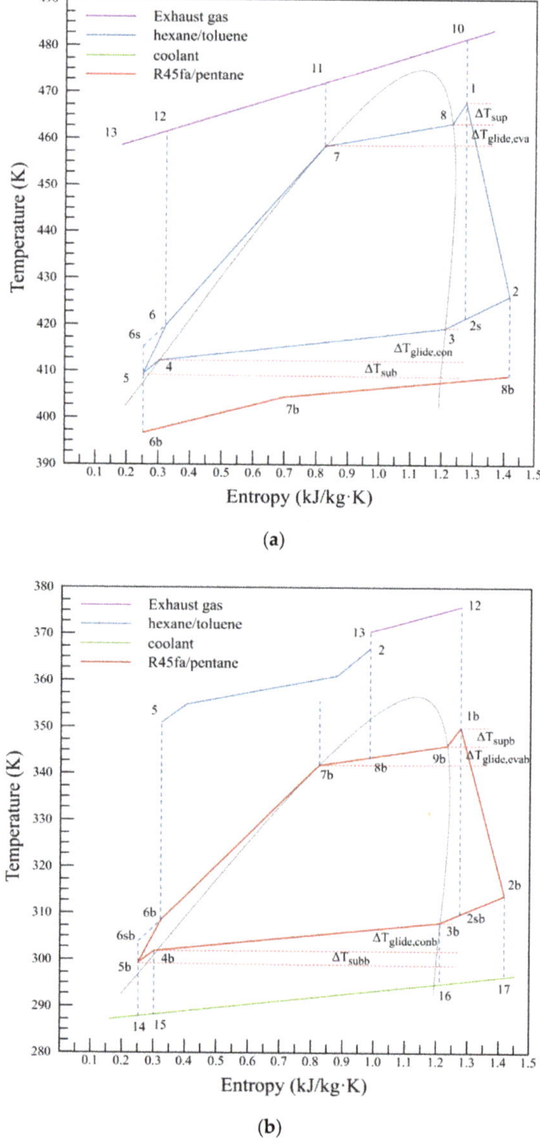

Figure 2. *T-s* diagram for the system: (**a**) The high-temperature loop; (**b**) the low-temperature loop.

Similar to HT loop, the low-temperature fluid from condenser (H4) is first pumped. Then the working fluid flows into heat exchangers (H2, H3) sequentially and absorbs residual heat of heat source and HT loop fluid, respectively. The saturated or superheated fluid enters the expander (T2) and expands to produce energy, and the pressure decreases to a prescribed condensation pressure. Eventually, the low-temperature loop fluid is cooled in condenser (H4) and flows into the pump for recycle.

2.2. Working Fluids Selection

Working fluids selection has a significant influence on ORC performance. Hydrocarbons (HCs) such as alkanes could perform great in high temperature ORCs. Their high critical temperature characteristics could result in excellent matching between working fluids and heat source. Toluene could get more power output than others in thermodynamic investigation, as introduced by Neto et al. [27]. Song et al. [28] selected cyclohexane, benzene, and toluene as working fluids for HT loop, while R123, R236fa, and R245fa for LT loop. Results revealed using cyclohexane and toluene in HT and R245fa in LT could increase total net power output. Shu et al. [29] used cyclohexane, cyclopentane, benzene, and two retardants for combination. Lecompte et al. [30] made exergy analysis and found that using R245fa/pentane could maximize the power output under low temperature environment (120–160 °C). Feng et al. [31] discussed the ORC performance using R245fa, pentane, and their mixture. They found that using 0.5 R245fa/0.5 pentane could obtain a low thermodynamic performance and a medium economic performance than pure fluids. Hence, hexane/toluene, cyclopentane/toluene, and R123/toluene are chosen as the working fluids for HT loop, while 0.5 pentane/0.5 R245fa is selected for LT loop. Table 1 lists the major thermodynamic properties of working fluids.

Table 1. Major properties of working fluids.

Fluids	Molar Mass (g/mol)	Triple Point Temperature (K)	Standard Boiling Point (K)	Critical Temperature (K)	Critical Pressure (MPa)
hexane	86.18	177.83	341.86	507.82	3.03
toluene	92.14	178	383.75	591.75	4.13
cyclopentane	70.13	179.72	322.4	511.69	4.51
pentane	72.15	143.47	309.21	469.7	3.37
R123	152.93	166	300.97	456.83	3.66
R245fa	134.05	171.05	288.3	427.2	3.65

2.3. Mathematical Modeling

Some assumptions based on the actual and literature are made to facilitate system analysis, including:

- The whole system operates in a steady state.
- Ignoring the pressure drop loss and thermal radiation in heat exchangers.
- The components and thermodynamic properties of mixtures do not vary with pressure or temperature.
- The isentropic efficiency of expanders and pumps are 0.8.
- The ambient temperature is set to be 10 °C.

The parameters of heat source used in this paper comes from the testing data of a six-cylinder turbocharged engine [32]. The major properties of the engine, the exhaust gas which is the heat source, are presented in Table 2. The mass fractions of main components of exhaust gas is calculated about: oxygen = 0.15, carbon dioxide = 0.05, water = 0.06, nitrogen = 0.74. The state parameters of all points are calculated by REFPROP 9.0 [33], and a simulation program combining MATLAB and REFPROP is established. Table 3 presents the major parameter assumptions of DORC.

Table 2. The main properties of engine.

Property	Power (kW)	Rotation Rate (r/min)	Torque (N·m)	Exhaust Temperature(K)	Exhaust Gas Mass Flow Rate (kg/s)
Value	1000	1500	6350	623	1

Table 3. Main assumptions for DORC system.

Parameter	Sign	Value
Environment temperature	T_0	283.15 (K)
Cold stream temperature	T_c	288.15 (K)
Superheating degree	ΔT_{sup}, $\Delta T_{sup,b}$	5 (K)
Supercooling degree	ΔT_{sub}, $\Delta T_{sub,b}$	0 (K)
Expander total efficiency	%	80
Pump total efficiency	%	80

2.3.1. Energy Modeling

The heat transfer rate of DORC system is:

$$\dot{Q}_g = \dot{m}_g(h_{10} - h_{13}) \tag{1}$$

where $\dot{m}_g$ represents the exhaust gas mass flow rate; where h_{10} and h_{13} represent specific enthalpies at heat source import and export, respectively.

There are two evaporators for DORC, one is for HT loop and the other is for LT loop. The heat released in high-temperature evaporator is:

$$\dot{Q}_{eva} = \dot{m}_g(h_{10} - h_{12}) \tag{2}$$

where h_{12} represents heat source specific enthalpy at high-temperature evaporator export. The high-temperature condenser is one of LT evaporators. The heat absorbed by the other evaporator is:

$$\dot{Q}_{evab2} = \dot{m}_g(h_{12} - h_{13}) \tag{3}$$

The working fluid mass flow rate in HT loop could be calculated by:

$$\dot{m}_r = \frac{\dot{m}_g(h_{10} - h_{12})}{h_1 - h_6} \tag{4}$$

where h_6 and h_1 are the specific enthalpies at high-temperature evaporator import and export, respectively.

The work capacity of high-temperature expander is:

$$\dot{W}_t = \dot{m}_r \cdot (h_1 - h_2) = \dot{m}_r \cdot (h_1 - h_{2s}) \cdot \eta_t \tag{5}$$

where η_t represents expander isentropic efficiency; h_{2s} and h_2 represent ideal and actual specific enthalpies at high-temperature expander export, respectively.

The working fluid mass flow in low-temperature loop can be calculated by:

$$\dot{m}_{rb} = \frac{\dot{m}_r(h_2 - h_5)}{h_{8b} - h_{6b}} \tag{6}$$

where h_5 represents specific enthalpy at high-temperature pump import; h_{8b} and h_{6b} represent specific enthalpies of low-temperature loop fluid at shared heat exchanger import and export, respectively.

Similarly, the power generated by low-temperature expander is:

$$\dot{W}_{tb} = \dot{m}_{rb} \cdot (h_{1b} - h_{2b}) = \dot{m}_{rb} \cdot (h_{1b} - h_{2sb}) \cdot \eta_t \tag{7}$$

where h_{1b} represents specific enthalpy at low-temperature expander import; h_{2sb} and h_{2b} are ideal and actual specific enthalpies at low-temperature expander export, respectively.

The pump power consumption is:

$$\dot{W}_{\text{p}} = \dot{m}_{\text{r}} \cdot (h_6 - h_5) = \frac{\dot{m}_{\text{r}} \cdot (h_{6s} - h_5)}{\eta_{\text{p}}} \tag{8}$$

$$\dot{W}_{\text{pb}} = \dot{m}_{\text{rb}} \cdot (h_{6b} - h_{5b}) = \frac{\dot{m}_{\text{rb}} \cdot (h_{6sb} - h_{5b})}{\eta_{\text{p}}} \tag{9}$$

where η_{p} represents pump isentropic efficiency; h_{5b} represents specific enthalpy at low-temperature expander import.

$$\dot{W}_{\text{T}} = \dot{W}_{\text{t}} + \dot{W}_{\text{tb}} \tag{10}$$

$$\dot{W}_{\text{P}} = \dot{W}_{\text{p}} + \dot{W}_{\text{pb}} \tag{11}$$

The net power output is the difference of expander work and pump consumption. Therefore, the net power output and thermal efficiency is:

$$\dot{W}_{\text{sys}} = \dot{W}_{\text{T}} - \dot{W}_{\text{P}} = \dot{W}_{\text{t}} + \dot{W}_{\text{tb}} - \dot{W}_{\text{p}} + \dot{W}_{\text{pb}} \tag{12}$$

$$\eta_{\text{sys}} = \frac{\dot{W}_{\text{sys}}}{\dot{Q}_{\text{g}}} \tag{13}$$

2.3.2. Exergy Modeling

Exergy analysis has proven to be an effective method for thermodynamic analysis. Exergy is the representative of energy available in thermodynamic processes. Exergy analysis considers not only the energy amount, but also the energy quality. The specific exergy of each steady point is:

$$e_i = h_i - h_0 - T_0(s_i - s_0) \tag{14}$$

where h_i and s_i represent the state point specific enthalpy and specific entropy; where h_0 and s_0 represent the specific enthalpy and specific entropy based on the environment. Exergy loss in each process is:

$$\dot{E} = \sum \dot{E}_{in} - \sum \dot{E}_{out} - \dot{W} \tag{15}$$

where $\sum \dot{E}_{in}$ and $\sum \dot{E}_{out}$ are the total exergies input and output, respectively. Therefore, the exergy loss in each component can be described as:

$$\dot{E}_{\text{eva}} = \dot{m}_{\text{r}}(e_6 - e_1) + \dot{m}_{\text{g}}(e_{10} - e_{12}) \tag{16}$$

$$\dot{E}_{\text{t}} = \dot{m}_{\text{r}}(e_1 - e_2) - \dot{W}_{\text{t}} \tag{17}$$

$$\dot{E}_{\text{con}} = \dot{m}_{\text{r}}(e_2 - e_5) + \dot{m}_{\text{rb}}(e_{6b} - e_{8b}) \tag{18}$$

$$\dot{E}_{\text{p}} = \dot{m}_{\text{r}}(e_5 - e_6) + \dot{W}_{\text{p}} \tag{19}$$

$$\dot{E}_{\text{evab2}} = \dot{m}_{\text{rb}}(e_{8b} - e_{1b}) + \dot{m}_{\text{g}}(e_{12} - e_{13}) \tag{20}$$

$$\dot{E}_{\text{tb}} = \dot{m}_{\text{rb}}(e_{1b} - e_{2b}) - \dot{W}_{\text{tb}} \tag{21}$$

$$\dot{E}_{\text{conb}} = \dot{m}_{\text{rb}}(e_{2b} - e_{5b}) + \dot{m}_{\text{c}}(e_{14} - e_{17}) \tag{22}$$

$$\dot{E}_{\text{pb}} = \dot{m}_{\text{rb}}(e_{5b} - e_{6b}) + \dot{W}_{\text{pb}} \tag{23}$$

Therefore, the system total exergy loss is:

$$\dot{E}_{\text{sys}} = \dot{E}_{\text{eva}} + \dot{E}_{\text{t}} + \dot{E}_{\text{con}} + \dot{E}_{\text{p}} + \dot{E}_{\text{evab2}} + \dot{E}_{\text{tb}} + \dot{E}_{\text{conb}} + \dot{E}_{\text{pb}} \tag{24}$$

2.3.3. Entropy Modeling

The entropy generation rate in entropy analysis is related to the irreversible loss and output power, and it represents the work ability. The concept of entropy was proposed by Clausius [34] in 1854. The larger entropy generation rate means greater irreversible loss in irreversible process. Therefore, entropy generation is always used for thermodynamic optimization. For any thermodynamic process, the entropy balance can be described as:

$$\dot{S}_{\text{g}} + \dot{S}_{\text{f}} = \Delta \dot{S} \tag{25}$$

where $\dot{S}_{\text{g}}$ represents entropy generation rate, while $\dot{S}_{\text{f}}$ represents entropy flow into the system. When the system keeps steady, the entropy change will not change with time, which means $\Delta \dot{S}$ is zero, hence, $\dot{S}_{\text{g}}$ could be expressed as:

$$\dot{S}_{\text{g}} = -\dot{S}_{\text{f}} = \dot{S}_{\text{f,out}} - \dot{S}_{\text{f,in}} \tag{26}$$

$$\dot{S}_{\text{f,in}} = \int_{T_0}^{T_{10}} \frac{c_g \dot{m}_g dT}{T} + \int_{T_0}^{T_{14}} \frac{c_c \dot{m}_c dT}{T} = c_g \dot{m}_g \ln \frac{T_{10}}{T_0} + c_c \dot{m}_c \ln \frac{T_{14}}{T_0} \tag{27}$$

$$\dot{S}_{\text{f,out}} = \frac{\dot{Q}_0}{T_0} \tag{28}$$

where $\dot{S}_{\text{f,in}}$ and $\dot{S}_{\text{f,out}}$ represent the entropies flow into and out of system, respectively; $\dot{Q}_0$ represents the heat flowing into the environment; c_g and c_c represent the specific heat capacities of heat and cold source, respectively. The heat released to the environment is:

$$\dot{Q}_0 = c_g \dot{m}_g (T_{10} - T_0) + c_c \dot{m}_c (T_{14} - T_0) - \dot{W}_{\text{sys}} \tag{29}$$

where $\dot{m}_c$ represents cooling water mass flow rate.

2.3.4. Entransy Modeling

Entransy concept is first proposed by Guo et al. [35] by thermoelectric comparison. They defined entransy as the total heat transfer capacity relative to 0 °C environment, which was similar to the electrical energy in a capacitor. Cheng et al. [36] discussed the thermodynamic process of Brayton cycles and found that calculating highest entransy could benefit parameter optimization design. Li et al. [37] conducted a model to optimize TSORC based on the entransy concept, and found that TSORC reduced the entransy dissipation because of the high average evaporation temperature. Zhu et al. [38] compared the thermodynamic performance using R123 and pentane. They found that entransy loss and entransy efficiency were suitable for optimization except entransy dissipation. For any object, the entransy could be expressed as:

$$G = \frac{1}{2} UT \tag{30}$$

where U and T represent internal energy and temperature, respectively. The internal energy could be expressed as:

$$U = cmT \tag{31}$$

where c represents the specific heat capacity at constant volume.

According to the first thermodynamics law, the heat-work conversion process is:

$$\delta Q = dU + \delta W \tag{32}$$

where δQ represents endothermic rate; dU and δW represents internal energy change rate and output power, respectively. The equation can be converted to:

$$\oint T\delta Q = \oint d\left(\frac{1}{2}cmT^2\right) + \oint T\delta W \tag{33}$$

For the ORC system, $\frac{1}{2}cmT^2$ represents working fluid entransy, which is a state quantity. After a cycle, the equation becomes as:

$$\oint T\delta Q = \oint T\delta W \tag{34}$$

The right part is work entransy outflow. The left part is net entransy flow into working fluid, which is:

$$\dot{G}_{net} = \oint T\delta Q \tag{35}$$

Part of entransy is used for heating or cooling, and some is used for expander work. The remaining is for low temperature source. Thus, from the perspective of heat source, the entransy loss in ORC system occurs in thermal cycle process, which could be expressed as:

$$\dot{G}_{loss} = \dot{G}_{net} + \dot{G}_{dis} = \dot{G}_h - \dot{G}_l \tag{36}$$

where $\dot{G}_h$ and $\dot{G}_l$ represent entransy inflow from high temperature source, and entransy outflow into low temperature source; $\dot{G}_{dis}$ represents the entransy dissipation because of the temperature difference between heat and cold source.

The entransy inflow could be calculated by:

$$\dot{G}_h = \frac{1}{2}c_g \dot{m}_g \left(T_{10}^2 - T_0^2\right) \tag{37}$$

The entransy outflow can be calculated by:

$$\dot{G}_l = \frac{1}{2}c_c \dot{m}_c \left(T_0^2 - T_{14}^2\right) + \dot{Q}_0 T_0 \tag{38}$$

where first part on right side is entransy flow into cold stream, and the other part is entransy loss due to heat flow released into the atmosphere. Therefore, the entransy loss rate can be expressed as:

$$\dot{G}_{loss} = \frac{1}{2}c_g \dot{m}_g \left(T_{10}^2 - T_0^2\right) + \frac{1}{2}c_c \dot{m}_c \left(T_{14}^2 - T_0^2\right) + \dot{W}_{sys} T_0 \tag{39}$$

Accordingly, the entransy efficiency is

$$G_x = \frac{\dot{G}_l}{\dot{G}_h} \tag{40}$$

3. Results and Discussion

There are many significant parameters affecting the system performance, including evaporation temperature, condensation temperature, superheating degree, pinch point temperature difference (PPTD), and working fluid mixture component. The parametric analysis of DORC using mixture fluids is conducted. The effects of evaporation temperature and condensation temperature of HT loop and LT loop are addressed in Sections 3.1 and 3.2, respectively. The effects of superheating degree and PPTD are analyzed in Section 3.3. Eventually, the effects of working fluid mixture component are addressed in Section 3.4.

3.1. Effects of Evaporation Temperature and Condensation Temperature of HT Loop

The effects of HT loop evaporation temperature on system performance are plotted in Figure 3. The HT loop evaporation temperature is in the range of 463–523 K, and LT loop evaporation temperature is set as 403 K. The HT and LT loop condensation temperatures are 413 K and 303 K, respectively. The working fluid mixture component in HT loop is assumed as 0.5/0.5.

As shown in Figure 3a, the net output power increases first and then decreases as HT loop evaporation temperature for all mixtures, which means there is an inflection point on the curve of the net output power. The optimal evaporation temperature is 488 K corresponding to maximum output power 51.58 kW for hexane/toluene. The net output power depends on the difference of expander shaft work and pump power consumption, and the expander work accounts for the main part which is determined by the enthalpy difference between expander import and export. The fluid mass flow rate decreases nonlinearly as HT loop evaporation temperature increases. The specific enthalpy difference in evaporator shows a linear growth. Therefore, HT loop expander work presents a parabola trend. The curve characteristics for three working fluid mixtures are same, while the optimal evaporation temperatures for cyclopentane/toluene and R123/toluene are 483 K and 473 K, which appear earlier than the case for hexane/toluene and the corresponding maximum output power is 48.64 kW and 45.61 kW, respectively.

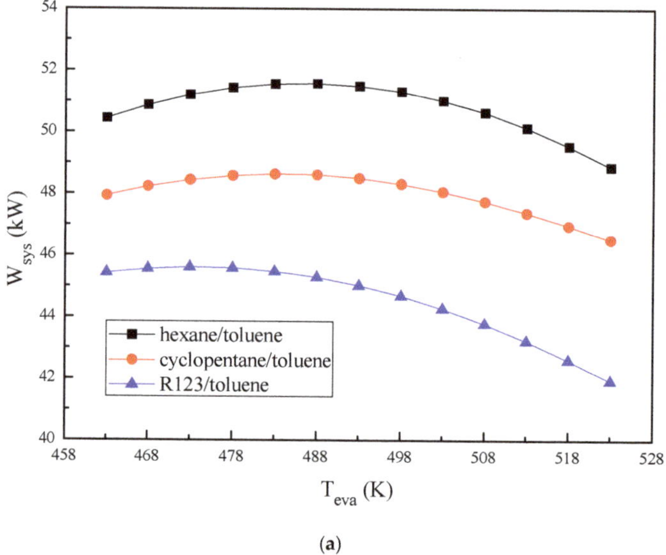

(**a**)

Figure 3. *Cont.*

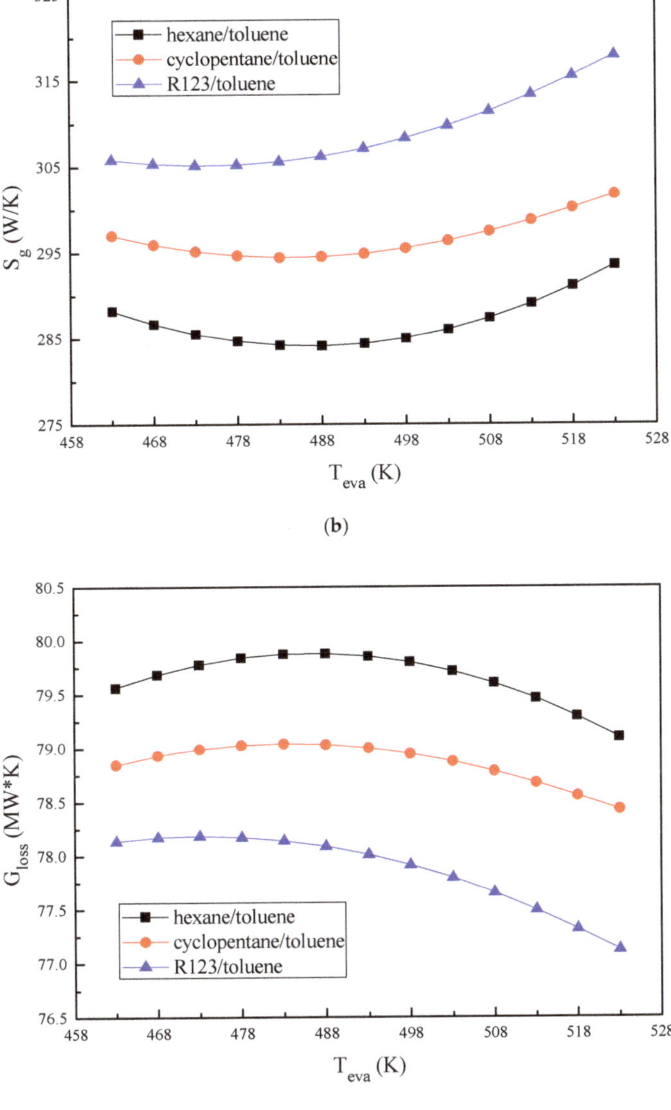

(b)

(c)

Figure 3. *Cont.*

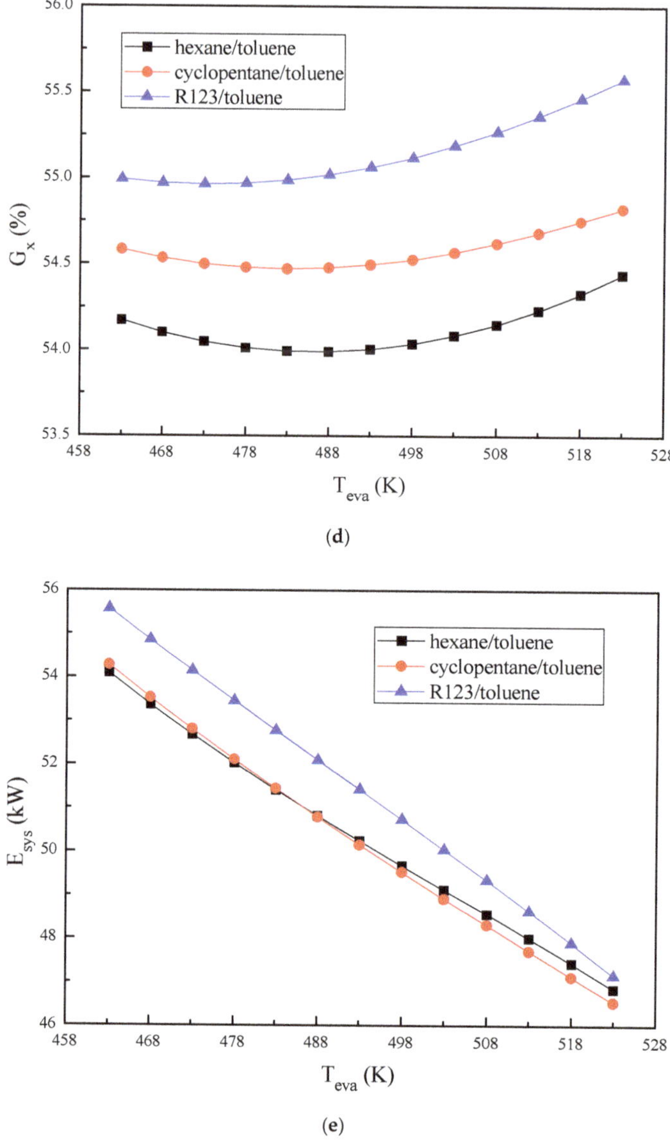

(**d**)

(**e**)

Figure 3. Effects of the high-temperature (HT) loop evaporation temperature on system performance: (**a**) Variation of net power output; (**b**) variation of entropy generation rate; (**c**) variation of entransy loss rate; (**d**) variation of entransy efficiency; (**e**) variation of exergy loss.

Entropy generation is indispensable in irreversible system. As shown in Figure 3b, the entropy generation rate for hexane/toluene decreases first and then increases as HT loop evaporation temperature increases, and reaches the minimum value 0.284 kW/K at the maximum output power point. The variation is same as the ones for cyclopentane/toluene and R123/toluene. Their minimum entropy generation rates are 0.294 kW/K and 0.305 kW/K respectively which are both higher than the one for hexane/toluene. Taking the system using hexane/toluene as example, when HT loop evaporation temperature falls below 488 K, net power output keeps increasing and the water mass flow requirement decreases. As a part of the entropy flowing out of system, the section $c_c \dot{m}_c (T_{14} - T_0)/T_0 - c_c \dot{m}_c \ln(T_{14}/T_0)$ decreases, leading to the decrease of entropy generation rate. When $T_{eva} \geq 488$ K, the net power output starts decreasing and water mass flow rate increase accordingly. The entropy generation rate yields a reverse trend because the growth rate of $\dot{W}_{sys}/T_0$ is larger than the decrement of $c_c \dot{m}_c (T_{14} - T_0)/T_0 - c_c \dot{m}_c \ln(T_{14}/T_0)$. The larger entropy generation rate means the greater irreversible loss. Hence, using hexane/toluene as the working fluid is better than others from the perspective of entropy analysis.

The entransy loss rate and entransy efficiency characterize the heat transfer capability from the perspective of hot and cold steam. Figure 3c indicates that the entransy loss rate increases to a maximum point and then reduces gradually with the rise of HT loop temperature. The maximum entransy loss rates are 79.88 MW·K (hexane/toluene), 79.04 MW·K (cyclopentane/toluene), and 78.19 MW·K (R123/toluene), respectively. According to the Figure 3c,d, entransy loss rate and entransy efficiency have adverse growth characteristics and the minimum entransy efficiency are 53.99% (hexane/toluene), 54.47% (cyclopentane/toluene), and 54.97% (R123/toluene), respectively. This proves that the system heat transfer capacity using hexane/toluene is optimal, followed by the one for cyclopentane/toluene and R123/toluene successively. It should be noted that the maximum entransy loss rates and highest power output occur under same temperature, which means there is a positive correlation between net power output and entransy loss. For theoretical study, when $T_{eva} < T_{eva,op}$, the entransy flowing into the system keeps constant under fixed exhaust gas inlet temperature. The water mass flow rate gradually reduces, leading to the decrease of the entransy flowing into water, while the decrement is lower than the increment of the entransy loss released into the environment. Therefore, the total entransy loss keeps rising. When $T_{eva} \geq T_{eva,op}$, the decline of the entransy flowing to the both water and environment results in the reduction of the total entransy loss.

Exergy analysis could study the energy loss during heat transfer process by comparing exergy destroyed in various components, as shown in Figure 3e. The system exergy loss decreases with the rise of HT loop temperature for all mixtures. Figure 4a–c reveal the variations of exergy loss rate and its distributions in the DORC system. As for hexane/toluene, the pump exergy loss is small enough to be neglected compared to other components, while the exergy destroyed in the evaporator and condenser accounts for the largest proportion, which are 22–31% and 33–35% respectively. The exergy loss in heat exchangers (H1, H2, H4) and LT loop expander decreases, while that of the heat exchanger H3 and the HT loop expander increases. When the HT loop evaporation temperature increases, the exergy loss in HT loop evaporator decreases. Although HT loop mixture mass flow rate is decreasing, the exergy change between HT loop expander import and export keeps increasing, and its increment is greater than the expander work, leading to the increase of expander exergy loss. The exergy analysis of other components can be verified in the same way. It can also be found that the exergy loss distributions for three mixtures have a similar trend.

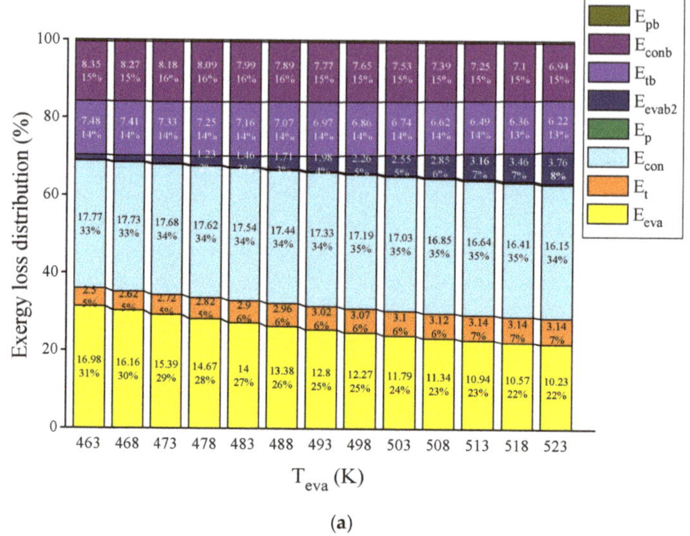

(**a**)

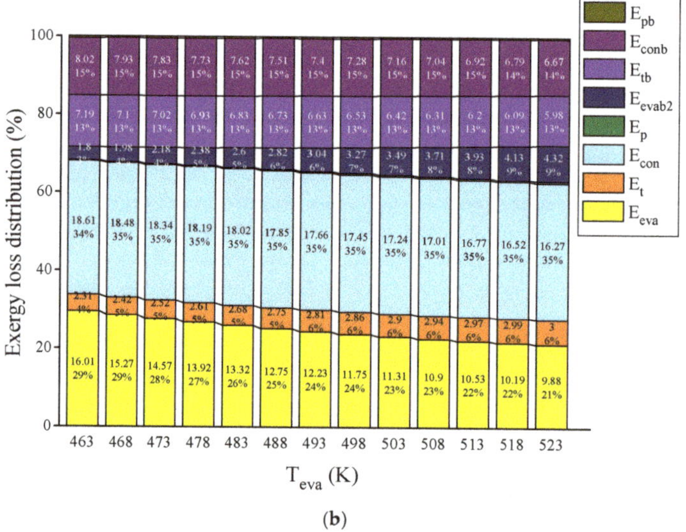

(**b**)

Figure 4. *Cont.*

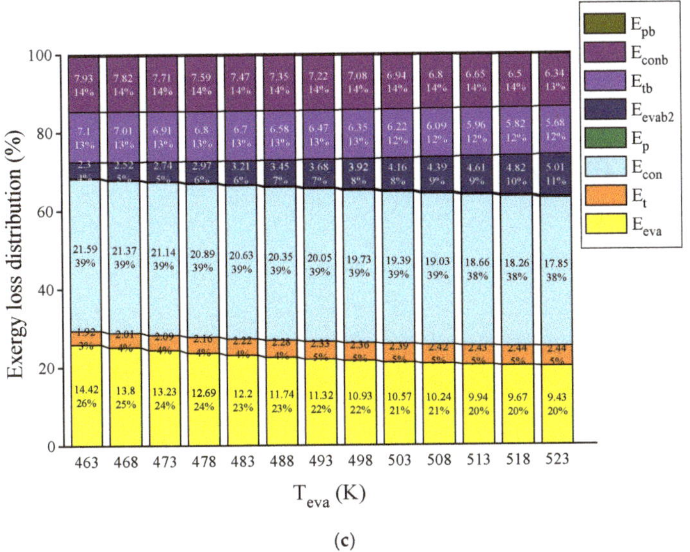

Figure 4. Variation of exergy loss distribution with HT loop evaporation temperature: (**a**) Hexane/toluene; (**b**) cyclopentane/toluene; (**c**) R123/toluene.

The simulation results for three mixture working fluids showed extremely similar characteristics. Thus, the researches below are only carried out for hexane/toluene as a typical case. Figure 5 presents the influence of HT loop condensation temperature on system performance. The HT loop evaporation temperature and LT loop condensation temperature are 513 K and 303 K, respectively. The net power output decreases as HT loop condensation temperature rises. This phenomenon can be explained as, the HT loop fluid mass flow rate keeps constant with the rise of HT loop condensation temperature, while the specific enthalpy difference between the expander import and export continuously decreases. Therefore, the expander shaft work decreases and the net power output decreases accordingly. When HT loop condensation temperature increases gradually, the water mass flow rate gradually decreases, resulting in the decline of the section $c_c \dot{m}_c \left(T_{14} - T_0\right)/T_0 - c_c \dot{m}_c \ln\left(T_{14}/T_0\right)$, but the dropping rate is less than that of $\dot{W}_{sys}/T_0$, so the entropy generation rate gradually improves. The entransy flowing into the system remains unchanged and water mass flow rate decreases, leading to the improvement of the entransy taken in from the water, while the portion is lower than the entransy loss released into the environment. Therefore, the entransy flow out of system increases, resulting in the decrease of total entransy loss rate. The curve characteristic of entransy loss rate is same as net power output and total exergy loss. The entropy generation rate is inversely proportional to the entransy loss rate as previous research results.

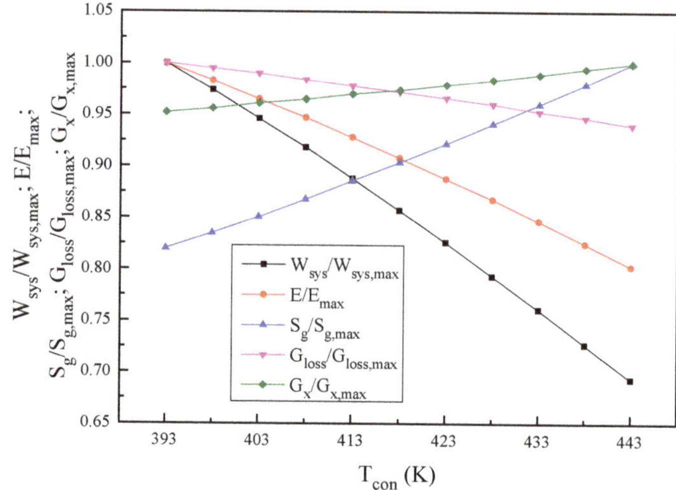

Figure 5. Effects of HT loop condensation temperature for hexane/toluene.

3.2. Effect of Evaporation Temperature and Condensation Temperature of LT Loop

For more discussion, some detailed 3-D coordinates are presented to describe visually the effects of the LT loop evaporation temperature and LT loop condensation temperature on system performance using hexane/toluene. The input data is set that T_{eva} = 513 K, T_{con} = 423 K, T_{evab} varies from 363 K to 413 K, T_{conb} varies from 293 K to 333 K.

Figure 6a shows that net power output increases as LT loop evaporation temperature increases. The specific enthalpy differences in LT loop expander and pump both increase as LT loop evaporation temperature increases, causing the increase of required mixture mass flow. The enthalpy difference in LT loop expander keeps rising under the comprehensive of mass flow rate and specific enthalpy. Besides, the LT loop pump power consumption is small enough to be ignored compared with the expander work; while the variation of the LT evaporation or condensation temperature could not affect the HT loop work capacity. Therefore, it can be approximated that there is similar variation between net power output and expander work, which means the net output power gradually increases as LT loop evaporation temperature and the maximum output power could reach 46 kW. In addition, the system entropy generation rate keeps falling as shown in Figure 6b. When LT loop evaporation temperature rises, the most parameters of HT loop could not be affected. The net power output increases and the water flow requirement gradually decreases, so the section $c_c \dot{m}_c \ln(T_{14}/T_0)$ and $c_c \dot{m}_c (T_{14} - T_0)/T_0$ both decreases while the decrement of latter is greater, which eventually leads to the decrease of entropy generation rate. The variation indicates that the system irreversible loss increases as LT loop evaporation temperature increases. Simultaneously, the entransy flowing into the water decreases, and its decrement is lower than that of entransy loss released into the atmosphere. Therefore, the entransy loss rate keeps rising, and the system heat transfer capability will get better with the improvement of LT loop evaporation temperature, as shown in Figure 6c.

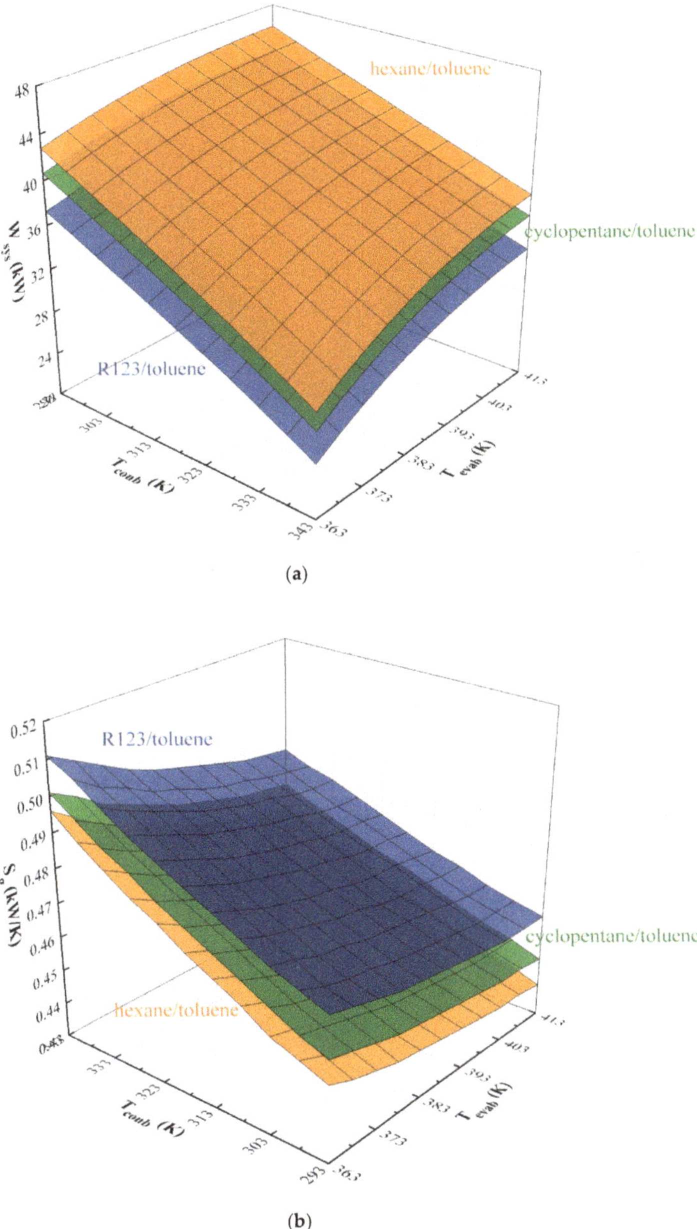

(**a**)

(**b**)

Figure 6. *Cont.*

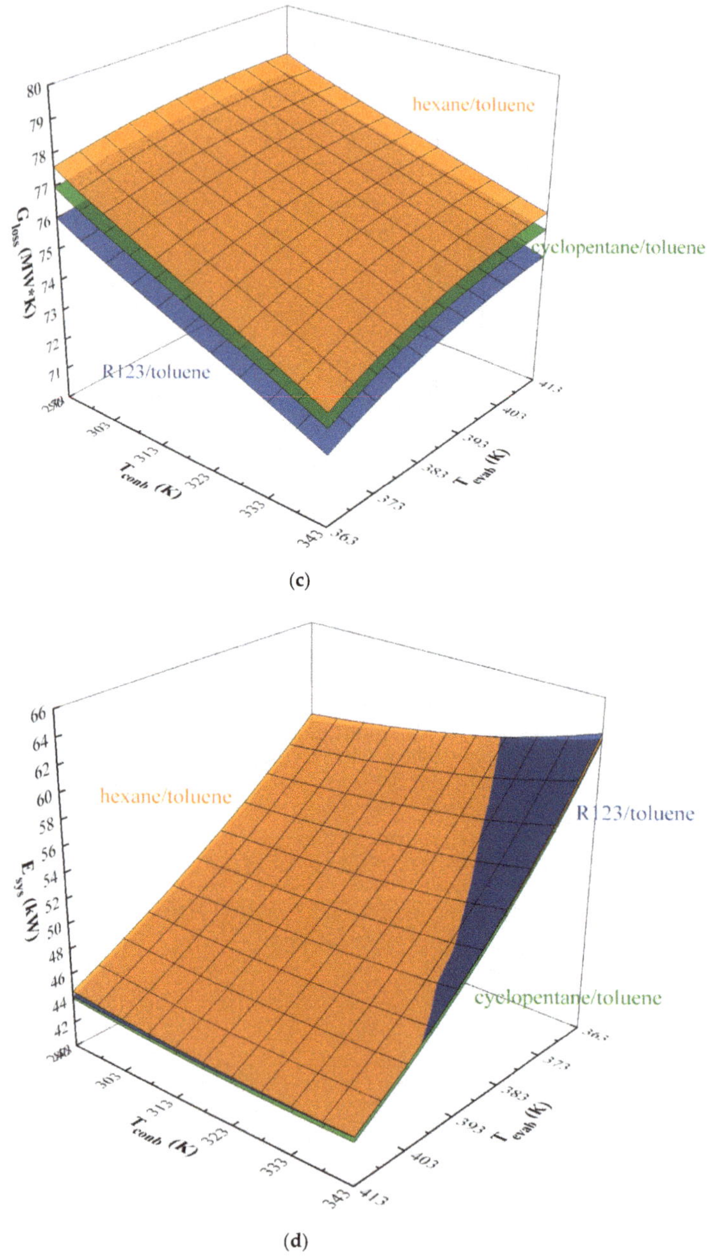

Figure 6. The effect of evaporation and condensation temperature of LT loop: (**a**) Power output; (**b**) entropy generation rate; (**c**) entransy loss rate; (**d**) exergy loss.

Figure 6 also presents the effects of LT loop condensation temperature on system performance. when the LT loop condensation temperature rises, the LT loop pump export enthalpy increases gradually, leading to the increase of LT loop fluid mass flow rate. The LT loop expander work decreases under the comprehensive effect of LT loop fluid mass flow rate and expander specific enthalpy difference.

Therefore, the net output power decreases. In addition, the enthalpy difference of R245fa/pentane in LT loop condenser keeps dropping, and the enthalpy difference of cooling water changes in a reverse trend. Therefore the water mass flow rate presents a downward trend under the combined influence of LT loop fluid mass flow rate and condenser enthalpy change, leading to a decline of the entropy flow out of system. Meanwhile, the section $c_c \dot{m}_c \ln(T_{14}/T_0)$ decreases entropy flow into the system both decrease accordingly, which eventually leads to the increase of entropy generation rate. There is a reverse relationship between system irreversible loss and LT loop condensation temperature. In addition, the entransy flowing into the water decreases and its portion is greater than the decrement of the entransy loss released into the environment, so the entransy outflow of the system keeping rising, and the total entransy loss rate keeps dropping accordingly, as shown in Figure 6c.

As presented in Figure 6b,c, the net power output has a position correlation with entransy loss rate, and is correlated with entropy generation rate inversely. Figure 6d indicates that the variation of exergy loss has an inflection point, and the various condensation temperatures corresponding to the lowest exergy loss are different. Therefore, the exergy analysis cannot be carried out for horizontal contrast under different environmental parameters.

3.3. Effect of Superheating Degree and PPTD of HT Loop

Figure 7a,b show the variation of entropy generation rate and entransy loss rate with the superheating degree and PPTD using hexane/toluene, cyclopentane/toluene, and R123/toluene. The temperatures are set that: T_{eva} = 513 K, T_{con} = 423 K, T_{evab} = 413 K, and T_{conb} = 303 K. The PPTD varies from 20 K to 40 K, and the superheating degree varies from 0 K to 20 K.

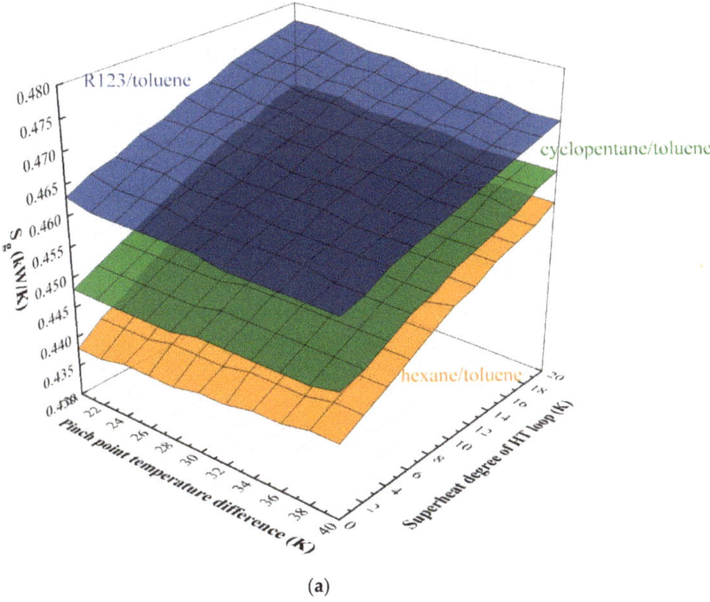

(a)

Figure 7. *Cont.*

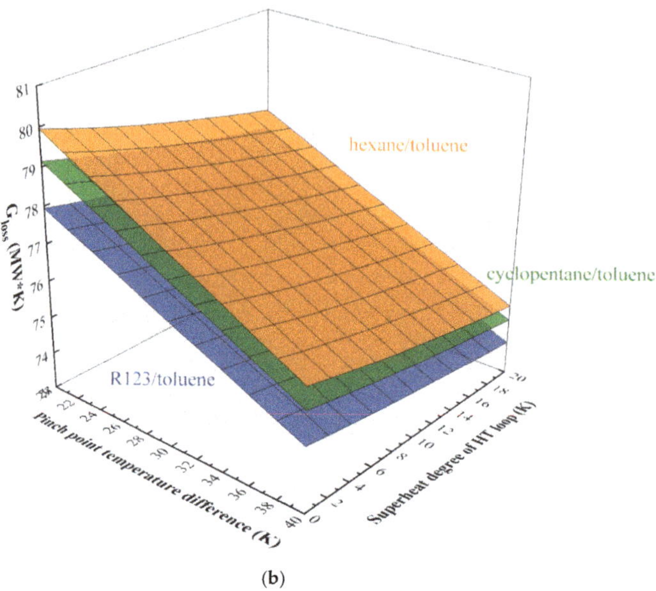

(b)

Figure 7. Effects of superheating degree and pinch point temperature difference of HT loop: (**a**) Entropy generation rate; (**b**) entransy loss rate.

The entropy generation rate and entransy loss rate for the three mixtures have a similar behavior to superheating degree and PPTD as Figure 7a. When the superheating degree increases, the expander import temperature increases, and the specific enthalpy at HT loop expander import and export both increase. The HT loop mixture mass flow rate gradually decreases, resulting in the decrease of specific enthalpy difference in HT loop expander. Thus, the expander work keeps decreasing, which eventually causes the dropping of system net output power. Meanwhile, the required mass flow of LT loop mixture would be less because of the decrease of the enthalpy change in LT loop expander. The required water mass flow rate will reduce accordingly, thus the entropy generation rate increases with the increase of both entropy inflow and outflow. The system using R123/toluene has the highest entropy generation which varies from 0.463 kW/K to 0.477 kW/K, while the lowest entropy generation is obtained by hexane/toluene in the range of 0.438–0.460 kW/K. In addition, when the PPTD increases, the entropy generation for hexane/toluene and cyclopentane/toluene increases and the one for R123/toluene yields a reverse trend. Despite that PPTD has an effect on the variation of exhaust gas heat in HT loop evaporator and then affects the subsequent parameters, the effect of PPTD on entropy generation rate could be negligible.

Figure 7b presents that the entransy loss rate decreases with the increase of both superheating degree and PPTD, and the highest entransy loss is obtained by the system using hexane/toluene, followed by ones using cyclopentane/toluene and R123/toluene successively. For the system using hexane/toluene, when the superheating degree increases at a specified PPTD of 20 K, the entransy loss rate decreases from 79.95 MW·K to 78.15 MW·K. The entransy flowing into the system keeps unchanged when the superheating degree varies, while HT loop mixture mass flow rate and the system output power both decrease. The heat released to the environment increases because $c_c \dot{m}_c (T_{14} - T_0)$ and the system output power vary in different degrees as Equation (29), leading to the increase of entransy flowing out of the system. The entransy outflow increases because that the flow absorbed from cooling water is less than that released to the environment. Therefore, the entransy loss rate decreases with the increase of superheating degree. Besides, when the PPTD increases gradually, the temperature and specific enthalpy of exhaust gas in HT loop evaporator export both increase, while the required mass flow of HT loop working fluid decreases, leading to the decrease of mass flow rate of LT loop working fluid and cooling water successively. The specific enthalpies in all state points will not be affected, so the net power output decreases due to the reduction in expander work. The expander work decrement is greater than the heat capacity absorbed by cold water from the environment, resulting in the increase of system heat flow released to the environment and entransy outflow. Therefore, the system entransy loss rate decreases with the increase of PPTD.

3.4. Effects of Working Fluid Mixture Component

In this section, the influence of working fluid mixture component on entransy loss and entropy generation rate is shown in Figure 8a,b. The evaporation temperature and condensation temperature of HT loop are 503 K and 403 K, respectively, and the evaporation temperature and condensation temperature of LT loop are 393 K and 303 K, respectively.

Figure 8a shows the variation of entransy loss rate with mass fraction of toluene. When the toluene mass fraction increases from 0 to 1, the entransy loss rate for hexane/toluene decreases from 81.09 MW·K to 71.20 MW·K, while the ones for cyclopentane/toluene and R123/toluene both decrease first and then increase. The entransy loss rate for cyclopentane/toluene decreases from 71.20 MW·K to 70.85 MW·K at the toluene mass fraction of 0.8 and then increases to 78.25 MW·K. Meanwhile, the toluene mass faction for the system using R123/toluene increases just from 0.4 to 1.0 restricted by the maximum evaporation temperature. The range of the entransy loss rate for R123/toluene is 70.46–71.76 MW·K and the minimum is 70.46 MW·K at the mass fraction of 0.7. The entransy loss rate for hexane/toluene is significantly higher than that for cyclopentane/toluene and R123/toluene because of the high specific enthalpy than other mixtures at the same evaporation temperature. Meanwhile, the system power output for hexane/toluene is greater than other mixtures. Therefore, the system using hexane/toluene has better heat transfer characteristic.

As observed in Figure 8b, the entropy generation rate yields a reverse trend to the entransy loss rate. The entropy generation rate for hexane/toluene increases gradually from 0.202 kW/K to 0.321 kW/K, while the ones for cyclopentane/toluene and R123/toluene both increase first and then decrease again. The entropy generation rate for cyclopentane/toluene is in the range of 0.236–0.325 kW/K with the maximum value of 0.325 kW/K at the mass fraction of 0.8. The entransy loss rate for R123/toluene increases from 0.315 kW/K to 0.321 kW/K with the maximum of 0.330 kW/K at the mass fraction of 0.7. The system using R123/toluene owns the highest entropy generation rate, while the lowest entropy generation rate is obtained by hexane/toluene.

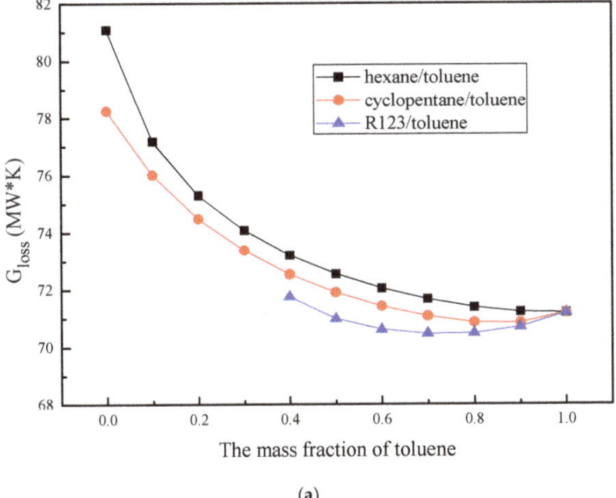

(a)

Figure 8. *Cont.*

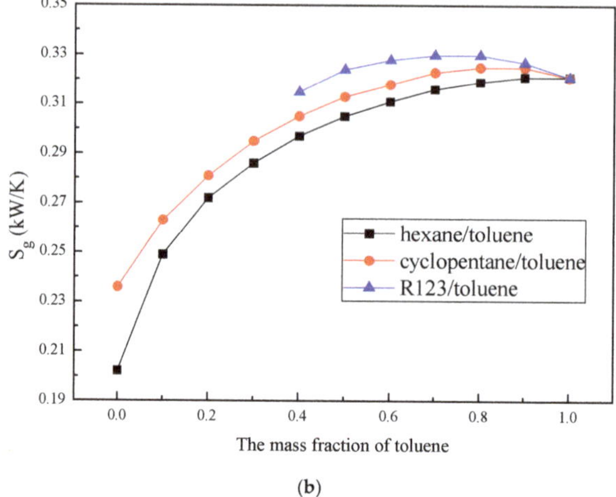

(**b**)

Figure 8. Effects of the working fluid mixture component: (**a**) Variation of the entransy loss rate; (**b**) variation of the entropy generation rate.

4. Conclusions

The exergy, entropy, and entransy analysis of DORC using mixture fluids (hexane/toluene, cyclopentane/toluene and R123/toluene) have been investigated. Effects of evaporation temperature and condensation temperature of HT loop, evaporation temperature and condensation temperature of LT loop, superheating degree, PPTD and working fluid mixture component on system exergy loss, entropy generation rate, entransy loss rate, and entransy efficiency are discussed. The analysis process is mainly through the combination of REFPROP and MATLAB. The major conclusions drawn from the simulation are:

- Hexane/toluene, cyclopentane/toluene, and R123/toluene are selected as high-temperature working fluids, and detailed analysis are proceeded under constant heat source condition. When the evaporator inlet temperature and condenser outlet temperature change, although the optimal evaporation temperature and maximum output power is different, the variation characteristics of entropy generation rate, entransy loss rate, entransy efficiency, and exergy loss are similar. Hence, the effects of operation temperature on these objects do not matter whatever the working fluid.
- The system net power output and entransy loss rate both increase first and then decrease as the HT evaporation temperature increases, while decrease as the HT or LT condensation temperature increases. The entropy generation rate and entransy efficiency show opposite variation.
- The curve characteristic of entransy loss rate is same as that of system power output, while the entropy generation rate yields a reverse trend to the entransy efficiency. The proposed method of the entropy and entransy analysis could be used for system optimization. Whereas the exergy analysis could be only utilized under fixed stream conditions.
- The entransy loss rate decreases with the increase of both superheating degree and PPTD. The entropy generation rate increases as superheating degree increases while PPTD has no effect on it. The entransy loss rate for hexane/toluene keeps dropping with the rising mass fraction of toluene, but that for cyclopentane/toluene and R123/toluene it decreases first and then increases. R123/toluene owns the lowest entransy loss rate and the highest entransy loss rate is obtained by hexane/toluene under same mass fraction. The entropy generation for three mixtures show a reverse trend with entransy loss rate.

Author Contributions: Conceived the presented idea, S.W. and Y.-Q.F.; theory and computation, Q.W., W.Z. and X.W.; analytical method verifying. Y.-Z.L., Y.W. and L.Y. All authors have read and agreed to the published version of the manuscript.

Funding: This research work has been supported by the National Natural Science Foundation of China (51806081), the Key Research and Development Program of Jiangsu Province, China (BE2019009-4), the Natural Science Foundation of Jiangsu Province (BK20180882), the China Postdoctoral Science Foundation (2018M632241), the 2019 Scholarship of the Knowledge Center on Organic Rankne Cycle (KCORC, www.kcorc.org), the Key Research and Development Program of Zhenjiang City, China (SH2019008), and the Key Project of Taizhou New Energy Research Institute, Jiangsu University, China (2018-20).

Conflicts of Interest: The authors declare no conflict of interest.

Nomenclature

c	specific heat capacity, J/kg·K
e	specific exergy, J/kg
E	exergy loss, W
G_{loss}	entransy loss rate, W·K
h	specific enthalpy, J/kg
m	mass flow rate, kg/s
Q	heat quantity, J
S_g	entropy generation rate, W/K
S_f	entropy flow rate, W/K
ΔS	entropy change rate, W/K
T	temperature, K
W	power, W
η	Efficiency
Hi	heat exchanger with number i, i = 1,2,3,4
HT	high temperature
LT	low temperature
Pi	pump with number i, i = 1,2
Ti	expander with number i, i = 1,2
subscripts	
b	parameters in LT cycle
con	condensation
dis	dissipation
eva	evaporation
glide	temperature slip
g, c	heat, cold source
op	optimal
p	pump
r	working fluid
sub	Supercooling
sup	Superheating
sys	System
t	expander
1-17	state point of T-s graph
2s,6s	Corresponding isentropic point

References

1. Dai, Y.P.; Wang, J.F.; Gao, L. Parametric optimization and comparative study of organic Rankine cycle (ORC) for low grade waste heat recovery. *Energy Convers. Manag.* **2009**, *50*, 576–582. [CrossRef]
2. Bombarda, P.; Invernizzi, C.M.; Pietra, C. Heat recovery from Diesel engines: A thermodynamic comparison between Kalina and ORC cycles. *Appl. Therm. Eng.* **2010**, *30*, 212–219. [CrossRef]

3. Feng, Y.Q.; Li, B.X.; Yang, J.F.; Shi, Y. Comparison between regenerative organic Rankine cycle (RORC) and basic organic Rankine cycle (BORC) based on thermoeconomic multi-objective optimization considering exergy efficiency and levelized energy cost (LEC). *Energy Convers. Manag.* **2015**, *96*, 58–71. [CrossRef]

4. Mago, P.J.; Chamra, L.M.; Srinivasan, K.; Somayaji, S. An examination of regenerative organic Rankine cycles using dry fluids. *Appl. Therm. Eng.* **2008**, *28*, 998–1007. [CrossRef]

5. Pei, G.; Li, J.; Ji, J. Analysis of low temperature solar thermal electric generation using regenerative Organic Rankine Cycle. *Appl. Therm. Eng.* **2010**, *30*, 998–1004. [CrossRef]

6. Roy, J.P.; Misra, A. Parametric optimization and performance analysis of a regenerative organic Rankine cycle using R-123 for waste heat recovery. *Energy* **2012**, *39*, 227–235. [CrossRef]

7. Franco, A. Power production from a moderate temperature geothermal resource with regenerative organic Rankine cycles. *Energy Sustain. Dev.* **2011**, *15*, 411–419. [CrossRef]

8. Li, M.Q.; Wang, J.F.; He, W.F.; Gao, L.; Wang, B.; Ma, S.L.; Dai, Y.P. Construction of preliminary test of a low-temperature regenerative organic Rankine cycle (ORC) using R123. *Renew. Energy* **2013**, *57*, 216–222. [CrossRef]

9. Li, T.L.; Wang, Q.L.; Zhu, J.L.; Hu, K.; Fu, W.C. Thermodynamic optimization of organic Rankine cycle using two-stage evaporation. *Renew. Energy* **2015**, *75*, 654–664. [CrossRef]

10. Sun, Y.R.; Lu, J.; Wang, J.Q.; Li, T.L.; Li, Y.J.; Hou, Y.S.; Zhu, J.L. Performance improvement of two-stage serial organic Rankine cycle (TSORC) integrated with absorption refrigeration (AR) for geothermal power generation. *Geothermics* **2017**, *69*, 110–118. [CrossRef]

11. Yuan, Y.; Xu, G.; Quan, Y.; Wu, H.; Song, G.; Gong, W.; Luo, X. Performance analysis of a new deep super-cooling two-stage organic Rankine cycle. *Energy Convers. Manag.* **2017**, *148*, 305–316. [CrossRef]

12. Xi, H.; Li, M.J.; Xu, C.; He, Y.L. Parametric optimization of regenerative organic Rankine cycle (ORC) for low grade waste heat recovery using genetic algorithm. *Energy* **2013**, *58*, 473–482. [CrossRef]

13. Shu, G.Q.; Liu, L.N.; Tian, H.; Wei, H.Q.; Yu, G.P. Parametric and working fluid analysis of a dual-loop organic Rankine cycle (DORC) used in engine waste heat recovery. *Appl. Energy* **2014**, *113*, 1188–1198. [CrossRef]

14. Choi, B.C.; Kim, Y.M. Thermodynamic analysis of a dual loop heat recovery system with trilateral cycle applied to exhaust gases of internal combustion engine for propulsion of the 6800 TEU container ship. *Energy* **2013**, *58*, 404–416. [CrossRef]

15. Panesar, A.S.; Morgan, R.E.; Miché, N.D.; Heikal, M.R. A novel organic Rankine cycle system with improved thermal stability and low global warming fluids. *Int. Conf. Prod.* **2014**, *13*, 6002. [CrossRef]

16. Boyaghchi, F.A.; Sohbatloo, A. Assessment and optimization of a novel solar driven natural gas liquefaction based on cascade ORC intergrated with linear Fresnel collectors. *Energy Convers. Manag.* **2018**, *162*, 77–89. [CrossRef]

17. Sun, Q.X.; Wang, Y.D.; Cheng, Z.Y.; Wang, J.F.; Zhao, P.; Dai, Y.P. Thermodynamic and economic optimization of a dual-pressure organic Rankine cycle driven by low-temperature heat source. *Renew. Energy* **2020**, *147*, 2822–2832. [CrossRef]

18. Oyewunmi, O.A.; Markides, C.N. Thermo-Economic and Heat Transfer Optimization of Working-Fluid Mixtures in a Low-Temperature Organic Rankine Cycle System. *Energies* **2016**, *9*, 448. [CrossRef]

19. Wang, E.H.; Yu, Z.B.; Collings, P. Dynamic control strategy of a distillation system for a composition-adjustable organic Rankine cycle. *Energy* **2017**, *141*, 1038–1051. [CrossRef]

20. Collings, P.; Yu, Z.B.; Wang, E.H. A dynamic organic Rankine cycle using a zeotropic mixture as the working fluid with composition tuning to match changing ambient conditions. *Appl. Energy* **2016**, *171*, 581–591. [CrossRef]

21. Oyewunmi, O.A.; Kirmse, C.J.W.; Pantaleo, A.M.; Markides, C.N. Performance of working-fluid mixtures in ORC-CHP systems for different heat-demand segments and heat-recovery temperature levels. *Energy Convers. Mang.* **2017**, *148*, 1508–1524. [CrossRef]

22. Oyewunmi, O.A.; Taleb, A.I.; Haslam, A.J.; Markides, C.N. On the use of SAFT-VR Mie for assessing large-glide fluorocarbon working-fluid mixtures in organic Rankine cycles. *Appl. Energy* **2016**, *163*, 263–282. [CrossRef]

23. Lecompte, S.; Ameel, B.; Ziviani, D.; van den Broek, M.; De Paepe, M. Exergy analysis of zeotropic mixtures as working fluids in Organic Rankine Cycles. *Energy Convers. Manag.* **2015**, *82*, 664–677. [CrossRef]

Energies **2020**, *13*, 1301

24. Feng, Y.Q.; Luo, Q.H.; Wang, Q.; Wang, S.; He, Z.X.; Zhang, W.; Wang, X.; An, Q.S. Entropy and Entransy Dissipation Analysis of a Basic Organic Rankine Cycles (ORCs) to Recover Low-Grade Waste Heat Using Mixture Working Fluids. *Entropy* **2018**, *20*, 818. [CrossRef]
25. Zhang, T.; Zhu, T.; An, W.; Song, X.; Liu, L.; Liu, H. Unsteady analysis of a bottoming Organic Rankine Cycle for exhaust heat recovery from an Internal Combustion Engine using Monte Carlo simulation. *Energy Convers Manag.* **2016**, *165*, 878–892. [CrossRef]
26. Baldi, F.; Larsen, U.; Garbrielii, C. Comparison of different procedures for the optimization of a combined Diesel engine and organic Rankine cycle system based on ship operational profile. *Ocean Eng.* **2015**, *110*, 85–93. [CrossRef]
27. Neto, R.D.O.; Sotomonte, C.A.R.; Coronado, C.J.R.; Nascimento, M.A.R. Technical and economic analyses of waste heat energy recovery from internal combustion engines by the Organic Rankine Cycle. *Energy Convers. Manag.* **2016**, *129*, 168–179. [CrossRef]
28. Song, J.; Gu, C.W. Parametric analysis of a dual loop Organic Rankine Cycle (ORC) system for engine waste heat recovery. *Energy Convers. Manag.* **2015**, *105*, 995–1005. [CrossRef]
29. Shu, G.Q.; Gao, Y.Y.; Tian, H.; Wei, Q.G.; Liang, X.Y. Study of mixtures based on hydrocarbons used in ORC (Organic Rankine Cycle) for engine waste heat recovery. *Energy* **2014**, *74*, 428–438. [CrossRef]
30. Lecompte, S.; Ameel, B.; Ziviani, D.; Broek, M.V.D.; Paepe, M.D. Exergy analysis of zeotropic mixtures as working fluids in Organic Rankine Cycles. *Energy* **2012**, *44*, 623–632. [CrossRef]
31. Feng, Y.Q.; Hung, T.C.; Zhang, Y.N.; Li, B.X.; Yang, J.F.; Shi, Y. Performance comparison of low-grade ORCs (organic Rankine cycles) using R245fa, pentane and their mixtures based on the thermoeconomic multi-objective optimization and decision makings. *Energy* **2015**, *93*, 2018–2029. [CrossRef]
32. Song, J.; Gu, C.W. Performance analysis of a dual-loop organic Rankine cycle (ORC) system with wet steam expansion for engine waste heat recovery. *Appl. Energy* **2015**, *56*, 280–289. [CrossRef]
33. Lemmon, E.W.; Huber, M.L.; Mclinden, M.O. NIST Standard Reference Database 23: Reference fluid thermodynamic and transport properties-REFPROP, version 9.1, Standard Reference Data Program, National Institute of Standards and Technology. *NIST NSRDS* **2010**, *22*, 1–49.
34. Zhao, K.H.; Luo, W.Y. *Thermal*; High Education Press: Beijing, China, 2002; pp. 1–222. (In Chinese)
35. Guo, Z.Y.; Zhu, H.Y.; Liang, X.G. Entransy-A physical quantity describing heat transfer ability. *Int. J. Heat Mass Transfer.* **2007**, *50*, 2545–2556. [CrossRef]
36. Cheng, X.T.; Liang, X.G. Entransy loss in thermodynamic processes and its application. *Energy* **2012**, *41*, 964–972. [CrossRef]
37. Li, T.L.; Yuan, Z.H.; Zhu, J.L. Entransy dissipation/loss-based optimization of two-stage organic Rankine cycle (TSORC) with R245fa for geothermal power generation. *Sci. China Technol. Sci.* **2016**, *59*, 1524–1536. [CrossRef]
38. Zhu, Y.D.; Hu, Z.; Zhou, Y.D.; Jiang, L.; Yu, L.J. Applicability of entropy, entransy and exergy analyses to the optimization of the Organic Rankine Cycle. *Energy Convers. Manag.* **2014**, *88*, 267–276. [CrossRef]

Article

Investigating the System Behaviors of a 10 kW Organic Rankine Cycle (ORC) Prototype Using Plunger Pump and Centrifugal Pump

Xin Wang [1], Yong-qiang Feng [1,*], Tzu-Chen Hung [2,*], Zhi-xia He [3], Chih-Hung Lin [4] and Muhammad Sultan [5]

1   School of Energy and Power Engineering, Jiangsu University, Zhenjiang 212013, China; strxinwang@163.com
2   Department of Mechanical Engineering, National Taipei University of Technology, Taipei 350118, Taiwan
3   Institute for Energy Research, Jiangsu University, Zhenjiang 212013, China; zxhe@ujs.edu.cn
4   Department of Refrigeration, Air Conditioning and Energy Engineering, National Chin-Yi University of Technology, Taipei 350118, Taiwan; feng200499@sina.com
5   Department of Agricultural Engineering, Bahauddin Zakariya University, Bosan Road, Multan 60800, Pakistan; muhammadsultan@bzu.edu.pk
*   Correspondence: hitfengyq@gmail.com (Y.-q.F.); tchung@ntut.edu.tw (T.-C.H.)

Received: 31 January 2020; Accepted: 20 February 2020; Published: 3 March 2020

Abstract: Based on a 10-kW organic Rankine cycle (ORC) experimental prototype, the system behaviors using a plunger pump and centrifugal pump have been investigated. The heat input is in the range of 45 kW to 82 kW. The temperature utilization rate is defined to appraise heat source utilization. The detailed components' behaviors with the varying heat input are discussed, while the system generating efficiency is examined. The exergy destruction for the four components is addressed finally. Results indicated that the centrifugal pump owns a relatively higher mass flow rate and pump isentropic efficiency, but more power consumption than the plunger pump. The evaporator pressure drops are in the range of 0.45–0.65 bar, demonstrating that the pressure drop should be considered for the ORC simulation. The electrical power has a small difference using a plunger pump and a centrifugal pump, indicating that the electric power is insensitive on the pump types. The system generating efficiency for the plunger pump is approximately 3.63%, which is 12.51% higher than that of the centrifugal pump. The exergy destruction for the evaporator, expander, and condenser is almost 30%, indicating that enhancing the temperature matching between the system and the heat (cold) source is a way to improve the system performance.

Keywords: organic Rankine cycle (ORC); plunger pump; centrifugal pump; pressure drop; temperature utilization rate; system generating efficiency

---

1. Introduction

The issues of energy high-speed consumption and safe supply have aroused widespread concern in society. In order to effectively solve the energy problem, countries around the world have proposed measures that focus on both energy development and conservation. According to statistics, 50% of the energy used by humans is directly emitted by low-temperature waste heat. If this energy can be used, it will not only solve some energy problems but also reduce environmental pollution. Among much of the low-grade thermal energy conversion and utilization technologies, the organic Rankine cycle is extensively applied on account of its wide temperature range and moderate power [1–3]. In addition, the organic Rankine cycle is often used in conjunction with other systems to achieve higher efficiency. Peris et al. [4] were interested in using ORC for combined heat and power (CHP) applications, and Capata et al. [5] recovered vehicle waste heat with small-scale ORC. In addition, solar energy, wind energy, and ocean temperature difference energy are of particular interest in small-scale ORC systems [6–8]. Therefore, extensive research studies have been conducted on the ORC.

For the industrial ORC prototype, the back work ratio is as high as 25%, indicating that improving the pump performance is a key for ORC commercial application. Numerous studies devoted main efforts on the pump improvement, including piston pumps, gear pumps, and centrifugal pumps. The vane pump delivers energy by the impeller, while the positive displacement pump is dependent on the periodic variation of the volume. Mathias et al. [9] conducted a comparison between the piston pump and gear pump, representing that the piston pump was preferred for the ORC system. Lei et al. [10] tested an ORC system using a roto-jet pump, stating that the pump efficiency of 11–23% was obtained. Bianchi et al. [11] conducted a test on an ORC system using a sliding vane pump, reporting that the shaft power has a significant influence on mass flow rate and pressure. Villani et al. [12] proposed two different ORC systems combined with the heavy diesel engines. One is that the pump and the expander are connected to achieve a fixed speed, and the other is that the pump and expander are separated, the pump optimal speed and expander optimal speed are selected by adjusting parameters. Zeleny et al. [13] applied a gear pump on the ORC system to evaluate the pump mechanical losses. Xu et al. [14] tested the operation characteristics of a piston pump on an ORC system, demonstrating that the low pump frequency was applicable to all expander torques. Carraro et al. [15] integrated a multi-diaphragm positive displacement pump into a 4-kW experimental prototype and found that the pump global efficiency was about 45–48%. Bianchi et al. [16] changed the pump speed to measure the performance of the pump and the overall system in a micro-ORC and found that the pump has a back-work ratio of 50–75% and causes a lot of power consumption. Therefore, special attention should be paid to the design of the pump in micro-ORC. Zhang et al. [17] experimentally studied the change in pump characteristics with evaporation and condensation temperatures. The pump consumption decreases with increasing condensation temperature and presents a non-linear relationship with increasing evaporation temperature. Xi et al. [18] experimentally tested the transient process for the sudden stop of the working fluid pumps in the ORC and regenerative ORC (RORC) systems and found that the expander showed good performance if the pumps were closed when the working fluid was overcharged. Abam et al. [19] analyzed the exergy performance for each component of the four different ORCs and showed that the exergy destruction of evaporators was the largest and that of the pumps was the smallest. Aleksandra et al. [20] integrated that the use of different refrigerants can produce different pump work. Wu et al. [21] utilized a booster pump instead of a common working pump to optimize system characteristics. Meng et al. [22] investigated the performance of the centrifugal pump applied to the engine exhaust recovery ORC device and found that the total efficiency of the pump was 15–65.7%. In addition, Yang et al. [23] and Sun et al. [24] compared a variety of pumps that can be applied to the ORC in order to find the most suitable pump for ORC systems, indicating that the hydraulic diaphragm metering pump was suitable for low heat capacity, whereas the multistage centrifugal pump was preferred for higher heat capacity. Feng et al. [25–27] experimentally compared the system behaviors on a 3kW ORC between pure working fluids and mixture working fluids using a scroll expander. In addition, when the working fluids are condensed in a condenser, the working fluids' temperature and pressure decrease, resulting in the working fluid pump cavitation. Cavitation affects the stability of the system, resulting in a decrease of system efficiency. D'amico et al. [28] introduced the thermodynamic model of a piston pump in the ORC and proposed the prediction of available head margin to avoid cavitation. Liu et al. [29] found that cavitation can occur when the working fluid was insufficient and proposed overcharging working fluid to avoid cavitation. Yang et al. [30] stated that a subcooling of 20 °C was needed to prevent the cavitation of the piston pump. Pei et al. [31] emphasized the use of a bypass tube to balance the pressure of the pump and the tank to solve the cavitation. Galindoe et al. [32] and Dumont et al. [33] used a liquid sub-cooler to prohibit the cavitation.

To lessen the influence of the pump on the ORC property, a novel concept was proposed—pumpless ORC. Gao et al. [34] raised the gravity-type pumpless ORC to ensure the stability and continuity of the system shaft power output. Bao et al. [35] believed that an ORC system without a pump can achieve a more compact and efficient arrangement, which can improve the system's net efficiency. Jiang et al. [36] raised a cascade cycle of power and refrigeration, applying the pumpless ORC to the upper cycle and the adsorption refrigeration cycle to the lower cycle. Within the range of experimental parameters, the maximum power is 232 W and the maximum cooling capacity is 4.94 kW.

As mentioned above, it is evident that several experimental investigations using different pumps have been performed. However, limited studies fulfilled the work on the experimental comparison on an ORC operation characteristic using different pumps. Simultaneously, micro- ORC is still in infancy, and more effort should be focused on the components' design and test. For the micro-ORC prototype, the considerably low efficiency of the pump may heavily affect the overall performance. Therefore, comparing the operation characteristics of small-scale ORC prototypes using different pumps is certainly of great interest. In the present study, a 10 kW R245fa-based ORC experimental prototype is used to study the operation characteristics. A plunger pump and centrifugal pump are adopted, which were widely used in previous experimental tests. The basic operation characteristics for the plunger pump and centrifugal pump are first analyzed. The components' behaviors are addressed, and the overall performance is examined.

2. Experimental Setup Description

2.1. System Design and Operating Method

A 10 kW ORC test platform is adopted. Figure 1 displays a schematic diagram of the experimental bench, and Figure 2 presents the pictures of the experimental equipment. To provide a simulated heat source, diathermic oil is adopted and heated by an electric heater. An electric heater has twelve electric heating rods, and each rod has a capacity of 10 kW. The heating rods are connected in series with an SCR power regulator to adjust the input electric. The heat source temperature is from 85–105 °C and each variation increases by about 5 °C. The heat input is in the range of 45–85 kW. The evaporator and condenser are both plate heat exchangers and the specific parameters for the heat exchangers are listed in Table 1.

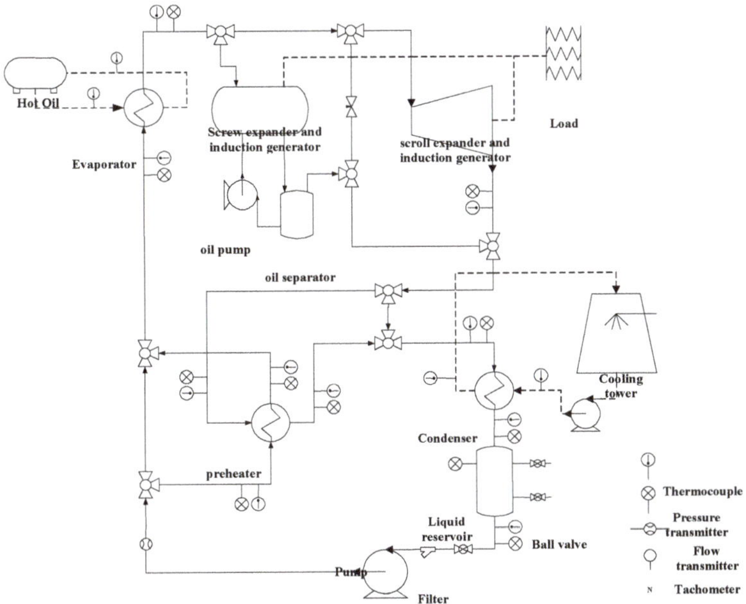

Figure 1. Schematic diagram of the experimental bench.

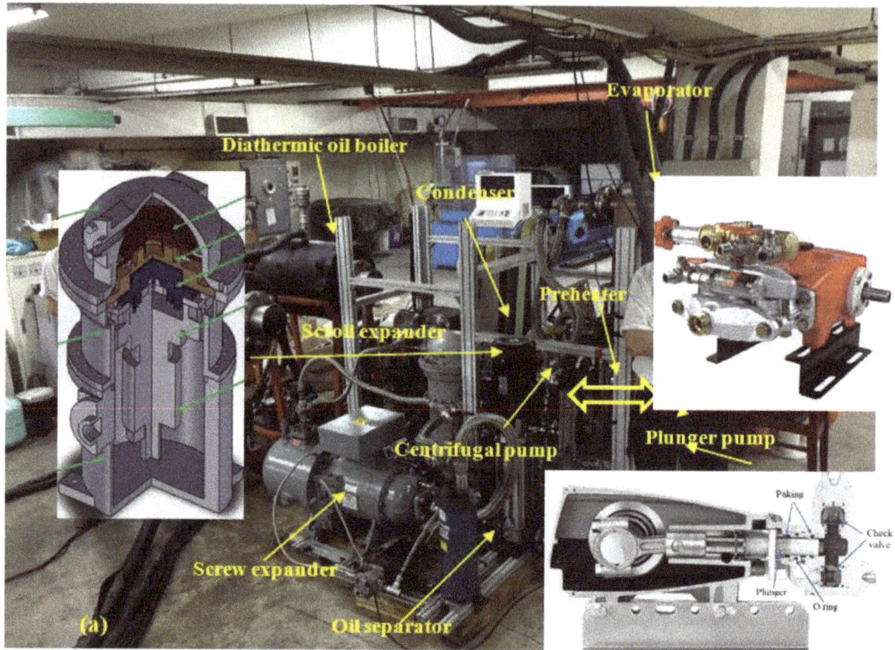

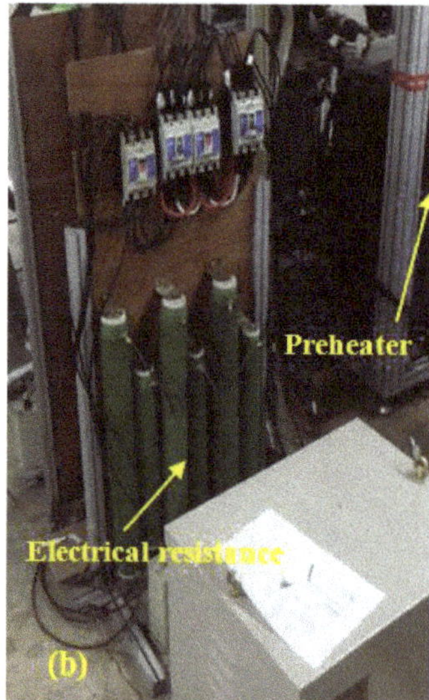

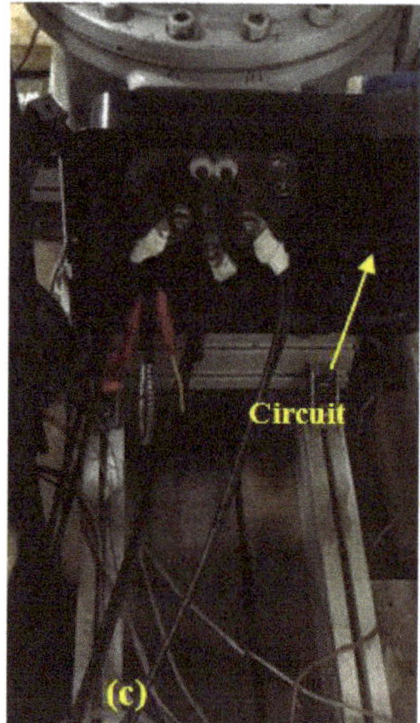

Figure 2. The pictures of (**a**) experimental setup; (**b**) the electrical load; (**c**) the circuit and (**d**) centrifugal pump.

Table 1. Parameters of plate heat exchangers.

Component	Maximum Temperature	Total Volume	Total Heat Transfer Area	Maximum Pressure
Heat exchanger	200 °C	8.848 L	4.175 m^2	30 bar

A centrifugal pump and a plunger pump are adopted in this experimental prototype. The mass flow rates are adjusted by the rotating speed of the pump and the pump rotating speed is controlled by a frequency converter. Meanwhile, the pump raises the refrigerant to reasonable working pressure and the refrigerant enters the evaporator to absorb heat. R245fa is chosen as the refrigerant because of its good thermodynamics and economic characteristics. The scroll expander is adopted, which is improved by a scroll air compressor that is operating in reverse. The product of the system is consumed by the electrical resistance and capacitance, which determines the expander speed. The speed of the expander is measured at 2500–2900rpm and the expander specifications are listed in Table 2. The steam refrigerant absorbed heat enters the expander which exports power. In Figure 2, the scroll expander is encapsulated with a generator, and the shaft power is hard to measure. Therefore, the expander shaft power is expressed by the mass flow rates and expander enthalpy difference. Mass flow rates are converted from volume flow rates directly measured by the flowmeter. The enthalpy difference is determined by checking the REFPROP(a software that can check physical properties) according to the temperature and pressure at both ends of the expander. An inductive generator is used as the power output device.

Table 2. Parameters of the scroll expander.

Component	Volume Ratio	Inspiratory Volume	Gear Height	Basic Circle Radius
Scroll expander	2.95	23.56 m^3/hr	48.2 mm	4.456 mm

A 3% lubricating oil is added in the expander to avoid the leakages and reduce the friction losses in the expander. In addition, the compatibility of the lubricating oil and R245fa is tested at first. After the expansion, the gaseous refrigerant enters the condenser and transfers the residual heat to the cooling water. The mass flow rate of cooling water is 4m/s, which is regulated by the cooling pump frequency. To better ascertain the effect of the pump on the system behavior, a plunger pump and centrifugal pump are tested and compared in this study.

2.2. Plunger Pump and Centrifugal Pump

The plunger pump has a high volume ratio, small flow rate, good characteristic curve, and low cost. It sucks and discharges the working fluid through the reciprocating motion of the plunger. The maximum flow rate of the reciprocating plunger pump is 15.5 L/min with the maximum pressure of 20 bar.

The centrifugal pump has a small area, less material consumption, less manufacturing and installation costs, and can run at high speeds. The centrifugal pump is driven by centrifugal force. The liquid is pumped out from the center to the periphery along the blade flow path and is sent to the discharge pipe through the volute. The centrifugal pump has a maximum flow rate of 36.7 L/min and the delivery pressure of 25 bar. More detailed information about the pumps is displayed in Table 3.

Table 3. Parameters of the plunger pump and centrifugal pump.

Components	Maximum Pressure(bar)	Maximum Flow(L/min)	Temperature Resistance (°C)	Nominal Speed (RPM)
Plunger pump	20	15.5	40	850
Centrifugal pump	25	36.7	120	3500

3. Measuring Device and Thermodynamic Analysis

The detailed operation parameters are measured, including temperature, pressure, and mass flow rate. A vortex flowmeter placed at the pump outlet is used to measure volume flow rates, and then the volume flow rates are converted into mass flow rates, with the detailed location shown in Figure 1. The heat source temperature for both pump experiments was 85–110 °C. The system performance can be obtained based on the measured operation parameters, while an uncertainty analysis is conducted [37].

$$\Delta Y = \sqrt{\sum_i \left(\frac{\partial Y}{\partial X_i}\right)^2 \Delta X_i^2}$$

(1)

where X and ΔY are the independent variable and uncertainty, respectively. Table 4 lists the measuring devices and the uncertainties for system parameters.

Table 4. The measuring devices and the uncertainties for system parameters.

Item	Type	Measurement Range	Device Uncertainty
Temperature	T-type	0–623.15 K	±0.3 K
Pressure	JPT-131S	0–30 bar	±0.5% F.S
Flow rate	GPI S050	1.9–37.9 L/min	±3% L/min
Rotation speed	UT-372	10–99,999 rpm	±3% rpm
Electrical power	PA310		0.5%
Pump shaft power			3.9%
Pump isentropic efficiency			4.5%
Evaporator heat transfer coefficient			1.9%
Condenser heat transfer coefficient			2.2%
Expander isentropic efficiency			4.7%
System generating efficiency			1.2%

Figure 3 shows the *T-s* diagram of the ORC system, the expander isentropic efficiency is calculated based on the ideal expansion process and actual expansion process. The expander isentropic efficiency ($\eta_{is,exp}$) and pressure difference (ΔP) can be calculated as follows:

$$\eta_{is,exp} = \frac{h_1 - h_2}{h_1 - h_{2s}}$$

(2)

$$\Delta P = P_1 - P_2$$

(3)

where h_1 and p_1 represent the enthalpy and pressure of the expander inlet, and h_2, p_2 and h_{2s} represent the enthalpy, pressure and isentropic enthalpy of the expander outlet.

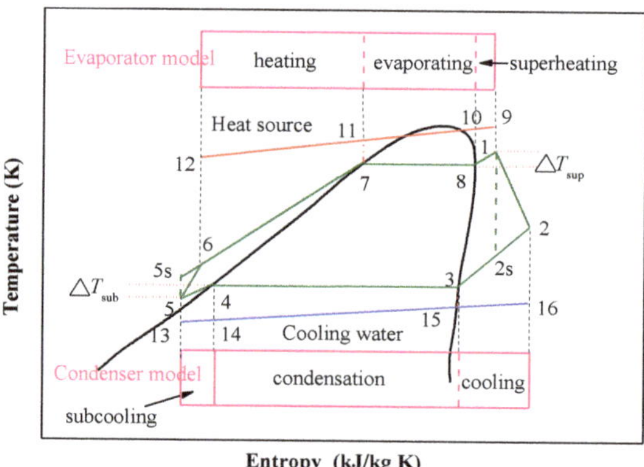

Figure 3. *T-s* plot of the organic Rankine cycle system.

The condenser heat transfer rate (Q_{cond}), the logarithmic mean temperature difference (LMTD) (ΔT_{cond}), heat transfer coefficient (U_{cond}) and pressure drop(ΔP_{cond}) can be calculated as follows:

$$Q_{cond} = m(h_2 - h_5)$$

(4)

$$\Delta T_{\text{cond}} = \frac{\Delta T_{\max} - \Delta T_{\min}}{\ln \frac{\Delta T_{\max}}{\Delta T_{\min}}} = \frac{(T_{13} - T_5) - (T_{16} - T_2)}{\ln \frac{T_{13}-T_5}{T_{16}-T_2}} \tag{5}$$

$$U_{\text{cond}} = \frac{Q_{\text{cond}}}{\Delta T_{\text{cond}} \cdot A} \tag{6}$$

$$\Delta P_{\text{cond}} = P_2 - P_5 \tag{7}$$

where h_5 is the outlet enthalpy of the condenser; $\Delta T_{\max}$ and $\Delta T_{\min}$ are the maximum and minimum temperature difference at the condenser, respectively; and A represents the surface area of the heat exchanger.

The pump isentropic efficiency is calculated by the actual compression process and the ideal compression process. The pump shaft power ($W_{\text{sh,pump}}$) and isentropic efficiency ($\eta_{\text{is,pump}}$) are expressed as:

$$W_{\text{sh,pump}} = m(h_6 - h_5) \tag{8}$$

$$\eta_{\text{is,pump}} = \frac{h_{5s} - h_5}{h_6 - h_5} \tag{9}$$

where h_{5s} and h_6 denote the pump outlet enthalpy and isentropic enthalpy.

Similarly, the evaporator heat transfer rate (Q_{eva}), LMTD (ΔT_{eva}), heat transfer coefficient (U_{eva}) and pressure drop (ΔP_{eva}) can be expressed as:

$$Q_{\text{eva}} = m(h_1 - h_6) \tag{10}$$

$$\Delta T_{\text{eva}} = \frac{\Delta T_{\max} - \Delta T_{\min}}{\ln \frac{\Delta T_{\max}}{\Delta T_{\min}}} = \frac{(T_{12} - T_6) - (T_9 - T_1)}{\ln \frac{T_{12}-T_6}{T_9-T_1}} \tag{11}$$

$$U_{\text{eva}} = \frac{Q_{\text{eva}}}{\Delta T_{\text{eva}} \cdot A} \tag{12}$$

$$\Delta P_{\text{eva}} = P_1 - P_6 \tag{13}$$

To better understand the heat source utilization, the temperature utilization rate of the heat source (θ) is proposed [38]. Assuming that the lowest heat source temperature can reach 60 °C (333.15 K), the denominator of temperature utilization rate θ indicates the heat that the system can use. The numerator denotes the actual heat used by the ORC system. So θ can be expressed as:

$$\theta = (TH, \text{in} - TH, \text{out}) / (TH, \text{in} - 333.15) \tag{14}$$

The generating efficiency can be expressed as:

$$\eta_{\text{ele}} = \frac{W_{\text{ele,exp}} - W_{\text{ele,pump}}}{Q_{\text{eva}}} \tag{15}$$

where $W_{\text{ele,exp}}$ and $W_{\text{ele,pump}}$ represent the electrical power and pump consumption power.

The exergy destruction of the four components including the pump (Ed, pump), evaporator (Ed, eva), expander (Ed, exp) and condenser (Ed, cond) can be calculated as follows:

$$E_{\text{d,pump}} = T_0 m(s_6 - s_5) \tag{16}$$

$$E_{\text{d,eva}} = T_0 m[(s_1 - s_6) - 2(h_1 - h_6)/(T_9 + T_{12})] \tag{17}$$

$$E_{\text{d,exp}} = T_0 m(s_2 - s_1) \tag{18}$$

$$E_{\text{d,con}} = T_0 m[(s_5 - s_2) - 2(h_5 - h_2)/(T_{13} + T_{16})] \tag{19}$$

4. Results and Discussion

To better compare the cycle behaviors of a 10-kW experimental prototype using two different pumps, the system operation parameters at different heat inputs are collected. The heat input for the plunger pump and centrifugal pump are in the range of 44.74–76.48 kW and 45.61–81.35 kW, respectively. The basic operating parameters using the plunger pump and centrifugal pump are displayed in Section 4.1. The detailed components' behaviors are described in Section 4.2, while the overall cycle characteristics, including system generating efficiency and exergy destruction are expressed in Section 4.3.

4.1. Basic Operating Parameters

In particular, the present data is collected at different time periods. It is difficult to keep the ambient temperature constant because of the fluctuating ambient conditions. However, the environment temperature has a great influence on the condensation process, so it has a strong guiding significance for explaining many basic operating parameters of ORC, such as pump inlet temperature and shaft work, etc. The environment temperatures for the ORC system using the plunger pump and centrifugal pump are listed in Tables 5 and 6, respectively. The environment temperature for the plunger pump is approaching 23 °C, which is 4 °C higher than that of the centrifugal pump. Figure 4 illustrates the relationship between mass flow rate and heat input using the plunger pump and centrifugal pump. When the heat input keeps rising, more working fluids are needed to absorb the heat from the evaporator. It also can be found that the centrifugal pump has a slightly higher mass flow rate than the plunger pump for the same heat input, which may be contributed to the centrifugal pump having the higher rotating speed and greater flow per revolution. The mass flow rate for the centrifugal pump is from 0.17 to 0.31 kg/s, which is 6.7% higher than that of the plunger pump.

Table 5. The environment temperature for using the plunger pump.

Heat input (kW)	38.7	48.8	58.2	63.8
Environment temperature (°C)	23.5	23.5	24.0	21.5

Table 6. The environment temperature for using the centrifugal pump.

Heat input (kW)	46.0	56.2	66.6	77.3
Environment temperature (°C)	17.5	17.3	18.9	18.5

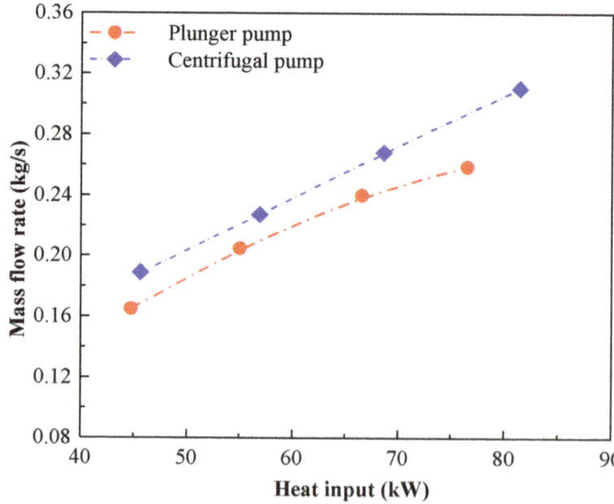

Figure 4. Mass flow rates with heat input using the plunger pump and centrifugal pump.

Figure 5 shows details of the temperature and pressure for the pump inlet and outlet with heat input using the plunger pump and centrifugal pump. In Figure 5a, the centrifugal pump inlet temperature and pressure have no obvious variation with the heat input. However, the plunger pump inlet temperature appears to suddenly decrease when the heat inputs exceed 66.53 kW, owing to the fluctuating environmental temperature. The environment temperature decreases slightly for heat inputs over 66.53 kW, resulting in a decrease of cooling water temperature and the pump inlet temperature. In Figure 5b, the pump outlet temperature has a similar trend to the state at the pump inlet, whereas the pump outlet pressure shows a sharp increase with the heat input. Because of the difficulty in repeating tests with similar environmental temperatures, the non-dimensional operating parameter (pump pressure ratio) is chosen to compare the characteristic of the two pumps, which is shown in Figure 5c. The centrifugal pump pressure ratio is much higher than the plunger pump. The pump outlet pressure for the plunger pump is in the range of 7.58–10.87 bar, which is 0.3 bar higher than that of the centrifugal pump ranging from 7.29 bar to 10.84 bar.

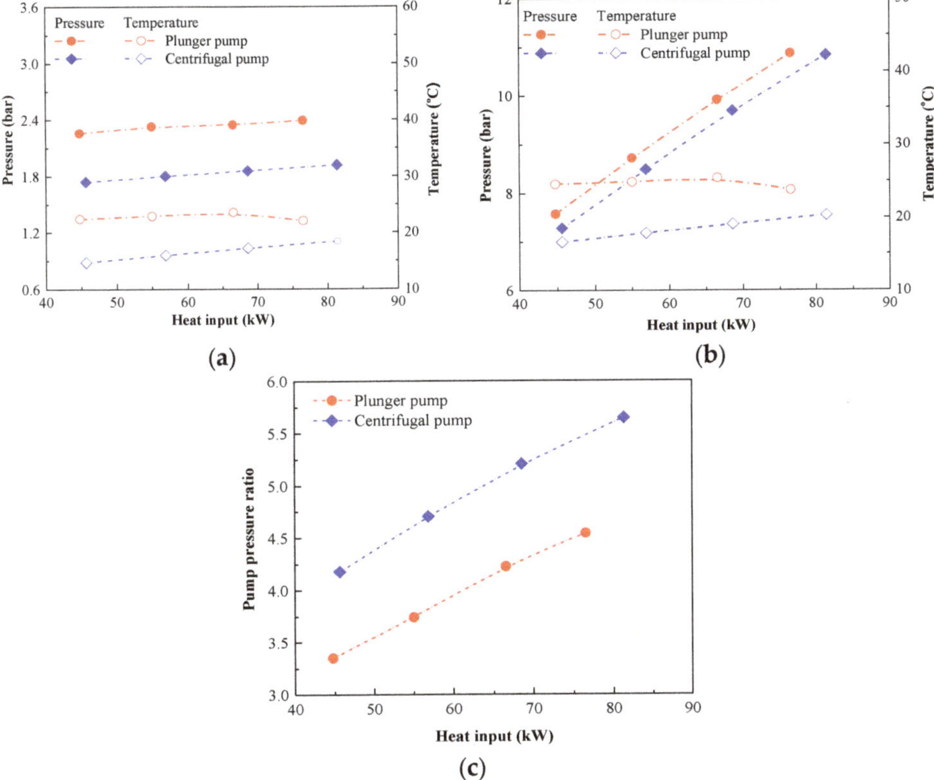

Figure 5. Pressure and temperature at the pump inlet and outlet with heat input using the plunger pump and centrifugal pump: (**a**) pressure and temperature at the pump inlet; (**b**) pressure and temperature at the pump outlet and (**c**) pump pressure ratios.

Figure 6 presents the comparison of expander inlet and outlet temperatures and pressure with heat input using the plunger pump and centrifugal pump. In Figure 6a, the expander inlet temperature and pressure increase monotonically with heat input. The expander inlet temperature of the plunger pump rises from 82.20 °C to 97.74 °C for heat inputs increasing from 45 kW to 85 kW, with the corresponding expander inlet pressure ranging from 7.13 bar to 10.23 bar. Meanwhile, the plunger pump demonstrates a higher expander inlet pressure and temperature than the centrifugal pump.

The expander outlet pressure and temperature keep rising in Figure 6b, which is similar to that at the expander inlet. Similarly, in order to more clearly compare the features of the expander using the two pump systems, the pressure ratio for the expander is shown in Figure 6c. The expander pressure ratio in the system using the centrifugal pump is more favorable than that using the plunger pump because the centrifugal pump can provide higher pressure to the system.

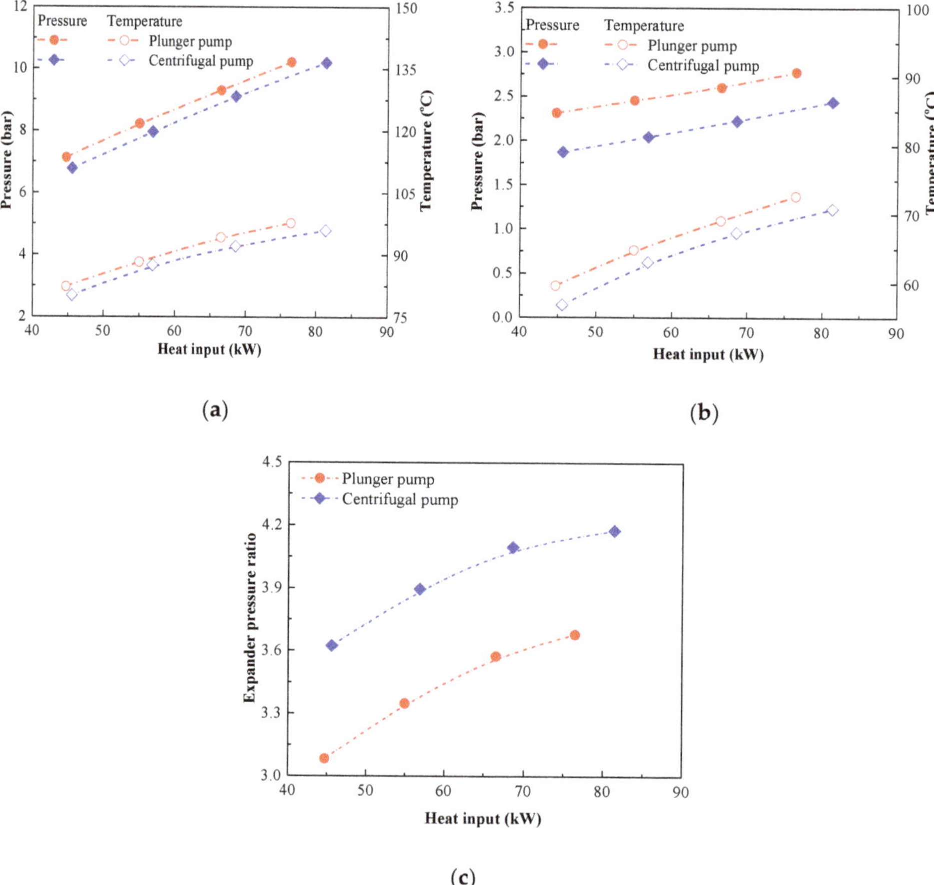

(a)

(b)

(c)

Figure 6. Temperature and pressure at the expander inlet and outlet with heat input using the plunger pump and centrifugal pump: (**a**) pressure and temperature at the expander inlet; (**b**) pressure and temperature at the expander outlet; (**c**) expander pressure ratios.

4.2. Detailed Components' Behavior

4.2.1. Pump Behavior

The pump shaft power cannot be tested and is expressed by Equation (8). Figure 7 demonstrates the details of shaft power using the plunger pump and centrifugal pump with heat input. Apparently, the pump shaft power of the centrifugal pump keeps increasing with heat input, which may be caused by the increase in mass flow rate and pump enthalpy difference. However, the shaft power of the plunger pump presents a slight decrease when heat inputs exceed 66.53 kW because of the higher sensitivity to the environmental temperature. The pump shaft power using the centrifugal pump increases from 0.51 kW to 0.85 kW, which is 37% higher than that of the plunger pump.

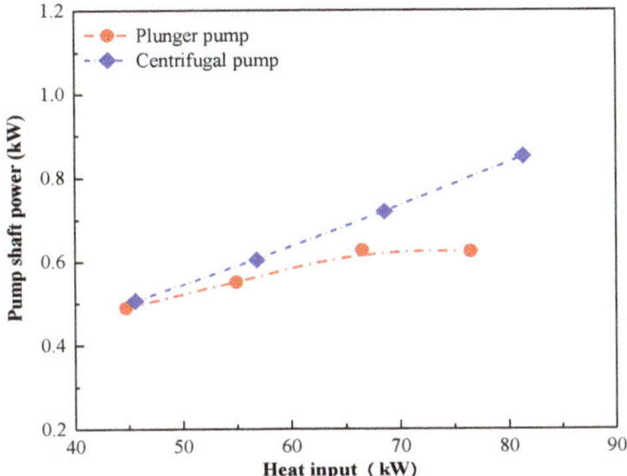

Figure 7. Pump shaft power with heat input using the plunger pump and centrifugal pump.

The pump isentropic efficiency can be considered as an important parameter to ascertain pump characteristics. Figure 8 displays the pump isentropic efficiency of the plunger pump and centrifugal pump. The isentropic efficiency keeps increasing with heat input because the pump compression process gets closer to the ideal isentropic process. The isentropic efficiency of the plunger pump and centrifugal pumps is in the range of 13.2–26.1% and 15.3–24.5%, respectively. The pump isentropic efficiency is really low because there is no specialized pump for the ORC system. Moreover, the centrifugal pump isentropic efficiency is higher than the plunger pump for low heat inputs, while a reverse trend for higher input heats. Meanwhile, the low pump isentropic efficiency reminds us that enhancing the pump's behavior is vital for the improvement in ORC performance.

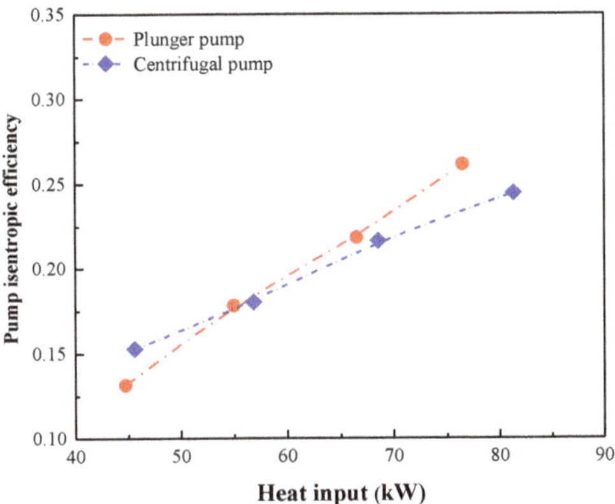

Figure 8. Pump isentropic efficiency with heat input using the plunger pump and centrifugal pump.

4.2.2. Heat Exchanger Behavior

The details of evaporator and condenser heat transfer coefficients with heat input using the plunger pump and centrifugal pump are plotted in Figure 9. In Figure 9a, the evaporator heat transfer coefficient of the centrifugal pump keeps increasing, whereas that of the plunger pump displays a parabolic trend with a maximum with heat input, which may be caused by both the evaporator heat transfer rate and LMTD. The condenser heat transfer coefficient keeps rising with heat input in Figure 9, this is because of an increasing expander outlet temperature. The increasing condenser heat transfer rate is greater than that of condenser LMTD, which causes an increasing condenser heat transfer coefficient. Meanwhile, the cycle using the centrifugal pump has a relatively higher heat transfer coefficient than that using the plunger pump (whether evaporator or condenser). The evaporator heat transfer coefficients for the cycle using the plunger pump and centrifugal pump are in the range of 116.45–147.01 W/m^2 °C and 126.95–149.58 W/m^2 °C, and the corresponding condenser heat transfer coefficients are 1003.18–1371.33 W/m^2°C and 1092.74–1608.27 W/m^2 °C, respectively.

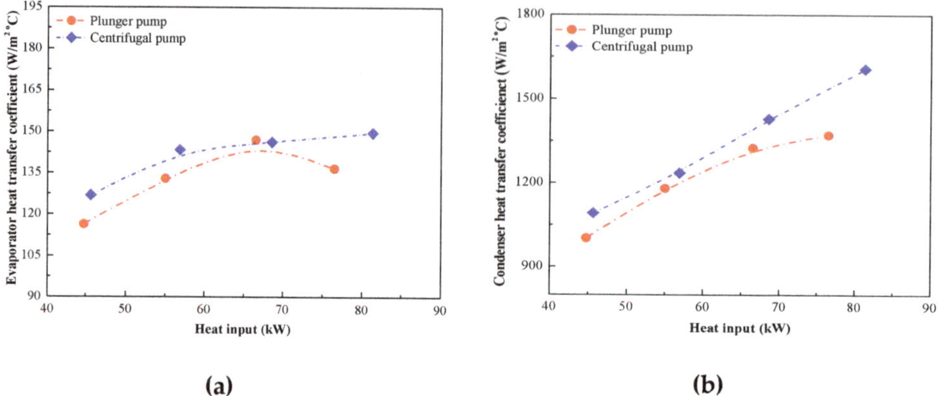

(a)   (b)

Figure 9. Evaporator and condenser heat transfer coefficients with heat input using the plunger pump and centrifugal pump: (**a**) evaporator heat transfer coefficients; (**b**) condenser heat transfer coefficients.

As for the ORC simulation, the pressure drop in the evaporation and condensation processes are usually ignored. However, for the actual cycle, having pressure drops can decrease the expander inlet pressure, and thus affect the overall system property. Figure 10 demonstrates the variation of condenser and evaporator pressure drop with heat input using the plunger pump and centrifugal pump. Obviously, the pressure drops for evaporator and condenser increase with the heat input, which may be caused by the increasing mass flow rate. A small difference in the evaporator pressure drop appeared between the plunger pump and the centrifugal pump. However, when the heat inputs raise from 40 kW to 82 kW, the condenser pressure drops using the centrifugal pump are 0.13–0.53 bar, which is 0.07 bar higher than that using the plunger pump. The evaporator pressure drops are in the range of 0.45–0.65 bar. It indicates that the pressure drop should be considered for the ORC simulation and decreasing the pressure drop is one way to improve the system performance.

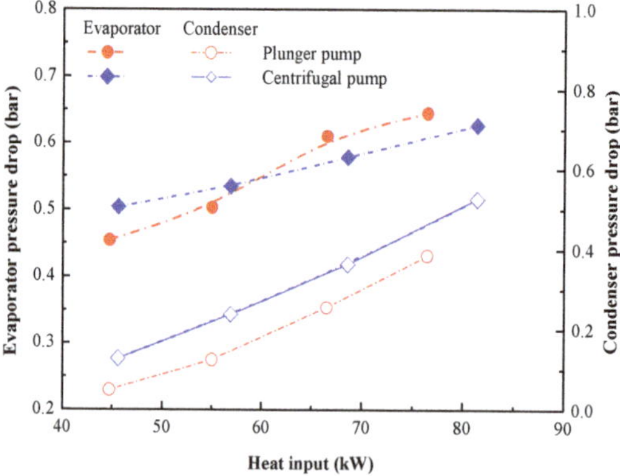

Figure 10. Condenser and evaporator pressure drop with heat input using the plunger pump and centrifugal pump.

The variation of temperature utilization rate with heat input using the plunger pump and centrifugal pump is plotted in Figure 11. For a specific heat source temperature, a higher thermal efficiency does not represent a higher net power output. Therefore, the temperature utilization rate is used to appraise the heat source utilization. As illustrated in Figure 11, the temperature utilization rate for the centrifugal pump has a slight decrease whereas the plunger pump has almost no change with the heat input. The average heat source temperature utilization rate for the centrifugal pump is about 30%, which is 5% higher than that of the plunger pump, indicating that the centrifugal pump absorbs more heat than the plunger pump at the same heat source condition.

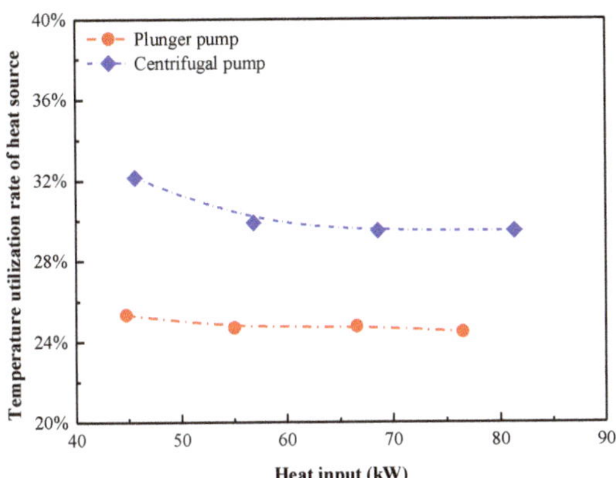

Figure 11. Temperature utilization rate with heat input using the plunger pump and centrifugal pump.

4.2.3. Expander Behavior

Figure 12 shows the expander isentropic efficiency with heat input using the plunger pump and centrifugal pump. It should be reminded that the scroll expander is designed with a nominal expander shaft power of 10 kW, indicating a heat input of 40–200 kW is needed. However, the heat input is set from 45 kW to 85 kW because of the power limitation. Therefore, the expander isentropic efficiency presents an apparent decrease trend from 58.8% to 39.1% with heat input because of the insufficient expansion. The expander isentropic efficiency for the plunger pump is in the range of 46.5–58.8%, which is 12.6% higher than that of the centrifugal pump.

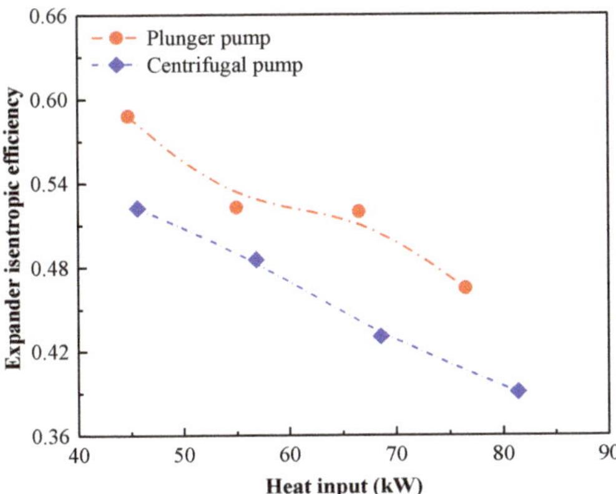

Figure 12. Expander isentropic efficiency with heat input using the plunger pump and centrifugal pump.

The variation of electrical power with heat input using the plunger pump and centrifugal pump is illustrated in Figure 13. The electrical power for the centrifugal pump rises from 1.83 kW to 3.01 kW, while that of the plunger pump is in the range of 1.76–2.87 kW. The reason is that the increment in pressure difference causes an increase in the expander rotational speed, resulting in an increase in electrical power. The electrical power has a small difference using the plunger pump and centrifugal pump, indicating that the electric power is insensitive on the pump types.

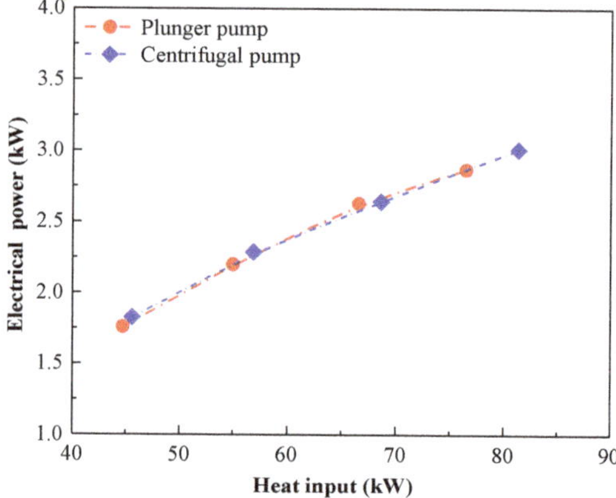

Figure 13. Electrical power with heat input using the plunger pump and centrifugal pump.

4.3. Overall System Performance

Figure 14 displays the system generating efficiency of the plunger pump and centrifugal pump. In this study, system generating efficiency is utilized as an evaluation criterion for this system, which is obtained based on the net electrical power and heat input. The system generating efficiency has no obvious change with heat input, demonstrating that the system generating efficiency has little effect on the heat input. The increasing net electrical power and the increasing heat input enable the almost unchanged system generating efficiency. The system generating efficiency for the plunger pump is approximately 3.63%, which is 12.51% higher than that of the centrifugal pump. One reason for the low overall system performance is that the pump consumed more power (as shown in Figure 7).

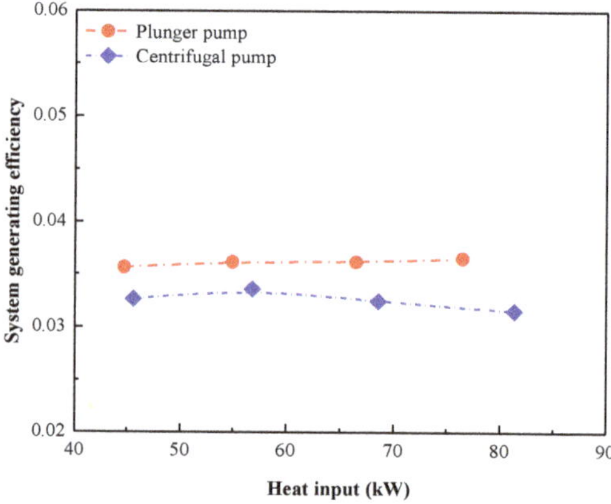

Figure 14. System generating efficiency with heat input using the plunger pump and centrifugal pump.

Figure 15 presents the details of the exergy destruction of the four important components with the heat input using the plunger pump and centrifugal pump. It is obvious that the exergy destruction for the evaporator, expander and condenser keep rising, whereas that of the pump almost has no change with the heat input. Besides, the exergy destruction of the piston pump is almost the same as that of the centrifugal pump. However, for the other three components, the exergy destruction using the centrifugal pump is higher than that using the plunger pump, because of the relatively higher mass flow rate for the centrifugal pump. For a specific heat input of 68.6 kW using the plunger pump, the exergy destruction for the pump, evaporator, expander, and condenser is 0.5 kW, 3 kW, 2.5 kW, and 3.2 kW, respectively.

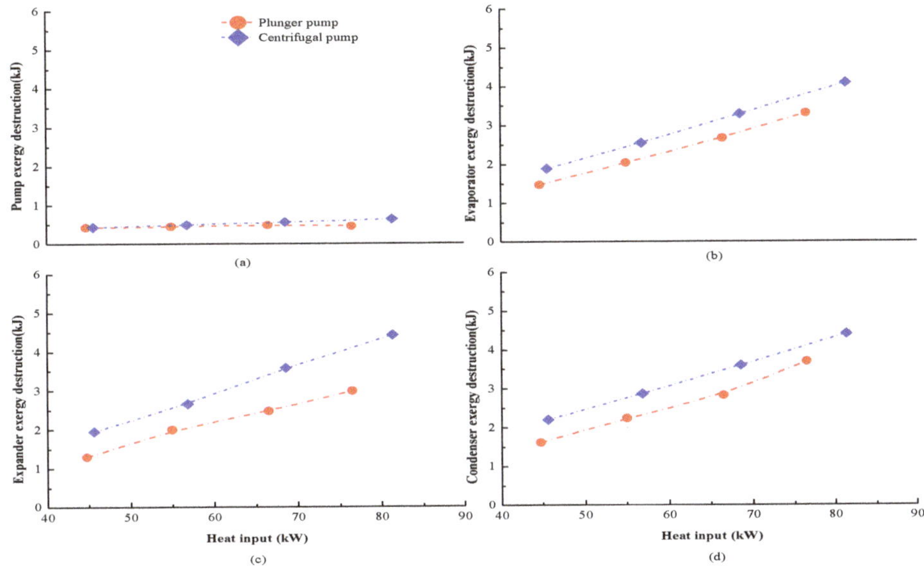

Figure 15. Exergy destruction of the four important components using the plunger pump and centrifugal pump: (**a**) pump exergy destruction; (**b**) evaporator exergy destruction; (**c**) expander exergy destruction; (**d**) condenser exergy destruction.

To ascertain which component contributes the maximum exergy destruction, the proportion of exergy destruction for each component using the plunger pump and centrifugal pump is shown in Figure 16. For the plunger pump and centrifugal pump, the exergy destruction for the evaporator, expander and condenser is almost 30%, indicating that improving the temperature matching between the cycle and the heat (cold) source is a way to improve the system property.

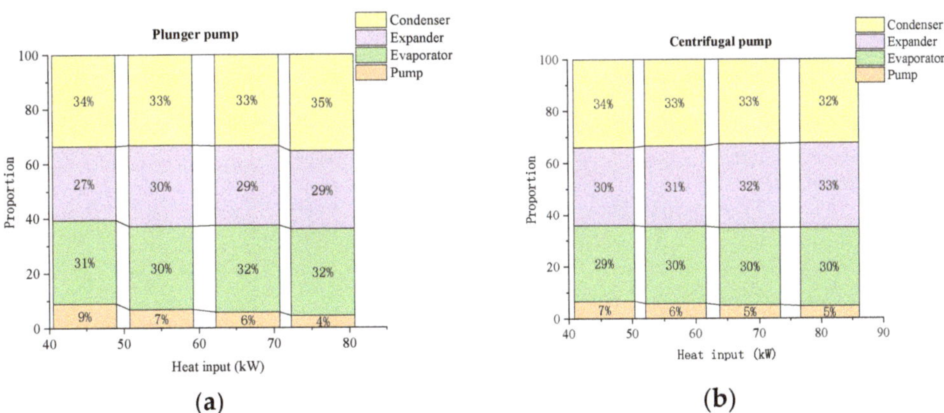

Figure 16. The proportion of exergy destruction for each component with heat input using the plunger pump and centrifugal pump: (**a**) the proportion of exergy destruction using the plunger pump; (**b**) the proportion of exergy destruction using the centrifugal pump.

5. Conclusions

The system behaviors using the plunger pump and centrifugal pump have been investigated experimentally. A 10 kW R245fa-based experimental prototype is adopted. The heat source temperature for both pump experiments was 85–106 °C. The heat inputs for the plunger pump and centrifugal pump are in the range of 44.74–76.48 kW and 45.61–81.35 kW, respectively. Simultaneously, the mass flow rates of the plunger pump are from 0.16–0.26kg/s and those of the centrifugal pump are in the range of 0.19–0.31 kg/s. The temperature utilization rate is used to appraise the heat source utilization. The detailed components' behaviors with the varying heat input are discussed, while the system generating efficiency is examined. The exergy destruction of the four main components is addressed. The conclusions are summarized below:

(1) The mass flow rates of the centrifugal pump are from 0.19–0.31 kg/s, which is 19% higher than that of the plunger pump. Compared with the plunger pump, the centrifugal pump owns a relatively higher mass flow rate and more pump shaft power.

(2) A small difference of evaporator pressure drop appeared between the plunger pump and the centrifugal pump. The condenser pressure drops using the centrifugal pump are 0.13–0.53 bar, while the evaporator pressure drops are in the range of 0.45–0.65 bar, demonstrating that the pressure drop should be considered for the ORC simulation.

(3) The average heat source temperature utilization rate for the centrifugal pump is about 30%, which is 5% higher than that of the plunger pump, indicating that the centrifugal pump absorbs more heat than the plunger pump at a same heat source condition.

(4) The electrical power for the centrifugal pump rises from 1.83 kW to 3.01 kW, while that of the plunger pump is in the range of 1.76–2.87 kW. The electrical power has a small difference using the plunger pump and centrifugal pump, indicating that the electrical power is insensitive of the pump types. The system generating efficiency for the plunger pump is approximately 3.63%, which is 12.51% higher than that of the centrifugal pump. It indicates that the plunger pump is more suitable for the ORC system in this study. The system generating efficiency is insensitive of the heat input.

(5) No matter which pump is used in the ORC, evaporator, expander, and condenser exergy destruction accounts for almost 30%.

(6) The exergy destruction for evaporators, expanders, and condensers is almost 30%, indicating that improving temperature matching between the system and the heat (cold) source is a way to improve the system property

Author Contributions: Y.-q.F., Q.W. and T.-C.H. conceived of the presented idea. X.W. and C.-H.L. developed the theory and performed the computations. Z.-x.H. and M.S. verified the analytical methods. All authors discussed the results and contributed to the final manuscript. All authors have read and agreed to the published version of the manuscript.

Funding: This research work has been supported by the National Natural Science Foundation of China (51806081), the Key Research and Development Program of Jiangsu Province, China (BE2019009-4), the Natural Science Foundation of Jiangsu Province (BK20180882), the China Postdoctoral Science Foundation (2018M632241), the 2019 Scholarship of the Knowledge Center on Organic Rankne Cycle (KCORC, www.kcorc.org), the Key Research and Development Program of Zhenjiang City, China (SH2019008), and the Key Project of Taizhou New Energy Research Institute, Jiangsu University, China (2018-20). The authors are grateful for the Ministry of Science and Technology, Taiwan under the grants of Contract No. MOST 107-2221-E-027-091, and by "Research Center of Energy Conservation for New Generation of Residential, Commercial, and Industrial Sector" from The Featured Areas Research Center Program within the framework of the Higher Education Sprout Project by the Ministry of Education (MOE) in Taiwan.

Conflicts of Interest: The authors declare no conflict of interest.

References

1. Johnson, I.; Choate, W.T.; Davidson, A. *Waste Heat Recovery: Technology and Opportunities in U.S. Industry*; BCS, Inc.: Laurel, MD, USA, 2008.

2. Feng, Y.Y.; Zhang, Y.Y.; Li, B.B.; Yang, J.J.; Shi, Y. Comparison between regenerative organic Rankine cycle (RORC) and basic organic Rankine cycle (BORC) based on thermoeconomic multi-objective optimization considering exergy efficiency and levelized energy cost (LEC). *Energy Convers. Manag.* **2015**, *96*, 58–71. [CrossRef]

3. Feng, Y.Y.; Hung, T.C.; Greg, K.; Zhang, Y.Y.; Li, B.B.; Yang, J.F. Thermoeconomic comparison between pure and mixture working fluids for low-grade organic Rankine cycles (ORCs). *Energy Convers. Manag.* **2015**, *106*, 859–872. [CrossRef]

4. Peris, B.; Navarro-Esbr, J.; Moles, F.; Martí, J.J.; Mota-Babiloni, A. Experimental characterization of an Organic Rankine Cycle (ORC) for micro-scale CHP applications. *Appl. Ther. Eng.* **2015**, *79*, 1–8. [CrossRef]

5. Capata, R.; Toro, C. Feasibility analysis of a small-scale ORC energy recovery system for vehicular application. *Energy Convers. Manag.* **2014**, *86*, 1078–1090. [CrossRef]

6. Freeman, J.; Hellgardt, K.; Markides, C.N. An assessment of solar-powered organic Rankine cycle systems for combined heating and power in UK domestic applications. *Appl. Energy* **2015**, *138*, 605–620. [CrossRef]

7. Pantano, F.; Capata, R. Expander selection for an on board ORC energy recovery system. *Energy* **2017**, *141*, 1084–1096. [CrossRef]

8. Atiz, A.; Karakilcik, H.; Erden, M.; Karakilcik, M. Investigation energy, exergy and electricity production performance of an integrated system based on a low-temperature geothermal resource and solar energy. *Energy Convers. Manag.* **2019**, *195*, 798–809. [CrossRef]

9. Mathias, J.J.; Johnston, J.J.; Cao, J.J.; Priedeman, D.D.; Christensen, R.N. Experimental testing of gerotor and scroll expanders used in, and energetic and exergetic modeling of, an organic Rankine cycle. *J. Energy Res. Technol. Trans. ASME* **2009**, *131*, 21–24. [CrossRef]

10. Lei, B.; Wang, J.-F.; Wu, Y.-T.; Ma, C.-F.; Wang, W.; Zhang, L.; Li, J.-Y. Experimental study and theoretical analysis of a Roto-Jet pump in small scale organic Rankine cycles. *Energy Convers. Manag.* **2016**, *111*, 198–204. [CrossRef]

11. Bianchi, G.; Fatigati, F.; Murgia, S.; Cipollone, R.; Contaldi, G. Modeling and experimental activities on a small-scale sliding vane pump for ORC-based waste heat recovery applications. *Energy Procedia* **2016**, *101*, 1240–1247. [CrossRef]

12. Villani, M.; Tribioli, L. Comparison of different layouts for the integration of an organic Rankine cycle unit in electrified powertrains of heavy duty Diesel trucks. *Energy Convers. Manag.* **2019**, *187*, 248–261. [CrossRef]

13. Zeleny, Z.; Vodicka, V.; Novotny, V.; Mascuch, J. Gear pump for low power output ORC—An efficiency analysis. *Energy Procedia* **2017**, *129*, 1002–1009. [CrossRef]

14. Xu, W.; Zhang, J.J.; Zhao, L.; Deng, S.; Zhang, Y. Novel experimental research on the compression process in organic Rankine cycle (ORC). *Energy Convers. Manag.* **2017**, *137*, 1–11. [CrossRef]

15. Carraro, G.; Pallis, P.; Leontaritis, A.D.; Karellas, S.; Vourliotis, P.; Rech, S.; Lazzaretto, A. Experimental performance evaluation of a multi-diaphragm pump of a micro-ORC system. *Energy Procedia* **2017**, *129*, 1018–1025. [CrossRef]

16. Bianchi, M.; Branchini, L.; Casari, N.; Pascale, D.; Melino, F.; Ottaviano, S.; Pinelli, M.; Spina, P.; Suman, A. Experimental analysis of a micro-ORC driven by piston expander for low-grade heat recovery. *Appl. Ther. Eng.* **2019**, *148*, 1278–1291. [CrossRef]

17. Zhang, H.H.; Xi, H.; He, Y.Y.; Zhang, Y.Y.; Ning, B. Experimental study of the organic rankine cycle under different heat and cooling conditions. *Energy* **2019**, *180*, 678–688. [CrossRef]

18. Xi, H.; He, Y.Y.; Wang, J.J.; Huang, Z.H. Transient response of waste heat recovery system for hydrogen production and other renewable energy utilization. *Int. J. Hydrogen Energy* **2019**, *44*, 15985–15996. [CrossRef]

19. Abam, F.F.; Ekwe, E.E.; Effiom, S.S.; Ndukwu, M.M.; Briggs, T.T.; Kadurumba, C.H. Optimum exergetic performance parameters and thermo-sustainability indicators of low-temperature modified organic Rankine cycles (ORCs). *Sustain. Energy Technol. Assess.* **2018**, *30*, 91–104. [CrossRef]

20. Aleksandra, B.-G. Pumping work in the organic Rankine cycle. *Appl. Ther. Eng.* **2013**, *51*, 781–786.

21. Wu, T.T.; Liu, J.J.; Zhang, L.; Xu, X.J. Experimental study on multi-stage gas-liquid booster pump for working fluid pressurization. *Appl. Ther. Eng.* **2017**, *126*, 9–16. [CrossRef]

22. Meng, F.F.; Zhang, H.H.; Yang, F.F.; Hou, X.X.; Lei, B.; Zhang, L. Study of efficiency of a multistage centrifugal pump used in engine waste heat recovery application. *Appl. Ther. Eng.* **2017**, *110*, 779–786. [CrossRef]

23. Yang, Y.Y.; Zhang, H.H.; Xu, Y.Y.; Yang, F.B.; Wu, Y.Y.; Lei, B. Matching and operating characteristics of working fluid pumps with organic Rankine cycle system. *Appl. Ther. Eng.* **2018**, *142*, 622–631. [CrossRef]

24. Sun, H.H.; Qin, J.; Hung, T.C.; Hua, H.H.; Ya, P.P.; Lin, C.H. Effect of flow losses in heat exchangers on the performance of organic Rankine cycle. *Energy* **2019**, *172*, 391–400. [CrossRef]
25. Feng, Y.Y.; Hung, T.C.; He, Y.Y.; Wang, Q.; Wang, S.; Li, B.B.; Lin, J.J.; Zhang, W.P. Operation characteristic and performance comparison of organic Rankine cycle (ORC) for low-grade waste heat using R245fa, R123 and their mixtures. *Energy Convers. Manag.* **2017**, *144*, 153–163. [CrossRef]
26. Feng, Y.Q.; Hung, T.C.; Wu, S.S.; Lin, C.C.; Li, B.B.; Huang, K.K.; Qin, J. Operation characteristic of a R123-based organic Rankine cycle depending on working fluid mass flow rates and heat source temperature. *Energy Convers. Manag.* **2017**, *131*, 55–68. [CrossRef]
27. Yang, S.S.; Hung, T.C.; Feng, Y.Y.; Wu, C.C.; Wong, K.K.; Huang, K.C. Experimental investigation on a 3 kW organic Rankine cycle for low grade waste heat under different operation parameters. *Appl. Ther. Eng.* **2017**, *113*, 756–764. [CrossRef]
28. D'Amico, F.; Pallis, P.; Leontaritis, A.D.; Karellas, S.; Kakalis, N.M.; Rech, S.; Lazzaretto, A. Semi-empirical model of a multi-diaphragm pump in an Organic Rankine Cycle (ORC) experimental unit. *Energy* **2018**, *143*, 1056–1071. [CrossRef]
29. Liu, L.L.; Zhu, T.; Wang, T.T.; Gao, N.P. Experimental investigation on the effect of working fluid charge in a small-scale Organic Rankine Cycle under off-design conditions. *Energy* **2019**, *174*, 664–677. [CrossRef]
30. Yang, X.X.; Xu, J.J.; Miao, Z.; Zou, J.J.; Yu, C. Operation of an organic Rankine cycle dependent on pumping flow rates and expander torques. *Energy* **2015**, *90*, 864–878. [CrossRef]
31. Pei, G.; Li, J.; Li, Y.Y.; Wang, D.D.; Ji, J. Construction and dynamic test of a small-scale organic rankine cycle. *Energy* **2011**, *36*, 3215–3223. [CrossRef]
32. Galindo, J.; Ruiz, S.; Dolz, V.; Royo-Pascual, L.; Haller, R.; Nicolas, B.; Glavatskaya, Y. Experimental and thermodynamic analysis of a bottoming Organic Rankine Cycle (ORC) of gasoline engine using swash-plate expander. *Energy Convers. Manag.* **2015**, *103*, 519–532. [CrossRef]
33. Dumont, O.; Quoilin, S.; Lemort, V. Experimental investigation of a reversible heat pump/organic Rankine cycle unit designed to be coupled with a passive house to get a Net Zero Energy Building. *Int. J. Refrig.* **2015**, *54*, 190–203. [CrossRef]
34. Gao, P.; Wang, Z.Z.; Wang, L.L.; Lu, H.T. Technical feasibility of a gravity-type pumpless ORC system with one evaporator and two condensers. *Appl. Ther. Eng.* **2018**, *145*, 569–575. [CrossRef]
35. Bao, H.H.; Ma, Z.Z.; Roskilly, A.P. Chemisorption power generation driven by low grade heat – Theoretical analysis and comparison with pumpless ORC. *Appl. Energy* **2017**, *186*, 282–290. [CrossRef]
36. Jiang, L.; Lu, H.H.; Wang, R.R.; Wang, L.L.; Gong, L.L.; Lu, Y.Y.; Roskilly, A.P. Investigation on an innovative cascading cycle for power and refrigeration cogeneration. *Energy Convers. Manag.* **2017**, *145*, 20–29. [CrossRef]
37. Feng, Y.Y.; Hung, T.C.; Su, T.T.; Wang, S.; Wang, Q.; Yang, S.S.; Lin, J.J.; Lin, C.H. Experimental investigation of a R245fa-based organic Rankine cycle adapting two operation strategies: Stand alone and grid connect. *Energy* **2017**, *141*, 1239–1253. [CrossRef]
38. Sun, H.H.; Qin, J.; Hung, T.C.; Lin, C.C.; Lin, Y.F. Performance comparison of organic Rankine cycle with expansion from superheated zone or two-phase zone based on temperature utilization rate of heat source. *Energy* **2018**, *149*, 566–576. [CrossRef]

Article

Modelling of Polymeric Shell and Tube Heat Exchangers for Low-Medium Temperature Geothermal Applications

Francesca Ceglia [1,*], Adriano Macaluso [2], Elisa Marrasso [1], Maurizio Sasso [1] and Laura Vanoli [2]

[1] Department of Engineering, University of Sannio, 82100 Benevento, Italy; marrasso@unisannio.it (E.M.); sasso@unisannio.it (M.S.)
[2] Department of Engineering, University of Study of Napoli "Parthenope", 80143 Naples, Italy; adriano.macaluso@uniparthenope.it (A.M.); laura.vanoli@uniparthenope.it (L.V.)
* Correspondence: fceglia@unisannio.it

Received: 30 March 2020; Accepted: 26 May 2020; Published: 29 May 2020

Abstract: Improvements in using geothermal sources can be attained through the installation of power plants taking advantage of low and medium enthalpy available in poorly exploited geothermal sites. Geothermal fluids at medium and low temperature could be considered to feed binary cycle power plants using organic fluids for electricity "production" or in cogeneration configuration. The improvement in the use of geothermal aquifers at low-medium enthalpy in small deep sites favours the reduction of drilling well costs, and in addition, it allows the exploitation of local resources in the energy districts. The heat exchanger evaporator enables the thermal heat exchange between the working fluid (which is commonly an organic fluid for an Organic Rankine Cycle) and the geothermal fluid (supplied by the aquifer). Thus, it has to be realised taking into account the thermodynamic proprieties and chemical composition of the geothermal field. The geothermal fluid is typically very aggressive, and it leads to the corrosion of steel traditionally used in the heat exchangers. This paper analyses the possibility of using plastic material in the constructions of the evaporator installed in an Organic Rankine Cycle plant in order to overcome the problems of corrosion and the increase of heat exchanger thermal resistance due to the fouling effect. A comparison among heat exchangers made of commonly used materials, such as carbon, steel, and titanium, with alternative polymeric materials has been carried out. This analysis has been built in a mathematical approach using the correlation referred to in the literature about heat transfer in single-phase and two-phase fluids in a tube and/or in the shell side. The outcomes provide the heat transfer area for the shell and tube heat exchanger with a fixed thermal power size. The results have demonstrated that the plastic evaporator shows an increase of 47.0% of the heat transfer area but an economic installation cost saving of 48.0% over the titanium evaporator.

Keywords: plastic heat exchanger; Organic Rankine Cycle; geothermal energy; shell and tube heat exchanger; fouling resistance

---

1. Introduction

Climate change, rising pollution, and the fossil fuel depletion encourage many countries to push towards renewable-based energy conversion systems [1]. In particular, the European Union energy policy gives a high priority to the increasing use of renewable energy sources (RESs), because of their strong contribution to the diversification of energy supply, the improvement of the security of energy systems, the minimization of greenhouse effects and social and economic cohesion. In 2014 the European Union (EU) agreed to implement strategies and targets (revised in 2018) to move toward a low carbon economy. One of the pillars of EU policy in energy and environmental matters consisted of the RESs share increase in final energy consumption by at least 32.0% in 2030 compared to the 1990 level [2]. The attainment of the EU renewables target is ensured by the governance system based on national and local energy planning. Indeed, in recent years, it can be observed that a proliferation of ambitious commitments and policies in the RESs field adopted by national, regional, and local governments aimed to achieve EU goals [3]. Moreover, the support policies and strategies offer useful instruments to encourage renewable energy use and to ensure the full exploitation of RES's potential in territorial areas where the energy policies' actions are in force [4].

According to Eurostat [5] renewable energies covered 17.5% of final EU energy consumption in 2017. It has been estimated that the main renewable energy options which will contribute to the additional EU potential in 2030 are: wind power (239 TWh), transport biofuels (218 TWh), solar thermal in industry and buildings (117 TWh), biomass in industry and buildings (105 TWh), and solar photovoltaic (93.0 TWh) [6]. Unfortunately, the widely used and most promising RESs (such as solar and wind) are not programmable. Their availability varies throughout hours and seasons and, as a consequence, their use has to be accurately designed in conjunction with management strategies such as load shifting and energy storage. Thus, one of the possible pathways to mitigate the criticalities of this uncertainty is the exploitation of more flexible and stable RESs such as geothermal and biomass energy sources. In particular, geothermal energy is an abundant renewable source with significant potential. The geothermal reservoirs could be employed in direct use such as in district heating systems, in indirect use to produce electricity, and in cogeneration systems for the combined production of power, heating, and cooling energy [7]. The hot water reservoirs in low deep sites are used for direct scopes and/or district heating systems. Native American, Chinese, and Ancient Roman people adopted hot mineral springs for cooking and bathing; currently these could be used for thermal scope, heating, and domestic hot water. Geothermal energy provides heat requirements to buildings by means of district heating systems. Surface hot water is directed to buildings in the intermediate circuit. For example, in Reykjavik (Iceland), a district heating system supplies thermal energy to most of the buildings. On the other hand, the use of geothermal energy in industrial fields concerns gold mining, food dehydration, and milk pasteurizing. One of the most widespread industrial application of geothermal energy is represented by the dehydration of vegetables and fruit drying [8]. As concerns the electricity production from geothermal sources, traditional geothermal power plants exploit steam or water at high temperatures (150 to 370 °C). Geothermal power plants are usually located near reservoirs within one or two miles of the Earth's surface.

Otherwise, geothermal sources are mostly available worldwide at medium and low enthalpy. In past years, geothermal energy has primarily been used in zones with a volcanic activity where there was a near sub-surface heat availability. Whereas, in recent decades geothermal energy usage has spread in many regions of the world thanks to the possibility of drilling to depths of several kilometres. In developed countries geothermal energy supply is especially widespread due to their flourishing economic means and industrial advances, as well as their availability of know-how and expertise. Besides, geothermal energy can be useful in developing countries to solve the energy access problems [9]. In Asia, only China owns 8.00% of world's geothermal resources and it has 27.8 MW of geothermal installed power generation [10]. In Canada and Australia there is a large availability of geothermal resources at 100 °C near the land surface [11,12]. In South Africa (Main Karoo Basin region) a geothermal potential has been recently addressed, finding a geothermal gradient ranging from 24.5 °C/km to 28.2 °C/km at 3500 m depth [13].

On the other hand, in many countries geothermal sources have remained unused because of the acceptability problems of geothermal installations. Meller et al. [14] have identified the precondition for public acceptability of geothermal power plants, and they have analysed the social response to novel geothermal activity. The aim of their study has been to develop technological and scientific options by means of sociological studies for the responsible use of geothermal resources. Indeed, the participation of the public into geothermal projects should go further with communication and awareness-raising measures. By respecting the research rules and the development steps of technological solutions it is necessary for an open approach aimed to include governance structures. In some European countries (such as Germany, France, and Italy) the social response to geothermal projects evidences the reticence of the public and relevant stakeholders to eventual geothermal development in all considered countries. In addition to these issues, the high cost of geothermal installations and the regulatory uncertainty have jeopardized the diffusion of energy conversion systems based on low and medium enthalpy geothermal sources.

This is an Italian case in which the massive presence of geothermal sources (at high, medium, and low temperatures) is not completely exploited even if its potential has been recognised since the 1980s [15]. Interesting geothermal areas had been found in the southern Alps region at a temperature of 80–120 °C and a depth of about 3000 m [16]. In the transition zone between the chain and the basin of the Po plain, a certain heat flow anomaly (80.0 mW/m^2) attributable to a geothermal gradient has been discovered. This heat flow is 21.3% higher than the Italian heat flow mean value (63.0 mW/m^2) [17]. Moreover, a great heat flow anomaly affects the Mediterranean and central/southern Tyrrhenian Sea (>150.0 mW/m^2). Intensive volcanic activity occurs on the Aeolian and Pantelleria islands (in Sicily) and in the Phlegraean Fields area (near Naples, South of Italy), resulting in geothermal fluid temperatures of 100–150 °C located near the Earth's surface [18]. Despite this large availability, the only Italian geothermoelectric power plants are installed in Tuscany, characterised by high-temperature geothermal sources covering a power equal to 915 MW. These power plants are all traditional flash steam technologies that can use only geothermal high temperature fluids letting the low-medium resources go unused [19,20].

The major possibility of employing the low-medium enthalpy reservoirs for geothermoelectric applications is represented by binary cycles such as the Rankine Cycle with organic working fluids (ORC) [21]. Its layout is the same as an ordinary Rankine cycle, with the main difference being that in the ORC an organic fluid is used instead of water. The liquid-vapour phase change of an organic fluid occurs at a temperature lower than the phase change of water-steam, allowing for the use of low-medium temperature sources. In addition, the use of an organic fluid results in further advantages related to the thermo-physical proprieties of the fluid. The major used fluids in ORC systems are characterized by a positive slope of the vapour saturation curve which permits the avoidance of both superheating at the inlet of the turbine and condensation during the expansion process.

Among the different sources for an ORC, the most used and promising are biomass, sun, geothermal brines, and exhaust gases from industrial processes and engines. ORC plants can be characterized by their wide size range from a few kW$_{el}$ (in residential cogeneration applications) up to tens of MW$_{el}$ (in large power plants uses). As regards geothermoelectric plants based on ORC systems, worldwide only 16.0% of geothermal power plants are based on ORC technology such as those in Italy [21]. In particular, ORC power plants activated by geothermal sources number about 337, and these represent 19.2% of total ORC plants feeding by different sources. Geothermal ORC power plants cover 2021 MW$_{el}$ representing 74.9% of the total power installed for the ORC systems. These data highlight that the geothermal ORC plants have higher size than other ORC plants fed by different RESs, and their medium power is 6.00 MW$_{el}$ for a single plant [22].

The rising interest in ORC power plants is demonstrated by several works conducted in recent years to investigate their performances in different applications.

In [23,24] a binary ORC power plant for the use of medium-low enthalpy geothermal reservoirs is analysed firstly in a thermodynamic investigation/optimization, and secondly from an economic aspect. In [23] a Matlab code is defined to find the optimal match cycle parameters, configuration, and fluid taking into account geothermal waters in the temperature range of 120–180 °C. Thermodynamic optimization results show that the highest plant efficiencies are obtained for fluids that present a range of 0.88–0.92 of the ratio between critical working fluid temperature and inlet geothermal temperature and a supercritical cycle with a reduced pressure between 1.10 and 1.60 bar. Meanwhile, in the economic optimization these values are slightly lower and the advantages of supercritical cycles usage are less present.

In [24] an ORC powered by diesel engine waste heat recovery with different selected working refrigerant fluids such as R123, R134a, R245fa, and R22 is modelled and optimized. The optimization results, conducted by means of a genetic algorithm, show that the R123 is the best working fluid in both an economical and a thermodynamic sense for a defined value of output power. The optimum result of R123 determines the 0.01%, 4.39%, and 4.49% improvement for the yearly cost compared to R245fa, R22, and R134a, respectively. As regards thermodynamic optimization, the percentages of efficiency increase using R123 and other fluids (R245fa, R22, and R134a) are 1.01%, 12.79%, and 10.6%, respectively.

Regarding the geothermal applications of ORC systems, many studies have been aimed to assess the performance of the cascade geothermal plants based on ORC systems [25], and the performance of this technology through numerical [26–29] and experimental [30] works has been evaluated.

The cascade geothermal plants based on ORC systems coupled with other components, such as absorption heat pumps and heat exchangers for heat recovery, were analysed in many studies. In [25] Pastor-Martinez et al. have compared different polygeneration systems activated by low and medium geothermal sources (80–150 °C). The systems consist of an ORC (to produce electricity), an absorption heat pump (to provide cooling energy), and a heat exchanger (for heat recovery) arranged in different cascade configurations, with parallel or series geothermal fluid use. The study considers two possible plants in two different temperature ranges. For the first case, from 80 to 111 °C, the polygeneration arrangement presents the highest exergetic performance in the hybrid arrangement for parallel-series configuration, and it reaches the exergy efficiencies varying from 42.8% to 50.1%. Instead, the second one ranges from 110–150 °C, and awards a series cascade configuration achieving exergy efficiencies from 51.4% to 52.9%. In [26–28] other simulation studies integrate the desalinization water system to improve the polygeneration system layout using the geothermal fluid in cascade allowing a re-injection of the fluid at a lower temperature. The hybrid multi-purpose plant consists of an ORC powered by a low-medium temperature geothermal source and by solar energy from parabolic collectors. The geothermal water is first employed to activate the ORC loop, then to supply heating needs at about 85–90 °C (in winter), or space cooling (in summer) using a single-effect absorption chiller. At the end of the cycle, the geothermal water is used to feed a multi-effect distillation system. In this process the seawater is turned into freshwater. The results determine an energy efficiency of the ORC plant equal to 11.6% and a simple payback of 4.47 years. Another simulation study [29] is based on a zero-dimensional model of an ORC that allows for the investigation of the effects on the main output parameters (heat exchangers efficiency, ORC power output, ORC first law efficiency) by only changing the heat transfer coefficients correlations of the vapour generator available in literature. Two ORC plants both activated by a medium-temperature geothermal source supplying a constant thermal load are analysed in the study. The first case study regards an ORC module with a size of 1.20 MW$_{el}$ fed by a geothermal fluid with a temperature of 160 °C and 7.00 bar and using n-pentane. The second ORC plant uses R245fa providing 8.00 kW$_{el}$ and it is activated by a geothermal fluid at 95.0 °C and 3.00 bar. The first plant shows an electric efficiency of about 14% for all correlations investigated, while the second plant obtains an electric efficiency ranging from 7.00–10.0%. In [30], a real ORC plant in Huabei (China) shows the possibility for providing geothermoelectric energy from different thermal fluids in an aggressive environment. The plant with 500 kW$_{el}$ employs fuel oil from an abandoned well using an intermediate circuit with an efficiency between 4.00–5.00% and consequently the output power of the turbine varies from 160–60.0 kW$_{el}$.

Even if ORC technology is very promising in low enthalpy geothermal applications, it is affected by a problem common to all technology exploiting geothermal sources. Geothermal fluids are often very aggressive (depending on geothermal sites) and they cause surface fouling, corrosion, and scaling over the heat exchanger (HEX) surface in which the heat transfer occurs through the working fluid evaporation. The scale formations threaten the heat transfer between geothermal and organic fluids, and they have also an impact on HEX performance in terms of flow velocity and pressure drop. Thus, they force the plant to turn off in order to clean or replace the HEX [31–33]. This issue causes an operating cost increase; indeed, it has been estimated that the impact of the HEX cost on the total plant investment costs ranges from 20.0% to 30.0% [34]. The HEX used in an ORC system is traditionally a metallic shell and tube type heat exchanger that has a high cost and low resistance to corrosion [35]. In [36] a mathematical model of the shell and tube heat exchanger (STHEX) is proposed for the capture and predicting fouling trends on both the shell and tube sides. The literature review is typically focused on fouling inside the tubes, and instead fouling on the shell side is not usually considered. In this work, instead, it has been demonstrated that fouling deposition on the shell side may be significant. It determines the not-equal heat transfer, growing pressure drops, and flow path modifications. Different solutions have been investigated to solve the fouling problems.

In [37] a study has been conducted to improve the corrosion-erosion features of carbon steel by using a multilayer composite coat for carbon steel plates. This solution is adopted to obtain better tribological behaviours and a lower wear rate of coatings than the untreated steel. In this application the nickel is linked with chromium to form Cr_2O_3 which allows us to use the material until 1200 °C. In particular the material use is ensured for temperatures lower than 800 °C thus, it could be used in both geothermal plant components and gas combustion applications. These coatings, however, produce very dense layers with a porosity under 0.500–1.00%. In [38] the performance of the HEX ceramic prototype has been evaluated. It is able to achieve a heat transfer of up to 6.00 kW and an effectiveness of up to 97.0%. It could operate at high temperatures and in harsh environments exceeding also the performance of a comparable metal HEX.

Another alternative path that can be followed is the replacement of the metallic HEX in geothermal ORC with a plastic (polymer) heat exchanger (PHEX). Indeed, a PHEX in ORC systems activated by geothermal sources ensures a higher inhibition of the scale depositions and the corrosion resistance limiting the maintenance periods for chemical or mechanical and/or hydrodynamic washing affecting the carbon steel HEX. Moreover, PHEX investments and maintenance costs are lower than those of a HEX with the same useful life cycle. Even if it is possible to consider the advantage of a PHEX for aggressive environment applications it is necessary to underline that the market for these components is thin for high working temperatures and pressures. The low operating temperature is compatible with geothermal sources, but the low pressure of exercise is not performing for organic fluids used in ORC systems. Thus, some advances are still required for the widespread use of plastic heat exchangers, but great improvements are expected in their market to justify the investigation on PHEX performance in innovative applications such as in ORC plants. This is the main topic of this work and in the following section the available literature information regarding PHEXs operating and design conditions in comparison with HEXs have been collected and summarised; in addition, the motivation and aims of the paper have been discussed too.

2. Polymeric Heat Exchanger: State of Art and Aim of the Study

Currently, the used materials for HEXs are metallic alloys, such as 70.0% Cu, 30.0% Ni, or titanium, with high thermal conductibility but with high corrosion and fouling resistances, especially in aggressive environments [39,40]. The types of fouling occurring in a HEX could be different, such as the precipitation of solid deposits in a fluid on the heat transfer surfaces (that can be cleaned by chemical treatment) or corrosion and other chemical fouling. The growth of algae in hot fluids can be a cause of contaminations in the heat exchangers. The biological fouling that the algae may determine can be avoided by chemical techniques. Moreover, in geothermal applications, the chemical aggressive composition of geothermal brines (strictly depending on geothermal site) determines the fouling on the HEX's surface. All these problems cause the deterioration of material, the worsening of heat transfer efficiency, the increase of pressure losses due to friction, the reduction of crossing flow section, and the need for HEX replacement and cleaning. As a consequence, it may be necessary to employ a larger, and thus more expensive, HEX to achieve the required heat transfer performance when fouling takes place. All of these categories of fouling can be reduced by using plastic pipes instead of metal ones or by coating metal pipes with glass. Regarding polymer heat exchangers, their main advantages are: low cost, low weight, anti-corrosion, antifouling, manufacturing simplicity, electrical insulation, high chemical resistance, high modelling, and excellent elasticity (depending upon the working temperatures) [41,42]. The low PHEXs' costs result in a decrease of up to 20% on power plant investment costs in which a titanium HEX evaporator is replaced by a PHEX one by considering the same life cycle [34]. Thus, the PHEX's opportunities are evident in small-size applications and when the specific unitary cost is high. Indeed, with reference to a metal HEX's cost in Figure 1, the investment cost of a common metal HEX against its heat transfer area is plotted starting from data available in literature for different heat exchangers' geometries [43]. In Figure 1, the x-axis is stopped at 120 m^2 because it is the maximum size of the heat transfer area for the plate heat exchanger available from the market data comparable to other models. In addition, a HEX's maintenance cost for 20 years of its life cycle is equal to 10.0–11.0% of the initial investment. Therefore, PHEXs offer an interesting alternative to the high cost of metal heat exchangers concerning both the investment and the maintenance costs, especially in aggressive operating conditions.

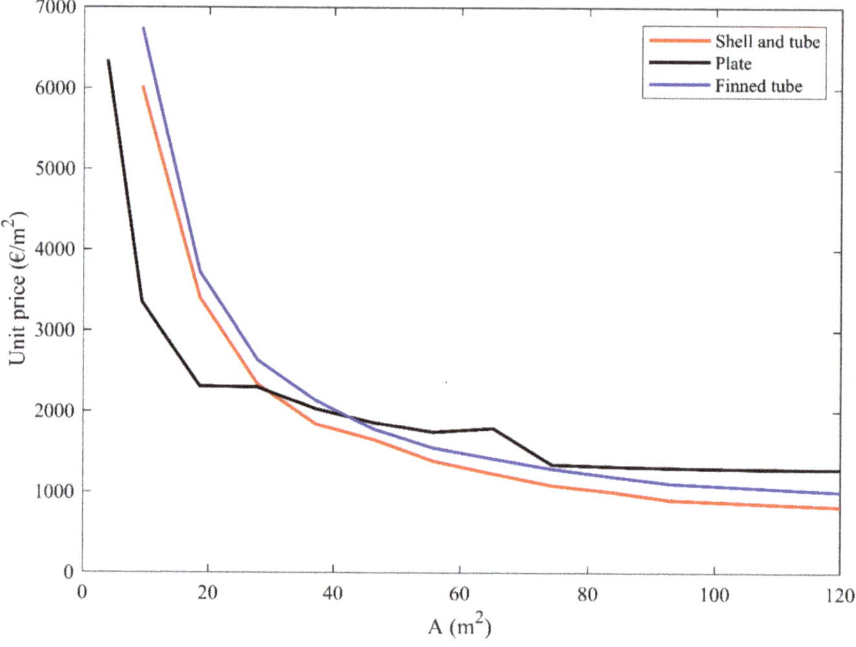

Figure 1. Metal heat exchangers' (HEXs) unit price as a function of the heat transfer area.

Another advantage of PHEXs is that they are already available on the market in multiple polymeric plastic materials (such as Ethilene Chloro TriFluoroEthylene (ECTFE); polyEthylenete TetraFluoroEthylene (ETFE); polyEthylene of high and low density (PEHD/PELD); PolyVinilChloride (PVC), etc.) and different manufacturing companies already produce PHEX worldwide: TMW (La Serre, France) [44], Polytetra (Bietigheim-Bissingen, Germany) [45], Aetna plastic (Valley View, United States of America) [46], HeatMatrix Group B.V. (Geldermalsen, Netherlands) [47], Fluorotherm (Parsippany, United States of America) [48], Ametek (Berwyn, United States of America) [49], Kansetu (Osaka, Japan) [50], and Calorplast (Krefeld, Germany) [51]. The most widespread geometries of PHEXs produced by the aforementioned companies are plate and shell and tube heat exchangers; only in particular cases is it possible to find immersion, hollow plates, or tube plate PHEXs. In addition to these benefits, PHEXs show also some critical issues that can be summarised as follows [52]:

- the most relevant parameter for HEXs' design is the global heat transfer coefficient (UA) that takes into account the convective, conductive, and fouling resistances. The plastic thermal conductivity is usually equal to 1.00% of the thermal conductivity of metallic materials [53,54]. This fact determines a penalization of the heat transfer performance; thus, a higher heat transfer surface area is requested.

- the PHEX has low structural strength and poor stability in terms of mechanical resistance at high temperatures or pressures. Indeed, the limits of PHEXs are the working couple of pressure and temperature that the material can support. At high temperatures, the maximum operating pressures of the commercial devices are often incompatible with the operating pressures of organic fluids defined for ORC market applications [34]. Table 1 shows a summary of the optimal working temperature and pressure values for each plastic material available for shell and tube exchangers. It is important to underline that the maximum operating temperatures could be higher when the pressure is reduced and vice versa. In general, it should be noted that the materials most resistant to high temperatures (>100 °C) are Perfluoroalkoxy alkanes (PFA) and polyvinylidene difluoride (PVDF) which, however, have lower thermal conductivity (about half) than PE and PP. The limit for the PE polymers is under 100 °C for the hot fluid considered. Over this value of temperature, it is necessary for the use of a polymer with a higher mechanical resistance. In particular, from market analysis, it has been found that it is easy to reach operating temperatures close to 100 °C by improving the composition of polymers; for example, by adopting high density PE (PEHD) as suggested by the manufacturer [51].

PHEX's problems have been investigated by researchers in recent years by means of simulative and experimental applications. For instance, in [55] the authors have demonstrated through an experimental analysis that PHEXs conductive resistance (that is proportional to the inverse of thermal conductivity) amounts only to 3.00% of the total thermal resistance (including fouling, convective, and conductive resistance). Moreover, the percentage weight of conductive resistance on total thermal resistance is strictly linked to the value assumed by convective and fouling resistances which depend on many factors such as the thermodynamic condition of the fluid and the Reynolds numbers (by velocity definition) of the fluid that crosses the PHEX [56]. Other authors have found that an enhancement of the polymer heat transfer coefficient can be obtained by filling the polymeric matrix with high conductivity additives or by controlling its crystallinity [57]. In support of these theories in [58] it is experimentally shown that the specific volumetric heat transfer coefficient of a PHEX in polyamide filled with carbon fibres is 1.65 times higher than that produced with a polyamide only. On the other hand, even if polymers constituting PHEXs have a low mechanical resistance, they can deform elastically meaning that they return to their original form after thermal or mechanical stress. Moreover, PHEXs' low structural strength results in a light weight. Indeed, in [59] it has been estimated that the metal HEXs have many drawbacks related to their weight, and PHEXs ensure up to 62.2% of weight reduction.

PHEXs' applications have been recognised as particularly suitable in different fields as seawater heat exchangers, solar water systems, heat recovery applications, and for the desalination industry in a corrosive environment [60]. PHEXs can also represent a good solution for ORC applications, since ORC systems are characterized by aggressive/low temperature heat sources and a high specific cost, so the heat exchanger's cost has a great weight in the overall cost of such technology. Despite this opportunity, there is a gap in the literature review concerning the PHEX use in ORC exploiting geothermal brine. To the best of authors' knowledge, only Gomez et al. [34] have conducted a simulative analysis to investigate the possible replacement of a titanium HEX with a PHEX in low size Organic Rankine Cycles for geothermal applications to reduce plant investment costs. They have conducted a thermodynamic analysis of a 20 kW$_{el}$ RORC plant according to the currently available data on market for shell and tube PHEXs. Then, the authors compared a PHEX and a metallic HEX from an economic point of view finding that for a plant availability for 5000 h of functioning in a year and a discount rate of 10%, a cost of the generated electricity equal to 94.8 $/MWh for a plastic solution and 118.9 $/MWh for a titanium HEX can be obtained.

In this context, the main aim of this work is to determine whether it is possible to employ PHEXs as an alternative to conventional metallic HEXs in an ORC-based plant using a geothermal source.

The work wants to define the advantages and disadvantages of HEX replacement considering the main design proprieties, such as the heat exchanger's surface area, the overall heat exchange coefficient, and also economic aspects. The novelty of this work concerns the development of a one-dimensional mathematical numerical model of a polymeric STHEX based on discrete elements for geothermal ORC applications. Thus, unlike other works in the literature review, PHEX is not modelled as a "black-box" accounting only for the inlet and outlet fluids' operating conditions, but the heat transfer inside tubes is modelled considering PHEXs' conductive, convective, and fouling resistances too. The study is based on a MATLAB algorithm with a REFPROP interface able to provide the value of design characteristics such as the heat transfer area, the length of tubes, and the heat transfer coefficient in a shell and tube heat exchanger configuration. A sensitive analysis is carried out varying the inlet temperature of geothermal brine using literature correlations in the worst fouling condition.

Table 1. Operating temperatures and pressures for different polymer materials.

Plastic Materials	Working Temperature (°C)	20	40	60	80	100	120	140
PFA	Rupture pressure (bar)	59.98	47.99	38.96	29.99	19.99	12.00	6.00
	Max working pressure (bar)	10.00	8.00	6.48	5.03	3.52	2.00	1.03
PVDF	Rupture pressure (bar)	79.98	55.02	49.99	39.99	29.99	22.41	17.44
	Max working pressure (bar)	12.00	10.00	7.52	6.00	4.48	3.52	2.96
PP/PE	Rupture pressure (bar)	24.96	18.00	14.00	8.00	-	-	-
	Max working pressure (bar)	8.00	6.00	4.00	2.00	-	-	-

3. ORC Configuration with PHEX in a Geothermal Plant and Working Fluid Selection

The most common applications provide the use of a STHEX in geothermal ORC systems because it can be cleaned easily. Among all heat exchangers, STHEXs are the most widespread heat exchangers covering 65.0% of the market in different applications [61]. The common practice considers the allocation of fluids with the maximum predisposition to scale in the side of tubes to facilitate the cleaning of the component. However, the fluid on the side of the shell can also be subjected to fouling. In such cases, the neglect of fouling effects in the shell-side may lead to great errors in the analysis of the data [36]. In this paper a shell and tube PHEX is considered as an evaporator in a geothermal ORC plant. The possibility of component cleaning is neglected due to its higher fouling resistance, whereas it is considered with the opportunity of PHEX replacement according to its low cost. Thus, this work considers the organic fluid flowing inside the tubes and the geothermal hot brine is confined in the shell, resulting in a counter-cross flow heat exchanger. This choice leads to two advantages:

- it is more economically convenient to build high pressure pipes (in which the organic working fluid with the higher pressure is flowing) than the high-pressure shell [56];
- the pipe-side heat transfer coefficient is superior to shell side coefficient thanks to the presence of boiling organic fluid in the pipes [62].

The considered plant, in a regenerative ORC configuration (RORC), is represented in Figure 2. It is composed of a pump, turbine, regenerator, condenser, and evaporator that provides the interaction between the two fluids: the hot vector fluid (geothermal brine) and the organic working fluid. Geothermal fluid is taken from the aquifer and then it is pre-treated. The geothermal fluid considered in this work refers to the Phlegrean Fields area (south of Italy). The choice of the hot fluid temperature derives from a previous analysis conducted by Carlino et al. in [63] from which it has been possible to evaluate the geothermal gradient. By analysing the census relating to the area of interest an average value of the well surface temperatures has been considered excluding temperatures below 70 °C (temperature typically used for reinjection). The temperature of the pre-treated hot resource used in this analysis is about 95 °C, compatible with the ORC manufacturer data request [64]. In particular, the Phlegrean Fields area is heavily affected by medium-low enthalpy geothermal sources [65]. It has been demonstrated that near Pozzuoli there are many zones interested by geothermal source availability with a temperature near 90 °C. Different studies have typically analysed the chemical composition of geothermal fluid at high deep sites and temperatures (~200–300 °C) [63,66], and have also presented data on flow rates and tempering available in the Campania region such as the geothermal brine chemistry characteristic. Furthermore, few literature data show the chemical composition at lower temperature conditions despite the studies evidencing the high variability of the geothermal source in Campania. However, in the Phlegrean Fields area the geothermal sites show various temperature gradients (20–180 °C/km) and temperature profiles, different flow rates, and heat flow density (from 20 to 140 mW/m^2) [67]. Some data about the chemical-physical composition of geothermal water at low temperatures in Phlegrean Fields are reported in Table 2. The data derive from studies conducted in Phlegraean Fields before 1980 [68], during the period 1990–1999 [69,70]. In some of these studies information about the possible change in geothermal fluid composition and temperature after the bradyseism both in the fumaroles and thermal waters are reported. In addition, it has been evidenced that the shallow water system is locally activated by a deep hydrothermal element, generating in the form of boiling pools and wells the hot waters at the surface with temperatures around 90 °C. In Table 2 the chemical composition of geothermal fluids suitable for our simulation data set according to their temperature and chemical composition compatible with a heat exchanger and ORC considered system is reported. Both the columns (H. Tennis #1 and H. Tennis #2) are referred to *H. Tennis well* for two different periods [64,69]. Nevertheless other geothermal sites at low-medium temperatures are available in Phlegrean Fields, but because of their extreme sulfate and ammonium composition prevalence, higher presence of Fe and SiO$_2$, and extremely low pH, they are not considered [66–68]. In fact, they could require further analyses for numerical simulation because of their large quantities of silica, iron, aluminum, and sulfate could induce massive precipitations of minerals in the PHEXs. According to the literature the prevalent elements are Na, Cl, and total dissolved solids (TDS). In this work the geothermal fluid in the water dominated condition is simulated by using the REFPROP library which allows us to evaluate the thermodynamic properties of pure water. This approximation is endorsed by a previous analysis which allowed for the estimation of the variation of the thermodynamic properties of geothermal fluid with respect to pure water for the Phlegrean Fields site in the case of the temperature of the source at 94 °C. The results showed that the thermodynamic parameters variation (estimated by means of a software capable of integrating the salinity of the water) is not appreciable and, therefore, does not affect the hot fluid heat transfer coefficient. The geothermal brine aggressive chemical composition determines corrosion and depositions of suspension solid particles on the heat exchanger surface. In order to consider these disadvantages in heat transfer efficiency, the fouling resistances are included in the design methodology of the heat exchanger. In addition, fouling conditions are closely related to the depth of the extraction site, the withdrawal temperature, the dominant water or steam condition, and the thermo-physical characteristics of the fluid in the geographical area of interest. Therefore, the treatment that the geothermal fluid undergoes before entering the evaporator has not been simulated.

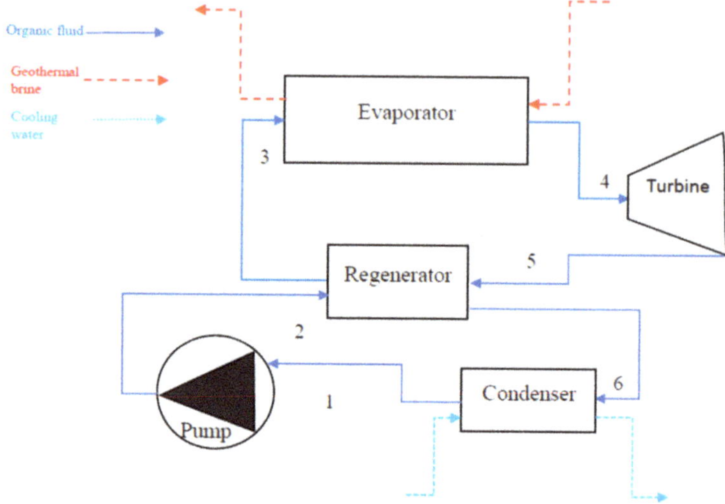

Figure 2. Organic Rankine Cycle (ORC) configuration.

Table 2. Chemistry composition of geothermal fluid for low-medium temperature sources.

Geothermal Fluid Properties	H. Tennis #1	H. Tennis #2
T (°C)	88.9	85
ph (–)	7.41	7.2
TDS (ppm)	5390	-
Na^+ (ppm)	1540	1295
K^+ (ppm)	455	407
Ca^{2+} (ppm)	38.7	58.1
Mg^{2+} (ppm)	1.3	4.9
Sr^{2+} (ppm)	0.4	0.9
Li^+ (ppm)	-	0.96
Rb^+ (ppm)	-	1.41
NH_4^+ (ppm)	-	23.5
Cl^- (ppm)	1536	1380
HCO_3^- (ppm)	753	699
SO_4^{2-} (ppm)	1040	1136
F^- (ppm)	-	3.5
SiO_2 (ppm)	-	183.5
Al^{3+} (ppm)	-	0.5
B (ppm)	-	44.8

Regarding the organic fluid, a lot of works have been conducted to choose the best fluid with optimum thermo-physical, economic, and environmental properties in coupling with a low-medium temperature source [71,72]. In particular, the working fluid selection depends on the heat source temperature and the condenser cooling fluid temperature as well as on the ORC size. Some analyses show that the organic fluids characterized by high critical temperature allow for the obtainment of a higher ORC cycle efficiency, but this condition is verified only for the fluids with high reduced pressure. Indeed, the high critical temperature determines substantial expansion ratios and low condensing pressures making these fluids suitable for low-temperature applications [73]. Besides, they allow for the avoidance of the technical problems related to low evaporation pressure [74], the obtainment of an acceptable pinch point temperature difference [75], and low costs associated with the exergy destruction even if they are strongly influenced by the amount of working fluid used [76,77]. Moreover, the use of fluids with a molecular weight greater than that of water can improve the isentropic efficiency of the turbine and it may allow for the use of single-stage expander reducing the plant costs [78].

Thus, the pentafluoropropane (R245fa) is the organic fluid chosen in this application. It is suitable for different heat transfer applications such as low-temperature refrigeration, passive cooling devices, ORC for heat recovery, and in sensible heat transfer. The R245fa is useful for its special compatibility with widespread materials of construction as metals, elastomers, and plastics. It is also assessed for its good properties concerning human safety, health, and environmental aspects as toxicity, flammability characteristics, exposure limits, global warming potential, and safe product handling [79]. Calise et al. [80] have demonstrated that R245fa is the fluid with the best performance (in terms of heat exchange efficiency and environmental pollution) among other organic fluids in geothermal applications when geothermal brine temperature is lower than 170 °C.

The design of the PHEX is based on the manufacturer data of an ORC system available on the market [51].

4. Methodology

This section aims to elaborate a numerical algorithm for the shell and tube ORC evaporator design carrying out a comparison among different materials for HEX construction such as PEHD, carbon steel, and titanium. The model has been developed considering the nominal thermal power ($\dot{Q}_{obj}$) of the STHEX provided by an ORC manufacturer [63]. The numerical model is based on physical equations developed in a MATLAB/REFPROP environment [81,82]. The physical equations presented here are equal for all the analysed materials while the parameters used in the equations vary from one STHEX material to another. Moreover, the correlations for evaluating the heat transfer coefficient used to determine the heat transfer rates are linked to the working fluid phase (single-phase or two phases) and to its position (shell or tube). The flow chart of the numerical algorithm developed and implemented in a Matlab code is reported in Figure 3. At the initial step the fixed parameters are allocated, and thermodynamic (Table 3) and geometric (Table 4) parameters are assigned as well as the thermal conductivity variable for each investigated material. At the first step of integration (j = 1), R245fa temperature ($T_{t,o}$, tube side) is set equal to its outlet temperature indicated by constructor, while the temperature of hot geothermal brine ($T_{s,i}$) is set equal to the hot source temperature required from the evaporator (since the evaporator is a HEX in counter-flow configuration). The mass flow rates in the tube side ($\dot{m}_t$) and shell side ($\dot{m}_s$) are allocated too, according to values reported in Table 3. Then, in each iteration step the heat transfer coefficients (Equations (3)–(15)), the overall heat transfer coefficient (Equation (2)), and the elementary thermal power (Equation (1)) are evaluated. At step j+1, the geothermal brine temperature in the shell side (T_s (j+1)) and the specific working fluid enthalpy in the tube side (h_t (j + 1)) are evaluated. Enthalpy at the tube side is obtained from REFPROP software.

This procedure must be repeated until the heat power target is achieved ($\dot{Q}_{obj}$). Finally, the total evaporator length L and the total heat exchanger surface (A) can be evaluated for each investigated material.

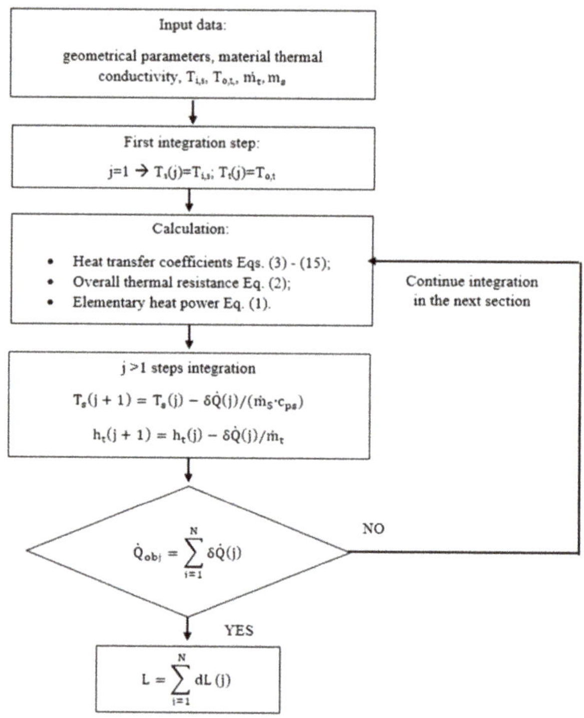

Figure 3. Flow chart of numerical algorithm.

Table 3. ORC nominal data [63].

Parameters	Value
Critical Temperature R245fa (°C)	154.05
Primary Thermal Power (kW$_{th}$)	450
Condensation water Temperature (°C)	33.0
Outlet Pressure Condensation (bar)	1.00
Flow rate working fluid R245fa (kg/s)	2.65
Flow rate geothermal brine (kg/s)	13.4
Inlet Evaporator Temperature hot fluid (°C)	94.0
Inlet Turbine Temperature working fluid (°C)	85.0
Outlet Turbine Temperature (°C)	60.0
Generator Cooling Water Temperature (°C)	<40.0

Table 4. Shell and tube heat exchanger data.

Parameters	Symbol	Value
Inlet diameter tube (m)	d_i	6.00×10^{-3}
Thickness diameter tube (m)	s_t	6.00×10^{-4}
Shell diameter (m)	D_s	0.180
Shell limit diameter (m)	$D_{limit,s}$	0.200
Max total length HEX (m)	L_t	6.00
Number of rows (-)	n_{row}	3.00
Number of tubes for each row (-)	$n_{t,bank}$	117
Fouling resistance tube side (R245fa) (m^{2}K/W)	$R_{t,f}$	1.00×10^{-4}
Fouling resistance shell side (water) (m^{2}K/W)	$R_{s,f}$	2.64×10^{-4}

4.1. Model Equations

The proposed configuration for counter-flow STHEX is shown in Figure 4. The working fluid (R245fa) flows in the tubes (with a total length named L_t and a diameter called d_i) while geothermal hot water crosses the shell with a diameter indicated as D_s and a length called L_s.

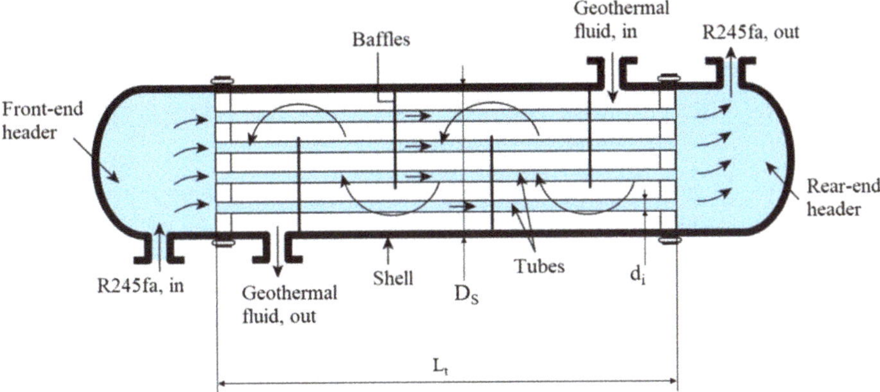

Figure 4. Shell and tube heat exchanger model.

The Matlab model is developed to define the STHEX area needed to obtain the fixed evaporator thermal power for each selected material.

The STHEX modelling depends on the following assumptions:

- steady-state conditions are considered for the the heat exchanger operations;
- the temperature of both working and hot fluids, at the exit and inlet of the section, is unvariant in the cross-sectional areas of the shell and the tube;
- the heat losses from the external heat exchanger surface to the surroundings are not considered;
- thermo-physical properties are integrated on a suitable infinitesimal element (dL) of tube with a fixed length of 5 mm;
- no variation in potential and kinetic energies of the flowing streams are taken into acoount;
- the plastic material able to withstand the operating pressures of the working fluid is considered;
- the pressure drops across tube side of heat exchanger are overlooked.
- the geometric condition fixes the tube banks rotated at 45°.

MATLAB model calculates the infinitesimal thermal power for each dL in every iteration using Equation (1). The calculation process stops when the sum of the infinitesimal thermal power (δQ) is equal to the nominal thermal power of the evaporator ($\dot{Q}_{obj}$). The infinitesimal temperature difference (dT) between geothermal and R245fa fluids in each dL is evaluated using REFPROP library starting from two thermodynamic properties of fluids. The overall heat transfer coefficient (U) is evaluated as the reciprocal of the thermal resistance (R) in each dL defined as expressed in Equation (2).

$$\delta \dot{Q} = \int_0^L \frac{dT \cdot dA}{R} dL \tag{1}$$

$$U = \frac{1}{R} = \frac{1}{R_s + R_{s,f} + R_w + R_t + R_{t,f}} = \frac{1}{\frac{1}{a_s} + \frac{1}{a_{s,f}} + R_w + \frac{1}{a_t} + \frac{1}{a_{t,f}}} \tag{2}$$

Thermal resistance considers the convective heat transfer process in the tube (R_t) and shell (R_s) side, fouling in the tube ($R_{t,f}$) and shell ($R_{s,f}$), and conductive heat transfer through wall material (R_w).

The correlations used to calculate the resistances in the shell and tubes are described in the following subsections. Additionally, the fouling resistances both in the shell and tube's side are constant and they depend upon the fluid.

4.2. Shell-Side Model Equations

The heat transfer convective coefficient into the shell side can be obtained considering the fluid in static condition assimilated to liquid water at a temperature range defined by the manufacturer. The convective coefficient for the shell side α_s depends on the Nusselt number (Nu), the diameter of shell (D_S) and thermal conductivity (k_s) as reported in following Equation (3):

$$\alpha_s = \frac{Nu \cdot k_s}{D_s} \tag{3}$$

where the Nusselt number is calculated according to Equation (4) in laminar condition [83] and to Equation (5) for turbulent regime [84].

$$Nu = 1.04 \cdot Re^{0.4} \cdot \left(\frac{Pr}{Pr_w}\right)^{0.36} \tag{4}$$

$$Nu = 0.023 \cdot Re^{0.8} \cdot Pr^{n} \tag{5}$$

In Equation (4) t Pr_w is the Prandtl number value on the wall. In Equation (5) n parameter assumes the value 0.300 if the shell side fluid is the cold one or 0.400 if it is the hot fluid. The Reynolds (Re) and Prandtl (Pr) numbers are defined as in Equations (6) and (7), respectively:

$$Re = \frac{\dot{m} \cdot D}{\mu \cdot A} \tag{6}$$

$$Pr = \frac{\mu \cdot c_p}{k} \tag{7}$$

where $\dot{m}$ represents the mass flow rate, D is the diameter, A is the area, μ is the dynamic viscosity, c_p is the specific heat at constant pressure, and k the thermal conductivity. Generally, α_s is multiplied by a corrective factor J that considers different parameters such as the number of baffles and their configuration, dispersion, and losses in shells and deflectors, and losses in the nozzles and temperature gradients that could cause accumulations. The presence of the baffles which cut the diameter by 20.0%–45.0% allows for support and prevents vibrations and vortices in the pipes. In this work as suggested by the literature, J is equal to 0.60 [56].

4.3. Tube Side Model Equations

For the tube side of STHEX the organic working fluid (R245fa) chosen for the simulation is in phase-transition; therefore, it is not advisable to consider it as a liquid for all tubes length. It is necessary to distinguish if the fluid is in two-phase or single-phase condition for convective heat transfer coefficient (α_t) evaluation. In a single phase condition, the equation used is Equation (5). In a two-phase condition α_t corresponds to α_{tp} expressed by the correlation of Gungor-Winterton [85] reported in Equation (8). It takes into account two contributions: a convective boiling term (α_{cb}) and a nucleate boiling term (α_{NB}).

$$\alpha_{tp} = S \cdot \alpha_{NB} + E \cdot \alpha_{cb} \tag{8}$$

where α_{NB} term is calculated from Cooper correlation [86] (Equation (9)):

$$\alpha_{NB} = 55 \cdot pr^{0.12 - 0.2 \cdot \log_{10} Rp} \cdot \left(-\log_{10} pr\right)^{-0.55} \cdot M^{-0.5} \cdot q^{0.67} \tag{9}$$

M is molecular weight of the substance, pr represents the ratio between the operating pressure and critical pressure and q is the heat flow.

α_{cb} term in Equation (8) is calculated from the Dittus–Boelter correlation that is employed to calculate the heat transfer of liquid only in the tubes (Equation (10)) [87]:

$$\alpha_{cb} = 0.023 \cdot \frac{k_t}{d_i} \cdot Pr_t^{1/3} \cdot Re_t^{0.8} \cdot \left(\frac{\mu_t}{\mu_{t,w}}\right)^{0.14} \tag{10}$$

where k_t is thermal conductivity, and μ_t and $\mu_{t,w}$ are the viscosity in the tube and wall, while Pr_t and Re_t are Prandt and Reynolds numbers, respectively, referred to the tube side as well, as k_t is the thermal conductivity of tube.

S (Equation (8)) shows the suppression factor of a nucleate boiling term that takes into account the decrease of fluid layer thickness as the vapour quality grows. E (Equation (8)) is the enhancement factor of the convective boiling process due to the increase of flow velocity when the vapour quality increases. The heat transfer coefficient is strongly and positively influenced by the vapour quality [88] and also by the frictional pressure gradient [89].

E and S factor are defined as in Equations (11) and (12), respectively:

$$E = 1 + 24000 \cdot \left(Bo^{1.16}\right) + 1.37 \cdot \frac{1}{X_{tt}}^{0.86} \tag{11}$$

$$S = \frac{1}{1 + 1.15 \cdot 10^{-6} \cdot E^2 \cdot Re_l^{1.17}} \tag{12}$$

In Equation (12) Reynolds number is referred to liquid condition (l).

In Equation (11) two dimensionless numbers are considered: the Martinelli parameter (X_{tt}) and the Boling number (Bo) defined in Equations (13) and (14), respectively:

$$X_{tt} = \left(\frac{1-x}{x}\right)^{0.9} \cdot \left(\frac{\rho_v}{\rho_l}\right)^{0.5} \cdot \left(\frac{\mu_v}{\mu_l}\right)^{0.1} \tag{13}$$

$$Bo = \frac{q}{G_t \lambda_{l,v}} \tag{14}$$

where x is the title, ρ_l, ρ_v are density in liquid and vapour condition, λ represents the latent heat and G_t is the flow rate in the tube.

4.4. Case Study

As before mentioned, the PHEX design is based on the manufacturer's data of an ORC system available on the market that delivers 30.0 kW_{el} with a primary power input equal to 450 kW_{th}.

In Table 3 the ORC module parameters that were considered for the modelling and simulation analysis are reported.

The geometrical construction data of the STHEX are referred to the schedule of PHEX company and they are listed in Table 4, while in Table 5 the thermal conductivity of simulated materials (kw) are showed.

Table 5. Thermal Conductivity of simulated Materials.

Material	k_w (W/mK)
PE	0.510
Titanium	16.0
Inox Steel	20.0
Carbon Steel	47.0

5. Results

The numerical algorithm has been employed with the aim of returning the heat transfer area (A) at the manufacturer's conditions for each considered material. The evaporator area needed to provide the required electric output, for the simulated ORC system is higher in the case of plastic material with respect to the other ones. In particular, the increase percentage between A of PHEX and HEX in carbon steel amounts to 48.5%. The percentage decreases to 47.5% and 47.0% if the PHEX's area is compared to area of a HEX made of inox steel and titanium, respectively. The difference among the area values is mainly due to the conductive and fouling resistance since the convective coefficients are similar for each material. In Figure 5 the values of the overall heat transfer coefficients and the heat transfer area of STHEX in plastic and metal materials are shown. As expected, PHEX provides the lowest value of U (88.5 W/m^{2}K) and its contemporary needs the highest A (524 m^2).

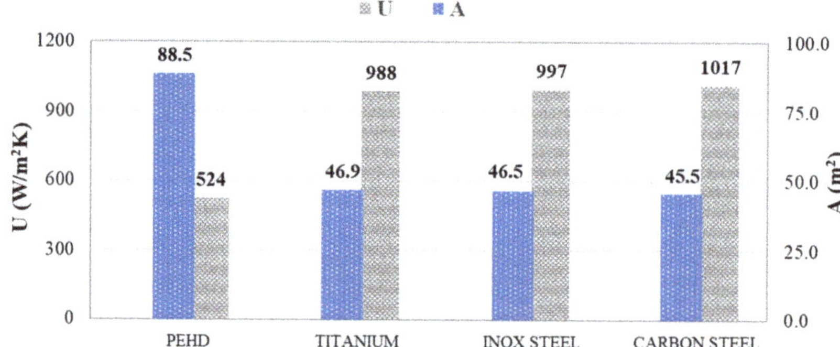

Figure 5. Overall heat transfer coefficient and heat transfer surface area.

STHEX length and the heat transfer coefficient are summarized in Table 6 for all considered materials. The length as well as the heat transfer area (Figure 5) are evaluated by summing the values in each iteration step, while UA and U values are calculated in each iteration. Table 6 and Figure 5 average values of UA and U weighted on the infinitesimal area are listed, respectively.

Table 6. Simulation results data.

STHEX Length and the Heat Transfer Coefficient	PEHD	Titanium	Inox Steel	Carbon Steel
L (m)	12.2	6.44	6.39	6.26
UA (W/K)	7.6	14.4	14.5	14.8

The purchase cost (*PC*) evaluation for STHEX made of common steel materials (inox steel and carbon steel) is carried out by means of a literature correlation (Equation (15)) suitable for evaporators at high-pressure conditions [90]:

$$PC = 190 + 310\,A \tag{15}$$

On the other hand, the evaluation of plastic and titanium STHEX costs have been conducted through a market analysis. The comparison is based on PC neglecting the maintenance and cleaning cost. In Figure 6 PCs for a PHEX (green bar) and metal materials (orange bars) are showed. The purchase cost of a STHEX in plastic material (27,597€) is significantly lower than a titanium STHEX (56,155€) while it is higher than a STHEX in inox steel (14,595€) and carbon steel (14,265€). Nevertheless, the economic comparison aimed to evaluate PC saving is relevant only between PHEX and HEX in titanium since in these two cases,the life cycle of two components are equal [91] and it is possible to neglect the cleaning cost for both STHEXs. Thereby, the percentage reduction of PC between PHEX and titanium STHEX is

48.0%. This reduction can have a significant weight for ORC plant investment costs since the unitary specific cost of the overall system is high.

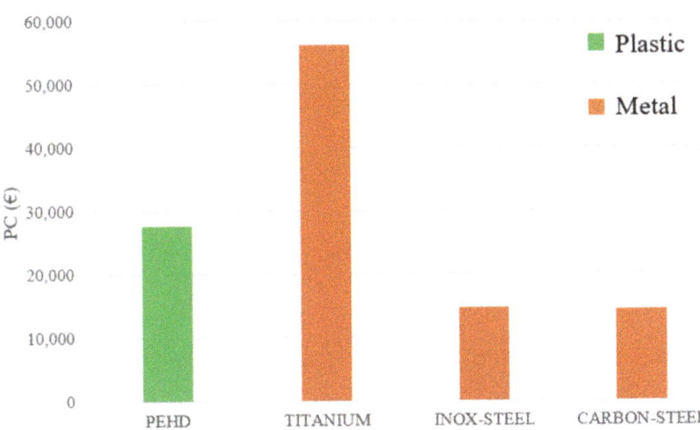

Figure 6. Length of tubes and purchase cost for different materials STHEXs.

In Figure 7 the convective heat transfer coefficient for the tube α_t, Figure 7a), for the shell α_s, Figure 7b) and overall heat transfer coefficient (U, Figure 7c) are shown as a function of heat exchanger length. These trends result from simulations in a MATLAB/REFPROP environment.

Since STHEX is in counter flow configuration, at L equal to 0 m (that represents the end of length integration) the geothermal brine is in inlet condition in the shell and the organic fluid is in the outlet condition of tubes side.

As reference to α_t (Figure 7a) at the integration start point of the evaporator (L = 0 m) the organic fluid in tubes is in satured vapour condition. Subsquently, α_t increases assuming a bell curve shape, and in this condition the organic fluid is in a two-phase state. As a matter of fact, in the two-phase condition the convective heat exchanger coefficient (α_{cb}) is characterized by two contributes (a convective boiling term α_{NB} and a nucleate boiling term α_{NB}) as reported in Equation (8). After that, α_t decreases and the curves becomes flat near the exit of the evaporator when the R245fais in the liquid condition (the condition of inlet state in STHEX) and the convective heat transfer coefficient is calculated through (Equation (10)). α_t curves for all metal materials have the same trend and are very similar; indeed, they are almost overlapping. α_t curve for plastic material has a larger bell, even if it shows the same trend of the metal materials' α_t curves. For this reason, if an L value is assigned, α_t value in metal STHEXs is very different from the α_t value of PHEX (for example for L = 5 m α_t is equal to 4024 W/m²K and 3601 W/m²K for PHEX and titanium, respectively). α_s (Figure 7b) shows a decreasing trend and it is proportional to the geothermal brine temperature; therefore, during the fluid cooling the efficiency of heat exchange is down. At the evaporator integration starting point (L = 0) geothermal fluid has a high temperature, and so also α_s is high. Then it decreases with the increase of length for all materials. α_s stops at L = 12.2 m for PHEX because it is the needed length to ensure the required ORC electric output. In Figure 7 R_w, $\alpha_{s,f}$, and $\alpha_{t,f}$ are not shown even if they are accounted for in the overall heat exchanger coefficient evaluation according to Equation (2). They are constant along whole STHEXs and they are calculated as following: R_w depends upon k_w of each material reported in Table 5, while $\alpha_{s,f}$ and $\alpha_{t,f}$ are the reciprocal of $R_{s,f}$ and $R_{t,f}$ evaluated for the geothermal brine and R245fa, respectively, according to values listed in Table 4. The overall heat transfer coefficient Figure 7) follows α_t trend for all materials since this term has the higher weight in Equation (2). Among metal materials the carbon steel STHEX has the highest U for a fixed L value since it shows the greatest k_w. PHEX has the lowest U for along whole evaporator.

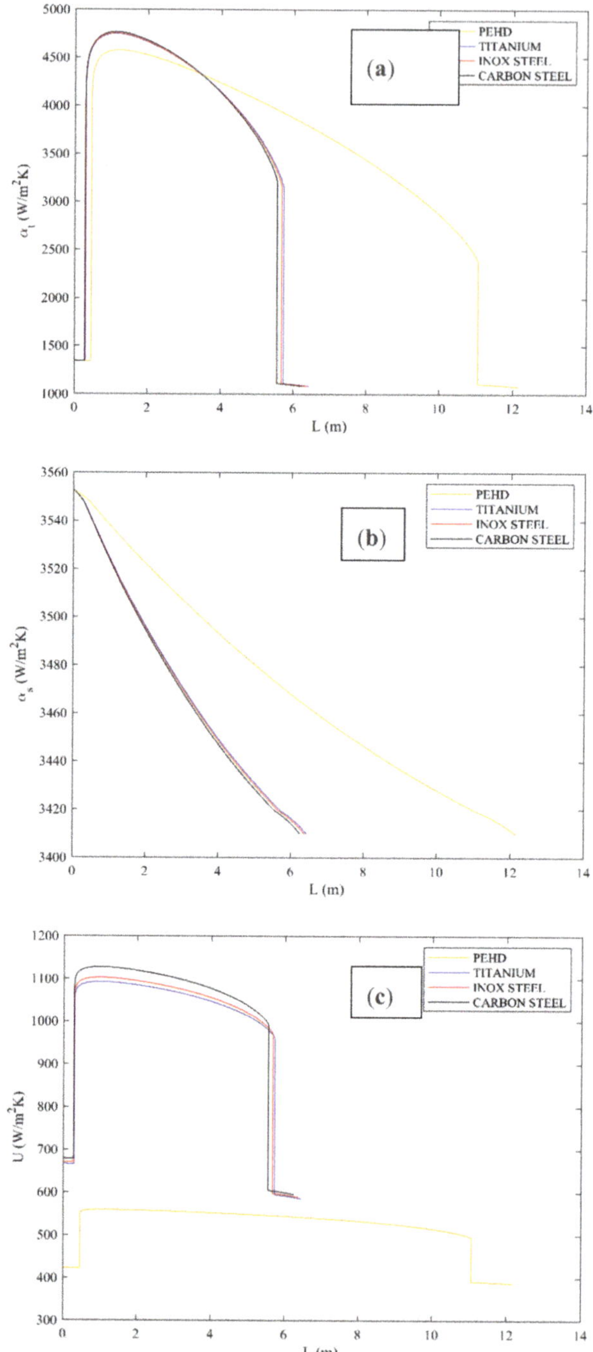

Figure 7. Convective heat transfer coefficient on tube side (**a**) and shell side (**b**) and overall heat transfer coefficient (**c**) as a function of evaporator length.

Sensitivity Analysis

The Figure 8 reports the purchase cost and the needed heat transfer area value for a STHEX made of PEHD and titanium by varying the inlet temperature of geothermal brine in the range 94–99 °C. The temperature range for this sensitivity analysis is chosen considering the temperature availability of geothermal brine in selected sites and the possible operating couple temperature-pressure in polymeric heat exchanger applications. These values of inlet temperature allow the use of the same correlation (Equations (3)–(7)) for the shell side of the main case analysed. In Figure 8 the values in red points and red bars correspond to the inlet geothermal fluid temperature provided by the manufacturer (94 °C). The PC and the heat transfer area diminishes with the increment of inlet temperature of geothermal sources in both PEHD and titanium evaporators.

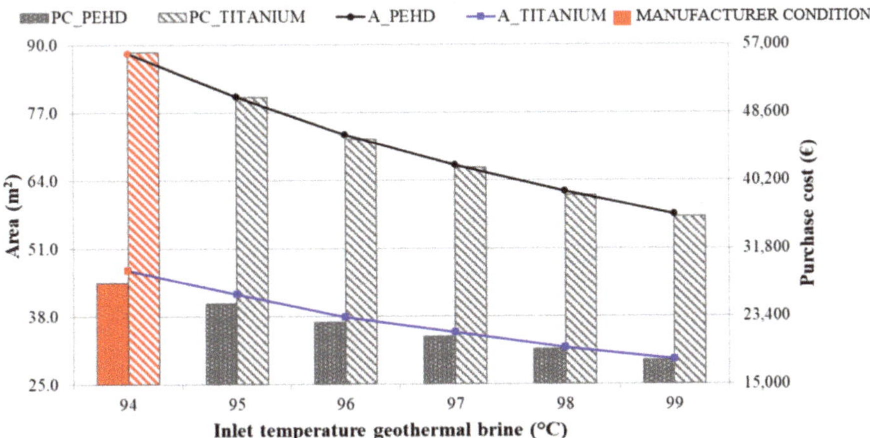

Figure 8. Purchase cost and area of evaporator in PEHD and titanium as a function of inlet temperature of geothermal brine.

In particular, the reduction of the heat transfer area for each degree of geothermal brine inlet temperature increase is higher for the PEHD evaporator than the titanium one. Vice versa, referring to PC the cost reduction is more rapid for the titanium than the PEHD evaporator. For example, by going from T = 95 °C to T = 96 °C the reduction of PC amounts to 2633€ for PHED and to 5612€ for titanium. Considering the same temperature range, the heat transfer area reduction is equal to 8.4 m^2 for the PHED evaporator and 4.7 m^2 for the titanium STHEX. In particular, the installation of the titanium STHEX evaporator in an ORC plant is responsible of an extra-cost equal to 28,558€ with respect to the PHED one when the inlet temperature of geothermal brine is 94.0 °C. This extra-cost ranges from 25,578€ to 17,799€ when the inlet temperature changes in the range of 95.0–99.0 °C. On the other hand, the heat transfer area growth by using a PHED evaporator instead of a titanium one varies from 41.6 m^2 to 29.8 m^2, when the inlet temperature of geothermal fluid goes from 94 °C to 99 °C.

6. Conclusions

Organic Rankine Cycle technologies are employed worldwide in different fields. Among other applications, this work is focused on ORC plants using a geothermal source. Such systems are affected by different issues: a corrosive heat source, low exercise temperatures, and high specific costs, thus evaporator cost has a great weight on the overall plant costs. These problems can be making possible the using of plastic evaporators instead of traditional ones made of metal materials with a high cost and a low corrosion resistance.

For these reasons in this work a one-dimensional model of a heat exchanger used as an evaporator in an ORC available on market, has been realized. The design modelling is grounded on physical equations describing the organic and geothermal fluid behaviours. The choice of installation site for the ORC plant (Phlegrean Fields, South of Italy) and the definition of working fluid have been conducted according to an analysis of the chemical composition of geothermal brine available in the chosen installation area.

The numerical model implemented in the MATLAB/REFPROP environments has been used to define the design properties of the plastic evaporator used to replace a metal one in the selected ORC plant. Furthermore, a comparison between plastic evaporator geometrical parameters and costs with those of evaporators made of titanium, inox steel, and carbon steel, has been carried out. The main outcomes of the simulation analysis can be summarised as following:

- it has been found that the plastic evaporator should have an area higher than 47% with respect to the titanium evaporator area to ensure the same ORC performance.
- As regards economic comparison between polymeric and metal evaporators, it has been highlighted that a plastic evaporator ensures a cost saving of EUR 28,558 with respect to titanium evaporator costs.
- The overall heat exchanger coefficients (depending upon convective, conductive and fouling resistance) have been evaluated for all considered evaporators and their variation as a function of evaporator length are shown. The plastic evaporator has an overall heat exchange coefficient lower than all metal evaporators along whole heat exchangers length. The carbon steel evaporator shows the higher overall heat exchanger coefficient due to its greater thermal conductivity.

As a general consideration, this work assesses that the plastic evaporators used in geothermal ORC plant can, therefore, be chosen nowadays as an encouraging technological option in order to reduce costs in small size plants. The analysis results can also be used to design the plastic evaporator for other ORC plants available on the market. Furthermore, the use of an ORC plant activated by a geothermal source allows us to address three goals:

- the use of the geothermal sources at low enthalpy that have not exploited yet according to their potential availability;
- the exploitation of geothermal brine to produce electric energy by means of an ORC system especially where it is not feasible to use a geothermal source for space heating because of the low heating demands in warm climatic zone (such as the Phlegrean Fields area);
- the diffusion of distributed systems activated by RESs that can reach a broad acclaim from society than the large plants, achieving, therefore, the RESs' sustainable exploitation linked to their social acceptability.

In future works, a sensitivity analysis varying the evaporator thermal power or its geometry and fluid distribution could be conducted in order to analyse the effect of the replacement of a metal evaporator with a plastic one in these new cases. In addition, the numerical model of a plastic heat exchanger can be used to design other ones exploited in direct geothermal ways too, such as the thermal heating networks.

Author Contributions: F.C., A.M., E.M., M.S. and L.V. contributed to the design and implementation of the research, to the analysis of the results and to the writing of the manuscript. All authors have read and agreed to the published version of the manuscript.

Funding: This research received no external funding

Acknowledgments: The authors gratefully acknowledge the financial support of GeoGrid project POR Campania FESR 2014/2020 CUP B43D18000230007.

Conflicts of Interest: The authors declare no conflict of interest.

Nomenclature

A	Heat transfer surface area (m^2)
Bo	Boling number (-)
c_p	Specific heat at constant pressure (J/kg K)
d	Tube diameter (m)
D	Shell diameter (m)
E	Enhancement factor of non-boiling forced heat thermal coefficient (-)
F	Correction factor for shell and tube thermal balance (-)
h	Specific Enthalpy (kJ/kgK)
J	Area corrective factor (-)
k	Thermal conductivity (W/mK)
L	Length (m)
$\dot{m}$	Fluid mass flow rate (kg/s)
M	Molar mass (g/mol)
Nu	Nusselt number (-)
p	Pressure (bar)
Pr	Prandtl number (-)
PC	Purchase cost (€)
q	Heat flow (W/m^2)
$\dot{Q}$	Heat power (W)
R	Thermal Resistance (m^2K/W)
Re	Reynolds number (-)
S	Suppression factor of nuclear boiled heat thermal coefficient (-)
T	Temperature (°C), (K)
U	Overall heat trasfer coefficient (W/m^2K)
X_{tt}	Martinelli number (-)
Greek symbols	
α	Convective heat transfer coefficient (W/m^2K)
δ	Thickness (m)
μ	Dynamic viscosity (Pa·s)
ρ	Density (kg/m^3)
λ	Latent heat (kJ/kg)
ΔT_{lm}	Logarithmic mean temperature (K)
Subscript	
cb	Convective booling
el	Electrical
f	Fouling factor
i	Inlet
l	Liquid
NB	Nucleate Boiled
o	Outlet
obj	Objective
s	Shell
t	Tube
th	Thermal
tp	Two-phase
v	Vapor
w	Wall material

Acronyms

EU	European Union
HEX	Heat Exchanger
MST	Maximum Service Temperature
ORC	Organic Rankine Cycle
PEHD	PolyEthylene high density
PHEX	Polymer Heat Exchanger
RESs	Renewable Energy Sources
RORC	Regenerative Organic Rankine Cycle
STHEX	Shell and Tube heat exchanger
TDS	Total Dissolved Solids

References

1. Dincer, I.; Acar, C. A review on clean energy solutions for better sustainability. *Int. J. Energy Res.* **2015**, *39*, 585–606. [CrossRef]
2. Directive (EU) 2018/2001 of the European Parliament and of the Council of 11 December 2018 on the Promotion of the Use of Energy from Renewable Sources. Available online: https://eur-lex.europa.eu/legal-content/EN/TXT/?uri=uriserv:OJ.L_.2018.328.01.0082.01.ENG&toc=OJ:L:2018:328:TOC (accessed on 10 December 2019).
3. Fruhmann, C.; Tuerk, A. Renewable Energy Support Policies in Europe" Climate Policy Info Hub, 2014. Available online: http://climatepolicyinfohub.eu/renewable-energy-support-policies-europe (accessed on 10 December 2019).
4. Suwa, A.; Sando, A. Geothermal energy and community sustainability. In Proceedings of the International Conference and Exhibition–Grand Renewable Energy 2018, Pacifico Yokaoma, Japan, 17–22 June 2018.
5. Eurostat, Statistical Office of the European Communities. Renewable Energy Statistics, 2019. Available online: https://ec.europa.eu/eurostat/statistics-explained/index.php/Renewable_energy_statistics#Share_of_energy_available_from_renewable_sources_highest_in_Latvia_and_Sweden (accessed on 11 December 2019).
6. Irena, International Renewable Energy Agency. Renewable Energy Prospects for the European Union, 2018. Available online: https://www.irena.org/-/media/Files/IRENA/Agency/Publication/2018/Feb/IRENA_REmap_EU_2018.pdf (accessed on 11 December 2019).
7. Van Der Zwaan, B.; Longa, F.D. Integrated assessment projections for global geothermal energy use. *Geothermics* **2019**, *82*, 203–211. [CrossRef]
8. U.S. Energy Information and Administration Independent Statistics & Analysis. Available online: https://www.eia.gov/energyexplained/geothermal/use-of-geothermal-energy.php (accessed on 11 December 2019).
9. Lund, J.W.; Boyd, T.L. Direct utilization of geothermal energy 2015 worldwide review. *Geothermics* **2016**, *60*, 66–93. [CrossRef]
10. Zhang, L.; Chen, S.; Zhang, C. Geothermal power generation in China: Status and prospects. *Energy Sci. Eng.* **2019**, *7*, 1428–1450. [CrossRef]
11. Harris, N.B.; Noyahr, C.; Renaud, E.; Banks, J.C.; Weissenberger, J.A. Geothermal Reservoir Characterization in Carbonate-Hosted Oil and Gas Fields from the Western Canada Sedimentary Basin. In Proceedings of the World Geothermal Congress 2020 Reykjavik, Reykjavik, Iceland, 21–26 May 2021.
12. Christ, A.; Rahimi, B.; Regenauer-Lieb, K.; Chua, H. Techno-economic analysis of geothermal desalination using Hot Sedimentary Aquifers: A pre-feasibility study for Western Australia. *Desalination* **2017**, *404*, 167–181. [CrossRef]
13. Campbell, S.A.; Lenhardt, N.; Dippenaar, M.; Götz, A.E. Geothermal Energy from the Main Karoo Basin (South Africa): An Outcrop Analogue Study of Permian Sandstone Reservoir Formations. *Energy Proced.* **2016**, *97*, 186–193. [CrossRef]
14. Meller, C.; Schill, E.; Bremer, J.; Kolditz, O.; Bleicher, A.; Benighaus, C.; Chavot, P.; Gross, M.; Pellizzone, A.; Renn, O.; et al. Acceptability of geothermal installations: A geoethical concept for GeoLaB. *Geothermics* **2018**, *73*, 133–145. [CrossRef]

15. Calamai, A.; Cataldi, R.; Locardi, E.; Praturlon, A. Distribuzione delle anomalie geotermiche nella fascia pre-appenninica Tosco-Laziale [Distribution of the geothermal anomalies in the pre-Apennine belt of Tuscany and Latium]. In Proceedings of the International Symposium on Geothermal Energy in Latin America, Ciudad Guatemala, Guatemala, 16–23 Oct 1976; pp. 189–229.

16. Calamai, A.; Ceppatelli, L.; Squarci, P. Summary of Italian experience in thermal prospecting for geothermal resources. *Zbl. Geol. Palaont.* **1983**, *1*, 156–167.

17. Cataldi, R.; Mongelli, F.; Squarci, P.; Taffi, L.; Zito, G.; Calore, C. Geothermal ranking of Italian territory. *Geothermics* **1995**, *24*, 115–129. [CrossRef]

18. Mormone, A.; Tramelli, A.; Di Vito, M.A.; Piochi, M.; Troise, L.; De Natale, G. Secondary hydrothermal minerals in buried rocks at the Campi Flegrei caldera, Italy: A possible tool to understand the rock-physics and to assess the state of volcanic system. *Int. J. Miner. Crystallogr. Geochem. Depos. Petrol. Volcanol.* **2011**, *80*, 385–406.

19. Pubblicazioni Statistiche. Available online: https://www.terna.it/it/sistema-elettrico/statistiche/pubblicazioni-statistiche (accessed on 3 November 2019).

20. Bertani, R. Geothermal Power Generation in the World 2010-2014 Update Report Enel Green Power. In Proceedings of the World Geothermal Congress 2015, Melbourne, Australia, 19–25 April 2015.

21. Ahmed, A.; Esmaeil, K.K.; Irfan, M.; Al-Mufadi, F.A. Design methodology of organic Rankine cycle for waste heat recovery in cement plants. *Appl. Therm. Eng.* **2018**, *129*, 421–430. [CrossRef]

22. Tartière, T.; Astolfi, M. A World Overview of the Organic Rankine Cycle Market. *Energy Procedia* **2017**, *129*, 2–9. [CrossRef]

23. Astolfi, M.; Romano, M.C.; Bombarda, P.; Macchi, E. Binary ORC (organic Rankine cycles) power plants for the exploitation of medium–low temperature geothermal sources–Part A: Thermodynamic optimization. *Energy* **2014**, *66*, 423–434. [CrossRef]

24. Hajabdollahi, Z.; Hajabdollahi, F.; Tehrani, M.; Hajabdollahi, H. Thermo-economic environmental optimization of Organic Rankine Cycle for diesel waste heat recovery. *Energy* **2013**, *63*, 142–151. [CrossRef]

25. Pastor, E.; Rubio-Maya, C.; Ambriz-Díaz, V.; Belman-Flores, J.; Pacheco-Ibarra, J. Energetic and exergetic performance comparison of different polygeneration arrangements utilizing geothermal energy in cascade. *Energy Convers. Manag.* **2018**, *168*, 252–269. [CrossRef]

26. Calise, F.; Di Fraia, S.; Macaluso, A.; Massarotti, N.; Vanoli, L. A geothermal energy system for wastewater sludge drying and electricity production in a small island. *Energy* **2018**, *163*, 130–143. [CrossRef]

27. Calise, F.; D'Accadia, M.D.; Macaluso, A.; Vanoli, L.; Piacentino, A. A novel solar-geothermal trigeneration system integrating water desalination: Design, dynamic simulation and economic assessment. *Energy* **2016**, *115*, 1533–1547. [CrossRef]

28. Calise, F.; D'Accadia, M.D.; Macaluso, A.; Piacentino, A.; Vanoli, L. Exergetic and exergoeconomic analysis of a novel hybrid solar–geothermal polygeneration system producing energy and water. *Energy Convers. Manag.* **2016**, *115*, 200–220. [CrossRef]

29. Calise, F.; Macaluso, A.; Pelella, P.; Vanoli, L. A comparison of heat transfer correlations applied to an Organic Rankine Cycle. *Eng. Sci. Technol. Int. J.* **2018**, *21*, 1164–1180. [CrossRef]

30. Yang, Y.; Huo, Y.; Xia, W.; Wang, X.; Zhao, P.; Dai, Y. Construction and preliminary test of a geothermal ORC system using geothermal resource from abandoned oil wells in the Huabei oil field of China. *Energy* **2017**, *140*, 633–645. [CrossRef]

31. Tan, M.; Karabacak, R.; Acar, M. Experimental assessment the liquid/solid fluidized bed heat exchanger of thermal performance: An application. *Geothermics* **2016**, *62*, 70–78. [CrossRef]

32. Zarrouk, S.J.; Woodhurst, B.C.; Morris, C. Silica scaling in geothermal heat exchangers ant its impact on pressure drop and performance: Wairakei binary plant, New Zealand. *Geothermics* **2014**, *51*, 445–459. [CrossRef]

33. Bott, T.R. *Fouling of Heat Exchangers*; Elsevier Science: Amsterdam, The Netherlands, 1995.

34. Aláez, S.G.; Bombarda, P.; Invernizzi, C.M.; Iora, P.; Silva, P. Evaluation of ORC modules performance adopting commercial plastic heat exchangers. *Appl. Energy* **2015**, *154*, 882–890. [CrossRef]

35. Cabral-Miramontes, J.A.; Gaona-Tiburcio, C.; Almeraya-Calderón, F.; Estupiñan-Lopez, F.H.; Pedraza-Basulto, G.K.; Salas, C.A.P.; Almeraya-Calderó, N.F.; Estupiñ An-Lopez, F.H. Parameter Studies on High-Velocity Oxy-Fuel Spraying of CoNiCrAlY Coatings Used in the Aeronautical Industry. *Int. J. Corros.* **2014**, *2014*, 1–8. [CrossRef]

36. Diaz-Bejarano, E.; Coletti, F.; Macchietto, S. Modeling and Prediction of Shell-Side Fouling in Shell-and-Tube Heat Exchangers. *Heat Transf. Eng.* **2018**, *40*, 845–861. [CrossRef]
37. Buzăianu, A.; Motoiu, P.; Csaki, I.; Ioncea, A.; Motoiu, V. Structural Properties Ni20Cr10Al2Y Coatings for Geothermal Conditions. *Multidiscip. Digit. Publ. Inst. Proc.* **2018**, *2*, 1434. [CrossRef]
38. Haunstetter, J.; Dreißigacker, V.; Zunft, S. Ceramic high temperature plate fin heat exchanger: Experimental investigation under high temperatures and pressures. *Appl. Therm. Eng.* **2019**, *151*, 364–372. [CrossRef]
39. Rodriguez, P. Selection of Materials for Heat Exchangers. In Proceedings of the 3rd International Conference on Heat Exchangers, Boilers and Pressure Vessels, Alexandria, Egypt, 5–6 April 1997.
40. Kakaç, S.; Liu, H.; Pramuanjaroenkij, A. *Heat exchangers Selection Rating and Thermal Design*; CRC Press: Boca Raton, FL, USA, 2002.
41. T'Joen, C.; Park, Y.; Wang, Q.; Sommers, A.; Han, X.; Jacobi, A. A review on polymer heat exchangers for HVAC&R applications. *Int. J. Refrig.* **2009**, *32*, 763–779. [CrossRef]
42. Zaheed, L.; Jachuck, R. Review of polymer compact heat exchangers, with special emphasis on a polymer film unit. *Appl. Therm. Eng.* **2004**, *24*, 2323–2358. [CrossRef]
43. Loh, H.P.; Lyons, J.; White, C.W. Process Equipment Cost Estimation, Final Report. *Process Equip. Cost Estim. Final Rep.* **2002**. [CrossRef]
44. TMW Technologies. Available online: www.tmwtechnologies.com (accessed on 10 October 2019).
45. Polytetra. Available online: www.polytetra.de (accessed on 10 October 2019).
46. Aetna Plastic. Available online: www.aetnaplastics.com (accessed on 10 October 2019).
47. HeatMatrix Group, B.V. Available online: www.heatmatrixgroup.com (accessed on 10 October 2019).
48. Fluorotherm. Available online: www.fluorotherm.com (accessed on 10 October 2019).
49. Ametek. Available online: www.ametekfpp.com (accessed on 10 October 2019).
50. Kansetu International. Available online: www.kansetuintl.com (accessed on 10 October 2019).
51. CALORPLAST Shell and Tube Heat Exchanger. Available online: http://www.calorplast-waermetechnik.de/wp-content/uploads/2013/09/rohrbuendel-waermetauscher-dokument.pd (accessed on 10 October 2019).
52. Deisenroth, D.C.; Arie, M.A.; Shooshtari, A.; Dessiatoun, S.; Ohadi, M. Review of most recent progresson development of polymer heat exchangers for thermal management applications. In Proceedings of the ASME 2015, International Technical Conference and Exhibition on Packaging and Integration of Electronic and Photonic Microsystems InterPACK2015, San Francisco, CA, USA, 6–9 July 2015.
53. Torres Tamayo, E.; Díaz Chicaiza, E.J.; Cedeño González, M.P.; Vargas Ferruzola, C.L.; Peralta Landeta, S.G. Heat transfer coefficients and efficiency loss in plate heat exchangers during the ammonia liquor cooling process. *Int. J. Theor. Appl. Mech.* **2016**, *1*, 55–60.
54. Shah, R.K.; Sekuli, D.P. *Fundamentals of Heat Exchanger Design*; Wiley: Hoboken, NJ, USA, 2003.
55. Arie, M.A.; Shooshtari, A.; Tiwari, R.; Dessiatoun, S.V.; Ohadi, M.M.; Pearce, J.M. Experimental characterization of heat transfer in an additively manufactured polymer heat exchanger. *Appl. Therm. Eng.* **2017**, *113*, 575–584. [CrossRef]
56. Çengel, Y.A. *Thermodynamics and Heat Transfe*; Springer: Berlin/Heidelberg, Germany, 2019; pp. 11–42.
57. Shen, S.; Henry, A.; Tong, J.; Zheng, R.; Chen, G. Polyethylene nanofibres with very high thermal conductivities. *Nat. Nanotechnol.* **2010**, *5*, 251–255. [CrossRef] [PubMed]
58. Cevallos, J.; Bar-Cohen, A.; Deisenroth, D.C. Thermal performance of a polymer composite webbed-tube heat exchanger. *Int. J. Heat Mass Transf.* **2016**, *98*, 845–856. [CrossRef]
59. Jabbour, J.; Russeil, S.; Mobtil, M.; Bougeard, D.; Lacrampe, M.-F.; Krawczak, P. High performance finned-tube heat exchangers based on filled polymer. *Appl. Therm. Eng.* **2019**, *155*, 620–630. [CrossRef]
60. Cevallos, J.G.; Bergles, A.E.; Bar-Cohen, A.; Rodgers, P.; Gupta, S.K. Polymer Heat Exchangers—History, Opportunities, and Challenges. *Heat Transf. Eng.* **2012**, *33*, 1075–1093. [CrossRef]
61. Erdoğan, A.; Colpan, C.O.; Cakici, D.M. Thermal design and analysis of a shell and tube heat exchanger integrating a geothermal based organic Rankine cycle and parabolic trough solar collectors. *Renew. Energy* **2017**, *109*, 372–391. [CrossRef]
62. Mastrullo, R.; Mauro, A.W.; Revellin, R.; Viscito, L. Modeling and optimization of a shell and louvered fin mini-tubes heat exchanger in an ORC powered by an internal combustion engine. *Energy Convers. Manag.* **2015**, *101*, 697–712. [CrossRef]
63. Carlino, S.; Somma, R.; Troise, C.; De Natale, G. The geothermal exploration of Campanian volcanoes: Historical review and future development. *Renew. Sustain. Energy Rev.* **2012**, *16*, 1004–1030. [CrossRef]

64. Zuccato Energia. Available online: https://www.zuccatoenergia.it/it/ (accessed on 20 May 2019).

65. Studio di Impatto Ambientale. Documento SCA-006-SIA-00-A01 Allegato 01 Relazione Geologico-Geotermica AMRA/INGV. Progetto per la Realizzazione di un Impianto Geotermico Pilota Nell'area del Permesso di Ricerca "Scarfoglio". Available online: https://docplayer.it/ (accessed on 16 April 2020).

66. Carlino, S.; Troiano, A.; Di Giuseppe, M.G.; Tramelli, A.; Troise, C.; Somma, R.; De Natale, G. Exploitation of geothermal energy in active volcanic areas: A numerical modelling applied to high temperature Mofete geothermal field, at Campi Flegrei caldera (Southern Italy). *Renew. Energy* **2016**, *87*, 54–66. [CrossRef]

67. Corrado, G.; De Lorenzo, S.; Mongelli, F.; Tramacere, A.; Zito, G. Surface heat flow density at the phlegrean fields caldera (SOUTHERN ITALY). *Geothermics* **1998**, *27*, 469–484. [CrossRef]

68. De Gennaro, M.; Franco, E.; Stanzione, D. Le alterazioni ad opera di fluidi termali alla solfatara di Pozzuoli (Napoli). Mineralogia e geochimica. *Period. Mineral.–Roma Anno* **1980**, *49*, 5–22.

69. Valentino, G.; Stanzione, D. Geochemical monitoring of the thermal waters of the Phlegraean Fields. *J. Volcanol. Geotherm. Res.* **2004**, *133*, 261–289. [CrossRef]

70. Valentino, G.; Cortecci, G.; Franco, E.; Stanzione, D. Chemical and isotopic compositions of minerals and waters from the Campi Flegrei volcanic system, Naples, Italy. *J. Volcanol. Geotherm. Res.* **1999**, *91*, 329–344. [CrossRef]

71. Györke, G.; Deiters, U.K.; Groniewsky, A.; Lassu, I.; Imre, A. Novel classification of pure working fluids for Organic Rankine Cycle. *Energy* **2018**, *145*, 288–300. [CrossRef]

72. Tchanche, B.F.; Lambrinos, G.; Frangoudakis, A.; Papadakis, G. Low-grade heat conversion into power using organic Rankine cycles–a review of various applications, Renew. *Sustain. Energy Rev.* **2011**, *15*, 3963–3979. [CrossRef]

73. Uusitalo, A.; Honkatukia, J.; Turunen-Saaresti, T.; Grönman, A. Thermodynamic evaluation on the effect of working fluid type and fluids critical properties on design and performance of Organic Rankine Cycles. *J. Clean. Prod.* **2018**, *188*, 253–263. [CrossRef]

74. Wang, J.; Zhao, L.; Wang, X. A comparative study of pure and zeotropic mixtures in low-temperature solar Rankine cycle. *Appl. Energy* **2010**, *87*, 3366–3373. [CrossRef]

75. Liu, X.; Zhang, Y.; Shen, J. System performance optimization of ORC-based geo-plant with R245fa under different geothermal water inlet temperatures. *Geothermics* **2017**, *66*, 134–142. [CrossRef]

76. Ghaebi, H.; Farhang, B.; Parikhani, T.; Rostamzadeh, H. Energy, exergy and exergoeconomic analysis of a cogeneration system for power and hydrogen production purpose based on TRR method and using low grade geothermal source. *Geothermics* **2018**, *71*, 132–145. [CrossRef]

77. Li, T.; Zhu, J.; Fu, W.; Hu, K. Experimental comparison of R245fa and R245fa/R601a for organic Rankine cycle using scroll expander. *Int. J. Energy Res.* **2014**, *39*, 202–214. [CrossRef]

78. Lê, V.L.; Kheiri, A.; Feidt, M.; Pelloux-Prayer, S. Thermodynamic and economic optimizations of a waste heat to power plant driven by a subcritical ORC (Organic Rankine Cycle) using pure or zeotropic working fluid. *Energy* **2014**, *78*, 622–638. [CrossRef]

79. Zyhowski, G.J.; Spatz, M.M.; Motta, S.M.Y. An Overview of the Properties And Applications of HFC-245fa. In Proceedings of the International Refrigeration and Air Conditioning Conference, West Lafayette, IA, USA, 9–12 July 2002.

80. Calise, F.; Capuozzo, C.; Vanoli, L. Design and parametric optimization of an organic rankine cycle powered by solar energy. *Am. J. Eng. Appl. Sci.* **2013**, *6*, 178–204. [CrossRef]

81. *MATLAB and Statistics Toolbox Release*; The MathWorks, Inc.: Natick, MA, USA, 2018.

82. Lemmon, E.W.; Huber, M.L.; McLinden, M.O. NIST Standard Reference Database 23: Reference Fluid Thermodynamic and Transport Properties-REFPROP, Version 9.1. Available online: https://www.nist.gov/publications/nist-standard-reference-database-23-reference-fluid-thermodynamic-and-transport (accessed on 10 December 2019).

83. Liang, Y.; Che, D.; Kang, Y. Effect of vapor condensation on forced convection heat transfer of moistened gas. *Heat Mass Transf.* **2006**, *43*, 677–686. [CrossRef]

84. Chien, N.B.; Vu, P.Q.; Choi, K., II; Oh, J.-T. A general correlation to predict the flow boiling heat transfer of R410A in macro/mini-channels. *Sci. Technol. Built Environ.* **2015**, *21*, 526–534. [CrossRef]

85. Gungor, K.; Winterton, R. A general correlation for flow boiling in tubes and annuli. *Int. J. Heat Mass Transf.* **1986**, *29*, 351–358. [CrossRef]

86. Cooper, M. Saturation nucleate pool boiling—A simple correlation. In *First U.K. National Conference on Heat Transfer*; Elsevier BV: Amsterdam, The Netherlands, 1984; Volume 86, pp. 785–793.

87. Dittus, F.; Boelter, L. Heat transfer in automobile radiators of the tubular type. *Int. Commun. Heat Mass Transf.* **1985**, *12*, 3–22. [CrossRef]

88. Citarella, B.; Lillo, G.; Mastrullo, R.; Mauro, A.W.; Viscito, L. Experimental Investigation on Flow Boiling Heat Transfer and Pressure Drop of Refrigerants R32 and R290 in A Stainless Steel Horizontal Tube. *J. Phys. Conf. Ser.* **2019**, *1224*. [CrossRef]

89. Lillo, G.; Mastrullo, R.; Mauro, A.W.; Viscito, L. Flow boiling data of R452A. *Energy Procedia* **2018**, *148*, 1034–1041. [CrossRef]

90. Bertsch, S.S.; Groll, E.A.; Garimella, S.V. A composite heat transfer correlation for saturated flow boiling in small channels. *Int. J. Heat Mass Transf.* **2009**, *52*, 2110–2118. [CrossRef]

91. Quoilin, S.; Declaye, S.; Tchanche, B.F.; Lemort, V. Thermo-economic optimization of waste heat recovery Organic Rankine Cycles. *Appl. Therm. Eng.* **2011**, *31*, 2885–2893. [CrossRef]

Article

Thermo-Economic Analysis of Hybrid Solar-Geothermal Polygeneration Plants in Different Configurations

Francesco Calise, Francesco Liberato Cappiello *, Massimo Dentice d'Accadia and Maria Vicidomini

Department of Industrial Engineering, University of Naples Federico II, 80125 Naples, Italy; frcalise@unina.it (F.C.); dentice@unina.it (M.D.d.); maria.vicidomini@unina.it (M.V.)
* Correspondence: francescoliberato.cappiello@unina.it

Received: 8 April 2020; Accepted: 5 May 2020; Published: 11 May 2020

Abstract: This work presents a thermoeconomic comparison between two different solar energy technologies, namely the evacuated flat-plate solar collectors and the photovoltaic panels, integrated as auxiliary systems into two renewable polygeneration plants. Both plants produce electricity, heat and cool, and are based on a 6 kWe organic Rankine cycle (ORC), a 17-kW single-stage H2O/LiBr absorption chiller, a geothermal well at 96 °C, a 200 kWt biomass auxiliary heater, a 45.55 kWh lithium-ion battery and a 25 m^2 solar field. In both configurations, electric and thermal storage systems are included to mitigate the fluctuations due to the variability of solar radiation. ORC is mainly supplied by the thermal energy produced by the geothermal well. Additional heat is also provided by solar thermal collectors and by a biomass boiler. In an alternative layout, solar thermal collectors are replaced by photovoltaic panels, producing additional electricity with respect to the one produced by the ORC. To reduce ORC condensation temperature and increase the electric efficiency, a ground-cooled condenser is also adopted. All the components included in both plants were accurately simulated in a TRNSYS environment using dynamic models validated versus literature and experimental data. The ORC is modeled by zero-dimensional energy and mass balances written in Engineering Equation Solver and implemented in TRNSYS. The models of both renewable polygeneration plants are applied to a suitable case study, a commercial area near Campi Flegrei (Naples, South Italy), a location well-known for its geothermal sources and good solar availability. The economic results suggest that for this kind of plant, photovoltaic panels show lower pay back periods than evacuated flat-plate solar collectors, 13 years vs 15 years. The adoption of the electric energy storage system leads to an increase of energy-self-sufficiency equal to 42% and 47% for evacuated flat-plate solar collectors and the photovoltaic panels, respectively.

Keywords: hybrid renewable polygeneration plant; micro organic Rankine cycle; evacuated solar thermal collectors; photovoltaic panels

---

1. Introduction

Polygeneration plants based on renewable energy sources (RES) (geothermal, solar, biomass, wind and hydro), represent a suitable solution to reach the long-term goals expected by 2050, i.e., a reduction of greenhouse gas emissions by 80%–95% with respect to 1990 levels and a 100% renewable electrical system. These targets are potentially achievable by considering that policies will continue to support renewable electricity worldwide, increasingly through competitive actions rather than feed-in tariffs, and by the transformation of the power sector amplified by rapid deployment of solar photovoltaic (PV) panels and wind turbine. An RES polygeneration plant can replace the existing conventional technologies based on fossil fuels and simultaneously produce several energy vectors (thermal, cooling and electric energy) and other outputs, by reducing significantly the primary energy consumption and CO_2 emissions [1].

Several studies presented in literature about renewable polygeneration plants investigate the integration of renewable systems with different conventional systems: electric heat pump, trigeneration, gas-fired boiler, district heating and cooling and combined heating, cooling and power CHCP plants [2] for different purposes and application. In the context of power plants, the organic Rankine cycles (ORCs) [3] represent an interesting opportunity when coupled with low-medium enthalpy energy sources (biomass products [4], geothermal [5], solar [6,7], waste heat [8], ocean thermal energy [9], etc.). ORC plants adopt specific organic fluids showing an energy performance considerably better than water used in the conventional Rankine cycles, due to a higher molecular weight, lower evaporation heat, positive slope of the saturated vapor curve in the T–s diagram and lower critical and boiling temperatures [6]. In addition, for small scale units, ORC turbines are today an interesting technology, demonstrating several advantages in terms of operation life, maintenance and part-load efficiency.

Although the rapid increase in PV application, it is well-known that solar electric production is extremely variable due to the fluctuations of the solar radiation. This leads to an important increase in the electricity bought from the grid during the night, when the PV production is null, and during low solar availability hours, with the consequent reduction of the plant profitability. Therefore, the hybridization of different power systems (PV panels, PV and thermal (PVT) panels, wind turbines, ORC power plants) and the adoption of suitable electricity storage systems (ESSs) [10] with the proper ESS size determination in renewable energy systems [11] to obtain a more stable availability of the electric energy production, is an attractive solution to be investigated.

The simulation of polygeneration plants based on ORC power plants supplied by low and medium-temperature energy sources is diffusely investigated in the literature. Particularly, the integration of solar and geothermal sources is one of the most attractive configurations, mainly in volcanic areas featured by high solar radiation availability [6]. Thermal energy sources at low and medium temperature, typically within 90–130 °C, obtainable by solar and/or geothermal ones, are often used as input for supplying absorption chillers (ACH), multi-effect distillation (MED) systems and ORC in a unique renewable polygeneration plant. In the following section, an overview of the renewable power plants based on ORC technology coupled with PV/PVT panels and/or low-temperature geothermal energy is provided. In addition, studies investigating the combination of several thermal activated technologies (ACH, MED, ORC, etc.) in different plant configurations are also reported.

A hybrid solar and geothermal polygeneration plant, producing thermal energy for solar space heating/cooling (SHC), domestic hot water (DHW), fresh water and electric energy was studied by Calise et al. [6]. The plant includes different technologies: geothermal wells at about 80 °C, concentrating photovoltaic and thermal (CPVT) collectors, producing heat at about 100 °C, a single-stage LiBr/H_2O ACH and a MED unit. The low-enthalpy geothermal energy, combined with the solar thermal energy, supplies the MED unit for fresh water production. Geothermal energy is also used to produce DHW at 45 °C. A dynamic simulation model is developed to simulate the whole plant performance and is applied to a suitable case study, the Pantelleria island and further volcanic Mediterranean islands. The pay back periods achieved for all the examined weather zones were extremely low, equal to about two years. Cheng Zhou [12] investigated a solar–geothermal hybrid plant, consisting of parabolic trough collectors, an ORC machine and a geothermal well at about 150 °C. Isopentane working fluid is adopted as working fluid and the ORC machine performance is evaluated by considering the subcritical and supercritical cycle. The studied ORC uses an air-cooled condenser for the condensation process and the exhaust fluid from the turbine by means of a recuperator. To perform the plant simulations, the Aspen HYSYS simulation tool is adopted. The supercritical ORC plant exhibits the better performance, producing from 4% to 17% more power and presenting from 4% to 19% lower solar-to-electricity cost with respect to the subcritical ORC plant.

The performance evaluation of several types of PV materials (silicon, gallium arsenide, indium phosphide, cadmium sulfide and triple-junction indium gallium phosphide/indium gallium arsenide

/germanium) in a system producing electricity using PV panels and utilizing the waste heat from the cells to drive an ORC is investigated by Tourkov and Schaefer [13]. Here, the performance of a variety of fluids as working fluids for the ORC is analyzed. It was found that n-butane is the optimal selection for the proposed application and that triple-junction cells at high concentration combined with an ORC were able to achieve over 45% solar efficiency. Kosmadakis et al. [14] carried out an experimental investigation of a small-scale low-temperature ORC machine coupled with CPVT collectors. R404A is selected as working fluid. The CPVT collectors produce electricity and heat and supply it to the ORC. The tests showed that such low-temperature ORC unit exhibits a fair efficiency and that its coupling with a solar field was feasible, increasing the power production of the whole system. The most important result from the laboratory tests is that the ORC machine with a capacity of 3 kW reached an adequate thermal efficiency, about 5%, when operated at a very low temperature. An energy and exergy analysis of a hybrid polygeneration plant based on solar collectors and medium-high enthalpy geothermal sources is presented by Bicer and Dincer [15]. The plant is designed for simultaneously producing electricity, drying air, hot water and space heating and cooling. The heat provided by the geothermal energy and an air PVT drives an ORC for producing power. The ORC waste heat is employed for the activation of a LiBr/H_2O ACH, which provides the energy for space cooling of a dairy farm. Whereas the energy for space heating is provided by an electric heat pump. Moreover, the outlet hot air provided by the PVT collectors was employed for the food drying process of the farm. The polygeneration plant here proposed achieved global exergy and energy efficiencies of 28% and 11%, respectively. The energy and exergy efficiencies of the ORC were 9% and 42%, respectively. The air PVT collectors reached an exergy efficiency of 12%. The COP and exergy of the ACH and electric heat pumps were 0.73 and 0.21, and 4.1 and 0.03, respectively. The optimal design of a hybrid solar power generation system for isolated zones, consisting of PV panels, diesel generators, batteries and an ORC machine is addressed in the work of Noguera et al. [16]. The novelty of this study is the adoption of the heat recovery of the exhaust gases from the diesel generator to supply the ORC machine. The selected objective function is the cost of power generation by considering as variables the nominal power of the diesel generator and the number of PV panels and batteries. Simulation results for the selected case study, the Cujubim city in Rondônia State, suggest that the optimized diesel-ORC-PV-battery hybrid system, including 6288 kW diesel generators, is able to obtain a generation cost of $0.301/kWh, reduced approximately of 38.15% in comparison with the generation cost of a diesel system.

The above-presented literature review shows that numerous studies have examined the use of the waste heat from PVT panels for various applications, whereas there is limited research about the optimization of a combined PV panels/ORC machine system [13]. This combination could present a potential benefit in terms of efficiency and electricity cost if compared with the one of concentrating PV panels. Also, the reported literature review shows hybrid renewable energy polygeneration plants supplied by geothermal and solar energy, based on thermally-activated technologies and ORC, are investigated in different and several plant configurations. These works often examined medium-high scale ORC, with the exception of the ORC machines presented in references [5,14]. Specifically, the work presented in reference [5] is developed by some of the authors of this paper. Here a 6 kW micro-scale ORC machine supplied by solar and geothermal energy is investigated. In particular, a 25 m^2 solar field consisting of flat-plate evacuated thermal collectors (ETCs) is coupled with a geothermal well at 96 °C to produce DHW, thermal energy, cooling energy by a single-stage LiBr/H_2O ACH and electricity by an ORC, for a hotel located in Ischia (South Italy). With respect to the paper presented in reference [5], where all the outputs produced by the plant are assumed to be fully consumed by the user, in this study the power production and the heat produced for space heating and cooling must match the real time-dependent loads of the investigated user, a commercial building located in Campi Flegrei, a famous volcanic area of Naples (South Italy). In addition, this paper also includes a further significant improvement, with respect to work reported in reference [5], since it presents a thermoeconomic comparison between two different solar layouts: in the first one, solar energy produced by ETC collectors is converted into electricity by an ORC; the second layout refers to a more mature and simple configuration where ETCs are replaced by PV panels, operating independently from the ORC, supplying additional electricity to the system. In other words, this work aims to compare an innovative complex solar geothermal plant with a simpler configuration including PV collectors, considering both energy and economic aspects. Finally, this paper also includes additional novelties: (i) in order to increase the renewable energy source utilization, both renewable polygeneration plants include a biomass auxiliary heater; (ii) an electric energy storage system based on lithium-ion technology is used to mitigate power production fluctuations; (iii) a ground-cooled condenser in order to provide the required cooling energy to the ORC and to the ACH.

Aim of the Study

In this paper, dynamic simulation models of two hybrid renewable polygeneration plants based on a micro-scale ORC machine coupled with a single-stage LiBr/H$_2$O ACH for producing power, heating and cooling are developed. The dynamic modeling involves the hybridization of geothermal, solar and biomass energy, thermal and electric energy storage systems. The thermoeconomic performance of two hybrid renewable polygeneration plants (the first based on ETCs and the second one on PV panels) are also compared. The dynamic energy models, developed by the means of the well-known TRNSYS software, include the modeling of complex operation control strategies and all the included technologies: ETCs, PV panels, micro-scale ORC, ACH, biomass auxiliary heater, lithium-ion energy storage, commercial building and all the other system components as storage tanks, heat exchangers, diverters, pumps, mixers, controllers and fan coils. From the achieved results interesting design and operating guidelines can be usefully provided for similar renewable polygeneration plants located in zones where both solar and geothermal energy are available.

2. System Layouts

The layouts of both the renewable polygeneration plants (Cases ETC and PV) are shown in Figure 1. They mainly consist of a low-temperature geothermal well (at 96 °C, depth 94 m), equipped with a submerged geothermal brine pump and a downhole heat exchanger, 25 m^2 of solar field (ETCs or PV panels), 6 kW ORC machine, 200 kW auxiliary biomass-fired heater (AH), 17 kW$_f$ H$_2$0/LiBr single-stage absorption chiller (ACH), stratified vertical storage tanks, ground-coupled heat exchangers, 45.55 kWh lithium-ion energy storage system (ESS), equipped with an inverter and a 308.5 m^2 commercial building. Note that the size of the PV and ETC field are assumed equal, in fact, the aim of the proposed analysis consists of studying the performance of these two layouts occupying the same surface. This is due to the fact that the considered zone is not too vast and, therefore, occupying less space is a remarkable positive aspect. The adoption of the ground-coupled heat exchangers allows one to enhance system efficiency since both ACH and ORC efficiency significantly increases when the temperature of the cold sink decreases. During the summer season, ground temperature is significantly lower than the air temperature. Conversely, in winter, ground temperature may be higher than that of the air. In this case, using outdoor air as a cold sink would be more efficient than the use of ground. However, for this specific case, the additional cost of an air cooler would not be balanced by the income determined by the higher ORC electrical production. Therefore, it is assumed to use the ground-coupled heat exchanger all year long. Note that in the ETC case, ORC is driven by the combination of geothermal energy and solar energy supplied by ETCs. Finally, additional auxiliary heat is supplied by the biomass auxiliary heater, AH, which is used in order to achieve the minimum ORC activation temperature. Conversely, in the PV Case, the ORC is only supplied by geothermal energy and by the biomass AH. The overall electrical production is due both to the ORC and to the PV panels. The geothermal source supplies energy for both building space heating and cooling. In particular, the geothermal heat is directly exploited for building space heating, whereas the cooling energy is provided by an absorption chiller driven by the geothermal heat. The power demand of the user is met by the power production of the ORC (Case ETC) or by the power production of the ORC and PV panels (PV Case), by the electric energy stored in the lithium-ion battery and by the electricity withdrawn from the grid.

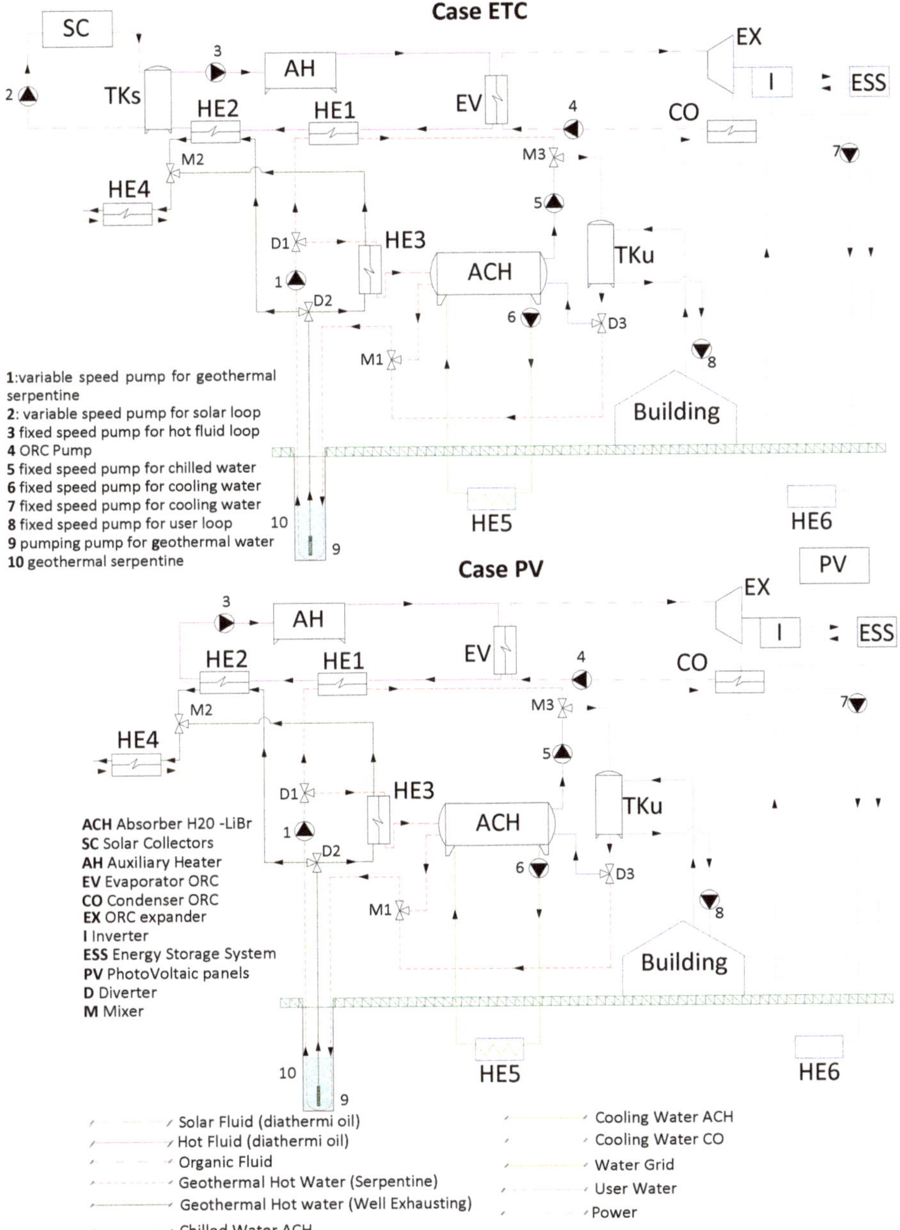

Figure 1. Systems layout: Case evacuated thermal collector (ETC, above) and Case photovoltaic (PV, below).

The layout of the examined plant consists of eleven main circuits:

- The solar fluid (only in Case ETC), which describes the diathermic oil flowing between the stratified vertical storage solar tank (TKs) and the evacuated flat-plate solar collectors (SC) and TKs and the ORC evaporator (EV);

- The hot fluid referred to the diathermic oil flowing from the stratified vertical storage solar tank (TKs) and the evaporator (EV) of the ORC (Case ETC) and referred to the diathermic oil flowing from the heat exchangers HE2 and the evaporator (EV) of the ORC (Case PV);
- The organic fluid, describing the working fluid of ORC machine, which is R245fa;
- The geothermal hot water describes the hot water flowing into the downhole heat exchanger transferring thermal flow rate from the geothermal well to the heat exchangers HE3 in the summer and HE1 in the winter;
- The geothermal hot water well exhausting (the geothermal brine) employed to feed the heat exchangers HE3 in the summer and HE2 in the winter. The well exhausting is designed with the aim of getting a suction of the high temperature geothermal brine from the geothermal ground to the well, in order to maintain a continuously high temperature in the well;
- The chilled water, describes the cold water produced by the ACH and stored in TKu;
- The cooling water, represents the cooling loop of the absorber and condenser of ACH, in particular, this loop exchange with the soil by means of the heat ground exchanger HE5;
- The cooling water of the condenser (CO), describes the cooling loop of the condenser of the ORC, which exchanges with the soil by the heat exchanger HE6;
- The water grid, the outlet geothermal hot water well exhausting (by the heat exchanger HE4) heats the grid water from 15 to 45 °C;
- The user water, described the heating/cooling water delivered to the fain coil unit inside the bar for building space heating/cooling purpose;
- The power circuit, which describes the power produced by the ORC expander (EX) (Case ETC) and by the PV panels and ORC expander (EX) (Case PV), the produced power is managed by the inverter (I) and sent to the system electric devices, building (Figure 2) and electric pumps, and subsequently employed for charging the energy storage system.

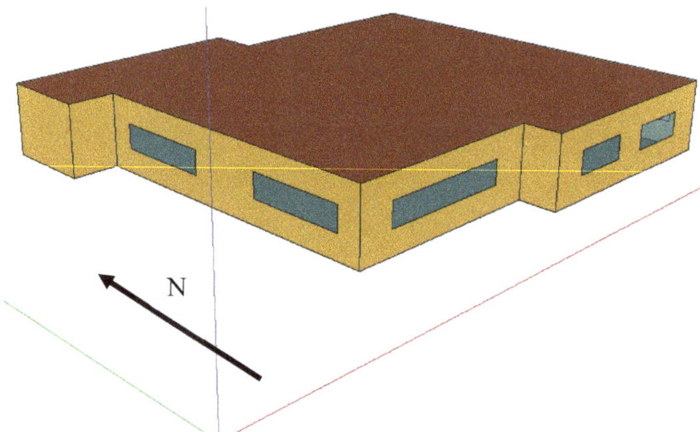

Figure 2. Building geometrical model.

Table 1 in detail reports the fluids adopted in each heat exchanger installed in the plant, the pumps and their seasonal scheduling.

According to the activation season of the several pumps and heat exchangers, during the winter season, D2 and D1 divert the geothermal brine and the geothermal hot water to HE2 and HE1, with the aim of heating the diathermic oil and obtain the minimum activation temperature of the ORC, i.e., $T_{min,act,ORC}$, equal to 90 °C. The outlet temperature of the geothermal hot water from HE1 is exploited for providing thermal energy for the building space heating. The geothermal hot water is delivered to the tank of the user circuit (TKu) by M3. Then, the water is supplied by the P8 constant speed pump from TKu to the fan coil unit with the aim of reaching the selected set point indoor temperature, $T_{set,heat}$, equal to 20 °C.

Table 1. Heat exchangers and pumps.

Heat Exchanger	Hot Side	Cold Side	Activation Season
HE1	Geothermal hot water	Diathermic oil *	Winter and summer *
HE2	Geothermal brine	Diathermic oil	
HE3	Geothermal brine	Geothermal hot water	Summer
HE4	Geothermal brine	Grid water	All year
HE5	Cooling water ACH	Ground	Summer
HE6	Cooling water CO of ORC		All year
Pumps	**Variable or Constant Speed**	**Fluid**	**Activation Season**
P1	variable	Geothermal hot water	All year
P2	variable	Diathermic oil	
P3		Diathermic oil	
P4		R245fa	
P5		Chilled water ACH	Summer
P6	Constant	Cooling water ACH	
P7		Cooling water CO of ORC	
P8		Chilled or hot water	All year
P9		Geothermal brine	

* The selected diathermic oil is a mixture consisted of biphenyl and diphenyl oxide. The analytical functions of the fluid properties are obtained by producer datasheets [17].

After the first preheating, performed by the geothermal thermal energy (HE1 and HE2), the diathermic oil is stored in the tank of the solar circuit. Then, the evacuated solar collector further heats the diathermic oil, to the selected set point temperature, $T_{set,SC}$, equal to 130 °C. The operation of the solar fluid loop is managed by a feedback controller, which varies the flow rate of the variable speed pump P2. Therefore, the controller tries to reach the outlet set point temperature $T_{set,SC}$, reducing the P2 flow rate. In addition, the controller stops pump P2 if the TKs bottom temperature is higher the outlet temperature of the solar collector, for preventing the dissipation of the thermal energy stored in the TKs. The P3 constant speed pump supplies the diathermic oil at a variable temperature to the ORC evaporator. In particular, the ORC feeding temperature ranges from 90 to 130 °C. When the top temperature of the tank TKs is lower than $T_{min,act,ORC}$, a biomass condensing boiler is activated, increasing the temperature of the diathermic oil to $T_{min,act,ORC}$.

During the summer season, the geothermal hot water is diverted to HE3 by D1, where the geothermal brine, diverted by the D2, further heats the geothermal hot water. This strategy allows the plant to reach a stable feeding temperature of the absorption chiller.

ACH is not in operation until the $T_{bottom,TKu}$ ranges between 6.5 °C and 15 °C (Table 2), then, in this case, also during the summer season, D2 and D1 respectively divert the geothermal brine and the geothermal hot water to HE2 and HE1, with the aim of increasing the thermal energy availability for the ORC machine. The constant speed pump P8 is activated for supplying the chilled water to the fan coil units, in order to achieve the selected set point indoor temperature, $T_{set,cool}$, equal to 26 °C.

Table 2. Design and operating parameters (1).

Component	Parameter	Description	Value	Unit
	l_{GHE}	Ground heat exchanger length	m	60
	d_{GHE}	Ground heat exchanger diameter	m	0.110
GHE (HE5)	A_{GHE}	Ground heat exchanger area	m^2	35
	$V_{backfill}$	Backfill material volume	m^3	7.2
	h_{conv}	Flow convection coefficient	$W/(K\,m^2)$	292
	l_{GHE}	Ground heat exchanger length	m	18
	d_{GHE}	Ground heat exchanger diameter	m	0.050
GHE (HE6)	A_{GHE}	Ground heat exchanger area	m^2	4
	$V_{backfill}$	Backfill material volume	m^3	0.02
	h_{conv}	Flow convection coefficient	$W/(K\,m^2)$	58
	k_{HDEP}	Thermal conductivity	$W/(m\,K)$	0.49
	ρ	Density of the material	kg/m^3	965
HDPE	c_p	Specific heat of the material	$J/(kg\,K)$	2.25
	ε	Roughness	mm	0.3
Sand (backfill material)	$k_{backfill}$	Thermal conductivity	$W/(m\,K)$	1.5
	$\rho_{backfill}$	Density of the material	kg/m^3	1500
	$c_{p,backfill}$	Specific heat of the material	$J/(kg\,K)$	1798
	k_{soil}	Thermal conductivity	$W/(m\,K)$	0.862
Clay (ground)	ρ_{soil}	Density of the material	kg/m^3	1430
	$c_{p,backfill}$	Specific heat of the material	$J/(kg\,K)$	1439
	SoC_{LIB}	High and low limit on fractional state of charge		0.95–0.05
I	$\eta_{I,AC,to,DC}$	Efficiency (AC to DC)	–	0.98
	$\eta_{I,DC,to,AC}$	Efficiency (DC to AC)		0.96
	η_R	Regulator efficiency		0.95
	C_{cell}	Cell energy capacity	Ah	63.27
	$V_{battery}$	Battery voltage	V	360
LIB	$C_{battery,aviable}$	Available capacity	kWh	41.00
	η_{LIB}	Battery efficiency	–	0.9
	$P_{LIB,discharge,max}$	Maximum allowed discharging power	kW	10
	$P_{LIB,discharge,max}$	Maximum allowed charging power		10
	P_{rated}	Rated cooling power	kW_{th}	17.1
ACH	COP	Rated coefficient of performance	–	0.7
	$T_{set,ACH}$	Set-point temperature for the chilled water	°C	6.5
AH	P_{rated}	Rated auxiliary heater power	kW	200
	$T_{set,AH}$	Set point temperature for AH	90	°C
TKu	H	Height	m	0.5
	V	Volume	m^3	2

Finally, M2 collects the geothermal brine exiting from HE3 and HE2, in order to exploit the waste heat of the geothermal brine for producing domestic hot water in HE4.

Note that due to the lack of a cold-water source the condenser of the ORC and ACH are cooled by ground-coupled heat exchangers HE6 and HE5.

The power produced by the ORC expander (Case ETC) and by the ORC expander and PV panels (Case PV) is delivered to the user through the regulator/inverter. When the power produced is greater than the power demand, including the power supplied to the auxiliary hydronic systems, the surplus power is employed for charging the lithium-ion battery. Note that the charge of the battery is allowed only if the battery state of charge (*SoC*) ranges between the low and the high safe limit, assumed to be equal to 0.05 (*SoC$_{inf}$*) and 0.95 (*SoC$_{sup}$*), respectively (Table 2). Note that if the lithium-ion battery is completely charged, and the building electric load is absent, the surplus power is delivered to the electric national grid. Finally, when the power produced and the stored energy in battery are not sufficient to match the power demand, the power is withdrawn from the electric national grid.

The adoption of a lithium-ion battery is affected by some criticism regarding the battery overheating as explained by many literature works [18–20]. The battery overheating is a dangerous aspect for the lithium-ion battery use, in fact, it may cause the battery to explode in the worst case but causes the degradation of the battery integrity and consequently the deterioration of its energy performances and the reduction of battery life-cycle [18–20]. This problem is caused by the high current intensity during the phase of charge and discharge of the battery, in fact, high current intensity leads to an increase of the temperature inside the battery. In conclusion, in order to prevent this problem, many literature works ([19,21,22]) suggest limiting the discharge/charge power with respect to the maximum value (assumed equal to the power required to fully charge/discharge the battery in one hour). Therefore, the maximum allowed discharging/charging power is assumed to be equal to the power that would discharge/charge the battery in 4.5 h, in order to achieve a battery life of 10 years [23].

3. System Model

The dynamic simulation models simulating the two renewable polygeneration plants shown in Section 2 are developed in the TRNSYS environment (version 17). This is a tool widely adopted both in academic and commercial areas. Some plant components (heat exchangers, ACH, building, energy storage system, controllers, mixers, diverters, pumps, tanks, fan coil units, inverter, etc.) are simulated by the "types"(i.e., libraries) included into the TRNSYS library, whereas the models of the ORC machine and the geothermal well are developed by the authors of this work and presented in reference [5]. In this section, the models of the main components of both compared cases (Cases ETC and PV), i.e., the flat-plate ETCs and the PV panels and thermoeconomic model developed for evaluating the economic and energy performance of both plants are reported in detail. Note that the reliability of the results achieved by the developed models is based on the fact that all the components adopted are validated vs experimental data [24], vs data available in literature and/or based on manufacturers' data.

The data concerning the design and operating parameters of the main components of both the plants and the building simulation data are summarized in Tables 2–4. Conversely, the data concerning the ORC machine and geothermal well are reported in reference [5].

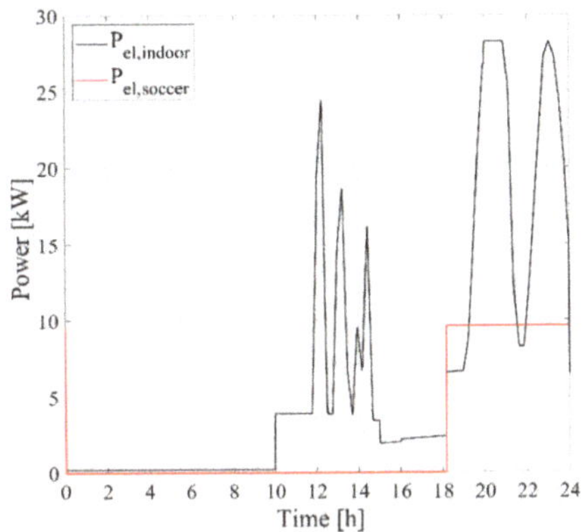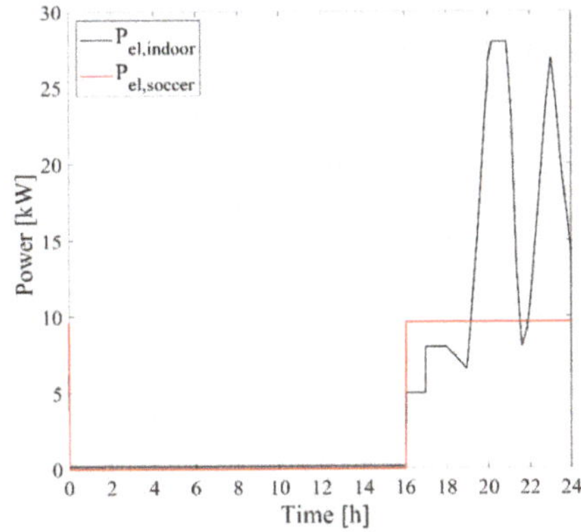

Figure 3. Daily electric load for a typical summer day (**left**) and for a typical winter day (**right**).

Table 3. Design and operating parameters (2).

Component	Parameter	Description	Value	Unit
ETC	a_1	Zero collector heat loss coefficient	0.399	$Wm^{-2}K^{-1}$
	a_2	Temperature difference dependence of the heat loss coefficient	0.0067	$Wm^{-2}K^{-2}$
	a_3	Wind speed dependence of the heat loss coefficient	0	$Jm^{-3}K^{-1}$
	a_4	Long-wave radiation dependence of the collector	0	-
	a_5	Effective heat capacity of the collector	7505	$Jm^{-2}K^{-1}$
	a_6	Wind speed dependence in zero loss efficiency	0	ms^{-1}
	A_{ETC}	Solar collector aperture area	25	m^2
	q_{P2}	P2 rated flow rate	3960	kg/h
	v_{TK}	Tank TK volume per unit SC aperture area	5	l/m^2
	η_0	SC zero loss efficiency at normal incidence	0.82	-
	c_f	Diathermic oil specific heat	1.8	kJ/kg K
	α	Collector slope	30	°
	β	Collector azimuth	0	
	$T_{set,ETC}$	SC outlet set point temperature	130	°C
PV	P_{max}	Maximum power	260	W_p
	V_{oc}	Open-circuit voltage	37.7	V
	I_{sc}	Short-circuit current	9.01	A
	V_{mpp}	Voltage at point of MPP	30.5	V
	I_{mpp}	Current at point of MPP	8.51	A
	N_s	Number of modules in series	2	-
	N_p	Number of modules in parallel	50	
	A	PV module area	1.6	m^2
	N_{cell}	Number of cells in series	15	-
	η_{PV}	Module efficiency	15.8	
	$P_{rated,PV}$	PV panel rated power	7.63	kW
	A_{tot}	PV field area	48.27	m^2

Table 4. Building simulation data.

Thermal Zone: Height (m), Volume (m³), Floor Area (m²), Glass Area (m²)		Height (m): 3	Volume (m³): 924	Floor Area (m²): 308	Glass Area (m²): 37
Building Element	U (W/m²K)	**Thickness (m)**	ρ_s (–)	ε (–)	
Roof and facades	0.828	0.300			
Wall	0.866	0.291	0.4	0.9	
Windows glass	2.89	0.004/0.016/0.004	0.13	0.18	
Rated Heating and Cooling Capacity of the Fan Coil Unit (kW)			Heating: $Q_{heat} = 15.5$ Cooling: $Q_{cool} = 17.1$		
Set Point Indoor Air Temperature (°C)			Heating: $T_{set,heat} = 20$ Cooling: $T_{set,cool} = 26$		
Heating and Cooling Season			Heating: 15 November–31 March Cooling: 1 May–30 September		
Occupancy Schedule (h)		Winter	Working Day 16:00–24:00 Weekend 10:00–24:00		
		Summer	10:00–24:00		
Number of Occupants per Zone			20 × 10:00–14:00; 30 × 14:00–22:00; 20 × 22:00–24:00		
People Heat Gain (W/p)			Sensible: 60 Latent: 40		
Light + Machineries Heat Gains Schedule (kW/h)			Figure 3		
Air Infiltration Rate (vol/h)			0.6		
Free Cooling Ventilation Rate (1/h)			2		
Average Daily DHW Demand (l/day)			30,240 (70 l/day/person)		
DHW Set Point Temperature (°C)			45		
Tap Water Temperature (°C)			15		

3.1. Evacuated Thermal Collector Model

The flat-plate evacuated solar collectors are modeled by considering the high-vacuum HT-Power collectors (version 4.0), designed and manufactured by the TVP Solar company [25]. In particular, the model, validated by several outdoors and indoors experimental campaigns, is based on a modified version of TRNSYS Type 132 [26], which considers the Hottel–Whillier equation integrated with the incidence angle modifier (IAM) coefficients (determined by the tests according to EN 12975 and EN 12976 [27]) and takes into account the wind effect on the zero loss efficiency, the wind influence on the heat losses and the long-wave irradiance dependence of the collector. Therefore, the heat transferred from the ETC per unit aperture at each time-step is:

$$
\begin{aligned}
Q_{th} = F' \cdot (\tau\alpha)_{en} \cdot \left\{ \left[1 - b_0 \cdot \left(\tfrac{1}{\cos\theta_b} - 1 \right) \right] \cdot G_b + K_{\theta d} \cdot G_d \right\} - a_1 \cdot (t_m - t_a) \\
- a_2 \cdot (t_m - t_a)^2 - a_3 \cdot u \cdot (t_m - t_a) - a_4 \cdot (E_L - \sigma T_a^4) - a_5 \cdot \tfrac{dt_m}{dt} - a_6 \cdot u \cdot G
\end{aligned}
\tag{1}
$$

where $F' \cdot (\tau\alpha)_{en}$ is the zero loss efficiency of the collector at normal incidence angle for the solar radiation onto the collector, $K_{\theta d}$ is the IAM for diffuse radiation, G_b is the beam of solar radiation, G_d is the diffuse radiation, the factor reported in square brackets is the incidence angle modifier (IAM) for beam radiation (b_0 is the IAM determined by the collector test, θ_b is the incidence angle for beam radiation onto the solar collector plane). The description of the coefficients of Equation (1) and their numerical values are reported in Figure 3.

3.2. PV Panel Model

The PV panel model is based on the so-called "four parameters" model, which is implemented by the Type 94 using the manufacturers' data and generating the IV curve every time step. The four parameters used are: (i) $I_{L,ref}$, the photocurrent of module at reference condition; (ii) $I_{0,ref}$, the diode reverse saturation current at reference condition; (iii) γ, the empirical PV curve-fitting parameter; (iv) R_s, the module series resistance. The main assumption of the model is that the slope of the IV curve at the short-circuit condition is zero. By considering R_s and γ to be constant, the current–voltage equation of the circuit is:

$$
I = I_{L,ref} \frac{G_T}{G_{T,ref}} - I_{0,ref} \left(\frac{T_c}{T_{c,ref}} \right)^3 \left[\exp\left(\frac{q}{\gamma k T_c} (V + I R_s) \right) - 1 \right]
\tag{2}
$$

where the current I is a linear function of G_T, the total incident solar irradiance on the PV panel and $G_{T,ref}$, the reference solar irradiance and depends on the temperature at the reference open-circuit condition T_c.

The current (I_{mpp}) and the voltage (V_{mpp}) at the maximum power point are evaluated by means of an iterative routine. Thus, the system of equations, that describes the four equivalent circuit characteristics, is solved. The first step is to substitute the voltage and current into Equation (2) at the short circuit, open-circuit and maximum power conditions. After some handling, one obtains the following three equations, that depend on $I_{L,ref}$ (Equation (3)), γ (Equation (4)) and $I_{0,ref}$ (Equation (5)).

$$
I_{L,ref} \approx I_{sc,ref}
\tag{3}
$$

$$
\gamma = \frac{q \left(V_{mp,ref} - V_{oc,ref} + I_{mp,ref} R_s \right)}{k T_{c,ref} \ln\left(1 - \frac{I_{mp,ref}}{I_{sc,ref}} \right)}
\tag{4}
$$

$$
I_{0,ref} = I_{sc,ref} \exp^{-\left(\frac{q V_{oc,ref}}{\gamma k T_{c,ref}} \right)}
\tag{5}
$$

Another equation is needed to determine the last unknown parameter, i.e., the temperature coefficient of open-circuit voltage. This parameter is obtained by the analytical derivate of voltage V_{oc} with respect to T_c:

$$\frac{\partial V_{oc}}{\partial T_c} = \mu_{voc} = \frac{\gamma k}{q}\left[\ln\left(\frac{I_{sc,ref}}{I_{0,ref}}\right) + \frac{T_c\mu_{isc}}{I_{sc,ref}} - \left(3 + q\varepsilon\left(\frac{\gamma}{N_s}kT_{c,ref}\right)^{-1}\right)\right] \tag{6}$$

where μ_{isc} is the temperature coefficient of short-circuit current, N_s is the number of individual cells in a module, q is the electron charge constant, k is the Boltzmann constant, ε is the semiconductor bandgap. The manufactures' specification about the open circuit temperature is equal to this analytical value. Therefore, a search routine is used iteratively to evaluate the equivalent open circuit characteristics.

3.3. Thermoeconomic Model

The energy and economic performance of both the renewable power plants are evaluated by the calculation of the primary energy saving (*PES*) and the simple pay back period (SPB) index. With this aim in mind, a suitable conventional reference system (RS) is selected to compare each plant with the same RS. The RS consists of a conventional vapor-compression chiller and a gas-fired heater for cooling and thermal energy production, respectively, whereas the national grid is the conventional system providing the electric energy to the user. The achievable *PES* by both renewable plants, in terms of electric, heating and cooling energy savings vs the conventional RS is calculated as in Equation (7).

$$PES = \sum_t \left[\frac{\left(\dfrac{E_{el,devices,t}+E_{el,chiller,t}}{\eta_{el,t}} + \dfrac{Q_{th,H,NG,spaceheating,t}+Q_{th,H,NG,DHW,t}}{\eta_{H,NG,t}}\right)_{RS} - \left(\dfrac{E_{el,fromGRID,t}-E_{el,toGRID,t}}{\eta_{el,t}}\right)_{PS}}{\left(\dfrac{E_{el,devices,t}+E_{el,chiller,t}}{\eta_{el,t}} + \dfrac{Q_{th,H,NG,spaceheating,t}+Q_{th,H,NG,DHW,t}}{\eta_{H,NG,t}}\right)_{RS}}\right] \tag{7}$$

where $E_{el,fromGRID,t}/E_{el,toGRID,t}$ are the electric energy withdrawn/sent from/to the national grid in the proposed system (PS), respectively, $Q_{th,AH,biomass,t}$ is the thermal energy supplied by the auxiliary biomass-fired heater in PS, $\eta_{H,NG,t}$ and $\eta_{el,t}$ are the natural gas-fired heater efficiency and conventional thermo-electric power plant efficiency, $E_{el,chiller,t}$ and $E_{el,devices,t}$ are the electric energy required by the compression chiller and by the electric devices of the building, respectively, in RS, $Q_{th,H,NG,spaceheating,t}$ and $Q_{th,H,NG,DHW,t}$ are the thermal energy supplied by the natural gas-fired heater for space heating and DHW, respectively, in RS.

The corresponding potential yearly economic savings ΔC achievable by both renewable plants are calculated by Equation (8). Here, the yearly operating costs of both PSs (Case ETC and Case PV) due to the yearly maintenance of an ORC machine m_{ORC} and yearly maintenance of both solar fields m_{SF} (ETCs and PV panels), are considered.

$$\Delta C = \sum_t \left[\begin{array}{l}\left(\left(E_{el,devices,t} + E_{el,chiller,t}\right)\cdot c_{Eel,fromGRID} + \dfrac{\left(Q_{th,H,NG,spaceheating,t}+Q_{th,H,NG,DHW,t}\right)\cdot c_{NG}}{LHV_{NG}\eta_{H,NG,t}}\right)_{RS} - \\[2ex] +\left(\dfrac{Q_{th,AH,biomass,t}\cdot c_{biomass}}{LHV_{biomass}\eta_{AH,biomass,t}} + E_{el,fromGRID,t}\cdot c_{Eel,fromGRID} - E_{el,toGRID,t}\cdot c_{Eel,toGRID} - \dfrac{f_{HE4}Q_{th,HE4}c_{NG}}{LHV_{NG}\eta_{H,NG,t}}\right. \\[2ex] \left. +m_{ORC} + m_{SF} + C_{ex}\right)_{PS}\end{array}\right]_i \tag{8}$$

where $c_{biomass}$ is the biomass cost for the auxiliary biomass-fired heater, C_{ex} is the fixed yearly cost due to the electric energy exchanged with the national grid, $c_{Eel,fromGRID}$ is the purchasing cost of the electric energy withdrawn from the national grid, $c_{Eel,toGRID}$ is the selling cost of the electric energy sent to the national grid. Note that if $f_{HE4} = 1$, the economic saving of the PS also takes into account the thermal energy recovered by heat exchanger HE4 for DHW production. Note that the same amount of DHW is produced in RS by a conventional gas-fired boiler.

The SPB is assessed as reported in Equation (9)

$$SPB = \frac{\sum_i J_i}{\Delta C} \tag{9}$$

Here, the capital costs of all the components J_i are considered. In particular, the capital costs for storage tanks, pumps and heat exchangers (HE1, HE2, HE3 HE4) are calculated by suitable polynomial equations [28] as a function of the rated volume, flow rate and heat exchange area, respectively. The unit capital costs assumed for the other components are summarized in Table 5. Concerning the heat exchangers HE5 and HE6, i.e., the ground-coupled heat exchangers, their capital cost J_{GHE} is obtained by the calculation reported in Equation (10).

$$J_{GHE} = c_{excavation} \cdot A_{GHE} + c_{lenght} \cdot l_{GHE} + c_{backfill} \cdot V_{backfill} \cdot \rho_{backfill} \tag{10}$$

where $c_{excavation}$ is the excavation cost assumed equal to 80 €/m^2, c_{lenth} is the specific cost of the horizontal pipe, assumed equal to 2.72 €/m and 14.47 €/m for HE6 and HE5, respectively, note that c_{length} depends on the diameter dimension, A_{GHE} is the area of the ground heat exchanger, l_{GHE} is the length of the ground heat exchanger (the length of the buried horizontal pipes), $\rho_{backfill}$ and $V_{backfill}$ are the backfill material density and volume respectively, and $c_{backfill}$ is the cost of the backfill material (sand), assumed equal to 14.45 €/t (Table 2).

Table 5. Thermoeconomic assumptions.

Parameter	Description	Value	Unit
J_{ACH}	ACH unit capital cost per kW of cooling capacity	310 [2]	€/kW
J_{ETC}	ETC unit capital cost per m^2 of solar field	300 [29]	€/m^2
J_{PV}	PV unit capital cost per kW$_{el}$	1000 [30]	
J_{ORC}	ORC unit capital cost per kW$_{el}$	583 [5]	€/kW$_{el}$
J_I	Inverter unit capital cost per kW$_{el}$	180 [30]	
J_{ESS}	ESS unit capital cost per kWh of capacity	346 [31]	€/kWh
$c_{Eel,fromGRID}$	Electric energy purchasing unit cost	0.17	€/kWh
$c_{Eel,toGRID}$	Electric energy selling unit cost	0.08	€/kWh
c_{NG}	Natural gas unit cost	0.88	€/Sm^3
$c_{biomass}$	Biomass gas unit cost	0.06	€/kg
C_{ex}	Energy exchange yearly cost	30	€/year
m_{ORC}	ORC machine maintenance yearly cost	1	%/year
$m_{SF,ETC}$	Solar field maintenance yearly cost (Case ETC)	2	%/year
$m_{SF,PV}$	Solar field maintenance yearly cost (Case PV)	2	%/year
$LHV_{biomass}$	Biomass lower heating value	3.7	kWh/kg
LHV_{NG}	Natural gas lower heating value	9.6	kWh/Sm^3
$\eta_{el,t}$	Conventional thermo-electric power plant efficiency	46	%
$\eta_{H,NG,t}$	Natural gas-fired heater efficiency.	95	%
$\eta_{AH,biomass,t}$	Auxiliary biomass-fired heater efficiency.	95	%

Table 5 summarizes the thermoeconomic assumptions for the yearly economic saving.

4. Case Study

The analyzed case study refers to a real commercial user, located in Campi Flegrei (near Naples) worldwide famous for its volcanic activity. The geothermal well is located close to a small bar serving five small soccer fields. The proposed polygeneration system will provide electricity, thermal and cooling energy to this user which is also selected to perform the dynamic simulations and the related thermoeconomic analysis.

Two innovative micro renewable power plants are presented, based on the exploiting of the solar and the geothermal energy source. Note that concerning the geothermal source of energy an existing low-temperature geothermal well at the selected user is considered.

The first one (Case ETC) uses this geothermal well and a small solar field for feeding a 6 kW$_e$ ORC machine and produce electricity and to supply a 17.1 kW$_f$ absorption chiller (supplied by the geothermal energy only), the data about the solar field are reported in Table 3. The second one (Case PV) only differs from Case ETC, a photovoltaic field of 48.27 m^2 is installed (design data displayed in Table 3) and no solar thermal plant is considered.

Note that ETC and PV areas are selected with the aim to obtain in both cases a similar cost. The scope of the analysis is to compare, at the same capital costs, PV and ETC configurations.

The geothermal well included in this case study was drilled several years ago for DHW production. However, due to its significantly high temperature compared to the other geothermal wells available in the selected zone, it is presently unused. Indeed, the selected zone is also rich in low temperature geothermal wells, which better suits the domestic hot water production. Whereas, the geothermal brine temperature is equal to about 96 °C. Therefore, this well is presently available for the research described in this work. The ORC machine is designed to be driven by the diathermic oil at an inlet temperature ranging between 90 °C and 130 °C, since the additional temperature increase can be provided by the solar field. Reference [5] in detail describes the ORC machine design parameters and operation. Note that the biomass condensing boiler is activated only if the temperature of the oil feeding ORC evaporator is below 90 °C (Figure 1).

Note that the ground heat exchangers (HE5 and HE6) are selected to cool the condenser and absorber of ACH and the condenser of the ORC machine because in the selected location no suitable cold-water source is available. In addition, the installation of cooling towers and/or dry coolers is not feasible due to space availability and noise constraints.

The ground heat exchanger for the ACH (HE5) consists of a pump (P6, Figure 1 and Table 1) which feeds a 60 m long tube of high-density polyethylene (HDPE). The heat exchange occurs in this tube, which is buried at 5 m depth. The tube diameter is chosen equal to 0.110 m (Table 2). Moreover, a layer of sand surrounds the tube, as backfill material, with the aim of enhancing the heat exchange between clay ground and the pipe. The ground heat exchanger for the ORC (HE6) is developed with the same approach, Table 2 in detail describes the characteristics and the thermodynamic features of both the ground heat exchangers.

The commercial area consists of a 308.5 m^2 small bar. Table 4 summarizes the opening hours, schedule of the people and machines inside the bar. The assumed machines installed inside the bar are an induction cooking professional plate, a coffee machine, a professional cooling table, an ice machine and a fryer. The five soccer fields are equipped with 48 lights of 200 W (Table 4), switched on during the same bar opening hours. Note that lights are turned on as a function of the solar radiation availability after 18:00 during summer and after 16:00 during winter. The soccer fields also include locker rooms, where the people may change and have a shower. In particular, the amount of DHW for the locker rooms showers is assumed to be equal to 30,204 l/day (Table 4).

Figure 3 displays the power demand of the selected user. The spiky shape of the power demand curve (Figure 3) is mainly due to the fact that the selected user is a small bar. Thus, the user is not able to serve many consumers at the same time. Then, the electric appliances installed into the bar are intermittently used.

The heating season is assumed to be from November 15th to March 31st with an indoor setpoint temperature equal to 20 °C, while the cooling season is assumed to be from May 1st to September 30th with an indoor setpoint temperature equal to 26 °C, according to Italian regulation (Table 4).

Finally, the proposed system is equipped with an ESS based on the lithium-ion technology. The ESS assumed for this model is the Renault Zoe ZE nickel manganese cobalt oxide (NMC) lithium-ion battery (LIB) [23], consisting of 63.35 Ah rated capacity cells [23]. In particular, the rated capacity of the battery is 45.56 kWh, while the rated voltage is 360 V (Table 3). Table 5 reports the assumptions made for the thermoeconomic analysis.

5. Results and Discussion

In this section, the results achieved by means of the dynamic simulation models of both renewable plant (Case ETC and Case PV) are presented. In particular, the dynamic, monthly and yearly results are displayed. Moreover, a parametric analysis is shown by varying the lithium-ion storage system capacity, and the PV and ETC solar field area.

5.1. Daily Results

Figure 4 displays the transient results for Case ETC and Case PV, above and below, respectively. In particular, Figure 4 plots the power produced by ORC ($P_{el,ORC}$), the electric load of the proposed layout ($P_{el,LOAD}$), the power sent and withdrawn to/from the grid ($P_{el,fromGRID}$ $P_{el,toGRID}$), the power sent/withdrawn to/from LIB ($P_{el,fromLIB}$ $P_{el,toLIB}$), the state of charge of the LIB battery (SoC_{LIB}) and the power produced by the PV panels ($P_{el,PV}$).

For Case ETC, SoC_{LIB} achieves its maximum value equal to 0.76 at 11:47, in fact, at this hour the load of the bar hugely increases, by reaching the value of 25.18 kW, due to the activation of the induction cooking plate (Figure 4, above). Consequently, LIB is discharged, from 11.45 to 14:31 by matching the electric load of the bar, along with the power produced by the ORC machine. Note that the residual electric load, defined as $P_{el,residual} = P_{el,LOAD} - P_{el,ORC}$, is greater than the maximum allowed discharging power ($P_{LIB,dicharge,max}$), which is equal to 10 kW (Table 2). Therefore, the power discharged from LIB is equal to $P_{LIB,discharge,max}$, and a rate of power equal to $P_{el,residual} - P_{el,fromLIB}$ is withdrawn from the grid (Figure 4, above). From 14:40 to 18:05 $P_{el,LOAD}$ decreases, while $P_{el,ORC}$ remains almost constant, consequently, SoC_{LIB} grows up from 0.40 to 0.53. Anyway, at 18:05 the increase of the bar activity and the turning on of the soccer field lights causes a dramatical growth in $P_{el,LOAD}$, reaching 38.47 kW. Consequently, LIB is totally discharged in 2h, and from 20:05 the load of the system is completely satisfied by the grid and ORC machine. The power produced by the ORC is averagely equal to 5.5 kW over the day (Figure 4, above).

For Case PV, SoC_{LIB} achieves the high limit, equal to 0.95 (Table 2), at 10:00, when the PV field power production increases (Figure 4, below). In fact, during the daylight hours, the photovoltaic power production contributes significantly to charge the battery by matching a larger amount of the system electric load. The surplus power is supplied to the grid, with a power value equal to 5.72 kW. When the bar activity increases, from 11.45 to 14:31, the ORC and PV production and LIB discharging supply the electric load of the bar. However, at 14:31 SoC_{LIB} rises again, because the overall electric production is higher than $P_{el,LOAD}$. Finally, at 18:05 when $P_{el,LOAD}$ hugely increases, as explained before, LIB is discharged and it is able to supply the system for 3h. Anyway, during this period $P_{el,fromGRID}$ is not null, indeed, $P_{el,residual} = P_{el,LOAD} - P_{el,ORC} - P_{el,PV}$ is greater than $P_{LIB,dicharge,max}$ (Table 2). Therefore, as explained before, $P_{el,fromLIB}$ is fixed to 10 kW and the remaining amount of power, i.e., $P_{el,residual} - P_{el,fromLIB}$, is withdrawn from the grid. Finally, from 21:45 the system is completely supplied by the grid and ORC machine. Obviously, the total electric production, sum of the ORC and PV production significantly increases in the middle of the day, by achieving the maximum value of 11.17 kW, due to the growth of the solar power production in these hours.

In conclusion, from Figure 4 it is clear that Case PV with respect to Case ETC produces a larger amount of power, consequently, the electric energy stored in the battery is greater. Thus, in the PV Case, LIB is able to cover a larger amount of the electric load of the system.

In Figure 5, the inlet oil temperature to the ORC evaporator (T_{toEV}) for typical summer and winter days are displayed. It is obvious that higher values of T_{toEV} are obtained in Case ETC. Obviously, the higher temperatures are reached in the middle of the day when the solar radiation is higher (the maximum value is 101.1 °C at 11:45). Conversely, for Case PV, T_{toEV} is averagely equal to 90–91 °C during winter and summer (Figure 5) days. Anyway, during the remaining part of the day, when the solar radiation is very low or absent there are no difference between the inlet oil temperatures in ETC and PV layouts.

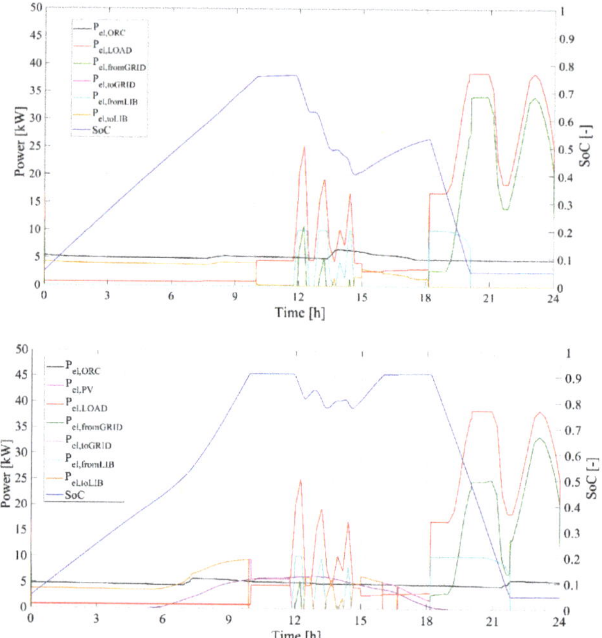

Figure 4. Dynamic results, typical summer working day: Case ETC (**above**) and Case PV (**below**).

5.2. Monthly Results

In this section, the monthly results for Case ETC and Case PV are reported. Figure 6 displays the ratios of the electric energy (i) produced by ORC $E_{el,ORC}$, (ii) self-consumed $E_{el,self}$; (iii) withdrawn from the grid $E_{el,fromGRID}$, (iv) sent to the grid $E_{el,toGRID}$, (v) discharged from LIB $E_{el,fromLIB}$, (vi) sent to LIB $E_{el,toLIB}$ and (vii) produced by PV $E_{el,PV}$ on the monthly electric load of the studied layout ($E_{el,LOAD}$). The total produced amount of renewable electric energy (for Case PV) $E_{el,renw}$ is also displayed.

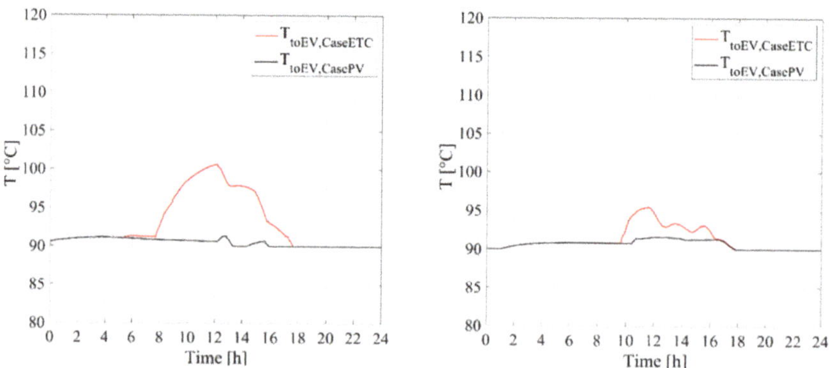

Figure 5. Feeding temperature of the evaporator of ORC: a typical summer day (**left**) and a typical winter day (**right**).

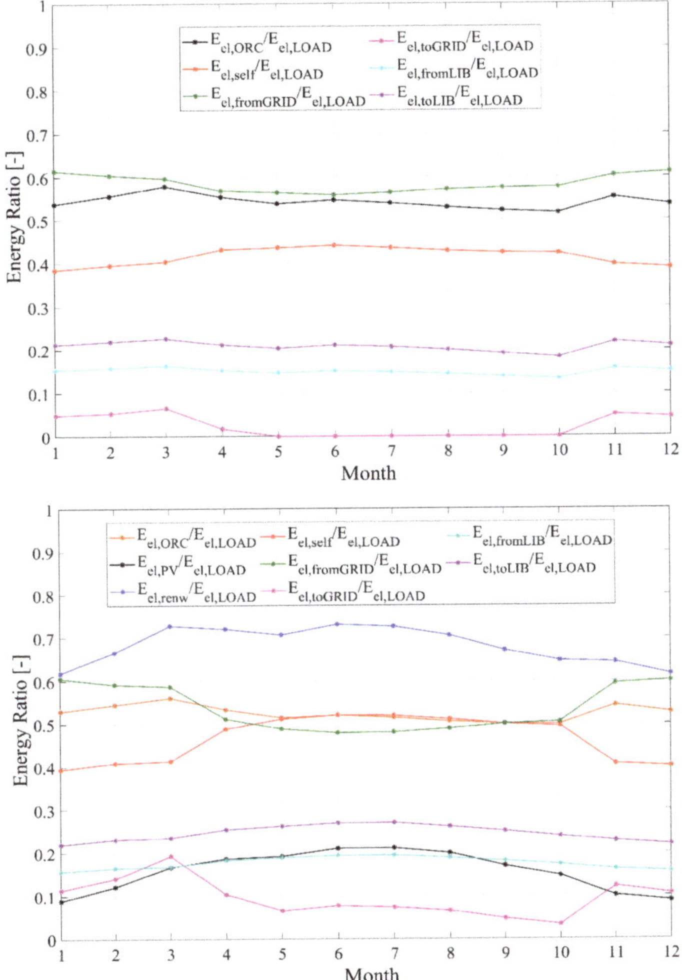

Figure 6. Monthly results, electric performance: Case ETC (**above**) and Case PV (**below**).

Case ETC does not achieve the energy self-sufficiency, indeed the ratio $E_{el,self}/E_{el,LOAD}$ constantly is lower than about 45% for all months of the year. LIB is able to cover about 14%–16% of the electric energy demand of the system. Note that LIB adoption causes the absence of surplus electric energy sent to the grid during the summer season (May–October), while during the remaining part of the year the $E_{el,toGRID}$ ratio is lower than 6.7% (Figure 6, above).

Although Case PV achieves self-consumed energy ratios $E_{el,self}/E_{el,LOAD}$ higher than Case ETC, ranging from 40% to 52%, the energy self-sufficiency is not reached (Figure 6, below).

Note that the electric energy produced by the renewable sources (by PV and ORC) achieves higher values during the summer season due to the higher energy production by the PV field during the months of higher solar radiation. In fact, during the months of July and August $E_{el,renw}/E_{el,LOAD}$ is equal to about 73%. Moreover, during these months (July and August) the proposed plant (Case PV) reaches the higher value of self-consumed energy, i.e., $E_{el,self}/E_{el,LOAD}$ equal to 52% and a lower value of electric energy withdrawn from the grid, i.e., $E_{el,fromGRID}/E_{el,LOAD}$ equal to 48% (Figure 6, below).

The adoption of PV panels instead of ETCs leads to a significant increase in the electric energy produced by the proposed renewable plant. However, this higher electric energy production is not fully self-consumed by the user, in fact, Case PV exhibits a limited increase in $E_{el,self}$, because the electric energy demand is not simultaneous with the electric energy production, mainly because the electric load is concentered mainly in a few hours of the day (evening hours). In addition, the battery (LIB) is not able to store all the surplus electric energy, that is delivered to the grid, see $E_{el,toGRID}/E_{el,LOAD}$ ratio in Figure 6. Therefore, it is possible to conclude that the energy self-sufficiency is not reached in both plants.

Figure 7 shows, for both ETC and PV Cases, the thermal energy supplied to the ORC evaporator ($Q_{toEV,CaseETC}$ and $Q_{toEV,CasePV}$), the electric energy produced by the ORC machine ($E_{el,ORC,CaseETC}$ and $E_{el,ORC,CasePV}$) and the efficiency of the ORC machine ($\eta_{ORC,CaseETC}$ and $\eta_{ORC,CasePV}$).

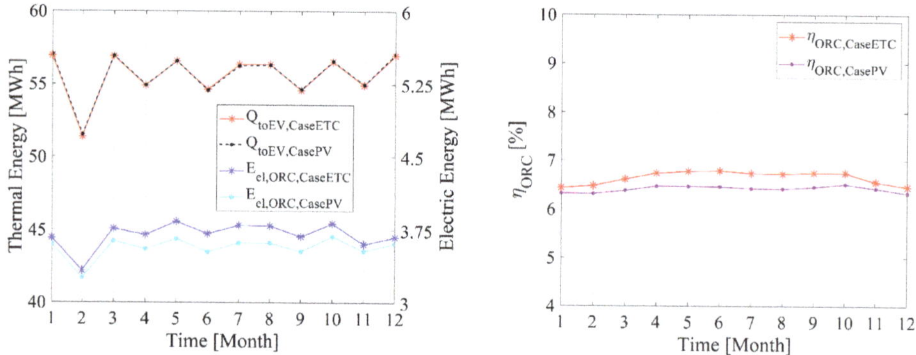

Figure 7. ORC performance: energy performance (**left**) and efficiency (**right**).

The thermal energy supplied to the ORC evaporator is about the same for both cases. The efficiencies are also similar but $\eta_{ORC,CaseETC}$ is slightly higher than $\eta_{ORC,CasePV}$, 6.7% vs 6.4%. This is due to the fact that the ORC inlet oil temperature is averagely higher in Case ETC than in Case PV (see Section 5.1 (Figure 5).

Figure 8 plots the ratios calculated by the following equations:

$$R_i = \frac{Q_i}{Q_{toEV}} = \frac{Q_i}{Q_{solar} + Q_{geoth} + Q_{biomass}} \tag{11}$$

where Q_{geoth} is the geothermal energy provided to the ORC evaporator by means of HE1 and HE2, Q_{solar} is the solar thermal energy provided by ETCs to the ORC evaporator and $Q_{biomass}$ is the thermal energy provided by AH to the ORC evaporator when the solar tank top temperature is lower than 90 °C (Table 2). In Case PV, the geothermal source provides almost the total amount of thermal energy for driving the ORC machine and reduces slightly in Case ETC due to the solar thermal energy production. Indeed, in Case ETC R_{solar} achieves the maximum value of 5.11%, while $R_{biomass}$ is lower than 1.1%. Without ETCs, $R_{biomass}$ does not significantly increase and achieves the maximum value of 1.73%. Thus, in Case PV, the geothermal source supplies more than 98% of the thermal energy needed for the ORC (Figure 8).

In both investigated cases, the slight reduction of R_{geoth} during the summer months (Figure 8), is due to the control strategy of the proposed system. In particular, during the summer season, the geothermal energy is used to supply the ACH producing the building space cooling, by reducing the geothermal energy sent to the ORC evaporator. In both plants, the biomass auxiliary heater could be removed without affecting significantly the overall plant performance.

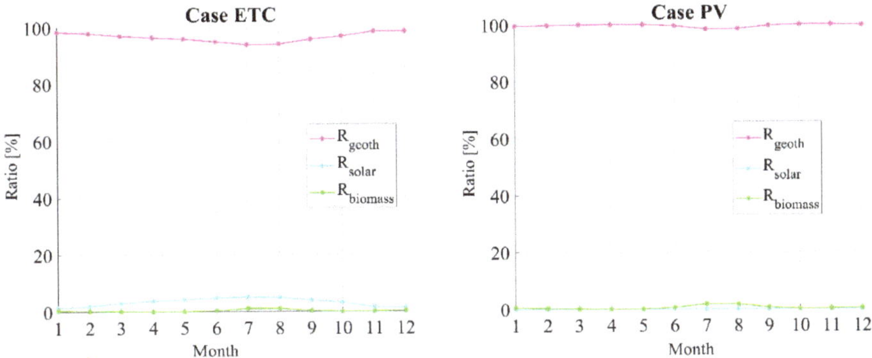

Figure 8. ORC evaporator energy ratios: Case ETC (**left**) and Case PV (**right**).

Figure 9 displays the thermal performance of the ground heat exchangers used to cool the condenser of ORC and ACH. In particular, in this figure, the thermal energy transferred from HE5 (Q_{HE5}) and HE6 (Q_{HE6}) to the ground for both the cases studied are represented. As mentioned before, the use of PV panels vs ETCs shows minor effects on ORC operation, and consequently, the thermal energy transferred from the ORC condenser to the ground is about the same for both the cases (Figure 9). ACH operates in the same way both in Case ETC and in Case PV, consequently, the values of Q_{HE5} are about similar (Figure 9). Anyway, the maximum values of energy transferred to the ground occur during the summer months, when the thermal energy required for the building space cooling increases. In fact, during the months of July and August Q_{HE5} achieves the values of 10.7 MWh and 10.3 MWh, for both cases (Cases ETC and PV, Figure 9).

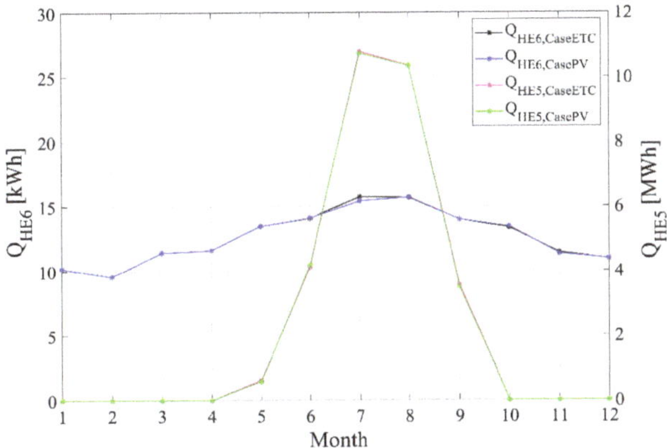

Figure 9. Ground heat exchanger energy performance.

5.3. Yearly Results

Considering the previously discussed results, the yearly thermo-economic and environmental results of the studied cases are discussed in Tables 6 and 7. The adoption of PV panels leads to a greater production of electric energy, in fact $E_{el,prod,CasePV}$ is equal to 55.77 MWh/year, whereas $E_{el,prod,CaseETC}$ is equal to 44.40 MWh/year (Table 6). Consequently, Case PV reaches a higher amount of self-consumed energy. Therefore, Case PV achieves a higher *PES* than Case ETC, 51.21% vs 37.81% (Table 7). The avoided equivalent CO_2 emissions obviously follow the same trend of *PES* (Table 7).

Table 6. Energy yearly results.

Case	$E_{el,ORC}$ [MWh/Year]	$E_{el,PV}$ [MWh/Year]	$E_{el,prod}$ [MWh/year]	$E_{el,fromGRID}$ [MWh/Year]	$E_{el,toGRID}$ [MWh/Year]
ETC	44.40	-	44.40	47.85	1.83
PV	42.86	12.91	55.77	43.74	7.63

Table 7. Yearly energy, economic and enviromental results.

Case	ΔPE [MWh/Years]	PES [%]	SPB [Years]	SPB_{inc} [Years]	J_{tot} [k€]	ΔC [k€/Year]	ΔCO_2 [t/Year]	ΔCO_2 [%]
ETC	60.94	37.81	14.71	7.36	111.64	7.59	13.37	37.70
PV	82.40	51.21	12.56	6.28	109.62	8.73	18.13	51.13

In conclusion from and energy and environmental point of view, Case PV exhibits better performance.

From an economic point of view, Case PV achieves a better SPB with respect to Case ETC. This result is mainly due to the higher electric energy produced, self-consumed and sold to the grid in Case PV. Note as $E_{el,toGRID}$ is equal to 1.83 MWh/year and 7.63 MWh/year for Cases ETC and PV, respectively (Table 6). Therefore, the yearly economic saving for Case ETC (7.59 k€/year) is lower than the one obtained for Case PV (8.73 k€/year), while the total capital costs of both cases are about similar (Table 7). In conclusion, SPB$_{CaseETC}$ was equal to 7.36 years, whereas SPB$_{CasePV}$ was equal to 6.28 years (Table 7).

Table 8 shows the performance indexes of the studied cases. For Case ETC, the solar thermal energy meets about 3% of the thermal energy delivered to the ORC, i.e., R_{solar} equal to 3.24%. $R_{biomass}$ passes from 0.31% for Case ETC to 0.42% for Case PV (Table 8). Thus, the absence of ETCs determines minor variations in the ORC electricity production. Anyway, the absence of ETC causes a lower inlet oil temperature to the ORC evaporator and consequently, ORC efficiency is slightly lower in Case PV, 6.45% vs 6.70% in Case ETC. Note that the value of the efficiency achieved by both the analyzed layouts (Case ETC and Case PV) is consistent with the values available in the literature [32]. For the reason above explained, the electric energy produced by ORC is slightly lower in Case PV, 42.86 MWh/year vs 44.40 MWh/year in Case ETC (Table 6). This difference does not affect the overall performance of Case PV, exhibiting better results from energy, environmental and economic points of view.

Table 8. Yearly performance index.

Case	R_{solar} [%]	R_{geoth} [%]	$R_{biomass}$ [%]	COP [-]	η_{ORC} [%]	η_{solar} [%]	$m_{biomass}$ [kg]
ETC	3.24	96.45	0.31	0.74	6.65	50.46	596.90
PV	-	99.58	0.42	0.74	6.42	15.43	906.38

Finally, the yearly COP of the ACH is also evaluated (Table 8). This value, equal to 0.74, is similar to the results available in literature about the performance of single-stage LiBr/H$_2$O ACH [33]. Note that the achieved COP of 0.74 is slightly higher than the rated value, due to the high activation temperature of the geothermal brine.

5.3.1. Parametric Analysis

A parametric analysis is carried out with the aim of analyzing the effects of the variability of LIB capacity and of ETC and PV area on the energy, economic and environmental performance.

5.3.2. Battery Capacity

By taking into account the proposed renewable plants, LIB capacity varied from 22.11 kWh to 227.77 kWh. Figure 10 displays the energy and environmental result of the parametric analysis for each case (Case ETC and Case PV). In particular Figure 10 (left) displays *PES* and ΔCO_2, while Figure 10 (right) points out the ratios: (i) $E_{el,fromGRID}$ on $E_{el,LOAD}$, (ii) $E_{el,toGRID}$ on $E_{el,LOAD}$ and (iii) $E_{el,self}$ on $E_{el,LOAD}$.

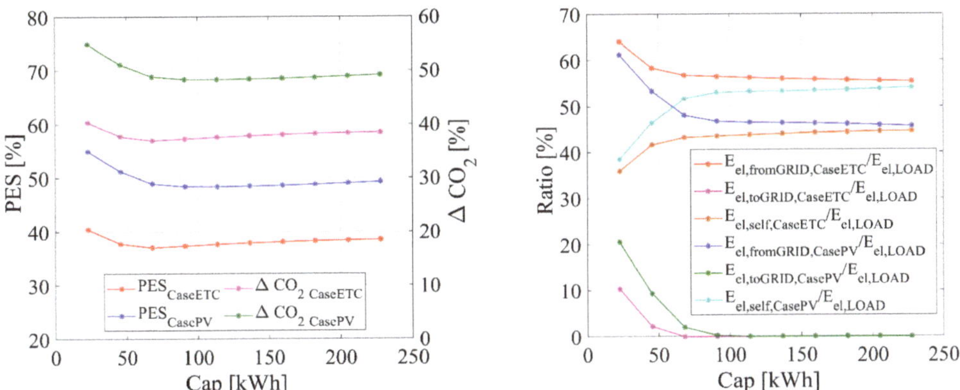

Figure 10. Parametric analysis: *PES* and ΔCO_2 (**left**) and energy ratios (**right**).

The increase in battery capacity from 22.11 to 227.77 kWh causes a reduction in the energy performance: $PES_{CaseETC}$ passes from 40.48% to 38.59% and PES_{CasePV} varies from 54.97% to 49.33% for Case PV. These trends are due to the battery discharge and charge efficiency. In fact, supplying the surplus electric energy to LIB and discharging the battery when needed, leads to a loss of electric energy, and therefore, a reduction of *PESs* (Figure 10). However, both the studied cases exhibit the same trend. In fact, after an initial decrease, the values of *PES* become almost constant by further increasing the battery capacity.

This behavior is well explained by Figure 10 (left), in fact, the $E_{el,self}/E_{el,LOAD}$ ratio initially increases but for capacity values higher than 113.9 kWh, the ratio is almost constant, to almost 53% and 44.2% for Case PV and Case ETC, respectively (Figure 10). In order to obtain a further increase of the ratio $E_{el,self}/E_{el,LOAD}$, an increase in the electric energy production by PV panels or ORC machine is needed because with the current installed electric capacity, the proposed renewable plant covers only 53% for Case PV and 44.2 % for Case ETC of the total electric energy demand (Figure 10).

Besides, the increase of LIB capacity leads to a limited worsening of the energy and environmental performance, for both the analyzed cases.

In Figure 11 the economic results are shown, in particular, Figure 11 (left) displays SPB, while Figure 11 (right) displays J_{tot} and ΔC. Figure 10 (left) also explains the economic results: by increasing the battery capacity, the values of electric energy sent/withdrawn to/from the grid initially decreases/increases but for capacity values higher than 113.9 kWh they become about constant (Figure 10). Consequently, the yearly economic savings initially increase and then become almost constant. In particular, for capacity values higher than 113.9 kWh, $\Delta C_{CaseETC}$ and ΔC_{CasePV} are equal to about 7.8 k€/year and to about 9.1 k€/year, respectively (Figure 11). By increasing the battery capacity from 22.11 kWh to 227.77 kWh, the capital cost hugely increases from 103.7 k€ and 101.7 k€ to 174.7 k€ and 172.7 k€, for Case ETC and Case PV, respectively. This is mainly due to the high cost of LIB. Consequently, SPB dramatically grows up: $SPB_{CaseETC}$ passes from 14.17 years to 22.25 years, and PB_{CasePV} varies from 12.17 years to 18.82 years.

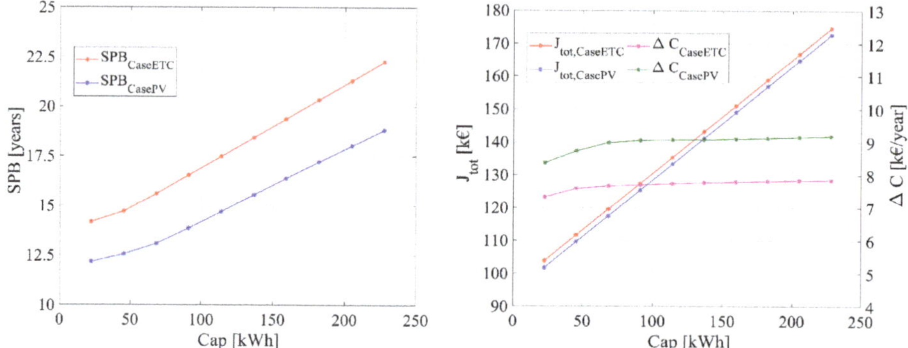

Figure 11. Parametric analysis: SPB (**left**) and capital cost and economic savings (**right**).

In conclusion, the LIB capacity increase causes a general reduction in the economic performance of the two studied cases.

5.3.3. PV and ETC Area

The area of ETCs and PV panels varied from 5 m^2 to 85 m^2. The energy and environmental results of this parametric analysis are displayed in Figures 12 and 13. In particular, Figure 12 (left) displays the electric energy produced by the ORC and the ORC efficiency, for both the studied cases, while Figure 12 (right) shows the electric energy withdrawn/sent from/to the grid, and the total electric energy produced by the proposed renewable power plants. Figure 13 (left) displays *PES* and Figure 13 (right) shows ΔCO_2.

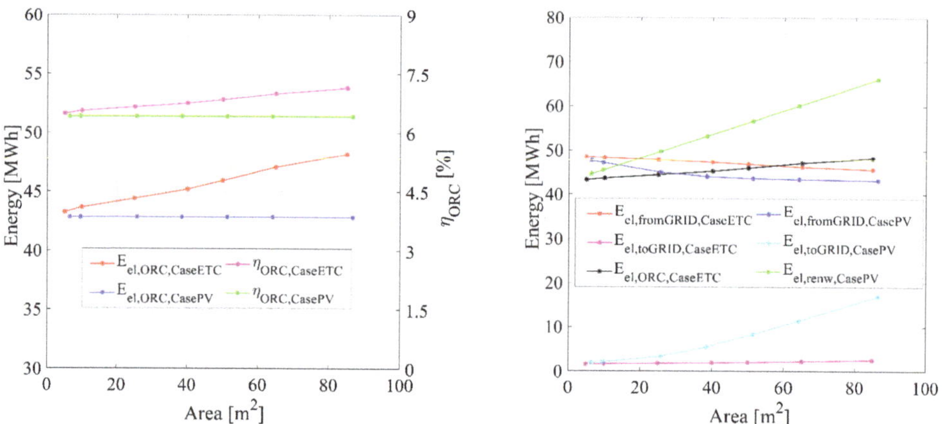

Figure 12. Parametric analysis: ORC performance (**left**) and electric energy performance (**right**).

The increase of ETCs area directly affects the ORC performance: the higher the ETCs area the higher the thermal energy provided to the ORC evaporator, and consequently, the higher the electric energy produced by the ORC. Conversely, the variation of PV field area obviously does not affect the ORC performance. Thus $E_{el,ORC,CaseETC}$ increases, by passing from 43.28 MWh/year to 48.21 MWh/year, while $E_{el,ORC,CasePV}$ remains constant (Figure 12).

Note that the growth of the ETCs area causes also an increase of η_{ORC}, because for higher ETCs areas, higher inlet oil temperatures to the ORC evaporator are reached. In particular, $\eta_{ORC,CaseETC}$ passes from 6.49% to 7.15%, by varying ETCs area from 5 m^2 to 85 m^2 (Figure 12).

Anyway, by considering the whole plant, the increase of PV field area with respect to the increase of ETCs area leads to a more significant enhancement of the energy performance of the power plant. In fact, the increase of the PV field causes a remarkable increase in the electric energy production, due to PV field energy production, thus $E_{el,renw}$ passes from 44.58 MWh/year to 66.09 MWh/year (Figure 12). This result affects the values obtained for $E_{el,toGRID,CasePV}$, significantly increasing from 1.96 MWh/year to 16.96 MWh/year (Figure 12).

Note that, although the electric energy production for Case PV significantly increases, the electric energy withdrawn from the grid exhibits a limited reduction. This is due to the selected LIB capacity (that is not able to store all the electric energy surplus), and because the electric load and production are not simultaneous. For all the reasons above explained, $PES_{CaseETC}$ values increase slightly, whereas PES_{CasePV} values increase remarkably. In particular, $PES_{CaseETC}$ and PES_{CasePV}, respectively, passes from 36.89% and 38.19% to 41.94% and 64.57% (Figure 13). The avoided CO_2 emissions follow the same trend as PES. Finally, the increase of the solar fields area enhances the energy and environmental performance of the proposed plants.

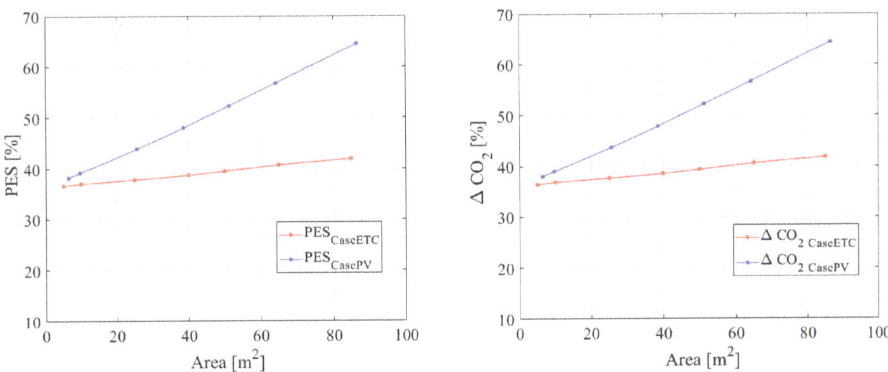

Figure 13. Parametric analysis: *PES* (**left**) and avoided equivalent CO_2 emissions (**right**).

Figure 14 displays the economic results, in particular this figure plots SPB (left) and J_{tot} and ΔC (right). $E_{el,fromGRID,CaseETC}$ exhibits a limited reduction without significative increasing of $E_{el,toGRID,CaseETC}$, whereas $E_{el,fromGRID,CasePV}$ exhibits a limited reduction but a remarkable increase in $E_{el,toGRID,CasePV}$. Therefore, $\Delta C_{CaseETC}$ remains about constant, conversely, ΔC_{CasePV} increases. In fact, $\Delta C_{CaseETC}$ is averagely equal to about 7.73 k€/year, whereas ΔC_{CasePV} passes from 7.73 k€/year to 9.44 k€/year. The capital cost of both the cases has a remarkable increase but $J_{tot,CaseETC}$ is higher than $J_{tot,CasePV}$ due to the higher capital cost of evacuated thermal solar collectors with respect to the PV panels.

Therefore, SPB$_{CaseETC}$ increases by passing from 13.69 years to 16.91 years. Conversely, SPB$_{CasePV}$ shows a limited decrease, by passing from 13.30 years to 12.27 years.

In conclusion, Case PV has better energy, environmental and economic performance. However, the increase of the ETCs and PV panels area leads to an enhancement of the energy and environmental performance of the proposed renewable power plans. Conversely, from the economic point of view, the better configuration suggests larger PV fields.

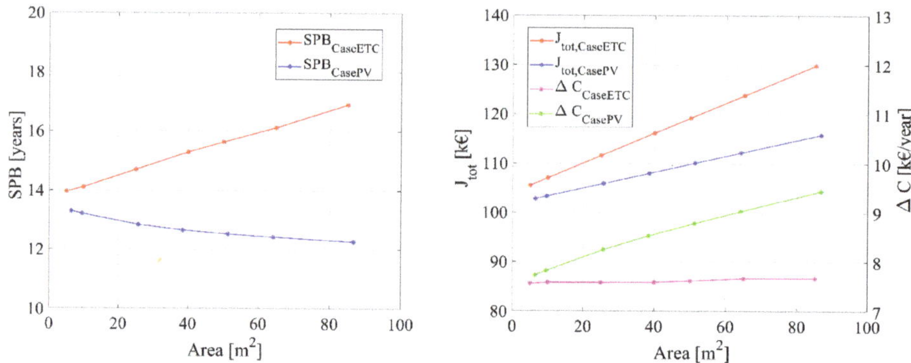

Figure 14. Parametric analysis: SPB (**left**) and capital and operative cost (**right**).

5.3.4. Optimization

In conclusion, an optimization analysis is performed in order to detect the optimal configuration of each analyzed layout. Primary energy savings, avoided CO_2 emissions and simple payback period are considered as object functions. For PV layout, the number of battery cells in parallel and the size of the PV field are simultaneously varied (Table 9). The results of these simulations are displayed in Figure 15. The best configuration achieves a PES equal to 107%, a ΔCO_2 equal to 107% and a SPB equal to 8.60 years. This configuration consists of a PV field area equal to 200 m^2 and a storage system capacity equal to 22.78 kWh. This layout exhibits the larger PV field and the smaller battery capacity. This trend is due to the high cost of the lithium-ion battery. Indeed, increasing the battery size the economic performance worsens. Figure 15 (left) displays the Pareto front for two objective functions: SPB and PES. Note that the higher the primary energy savings, the lower the payback period. These configurations are that in which the PV field is larger, and the battery capacity is smaller.

Table 9. Parameter varied during the optimization analysis, PV layout and ETC playout.

Case PV	PV Area [m^2]		Battery Capacity [kWh]	
	min	max	min	max
	5	200	22.78	683.31

Case ETC	ETC Area [m^2]		Battery Capacity [kWh]		Specific Tank Parameter [l/m^2]	
	min	max	min	max	min	max
	5	85	22.78	341.66	2.5	10

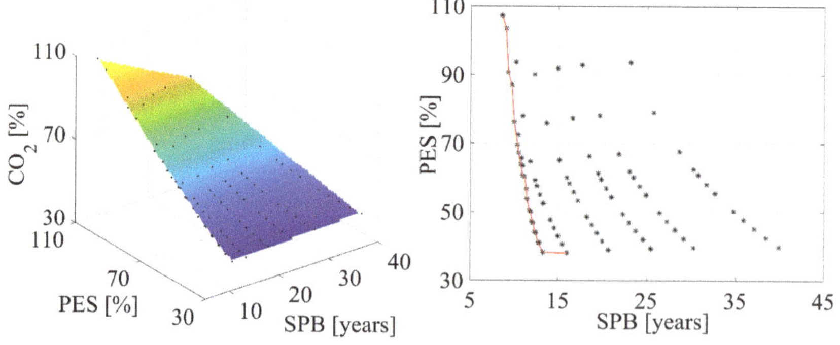

Figure 15. 2D Pareto frontier (**left**) and Optimal configuration research for Case PV (**right**).

For ETC layout, the number of the battery cells in parallel, the size of the ETC field and the specific tank parameter are simultaneously varied (Table 9). It is clear that the increase of the avoided CO_2 emissions and primary energy savings lead to a remarkable increase of the payback period, reducing the economic feasibility of the ETC layout (Figure 16). This trend is related to the high cost of the ETCs, indeed increasing the ETCs field area PES and ΔCO_2 increase while SPB worsens. Moreover, as deeply explained in the previous section, the increase in the ETC area leads to a not so significant improvement of the ORC efficiency. Therefore, the increase in the ETC area causes a huge rise in the capital cost and a limited increase in the yearly savings.

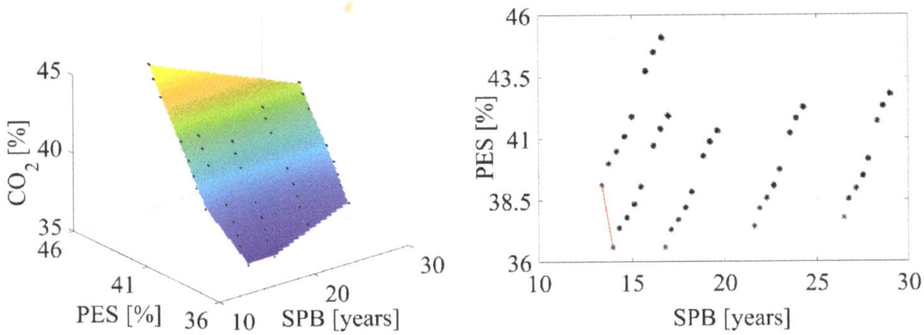

Figure 16. 2D Pareto frontier (**left**) and Optimal configuration research for Case ETC (**right**).

The optimal configuration, consisting of an ETC field equal to 5 m^2, a specific tank parameter equal to 7.5 l/m^2 and a battery capacity equal to 22.78 kWh, achieves a payback period of 13.36 years, a ΔCO_2 equal to 38.91% and a PES equal to 39.12% (Figure 16). Note that the system is not too sensitive to the variation of the specific tank parameter, in fact, the overlapped points (Figure 16) are due to the configurations among which varies only this parameter.

6. Conclusions

In this work, two innovative micro renewable polygeneration plants, both consisting of a micro organic Rankine cycle, single-stage H_2O/LiBr absorption chiller, geothermal well, biomass auxiliary heater and lithium-ion battery are presented. The main aim of this work is the thermoeconomic comparison of two alternative solar technologies integrated as auxiliary systems into two polygeneration plants, namely the evacuated solar collectors and photovoltaic panels. Both plants produce power for, heat and cool a small bar, located in Naples (South Italy), in a weather zone famous for its volcanic activity and high solar availability. Plant layouts are dynamically simulated in the TRNSYS environment, by developing comprehensive models suitable for evaluating the transient energy performance (temperatures, heat, power and efficiency) of all the plant components on hourly, daily, weekly, monthly and yearly time basis. A parametric analysis of the design parameters of the key units of the plant is also performed. The main findings of the simulations are summarised in the following:

- During the daylight hours, evacuated solar collectors rise the inlet oil temperatures to the organic Rankine cycle evaporator on average by of 5–10 °C with respect to the plant including photovoltaic panels, consequently, a higher organic Rankine cycle efficiency, 6.7% vs 6.4%, is obtained;
- The polygeneration plant including photovoltaic panels showed better performance from an energy, environmental and economic point of view with respect to the plant including evacuated solar collectors. In particular, the primary energy saving, payback period, and avoided CO_2 emissions are 51% and 38%, 15 years and 13 years and 51% and 38%;
- The lithium-ion battery capacity increasing causes an increase in the energy-self-sufficiency but a worsening of the economic, energy and environmental performance of two studied plants;

- From the economic point of view the better configuration suggests larger photovoltaic fields and smaller evacuated solar collectors fields due to the higher capital cost of evacuated solar collectors than photovoltaic panels and the achievable economic saving for the higher amount of selling electric energy.

Finally, an optimization analysis is carried out for both the analyzed layouts, the main findings are listed in the following:

- For the layout including photovoltaic panels (Case PV) the larger the photovoltaic field and the lower the battery capacity the better the energy, environmental and economic indices of the plant;
- For the layout including evacuated solar collectors (Case ETC) the tank size does not affect the performance of the plant;
- The optimal layout based on evacuated solar collectors (Case ETC) consists of the lower collector area and the lower battery capacity due to the high costs of both the battery and evacuated solar collector.

Author Contributions: Conceptualization, F.C., F.L.C., M.D.d. and M.V.; Methodology, F.C., F.L.C., M.D.d. and M.V.; Software, F.C., F.L.C., M.D.d. and M.V.; Writing—review and editing, F.C., F.L.C., M.D.d. and M.V. All authors have read and agreed to the published version of the manuscript.

Funding: It is important to stress that this work is developed in the framework of the project "GEOGRID", aiming at adopting suitable technologies and methods in order to use efficiently the geothermal energy of the Campania Region (South Italy). This project is funded by POR CAMPANIA FESR 2014/2020.

Conflicts of Interest: The authors declare no conflict of interest.

Nomenclature

A	area (m^2)
c	specific cost-price (€/kWh or €/m^2 or €/m or €/t)
c_p	specific heat at constant pressure (kJ kg^{-1} K^{-1})
d	pipe diameter (m)
e	open circuit voltages at full charge, extrapolated from V-I curve (V)
F_{SOC}	fractional state of charge (-)
G	incident solar total radiation (W m^{-2})
g	small-valued coefficients of H in voltage-current-state of charge formulas (V)
h	heat transfer coefficient (W m^{-2} K^{-1})
I	current (A)
J	capital cost (€)
k	conductibility (W m^{-1} K^{-1})
l	length (m)
LHV	lower heating value (kWh Sm^{-3})
m	mass flow rate (kg s^{-1})
m	cell-type parameters for the shape of the I-V-Q characteristics (-)
m_{ORC}	ORC yearly maintenance (%/year)
m_{SF}	solar field yearly maintenance (%/year)
N_p	number of modules in parallel (-)
N_{pipe}	number of pipe (-)
N_s	number of modules in series (-)
P	electric power (kW)
PE	primary energy (kWh/year)
PES	primary energy saving (-)
Q	thermal power (kW)
Q	electric charge (Ah)
Q_m	rated capacity of cell (Ah)
r	internal resistances at full charge (Ω)
R	thermal resistance (k W^{-1})

$r_{backfill}$	backfill material radius (m)
r_{pipe}	pipe radius (m)
SoC	state of charge (-)
SPB	simple pay back (years)
T	temperature (°C)
U	overall heat transfer coefficient (W m^{-2} K^{-1})
v	velocity (m s^{-1})
V	volume (m^3)
V_{el}	voltage (V)

Greek Symbols

Δ	difference (-)
ε	long wave emissivity (-)
η	efficiency (-)
θ	time step (s)
ρ	density (kg m^{-3})
ρ_s	solar reflectance (-)

Subscripts

a	ambient
act	activation
avg	average
c	referred to battery charge
$conv$	convective
$cool$	cooling
d	referred to battery discharge
DHW	domestic hot water
E	energy
el	electric
$el,devices$	electric devices of the building
ex	exchange
$FromLIB$	electric energy withdrawn from lithium-ion battery
$fromGRID$	electric energy withdrawn from national electric grid
$heat$	heating
i	number of nodes of ground-coupled heat exchanger
I	inverter
in	inlet
inf	inferior
min	minimum
NG	natural gas
ORC	organic Rankine cycle
out	output
p	primary energy
PS	proposed system
PV	photovoltaic field
$renw$	the renewable energy produced
RS	reference system
s	soil
sup	superior
t	the value of a parameter in time step
th	thermal
$toBUILD$	electric energy supplied to building
$toEV$	to evaporator of ORC machine
$toGRID$	electric energy sent to national electric grid
$toLIB$	electric energy sent to the lithium-ion battery
u	user

References

1. Calise, F.; di Vastogirardi, G.D.N.; d'Accadia, M.D.; Vicidomini, M. Simulation of polygeneration systems. *Energy* **2018**, *163*, 290–337. [CrossRef]
2. Calise, F.; d'Accadia, M.D.; Libertini, L.; Quiriti, E.; Vanoli, R.; Vicidomini, M. Optimal operating strategies of combined cooling, heating and power systems: A case study for an engine manufacturing facility. *Energy Convers. Manag.* **2017**, *149*, 1066–1084. [CrossRef]
3. Calise, F.; d'Accadia, M.D.; Vicidomini, M.; Ferruzzi, G.; Vanoli, L. Design and Dynamic Simulation of a Combined System Integration Concentrating Photovoltaic/Thermal Solar Collectors and Organic Rankine Cycle. *Am. J. Eng. Appl. Sci.* **2015**, *8*, 100–118. [CrossRef]
4. Uris, M.; Linares, J.I.; Arenas, E. Feasibility assessment of an Organic Rankine Cycle (ORC) cogeneration plant (CHP/CCHP) fueled by biomass for a district network in mainland Spain. *Energy* **2017**, *133*, 969–985. [CrossRef]
5. Buonomano, A.; Calise, F.; Palombo, A.; Vicidomini, M. Energy and economic analysis of geothermal–solar trigeneration systems: A case study for a hotel building in Ischia. *Appl. Energy* **2015**, *138*, 224–241. [CrossRef]
6. Calise, F.; Cipollina, A.; d'Accadia, M.D.; Piacentino, A. A novel renewable polygeneration system for a small Mediterranean volcanic island for the combined production of energy and water: Dynamic simulation and economic assessment. *Appl. Energy* **2014**, *135*, 675–693. [CrossRef]
7. Calise, F.; d'Accadia, M.D.; Vicidomini, M.; Scarpellino, M. Design and simulation of a prototype of a small-scale solar CHP system based on evacuated flat-plate solar collectors and Organic Rankine Cycle. *Energy Convers. Manag.* **2015**, *90*, 347–363. [CrossRef]
8. Patel, B.; Desai, N.B.; Kachhwaha, S.S.; Jain, V.; Hadia, N. Thermo-economic analysis of a novel organic Rankine cycle integrated cascaded vapor compression–absorption system. *J. Clean. Prod.* **2017**, *154*, 26–40. [CrossRef]
9. Tchanche, B.F.; Pétrissans, M.; Papadakis, G. Heat resources and organic Rankine cycle machines. *Renew. Sustain. Energy Rev.* **2014**, *39*, 1185–1199. [CrossRef]
10. Olabi, A.G. Renewable energy and energy storage systems. *Energy* **2017**, *136*, 1–6. [CrossRef]
11. Yang, Y.; Bremner, S.; Menictas, C.; Kay, M. Battery energy storage system size determination in renewable energy systems: A review. *Renew. Sustain. Energy Rev.* **2018**, *91*, 109–125. [CrossRef]
12. Zhou, C. Hybridisation of solar and geothermal energy in both subcritical and supercritical Organic Rankine Cycles. *Energy Convers. Manag.* **2014**, *81*, 72–82. [CrossRef]
13. Tourkov, K.; Schaefer, L. Performance evaluation of a PVT/ORC (photovoltaic thermal/organic Rankine cycle) system with optimization of the ORC and evaluation of several PV (photovoltaic) materials. *Energy* **2015**, *82*, 839–849. [CrossRef]
14. Kosmadakis, G.; Landelle, A.; Lazova, M.; Manolakos, D.; Kaya, A.; Huisseune, H.; Karavas, C.-S.; Tauveron, N.; Revellin, R.; Haberschill, P.; et al. Experimental testing of a low-temperature organic Rankine cycle (ORC) engine coupled with concentrating PV/thermal collectors: Laboratory and field tests. *Energy* **2016**, *117*, 222–236. [CrossRef]
15. Bicer, Y.; Dincer, I. Analysis and performance evaluation of a renewable energy based multigeneration system. *Energy* **2016**, *94*, 623–632. [CrossRef]
16. Galindo Noguera, A.L.; Castellanos, L.S.M.; Lora, E.E.S.; Cobas, V.R.M. Optimum design of a hybrid diesel-ORC / photovoltaic system using PSO: Case study for the city of Cujubim, Brazil. *Energy* **2018**, *142*, 33–45. [CrossRef]
17. Dow Chemical Company. *Dowtherm A—Heat Transfer Fluid—Product Technical Data*; Dow Chemical Company: Midland, MI, USA, 1997; Available online: https://www.dow.com/content/dam/dcc/documents/en-us/app-tech-guide/176/176-01407-01-dowtherm-q-heat-transfer-fluid-technical-manual.pdf?iframe=true (accessed on 5 May 2019).
18. Drake, S.J.; Martin, M.; Wetz, D.A.; Ostanek, J.K.; Miller, S.P.; Heinzel, J.M.; Jain, A. Heat generation rate measurement in a Li-ion cell at large C-rates through temperature and heat flux measurements. *J. Power Sources* **2015**, *285*, 266–273. [CrossRef]
19. Khandelwal, A.; Hariharan, K.S.; Gambhire, P.; Kolake, S.M.; Yeo, T.; Doo, S. Thermally coupled moving boundary model for charge–discharge of LiFePO4/C cells. *J. Power Sources* **2015**, *279*, 180–196. [CrossRef]

20. Grandjean, T.; Barai, A.; Hosseinzadeh, E.; Guo, Y.; McGordon, A.; Marco, J. Large format lithium ion pouch cell full thermal characterisation for improved electric vehicle thermal management. *J. Power Sources* **2017**, *359*, 215–225. [CrossRef]

21. Ahmadian, A.; Sedghi, M.; Elkamel, A.; Fowler, M.; Golkar, M.A. Plug-in electric vehicle batteries degradation modeling for smart grid studies: Review, assessment and conceptual framework. *Renew. Sustain. Energy Rev.* **2018**, *81*, 2609–2624. [CrossRef]

22. Dong, T.; Peng, P.; Jiang, F. Numerical modeling and analysis of the thermal behavior of NCM lithium-ion batteries subjected to very high C-rate discharge/charge operations. *Int. J. Heat Mass Transf.* **2018**, *117*, 261–272. [CrossRef]

23. Calise, F.; Cappiello, F.L.; Cartenì, A.; d'Accadia, M.D.; Vicidomini, M. A novel paradigm for a sustainable mobility based on electric vehicles, photovoltaic panels and electric energy storage systems: Case studies for Naples and Salerno (Italy). *Renew. Sustain. Energy Rev.* **2019**, *111*, 97–114. [CrossRef]

24. Klein, S.A. *Solar Energy Laboratory, TRNSYS. A Transient System Simulation Program*; University of Wisconsin: Madison, WI, USA, 2006.

25. Palmieri, V. European Patent Application. European Patent 2 672 194, 23 March 2012.

26. Perers, B.; Bales, C. Report of IEA SHC—Task 26, Solar Combisystems. 2002. Available online: https://www.iea.org/ (accessed on 8 May 2019).

27. Institut für Thermodynamik und Wärmetechnik (ITW). Test report n. 11COL1028. 2012. Available online: https://www.igte.uni-stuttgart.de/ (accessed on 8 May 2019).

28. Calise, F.; Dentice d'Accadia, M.; Piacentino, A. A novel solar trigeneration system integrating PVT (photovoltaic/thermal collectors) and SW (seawater) desalination: Dynamic simulation and economic assessment. *Energy* **2014**, *67*, 129–148. [CrossRef]

29. Calise, F.; d'Accadia, M.D.; Vanoli, R.; Vicidomini, M. Transient analysis of solar polygeneration systems including seawater desalination: A comparison between linear Fresnel and evacuated solar collectors. *Energy* **2019**, *172*, 647–660. [CrossRef]

30. Buonomano, A.; Calise, F.; d'Accadia, M.D.; Vicidomini, M. A hybrid renewable system based on wind and solar energy coupled with an electrical storage: Dynamic simulation and economic assessment. *Energy* **2018**, *155*, 174–189. [CrossRef]

31. Tervo, E.; Agbim, K.; DeAngelis, F.; Hernandez, J.; Kim, H.K.; Odukomaiya, A. An economic analysis of residential photovoltaic systems with lithium ion battery storage in the United States. *Renew. Sustain. Energy Rev.* **2018**, *94*, 1057–1066. [CrossRef]

32. Santos, M.; André, J.; Costa, E.; Mendes, R.; Ribeiro, J. Design strategy for component and working fluid selection in a domestic micro-CHP ORC boiler. *Appl. Therm. Eng.* **2020**, *169*, 114945. [CrossRef]

33. Iranmanesh, A.; Mehrabian, M.A. Dynamic simulation of a single-effect LiBr–H2O absorption refrigeration cycle considering the effects of thermal masses. *Energy Build.* **2013**, *60*, 47–59. [CrossRef]

 *energies*

Article

Energy, Environmental, and Economic Analyses of Geothermal Polygeneration System Using Dynamic Simulations

Francesca Ceglia [1], Adriano Macaluso [2],[*], Elisa Marrasso [1], Carlo Roselli [1] and Laura Vanoli [2]

[1]   Department of Engineering, University of Sannio, 82100 Benevento, Italy;
      francesca.ceglia@unisannio.it (F.C.); marrasso@unisannio.it (E.M.); carlo.roselli@unisannio.it (C.R.)
[2]   Department of Engineering, University of Naples Parthenope, 80143 Naples, Italy;
      laura.vanoli@uniparthenope.it
[*]   Correspondence: adriano.macaluso@uniparthenope.it

Received: 18 June 2020; Accepted: 31 August 2020; Published: 4 September 2020

Abstract: This paper presents a thermodynamic, economic, and environmental analysis of a renewable polygeneration system connected to a district heating and cooling network. The system, fed by geothermal energy, provides thermal energy for heating and cooling, and domestic hot water for a residential district located in the metropolitan city of Naples (South of Italy). The produced electricity is partly used for auxiliaries of the thermal district and partly sold to the power grid. A calibration control strategy was implemented by considering manufacturer data matching the appropriate operating temperature levels in each component. The cooling and thermal demands of the connected users were calculated using suitable building dynamic simulation models. An energy network dedicated to heating and cooling loads was designed and simulated by considering the variable ground temperature throughout the year, as well as the accurate heat transfer coefficients and pressure losses of the network pipes. The results were based on a 1-year dynamic simulation and were analyzed on a daily, monthly, and yearly basis. The performance was evaluated by means of the main economic and environmental aspects. Two parametric analyses were performed by varying geothermal well depth, to consider the uncertainty in the geofluid temperature as a function of the depth, and by varying the time of operation of the district heating and cooling network. Additionally, the economic analysis was performed by considering two different scenarios with and without feed-in tariffs. Based on the assumptions made, the system is economically feasible only if feed-in tariffs are considered: the minimum Simple Pay Back period is 7.00 years, corresponding to a Discounted Pay Back period of 8.84 years, and the maximum Net Present Value is 6.11 M€, corresponding to a Profit Index of 77.9% and a maximum Internal Rate of Return of 13.0%. The system allows avoiding exploitation of 27.2 GWh of primary energy yearly, corresponding to $5.49 \cdot 10^3$ tons of CO_2 avoided emissions. The increase of the time of the operation increases the economic profitability.

Keywords: heating and cooling network; polygeneration system; geothermal energy community; ORC; geothermal energy; energy district

1. Introduction

Industrialization has promoted the use of oil, natural gas, coal, and other conventional energy sources causing the risks of stock depletion and environmental pollution [1]. Thus, the environmental emergency is a priority on the policy agenda of different countries. Indeed, in the framework of the Conference of the Parties in Paris [2], a group of countries has signed and ratified different documental acts with the common aim to solve the climate change problems [3]. Moreover, the European Commission developed the model to convert the energy and production economic panorama into a low-carbon model within 2050 [4]. The crucial topic of these legislations is the necessity to obtain a climate-friendly European economy encouraging renewable energy against long-term energy consumption.

The concerns regarding global warming and well-being targets allowed defining the elementary energy objectives to be reached [5], such as the almost exclusive use of renewable energy sources (RESs) and, in particular, the increase of local RESs exploitation. The objective of a completely renewable energy production panorama cannot disregard adoptable city strategies to enhance their sustainability to a globally competitive level [6,7].

Different studies [8,9] about energy districts fed by local energy sources demonstrate that the development of energy systems depends on the reciprocal connection of all the energy vectors such as the electric, heating, and cooling. Thus, the correct combination in energy polygeneration model permits to obtain a territorial energy planning strategy to optimize the use of the local energy resources and satisfy the energy needs of a defined territory [10,11].

Ecological policies, aimed to define energy-independent areas by use of the availability of RESs in the area, have increasingly become a branding factor of cities [12]. Avant-garde cities such as the Danish capital of Copenhagen, the Swedish city of Malmö, and the German city of Freiburg have all invested significant resources into the development of a green image via ambitious sustainability policies [13]. These cities attract thousands of foreign politicians interested in learning about ways to optimize the concept of a common green economy. Finally, a "green image" is often associated with a high degree of livability [14], which can attract new citizens [15]. Therefore, a requalification of a zone through a green imagine could be a way to develop and recover an interesting area with the promotion of social acceptability of renewable plants by encouraging citizenship in both social and economic aspects [16].

Sometimes the employment of renewable energy is limited by the uncertainty of RESs with higher impact of diffusion (such as solar or wind energy). The possible strategy can be the use of higher stable RESs such as geothermal and biomass. In addition, to reach energy independence, the use of local RESs is incentivized to obtain energy districts and communities [17]. In this context, the use of RESs to satisfy the energy requirements of an entire area of a city is crucial for achieving the energy and environmental targets associated with ecological policies. In worldwide panorama, the geothermal power plants represent the higher reliable RESs in terms of operating hours (this value is averagely the 62% of total yearly hours) [18]. The use of geothermal reservoirs could be a good solution to obtain a more flexible and stable energy system. In 2016, renewable energy was used to meet 13.7% of the worldwide electricity demand, and only 4.3% was covered by geothermal sources [19]. Concerning geothermal power plants, in 2014, the total worldwide installed capacity was 12.6 GW$_{el}$. Among the World's countries, the United States had the largest installed capacity (3.5 GW$_{el}$, 28% of the world total), followed by the Philippines (1.9 GW$_{el}$, 15%), Indonesia (1.4 GW$_{el}$, 11%), and New Zealand (1 GW$_{el}$, 8%) [20]. Although geothermal power generation accounted for only 0.3% of the total electricity production, it increased significantly in 2014, representing a substantial proportion of the total electricity generation in countries such as Kenya (32%), Iceland (30%), El Salvador (25%), and New Zealand (17%) [20].

As regards thermal direct geothermal applications, the installed geothermal power is ~107,727 MW$_{th}$ worldwide, with an annual increment of 8.73% from 2015. Geothermal energy is mainly exploited in ground-source heat pumps (58.8%), bathing and swimming (18.0%), and space heating (16.0%); only 3.50% is employed in greenhouse heating and 3.70% for all the other applications. China, Iceland, Turkey, France, and Germany are the best countries for geothermal district heating exploitation [21]. District heating systems coupled with renewable energy sources can save fossil fuels and reduce greenhouse gas emissions. Indeed, the interest in RES based-district heating and cooling systems has increased during the last years and different applications can be found worldwide. In [22], an application based on low-temperature geothermal district heating feeding the municipality of Aalborg has been analyzed. The simulation results highlighted that RES was not able to cover the total space heating request for the municipality. In Germany, during 2017 geothermal plants provided 1.3 TWh of heat on yearly based, which was generally used for district heating purposed [23]. Moreover, in [24], it was compared a district heating system based on deep geothermal energy with a conventional district heating system based on natural gas. The innovative RES system was able to increase the exergy efficiency by ~12% and to reduce heating costs by ~25%.

Furthermore, the geothermal reservoirs allow the cascade use of geothermal hot fluids [25,26] to provide different energy vectors such electricity and heat exchanging fluids to meet space heating and cooling, as well as domestic hot water (DHW) demands. From literature analysis about the future employment of geothermal uses, it has been calculated that this source could satisfy 5% of the global heating demand by 2050 [27].

Despite the encouraging data, the geothermal reservoirs are often not used in many areas such as Italy [28]. The employing of the geothermal source is threatened by the impossibility to generalize the plant configuration that strictly depends on chemical, enthalpy, deep, and quantity of geothermal site availability. In addition, to define a basic model compatible with all sites, a limit is the social acceptability of geothermal plants by citizens often caused by incomplete and inaccurate environmental information [29]. In the actual RESs, world panorama America and Asia exhibit the largest geothermoelectric installed capacity, followed by Europe. Italy is the leading country in Europe, with installed geothermal power plants around 916 MW, but all these plants are in Toscana. The Italian electricity generation from geothermal sources amounts to 5.9% of the total renewable-based electricity production, and up to 0.9% of thermal energy produced from renewables is from geothermal sources [30,31]. The usage of geothermal sites in electric applications in Italy regards the higher temperature sources (T > 150 °C) by traditional Rankine or Kalina Cycle. These plants are not suitable for the exploitation of the most widespread low/medium temperature reservoirs. To use the low/medium geothermal energy sources in power plants the literature considers the employment of Organic Rankine Cycle (ORC). This technology represents one of the most satisfying strategies to exploit renewable energies and low/medium temperature thermal cascades that permits in addition to electricity the thermal energy recovering if opportunely designed [32,33]. The ORC module consists of a Rankine cycle in which the water was replaced by an organic fluid that, despite the cons represented by high cost, toxicity, and flammability, it presents the vantages such as low critical temperature, low latent heat of evaporation, positive slope of the vapor saturation curve and high molecular weight.

The share of ORC installations in geothermal applications worldwide is 19% of the total number of ORC plants for all feeding typologies, but they cover 74.8% of the installed ORC capacity owing to their high size availability (typically under set of ten MW_{el}) [34]. In many zones, ORC installations at low/medium temperatures are available, allowing the use of the well-known ORC technology, which is particularly suitable for these applications. The renewed interest in the academic study of the ORC is due to its growing technological adoption in the energy conversion of thermal sources at low/medium temperatures ranging from 80 to 300 °C [35]. The results of ORC applications in this field are very promising [36] when coupled with different renewable sources as solar, geothermal, and biomass [27,37], but its diffusion is nowadays complexed because of the high specific cost of investment, low conversion efficiencies, high production costs, and maintenance cost linked to aggressivity of geothermal fluids [38].

Currently, only large power plants can compensate for the technological concerns in exploiting low-temperature heat sources for electricity production [39]. Globally, only 16.0% of geothermal power plants use ORC technology [39].

Typically, the ORC modules are suitable for large power range applications in particularly those with low/medium enthalpy, e.g., geothermal reservoirs, biomass combustion, industrial waste heat, waste heat from reciprocating engines [40–42] and gas turbines [43], biomass gasification [44,45], concentrated solar radiation [46–48], exploiting the recovery of residual heat from diesel engines [49], and drying [50,51]. Simulation studies on geothermal ORC plants indicated that the first-law efficiency for the ORC ranges from 7 to 15% with a geothermal source temperature of 160 °C. Sometimes to improve the system the integrations with solar collectors are considered to keep the working fluid under the desired conditions at the turbine inlet. In addition to trigeneration use, the geothermal brine sometimes could be used for the desalinization of water [52–54].

In the present study, an ORC module for geothermal applications was designed and simulated to supply the energy demand of a district in Monterusciello, a district of Pozzuoli (Naples), in the geothermal area of Phlegraean Fields, South Italy. The ORC module uses a geothermal source in cascade application to supply thermal demands for space heating and cooling and domestic hot water demands of the district in a polygeneration application; otherwise, the electric output energy is partly used to supply the energy request of thermal network and the remaining part is sold to electric national grid. The selection of organic fluids is based on a literature analysis and depends on the thermophysical, economic, and environmental properties [55,56]. An appropriate selection of the working fluid of the cycle is crucial for optimizing the efficiency of the binary plant, to maximize the conversion efficiency or to determine the best configuration for a given plant capacity. The selection of the working fluid also significantly affects the costs of the heat exchanger.

Considering the previous literature [57], the choice of the working fluid depended on the source temperature and critical temperature of the fluid. The working fluid must satisfy the general technological and environmental criteria, which have been widely discussed in the literature, for example, suitable thermodynamic fluid properties, no toxicity, no or low flammability, low cost, a low Global Warming Potential, and no Ozone Depletion Potential impact.

By using the previous analysis [58–60], the organic fluid R245fa was selected because of its environmental and thermodynamic efficiency in the case of heat sources with temperatures close to 150 °C for geothermal applications and/or other renewable sources for ORC applications. The use of R245fa is also effective for temperatures below 170 °C [61]. Other studies that optimize the ORC efficiency, consider the zeotropic mixtures such as the combination of 70% R245fa and 30% R125a [62].

The analyzed polygenerative plant application in Monterusciello is used to satisfy the thermal loads of a small local energy community through a mini-grid system. The geothermal district was inspired by energy community based on geothermal activity in Japanese real cases [63]. Its simulative model is created to evaluate the economic and social benefits of a district energy system as suggested in the literature [64], getting closer to meet economic and environmental needs. The study wants to encourage the energy district diffusion. It defines the limits and advantages of small independent energy districts fed by renewables reservoirs.

The keys for comprehensively developing such district contexts and making them economically profitable are the integration of different sources, the simultaneous production of different energy vectors (polygeneration), and the "load plant sharing" approach, which maximizes the source exploitation. In this context, a pre-feasibility study cannot be neglected. The authors propose a district system based on the exploitation of geothermal sources integrated with an auxiliary biomass backup system for the simultaneous production of electricity, heating, and cooling for domestic air conditioning (using a district heating and cooling network (DHCN) and domestic hot water network (DHWN). The plant system was simulated in TRNSYS environment [65], and all the heat exchangers and the ORC were first developed by previous analysis in the AspenEDR [66] and AspenONE [66] environments, respectively. AspenEDR allows a detailed and in-depth design; according to its simulation results, the heat-exchanger geometry is developed in the TRNSYS environment, creating user-defined components.

Because the ORC built-in model is absent in the TRNSYS library, it was first developed and simulated in AspenONE to extrapolate working maps as functions of the inlet heat source temperature and mass flow rate and after inserted in TRNSYS environment to create a user-defined component following the approach adopted in [52–54]. A control strategy is developed in this study to manage the layout and the simultaneous production of multiple energy vectors.

The dynamic model of the entire system includes a typical building configuration in the real considered residential district realized by TRNBUILD (a TRNSYS tool).

The novelty of the study is related to defining a possible geothermal energy district in a not yet used geothermal field. In particular, a combination of two flexible and stable RESs is considered (biomass and geothermal energy). A sustainable energy district in load sharing configuration is analyzed by using low-temperature geothermal source, to requalify a district of Naples by improving the local energy sources and the local energy network. Moreover, a sensitivity analysis is performed by varying the depth of the geothermal well, to take into account the uncertainty related to the depth where a suitable geothermal source temperature is available. An economic analysis, which is the most central aspect of the pre-feasibility study and based on accurate market surveys, is performed by taking into account different scenarios of the Italian market for electricity production (with and without feed-in tariffs).

2. Area of Interest

This study should be referred to the improvement of geothermal reservoirs exploitation in Italy. Considering the general Italian geothermal context, it has been highlighted from the geothermal maps (available at depths of 1000, 2000, and 3000 m) that the areas of interest concerning the temperature and geothermal flow are located in Tuscany, Aeolian Islands, and Neapolitan area [67,68]. In [69,70], it is demonstrated that the geothermal anomalies also in different Italian areas such as the Alps, Sicily, and the central Tyrrhenian and the Mediterranean Sea with a heat flux of 80, 40, and higher than 150 mW/m^2, respectively.

The area of interest in this study is the Neapolitan area (Phlegraean Fields), which (similar to numerous wells already surveyed in the 1980s) has a highly aggressive geothermal fluid and temperatures of approximately $100 \div 150$ °C at the land surface [71–73]. Interest in the geothermal area of Naples started in 1930 and grew rapidly in the mid-1980s. A total of 117 geothermal wells were investigated, with a maximum depth of 3046 m. The results were particularly encouraging for the Phlegraean Fields [74] and Ischia, where high geothermal gradients have been recorded, owing to the presence of localized high enthalpy fluids (T > 150 °C) at low depths (hundreds of meters), and both steam and water dominated [75–77].

In the Neapolitan contest, the Phlegraean Fields represent the area of greatest interest. According to the previous investigation for some reservoirs of these sites, the average geothermal gradient is 0.170 °C/m (average reference value of 0.03 °C/m) [68] and the average geothermal flow is 149 mW/m^2 (average reference value of 63.0 mW/m^2) [68]. In this area, buildings with different intended uses (such as residential, commercial, and also industrial buildings) can be found and they all can be connected by a thermal grid fed by geothermal sources.

In the past, the local geothermal applications in the Neapolitan area provided to direct use of the source in thermal employment using available reservoir sites in medium-low enthalpy. Nowadays, the technological maturity of the ORC component and the large diffusion of polygeneration systems to define energy district can be made usable these low/medium enthalpy sources. This study analyses the possibility to employ the large low/medium temperature geothermal sources available in Phlegraean Fields in polygeneration approach to satisfy the energy loads of a district. In addition, this work focuses on the replacement of pre-existent wells realized from 1930 until 1985, during which a large amount of data were collected firstly by the SAFEN Company and successively in ENEL-AGIP Joint Venture for geothermal exploration [77].

In a previous study about Monterusciello [78], it was analyzed only a system, without ORC module, meeting thermal loads using geothermal fluid at a temperature of 50 °C available at a depth of 100 m. In the current analysis, the geothermal fluid could be obtained from two wells of extraction (really investigated from the previous AGIP campaigns [77]) at the desired temperature and flow rate. In this upgrade study, an analysis of the existing geothermal wells of Phlegraean Fields was conducted as reported in Table 1. According to geological maps, the geothermal analyzed wells are collected and geolocated. For each well, the geothermal gradient is estimated by considering a linearity approximation as reported in the last column of the table. For Monterusciello, the gradient considered is 0.1 °C/m in the simulation at depth of 1500 m that guarantees a geothermal brine temperature of 150 °C. This value represents a possible real gradient in the Monterusciello by considering the near wells.

Table 1. Collection data of wells in Phlegraean Fields.

Well	Latitude	Longitude	Temperature (°C)	Depth (m)	Geothermal Gradient (°C/m)
Cigliano1D	40°50′54″	14°07′15″	342	2785	0.123
San Vito 1	40°50′54″	14°07′16″	419	3046	0.138
San Vito 3	40°51′48″	14°07′13″	385	2360	0.163
San Vito 8D	40°51′48″	14°07′17″	375	2868	0.131
Romano	40°49′46″	14°08′57″	97	35	2.771
Lopez	40°49′35″	14°07′17″	29	20	1.450
Regina	40°50′19″	14°06′25″	37.2	57	0.653
Sardo	40°51′06″	14°06′20″	29.6	60	0.493
Caruso	40°50′18″	14°05′34″	37	39	0.949
Damiani	40°50′21″	14°05′24″	48	40	1.200
Viola	40°50′51″	14°05′13″	30	60	0.500
Enel	40°50′45″	14°05′15″	29	45	0.644
Purziano	40°50′15″	14°04′51″	29	10	2.900
Giannotti	40°50′08″	14°04′53″	40	21	1.905
Babbo	40°50′02″	14°04′50″	32	11	2.909
Licola1	40°51′23″	14°03′12″	280	2653	0.106
Lubrano	40°50′23″	14°06′22″	41.8	120	0.348

3. Buildings and Heating and Cooling Network Characterization

A previous analysis for the building modeling was performed based on a field investigation to define both for the envelope characteristics and users/consumers typologies. The definition and modeling of the building are set up using a tool of TRNSYS software (TRNBUILD). Each building of the district, which is an ensemble of social housing, consists of four floors and eight apartments and is in a residential zone. In the SKETCHUP environment [79], the rendering of the building is represented in Figure 1. Each building consists of four floors with eight thermal zones of 198 m^2 and two apartments for each thermal zone; such a configuration was implemented by considering, on one hand, a plausible reproducibility of the considered context and, on the other hand, the computational burden of such a complex layout. The building model reproduces the typical council housing of 1960s. The stratigraphy respects the Italian building regulations for its specific age of construction [80]. All 90 buildings of the district led to a total of 1440 apartments and 1980 habitants, covering ~6% of the Monterusciello population. The opaque and transparent building envelope characteristics and the geometrical parameters for each building are reported in Table 2. The transmittance of windows is referred to a single glass component with aluminum frame. Space heating and cooling loads were evaluated considering an occupancy schedule, represented in Figure 2, based on four different cases (a, b, c, and d):

- case a: two working adults and two children,
- case b: two adults, a child, and an elderly person,
- case c: two elderly habitants, and
- case d: two working adults.

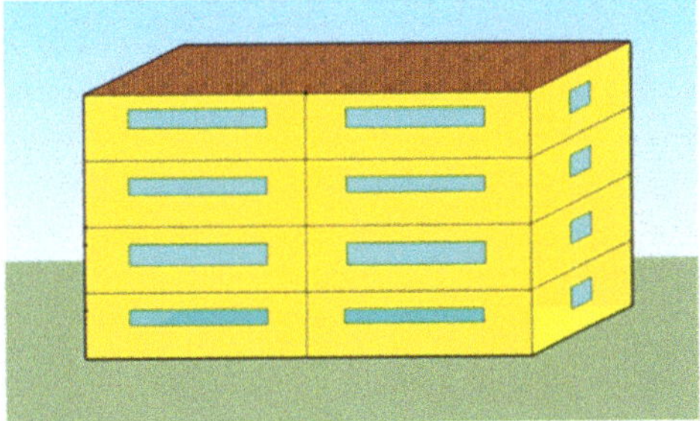

Figure 1. Single building representation in the SKETCHUP [79] environment.

Table 2. Building envelope and geometric parameters.

Structure Type	U (W/(m^2 K))	Geometrical Parameter for Each Building	
External wall	2.10	Building length (m)	39.8 m
		Building width (m)	9.95 m
Roof	1.21	Building height (m)	15.3 m
		Number of floors (-)	4
Adjacent wall	1.32	Number of the apartment per floor (-)	4
		Number of rooms for apartmen (-)	4
Common ceiling	1.06	Floor area (m^2)	396
		Floor height (m)	3.24
Window	5.68	Thermal-zone volume (m^3) (m^2)	574

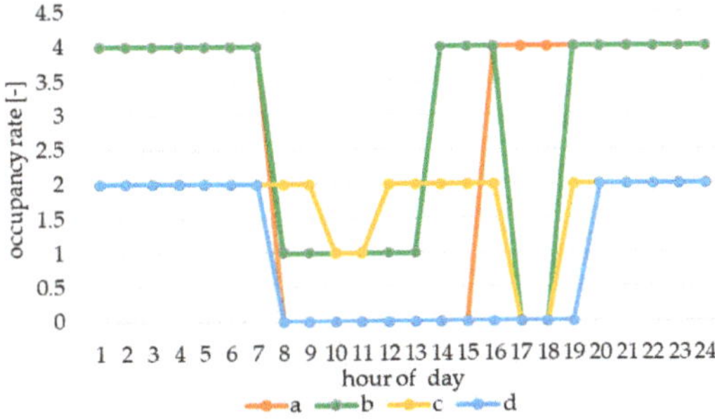

Figure 2. Schedule occupation.

Each one of these schedules refers to two apartments of each considered building.

For the domestic hot water, load is defined as a normalized hourly profile for different months of the year [81] and a daily volume consumption of 50 L per district resident [79].

According to Italian normative [82], the heating period for Naples in climate zone C goes from 15 November to 31 March. While the cooling period is from 1 June to 30 September. The maximum number of heating and cooling systems operating hours is established to 10 h per day.

All the electric energy produced by the plant is sold to the power grid excluding the part needed for auxiliaries of the thermal district network.

While during the winter period the heating load depends on the desired comfort condition (indoor temperature at 20 °C) required in the whole apartment, during the summer season the energy delivered by cooling network depends by activation of terminal cooling units that commonly serve only the rooms of the apartment effectively occupied. The yearly trends of heating and cooling load for the whole building are presented in Figure 3. The 1-year simulation was based on the weather file available in Trnsys library, which refers to [83].

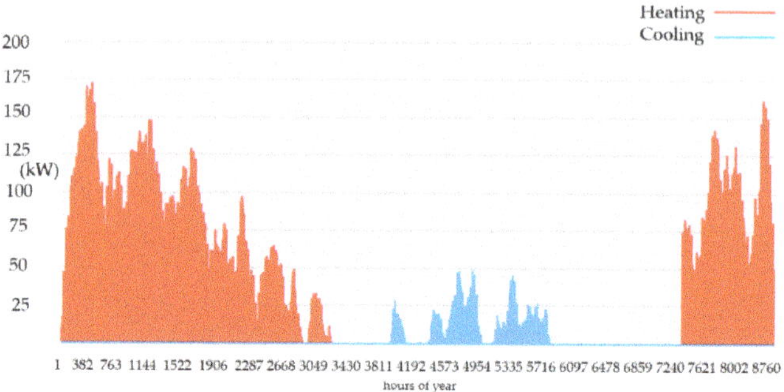

Figure 3. Yearly trend of the building load.

The buildings of the Monterusciello District area, modeled through TRNBUILD tools and based on the previous information, are connected by means of two energy networks: DHCN and DHWN.

The district area is divided by considering the constraints of the DHCN and DHWN network installation. Figure 4 reports the pipelines segments (in light blue color) running along the main roads, the spatial distribution of the buildings and a viable energy conversion system plant location (in red color). The pipelines segment are labeled as "A" or "B", referring to different blocks of the district or "MAIN". The lengths of the pipeline are consequently calculated by considering the real building location and the plausible location of the plant. The diameters sizing takes into account the thermal demand distribution and a fluid velocity of 2.0 m/s is considered. The DHWN sizing was performed similarly to the DHCN one. The geometric features of piping are listed in Table A1 in Appendix. The DHCN considers a time operation of ten hours (from 10 a.m. to 8 p.m.). A parametrical analysis was also conducted to consider the cases in which DHCN is turned off 2 and 4 h later than the base case (10 p.m. and 12 p.m.). The DHW is guaranteed for all 24 h. The set point of DHW temperature is ensured by the biomass boiler. The overall head and thermal losses of the distribution networks takes into account the ones related to the DHCN and the one related to the DHWN.

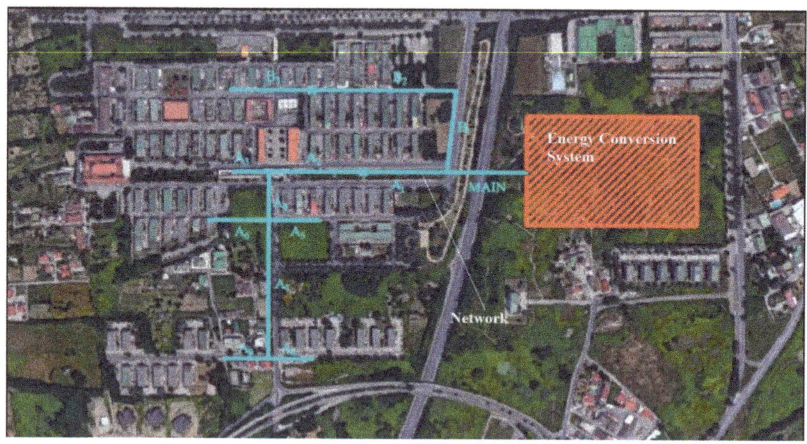

Figure 4. DHCN layout. Map data © 2019 Google.

4. System Configuration and Layout

The simplified system layout is shown in Figure 5, where the main components are presented but the DHCN and the DHWN are omitted. For simplicity of the scheme, only one production well is represented, even if such an analysis two production wells are supposed to be exploited.

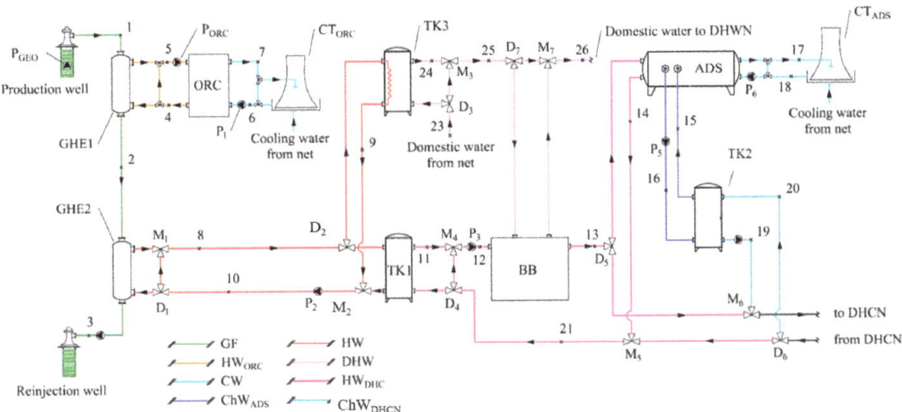

Figure 5. System layout.

First, geofluid powers an ORC module through a heat exchanger (GHE1) heating the hot water (HW$_{ORC}$); the HW$_{ORC}$ heats the ORC working fluid in the evaporator.

The ORC is calibrated to ensure a constant power production; it is condensed with cooling water (CW), which is cooled through the cooling tower (CT$_{ORC}$) and whose mass flow rate is adjusted using a variable speed pump to keep the temperature difference at the condenser constant. The ORC module is intended to operate under steady-state conditions with stable evaporation and condensation pressures. This condition at the evaporator is guaranteed by fixed temperature and flow rate of geothermal brine. While to assure the stationary condition at the condenser a variable cooling water flow rate from the water supply net is considered.

A little part of the electricity available from the ORC plant is used to cover the network auxiliaries (pumps, etc.), while the most is sold to the national grid. After delivering thermal energy at ORC evaporator the geothermal fluid is used to provide space heating, cooling, and DHW by means of DHCN and DHWN. A brief description of the system operation and the control strategy is presented herein following a previous study [78].

The Hot Water (HW, red line in the layout) feeding tanks (TK1 and TK3) are produced through the GHE2 heat exchanger. One part of HW is stored in a tank (TK1) to recover thermal energy for space heating and cooling. Through a control strategy, HW is alternatively sent to storage tank TK3, and it is used to store and heat the DHW. Then, a dedicated subsystem is used to ensure a constant temperature.

GHE2 is designed to provide a suitable HW mass flow rate basing on the temperature approach. If the temperature of the HW is higher than that of the geothermal source entering GHE2, the thermostatic valve (through D1/M1) diverts the HW bypassing GHE2, to prevent heat dissipation.

Diverter D2 diverts the HW flow to keep TK1 and TK3 thermally loaded, giving priority to TK1. Different opportune set temperatures on the top of the tanks are provided for both the heating and cooling operation modes. A thermostatic valve (through D3/M3) ensures a constant temperature (45 °C) of the DHW for the networks.

A thermostatic valve (through D4/M4) ensures a constant temperature of the HW$_{DHCN}$ for the DHCN, which is sent directly to the network during the heating operation while it is sent to the adsorption chiller (ADS) during the cooling operation. The HW temperature is set to 50.0 °C during the heating operation and 100 °C during the cooling operation for feeding the ADS. If the set point temperatures cannot be ensured, the biomass boiler (BB) fed with wood chip, is activated until the indicated temperature is reached and until the thermal storage TK1 is thermally loaded. Similarly, if the DHWN set point temperature cannot be ensured, the DHW flow is diverted to the biomass boiler trough thermostatic valve D7/M7.

During the cooling operation, the ADS is activated (by cooling tower, CT_{ADS}), and the chilled water (ChW_{ADS}) produced is sent to storage tank TK2, where the water ChW_{DHCN} is stored at 10.0 °C and is sent to the DCHN.

To fed DHCN and DHWN the GHE2 is designed to provide a suitable mass flow rate of HW in a temperature range of 40 ÷ 75 °C.

The modeling of the plant has considered an ad hoc defined flow rate for geothermal fluid and a recommended minimum rejection temperature of 70.0 °C as the literature suggests [84,85]. This value represents a good compromise for avoiding excessive depletion of the geothermal source and moderately mitigating problems involving scaling at the heat exchangers and the mechanical apparatus of the rejection process. Once the geothermal source is set, according to the mass flow rate and temperature range, the available thermal energy is calculated, and the entire system is calibrated accordingly. The ORC is calibrated to ensure constant power production. It is intended to operate under steady-state conditions of working fluid (R245fa); the evaporation and condensation pressures are stable.

5. Models

The development of the whole simulation model has been carried out in three steps, namely, the development of the ORC module in AspenPlus environment, the heat exchanger in AspenEDR environment, and the dynamic simulation model in TRNSYS environment. The TRNSYS library does not include the ORC module; the ORC was implemented in AspenPlus software and simulated by varying the inlet temperature of the source; there were obtained working maps reporting the main output parameters as a function of the inlet temperature of the thermal source, given a constant mass flow rate.

These functions have been implemented in TRNSYS environment adding a user-defined component. The entire dynamic simulation model also includes the ORC heat exchangers modules (evaporator and condenser) and geothermal heat exchangers (GHE1 and GHE2) designed, as previously defined in the introduction section, in the AspenPlus and AspenEDR environments, respectively.

All the heat exchangers of the plant have been developed and designed in AspenEDR environment.

After this, the geometry and suitable heat transfer coefficient correlations have been implemented in TRNSYS software to create a user-defined component: this approach is more rigorous with respect to using the built-in ones since it takes into account the real instant operation of the heat exchanger in terms of overall heat transfer coefficient and efficiency. In the Table A2 in Appendix A all the input parameters of the dynamic model are listed for ORC, working fluid, and heat exchangers. In particular, the heat exchanger GHE1 and GHE2 are simulated considering shell and tube heat exchanger model in titanium material. The resulted characteristics of GHE1 and GHE2 show an external diameter of tubes of 19.05 mm, thickness of tube of 1.2 mm, pitch of 23.8 mm. For the ORC module, the isentropic efficiencies are fixed: 70% and 85% for the pump and turbine, respectively. The ORC pinch point temperature differences are 7 and 5 °C respectively at the evaporator and the condenser. The working fluid used for the ORC plant is the R245fa. The evaporator and condenser for ORC module are two AISI306 shell and tube heat exchangers. The output parameters of the heat-exchanger design and the ORC module obtained as results are presented in Tables A3 and A4, respectively. In the ORC evaporator (not showed in the layout figure), the inlet and outlet temperatures of R245fa are, respectively, 51.08 and 120.1 °C at 19.35 bar, while at the condenser, they are 66.05°C for the inlet and 50 °C in the outlet section at 3.43 bar. The thermal power exchanged in the evaporator and condenser is 4698 and 3835 kW, respectively.

The nomenclature used to define each parameter in Tables A2–A4 is referred to system layout scheme represented in Figure 5.

The ORC is calibrated considering a nominal power production of 500 kW_{el}. All the heat exchangers are simulated by calculating the outlet streams' conditions using the heat and mass balances and the surface area. The inlet and the outlet temperatures of the condensation process are fixed. The mass flow rate is determined accordingly.

Once the cycle is completely defined and simulated in the AspenONE environment, the evaporator and the condenser are designed in AspenEDR.

Both the geothermal heat exchangers (GHE1 and GHE2) are designed by considering plausible values from an ORC market survey of the cold-side mass flow rate for feeding the ORC module, a plausible value of the cold-side temperature difference, and a hot inlet–cold outlet temperature approach. The correlations adopted for the overall heat-transfer coefficient to define heat exchangers parameters are based on the fully developed laminar flow inside the duct with an isothermal wall [86] and the fully developed turbulent flow inside the duct with an isothermal wall [87].

The calibration of the overall system is based on a preventive dynamic simulation of the building equipped with a plausible number of fan coils per apartment, which are also implemented in the TRNSYS environment. Then, according to the simulation results, the heat-exchanger geometry is developed in the TRNSYS environment (with the creation of user-defined components using macros and Calculator blocks) to dynamically simulate the real heat-transfer performance with regard to the instant value of the overall heat-transfer coefficient, efficiency, and number thermal Unit (NTU).

The simulation model of the building is linked to the simulation model of the fan coil in the same TRNSYS environment. The calibration of the fan-coil simulation model is based on a market survey.

The seasonal data about fan coil are listed in the Table 3, such as the nominal power of fan coil ($P_{nom,fc}$), flow rate water and air of fan coil ($m_{fc,w}$, $m_{fc,a}$), the comfort temperature set ($T_{set,amb}$), and the ratio between sensible and total thermal power ($P_{th,sens}/P_{th,TOT}$).

Table 3. Fan coils data set.

Operating Mode	$P_{el,nom,fc}$ (kW)	$M_{fc,w}$ (kg/h)	$M_{fc,a}$ (kg/h)	$T_{set,amb}$ (°C)	$P_{th,sens}/P_{th,TOT}$ (-)
Heating mode: 15 November/31 March	1.45	168	224	20	-
Cooling mode: 1 June/30 September	1.08	168	224	26	0.78

This approach allows us to determine the effective thermal and cooling loads under dynamic operating conditions and then take into account variable weather conditions, the thermal inertia of the building envelope, and the variability of the space occupation.

Once the time-dependent thermal and cooling loads are extrapolated and the maximum thermal energy available from the geothermal source is considered, it is possible to determine the maximum number of apartments served by the system.

Finally, the variable parameters are calibrated in the TRNSYS environment such as mass flow rates, set point temperatures, and characteristics of components under nominal conditions.

The system plant simulation model is linked to the DHCN one, which takes into account the heat losses occurring in the network piping. In turn, the DHCN simulation model is linked to the thermal and cooling loads, allowing the return water conditions to be determined.

6. Methodologies

The system described in the previous sections has been analyzed from an economic an environmental point of view according to the following methodology.

As regards to economic analysis, the total plant investment cost, Z_{TOT}, is given by the sum of the costs of all the modules composing the plant, the cost of the distribution networks, and the BOP (Balance of the Plant) cost Z_{BOP}, to take into account all the auxiliary systems and supporting components of the plant. Table 4 reports the parameters used in the economic analysis, as well as the main parameters of the reference scenarios for each energy vector and emission factor. The performances of the renewable ORC coupled with DHCN system are evaluated by assessing the energetic and economic savings, and by comparing the proposed system (PS) with the reference one (RS). In particular, in case of the reference scenario

- a reversible electric-driven air-to-water heat pump for space heating and cooling is supposed to be installed in each building; the coefficient of performance (COP) in heating mode is supposed to be 3.0, while in cooling mode it is supposed to be 2.5 [88];
- a natural gas boiler for DHW production is supposed to be installed in each dwelling; its efficiency is supposed to be 0.85;
- power production is given by the national grid characterized by an overall average efficiency, η_{grid}, equal to 46% [89]

The same space heating and cooling terminal units (fan coils) are adopted both in case of PS and RS.

All the electricity prices (for the cases with and without feed-in tariffs) are based on the Italian real market trends from January 2018 to August 2019 [31]. Different prices are provided for the three time-dependent partitions of the Italian market (F1, F2, and F3). The prices of heating and cooling energy are based on market surveys.

The Z_{BOP} is calculated as 3.00% of the overall plant cost. An economic analysis is performed by taking into account two different scenarios for electricity sales: with and without feed-in tariffs provided by the Italian market [31].

In the case of feed-in tariffs, the revenue related to electricity sales $R_{el,feed\text{-}in}$ is defined as follows,

$$R_{el,feed\text{-}in} = E_{el}c_{el,feed\text{-}in} \tag{1}$$

where E_{el} represents the amount of yearly electricity produced, and $c_{el,feed\text{-}in}$ represents the price of electricity under feed-in tariff market conditions. In the case where no feed-in tariffs are provided, a guarantee minimum price tariff system is assumed: regardless of the effective electricity market price, the plant always sells energy above a certain threshold price established by the GSE (Gestore dei Servizi Energetici, i.e., Italian energy services management institution).

Moreover, three selling prices are considered based on the hourly partition of the time-dependent tariff system. Given the uncertainty of the results due to the variability of prices, a plausible range of the revenue (i.e., $R_{el,min}$ and $R_{el,max}$) is calculated according to the minimum and maximum values of the sale price. The minimum and maximum sale price values are established with consideration of the real price trend from January 2018 to August 2019 (reported in Figure A1 in Appendix A).

Then, the minimum and the maximum revenues are calculated as follows,

$$R_{el,min} = E_{el}f_{F1}c_{el,F1,min} + E_{el}f_{F2}c_{el,F2,min} + E_{el}f_{F3}c_{el,F3,min} \tag{2}$$

$$R_{el,max} = E_{el}f_{F1}c_{el,F1,max} + E_{el}f_{F2}c_{el,F2,max} + E_{el}f_{F3}c_{el,F3,max} \tag{3}$$

where the subscript F_i indicates the specific partition of the time-dependent tariff system and the subscript f_{Fi} indicates the yearly fraction of the hours belonging to the specific F_i partition.

The revenue related to the selling of thermal energy for space heating R_{th} is calculated as follows,

$$R_{th} = E_{th}c_{th} \tag{4}$$

where E_{th} represents the amount of thermal energy provided to the network yearly, and c_{th} represents the price of thermal energy based on an Italian market survey.

Similarly, the revenue related to the selling of cooling energy R_{cool} is defined as

$$R_{cool} = E_{cool}c_{cool} \tag{5}$$

where E_{cool} represents the amount of cooling energy provided to the network yearly, and c_{cool} represents the price of cooling energy based on an Italian market survey.

The simple payback (SPB) is defined as follows,

$$SPB = \frac{Z_{TOT}}{CFS} \tag{6}$$

where CFS represents the yearly cash-flow statement, which is defined as

$$CFS = R_{TOT} - C_{TOT} \tag{7}$$

R_{TOT} represents the sum of all the revenues, and C_{TOT} represents the sum of the yearly maintenance cost $C_{O\&M}$ and the yearly operational cost C_{Op}, which are defined as follows,

$$C_{O\&M} = 0.05 \times Z_{TOT} \tag{8}$$

$$C_{op} = \frac{P_{AUX,TOT}}{LHV_{biom}} \times c_{biom} \tag{9}$$

where $P_{AUX,TOT}$ represents the total thermal energy provided by the auxiliary boiler, LHV_{biom} represents the lower heating value of the biomass (wood chips), and c_{biom} represents the unit cost of the biomass.

Finally, the main economic indicators are calculated to assess the economic profitability of the system, i.e., the discounted payback period (DPB), net present value (NPV), profit index (PI), and internal rate of return (IRR):

$$DPB = \frac{\ln(1 - SPBa)}{\ln(1 + a)} \tag{10}$$

$$NPV = (CFS \times AF) - Z_{TOT} \tag{11}$$

$$PI = \frac{NPV}{Z_{TOT}} \tag{12}$$

where a represents the discount rate, AF represents the annuity factor, and N represents the service life.

$$AF = \frac{1}{a} \times \left(1 - \frac{11}{(1+a)^N}\right) \tag{13}$$

Regarding the environmental analysis, the saved primary energy source (PE) is calculated by considering a specific reference scenario for each produced energy vector (electric energy and thermal energy for air conditioning).

Then, the PE is calculated as follows,

$$PE = \frac{E_{el}}{\eta_{grid}} + \frac{E_{th}}{\eta_{boil,ref}} + \frac{E_{cool}}{COP_{ref,HP}\eta_{grid}} \tag{14}$$

where $\eta_{boil,ref.}$ represents the reference value for the efficiency of the traditional natural gas boiler, and $COP_{ref,HP}$ represents the reference value for the coefficient of performance (COP) of a traditional chiller.

Finally, the avoided CO_2 emissions EM_{CO_2} are calculated as follows,

$$EC_{CO2}\left(E_{el} + \frac{E_{cool}}{COP_{ref,HP}}\right) \times EF_{grid} + \frac{E_{th}}{\eta_{boil,ref} + LHV_{nat.gas}}EF_{nat.gas} \tag{15}$$

where EF_{grid} represents the emission factor related to the national grid and $EF_{nat,gas}$ represents the emission factor of the natural gas.

The implemented economic model calculates the investment and the operating costs of both PS and RS. The cost functions adopted for PS and RS components are taken from scientific and technical literature. The cost of GHE1 and GHE2 are obtained by previous literature studies as a function of the heat exchanger area [51]. The tank cost is a function of occupied volume [90]. The cost of line networks for DHCN and DHWN is the functions of diameters and length of pipes [91]. The cost of pumping depends on flow rates which cross the pumps [92].

Table 4. Main input parameters for the economic and environmental analysis.

Parameter	Value
Reference scenario	
$\eta_{boil,ref}$ (-)	0.85
$c_{nat \cdot gas}$ (€/m^3) [93]	0.80
$LHV_{nat.gas}$ (kWh/m^3)	9.59
$COP_{ref,\,HP}$ (-)	2.5
η_{grid} (-) [89]	0.46
LHV_{biom} (MWh/t)	3.70
c_{biom} (€/t)	60.00
Price of electricity	
Feed-in tariffs [31]	
$c_{el,feed-in}$ (€/MWh)	165.00
No feed-in tariffs [31]	
$c_{el,F1,MIN}$ (€/MWh)	53.60
$c_{el,F1,MAX}$ (€/MWh)	82.82
$c_{el,F2,MIN}$ (€/MWh)	52.49
$c_{el,F2,MAX}$ (€/MWh)	79.89
$c_{el,F3,MIN}$ (€/MWh)	52.30
$c_{el,F3,MIN}$ (€/MWh)	69.84
f_{F1} (-)	0.32
f_{F2} (-)	0.23
f_{F3} (-)	0.45
Thermal unit price	
c_{th} (€/MWh)	81.17
c_{cool} (€/MWh)	90.00
Economic parameters	
a (-)	0.05
N (years)	20
AF (-)	12.46
Environmental parameters [90]	
EF_{grid} (kgCO$_2$/MWh$_{el}$)	445.50
$EF_{nat.gas}$ (t$_{CO_2}$/TJ)	55.80

The study realizes a parametric analysis with different depths of the geothermal well. The depth of the geothermal well affects both the operation of the system plant and economic performance. The aim of the analysis is the prefeasibility study by considering the uncertainty related to the depth where the geothermal source is available and assessing a range for each main performance indicator where the system can be considered feasible and profitable.

A second parametric analysis is performed with different operation times of the DHCN, without changing the operation time of the DHWN; this ensured a constant temperature of 45 °C.

When the DHCN operation time is reduced, a larger amount of electric energy is available, because all the auxiliaries of the DHCN subsystem are off. Consequently, higher revenue related to electricity production is expected.

Additionally, the DHWN subsystem is forced to use the auxiliary biomass boiler to ensure the appropriate temperature of the DHWN; thus, a higher operational cost related to boiler fuel (wood chip) is expected.

The objective is to evaluate the effects of the production strategy on profitability and to determine the optimum operation point.

7. Results

7.1. Thermodynamic Analysis

First, the daily results are presented and discussed through two days representative of the system functioning for each operation mode: one for heating (winter day) and one for cooling (summer day).

Figures 6 and 7 show the main system outputs, respectively, for the winter day where the nomenclature and numeration refer to Figure 5.

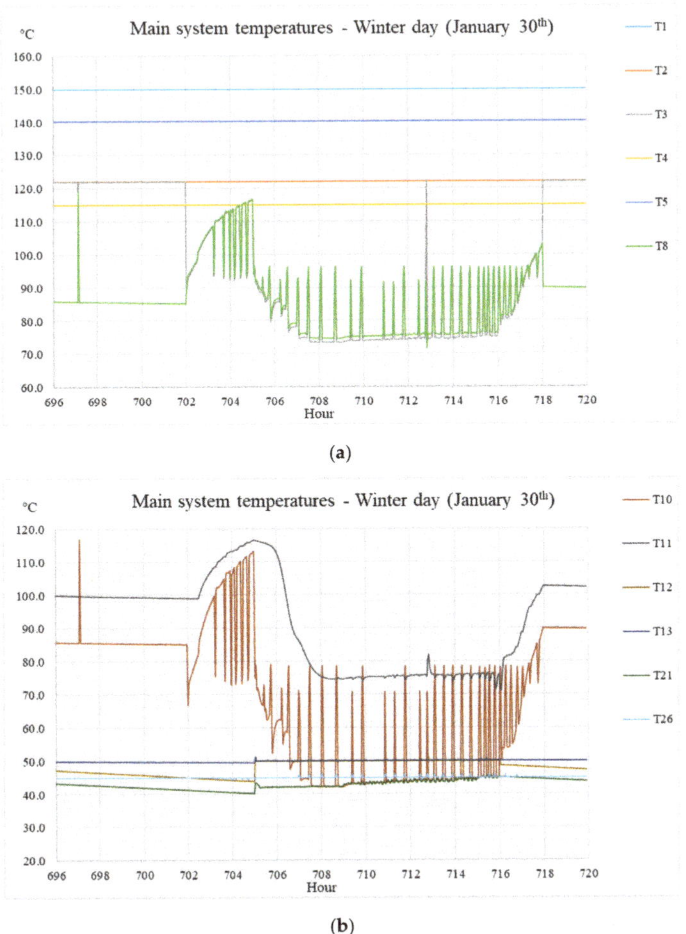

(a)

(b)

Figure 6. Main system temperatures: winter day (30 January). from T_1 to T_8 (**a**); from T_{10} to T_{26} (**b**).

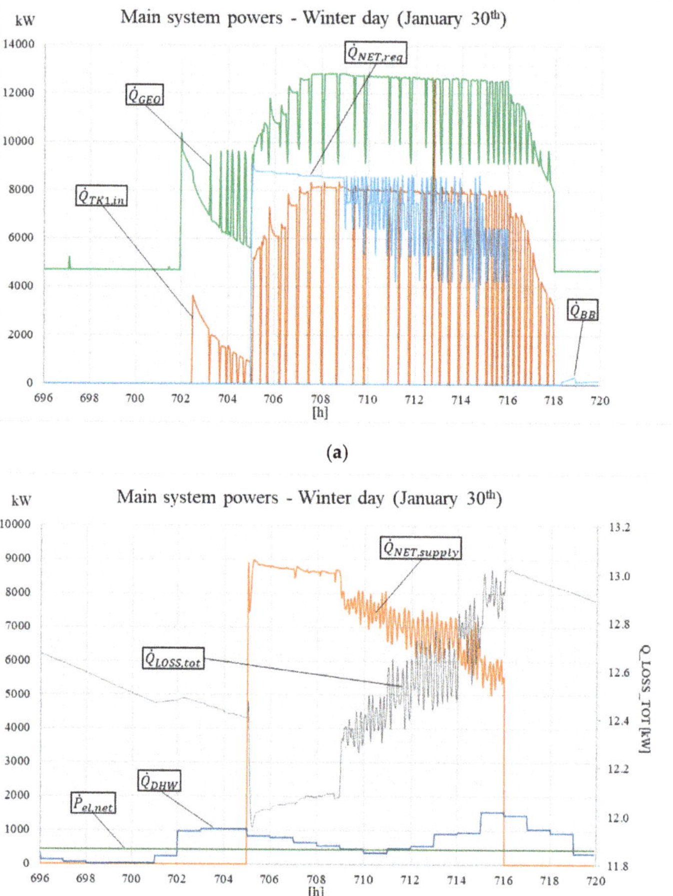

Figure 7. Main system powers: winter day (30 January). $\dot{Q}_{geo}$, $\dot{Q}_{NET.req}$, $\dot{Q}_{TK1,in}$ and $\dot{Q}_{BB}$ (**a**); $\dot{Q}_{DHW}$, $\dot{Q}_{NET.supply}$, $\dot{Q}_{loss,tot}$ and $\dot{P}_{el,net}$ (**b**).

Figure 6 presents the temperatures of the geothermal system for the winter day. T_1 and T_2 are not reported on the figure. T_1 is constant at 150 °C, according to the input parameters, and T_2 is constant at 122 °C due to the steady working conditions of the ORC module. T_3 varies depending on the time operation of the DHCN and the operation of TK1.

When the DHCN is off, T_3 coincides with T_2, whereas it continuously varies depending on the inlet condition of the cool side at GHE2, i.e., T_{10}.

T_3 never decreases to 70.0 °C, due to the calibration of the overall system and the control strategy.

T_{10} and T_8 exhibit the trends of TK1 to and from GHE2, respectively; T_{11} and T_{21} represent the trends of the water temperatures to and from the DHCN, respectively. The system can ensure a constant temperature of 50.0 °C to the DHCN and the set point temperature of the fan coils installed in the apartments taking into account the small heat losses occurring in the piping.

Moreover, the system is capable to give the DHW at a constant temperature value of 45.0 °C, as shown in Figure 6, independently by the load, shown in Figure 7.

From 6:00 to 10:00, the geothermal source is mainly employed to thermally load TK3, which belongs to the DHW subsystem. From 10:00 to 20:00, the source is mainly employed to thermally load TK1, which belongs to the DHCN.

Simultaneously, the geothermal source is constantly employed to feed the ORC module.

The foregoing is clearly shown in Figure 7, which presents the main system powers.

When the DHCN is off, the amount of achievable thermal power by the geothermal fluid is equivalent approximately to 4.50 MW, whereas it increases to 13.0 MW when the load required by the network is maximized.

The ORC net power production depends on the operation time of the DHCN. When the DHCN is off, the ORC power (approximately 430 kW_{el}) is employed for geothermal fluid pumping and the ORC cooling tower auxiliaries, and the remaining part is sent to the grid.

When the DHCN is on, ORC power is employed to feed all the system auxiliaries (pumps and the overall control and monitoring system). Then, the net power sent to the grid decreases to 380 kW_{el}.

Figures 8 and 9 show the main system temperatures for the characteristic representative summer day, and Figure 10 presents the main system powers.

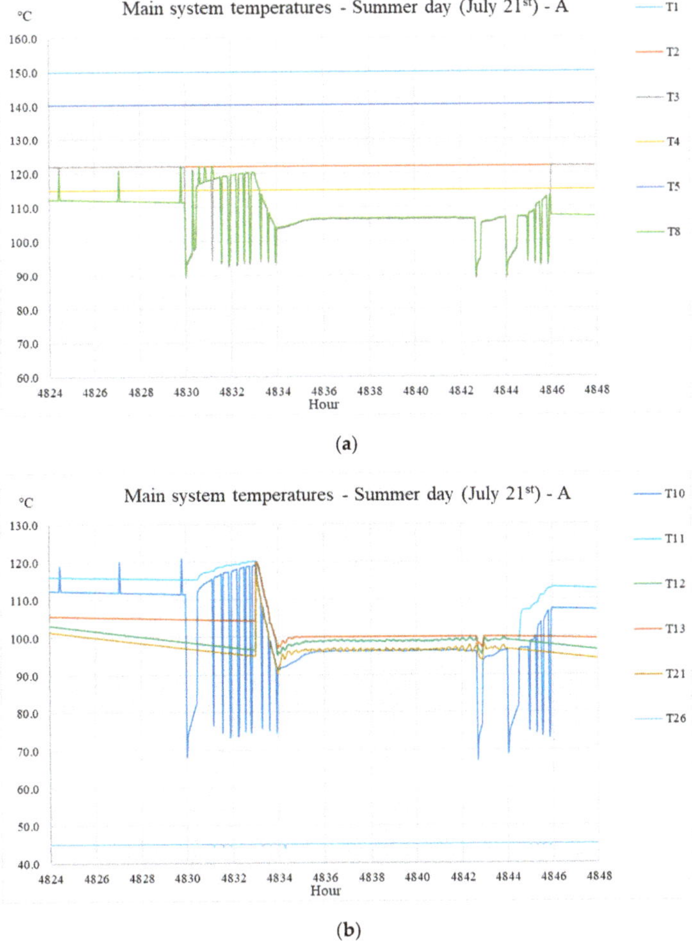

Figure 8. Main system temperatures: summer day (21 July—A). from T_1 to T_8 (**a**); from T_{10} to T_{26} (**b**).

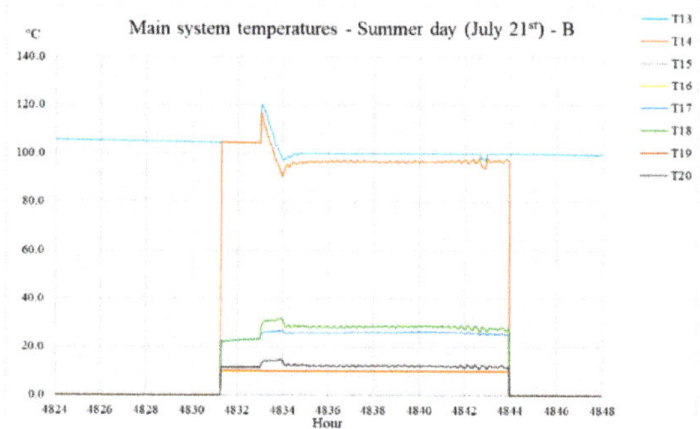

Figure 9. Main system temperatures: summer day (21 July—B).

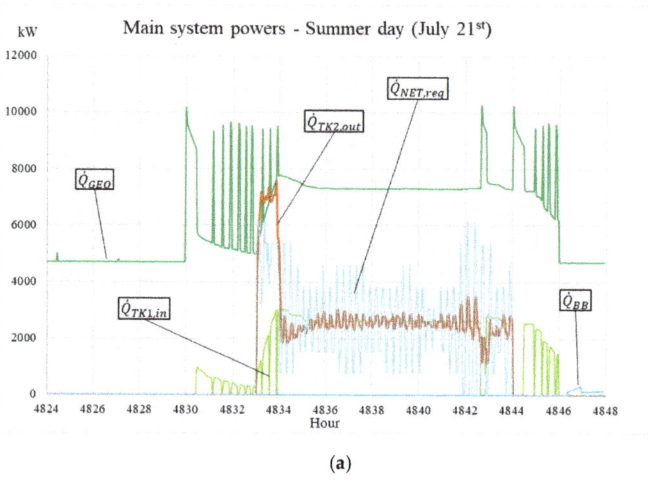

(**a**)

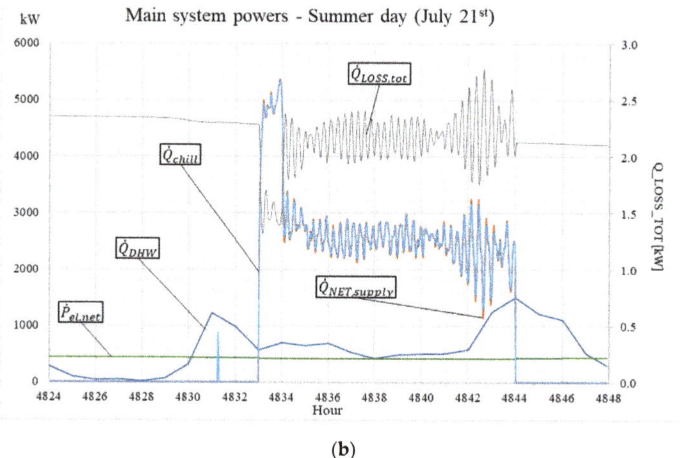

(**b**)

Figure 10. Main system powers: summer day (21 July). $\dot{Q}_{geo}$, $\dot{Q}_{NET.req}$, $\dot{Q}_{TK1,in}$, $\dot{Q}_{TK2,out}$ and $\dot{Q}_{BB}$ (**a**); $\dot{Q}_{DHW}$, $\dot{Q}_{NET.supply}$, $\dot{Q}_{chill}$, $\dot{Q}_{loss,tot}$ and $\dot{P}_{el,net}$ (**b**).

In particular, Figure 9 indicates the temperature trend of TK3 (T_{15}, T_{16}, T_{19}, T_{20}) and the ADS module (T_{13}, T_{14}, T_{15}, T_{16}, T_{17}, T_{18}). As shown, the feed temperature T_{13} of the ADS module is constant at 100 °C, except at the initial time, where no cooling load is observed (Figure 10) and the feed water is available at 120 °C. Moreover, the results indicate that the system is perfectly capable of ensuring a constant temperature T_{19} (10.0 °C) of the cooled water to be sent to the network, which corresponds (taking into account the losses in the network piping) to the set temperature of the fan coil in the cooling mode.

Table 5 presents the main results of the thermodynamic analysis on a monthly and yearly basis.

Table 5. Main results of the thermodynamic analysis (on a monthly and yearly basis).

Thermal/Electrical Energy	Jan	Feb	Mar	Apr	May	June	July	Aug	Sept	Oct	Nov	Dec	Year
E_{geo} (MWh)	6438	5850	5853	3877	3992	3932	4419	4297	3932	3965	4827	6339	57,720
$E_{TK1,in}$ (MWh)	2508	2278	1879	22.29	16.99	96.02	499.6	363.6	113.5	14.14	1007	2397	11,195
$E_{TK1,out}$ (MWh)	2486	2271	1869	7.406	2.671	81.46	483.9	347.4	98.12	2.531	997.2	2386	11,029
$E_{TK2,in}$ (MWh)	0.000	0.000	0.000	4.158	0.984	51.63	403.7	270.8	62.55	0.000	0.000	0.000	793.9
$E_{TK2,out}$ (MWh)	0.000	0.000	0.000	3.744	0.152	50.46	403.3	269.2	61.37	0.000	0.000	0.000	787.2
$E_{heat,ADS}$ (MWh)	0.000	0.000	0.000	6.054	1.77	81.62	601	408.8	97.45	0.000	0.000	0.000	1197
$E_{cool,ADS}$ (MWh)	0.000	0.000	0.000	4.188	1.037	51.71	403.7	270.9	62.61	0.000	0.000	0.000	794.2
$E_{req,WINT,DHCN}$ (MWh)	2518	2294	1860	0.000	0.000	0.000	0.000	0.000	0.000	0.000	997.7	2402	10072
$E_{req,SUMM,DHCN}$ (MWh)	0.000	0.000	0.000	0.000	0.000	48.8	399.8	265.5	57.58	0.000	0.000	0.000	771.6
$E_{SUPPLIED,DHCN}$ (MWh)	2527	2304	1870	0.000	0.000	50.19	402.3	268.90	61.19	0.000	1003	2412	10,900
$E_{TK3,in}$ (MWh)	440.9	417.8	482.2	458.3	466.1	440.8	413.8	426.9	422.7	440.7	432.1	449.8	5292
$E_{TK3,out}$ (MWh)	434.1	408.5	473.5	457.2	464.8	439.5	412.4	425.5	421.9	439.5	427.5	440.8	5245
$E_{req,DHW}$ (MWh)	468.1	428.7	476.4	457.9	465.3	440.4	445.3	438.4	422.5	440	433.5	458.2	5375
$E_{SUPPLIED,DHW}$ (MWh)	469.4	429.9	477.8	459.2	466.5	441.4	446.1	439.1	423.1	440.7	434.3	459.3	5387.0
$E_{AUX,DHCN}$ (MWh)	45.78	40.48	4.294	0.08738	0.000	1.5	117.7	64.99	2.091	0.000	7.869	30.03	314.8
$E_{AUX,DHW}$ (MWh)	34.22	20.31	2.854	0.75677	0.5793	0.9461	32.69	12.73	0.6141	0.6358	5.992	17.60	129.9
$E_{LOSS,DHCN}$ (MWh)	7.368	7.296	8.894	0.5708	0.3691	1.386	2.476	3.402	3.615	1.212	2.889	6.170	45.65
$E_{LOSS,DHW}$ (MWh)	1.252	1.248	1.419	1.31	1.197	0.961	0.7956	0.6573	0.5994	0.6896	0.8207	1.053	12
$E_{el,ORC}$ (MWh)	372.0	336.0	372.0	360.0	372.0	360.0	372.0	372.0	360.0	372.0	360.0	372.0	4380
$E_{el,net}$ (MWh)	327.2	295.7	328.6	311.3	321.7	311.1	321.1	321.1	311.4	329.8	318.3	328.0	3825
COP_{AVG} (-)	0.000	0.000	0.000	0.6917	0.586	0.6335	0.6717	0.6627	0.6425	0.000	0.000	0.000	0.6636
η_{ORC} (%)	10.64	10.64	10.64	10.64	10.64	10.64	10.64	10.64	10.64	10.64	10.64	10.64	10.64

As shown, the system is perfectly capable of ensuring the temperature levels in each subsystem and it is perfectly capable of covering the total network load required (sum of thermal total power required by DHCN and DHWN) and the losses occurring in the network.

During the heating season, the overall thermal energy provided by the system for district heating, $E_{SUPPLIED,DHCN}$, is higher than the thermal energy required, $E_{req,WINT,DHCN}$; the difference is caused by the losses occurring at the network piping.

Moreover, the sum given by the thermal energy exiting the TK1, $E_{TK1,out}$, and the one coming from the auxiliary boiler for the DHCN, $E_{AUX,DHCN}$, is higher than $E_{SUPPLIED,DHCN}$: this is due to the thermal losses occurring at the plant piping.

Thermal losses occurring at the thermal storages are given by the difference between the thermal energy entering and exiting the tanks.

What has just been discussed can be referred to the domestic hot water production system.

During the cooling season, the overall cooling energy provided by the system for district cooling, $E_{SUPPLIED,DHCN}$ is higher than the cooling energy required $E_{req,SUMM,DHCN}$; even in this case the difference is caused by the losses occurring at the network piping.

Similarly to the heating season, during the cooling season the cooling energy exiting the TK2 $E_{TK2,out}$ is higher than the supplied one $E_{SUPPLIED,DHCN}$ because of the losses occurring at the system plant.

The gross amount of electricity produced, $P_{el,ORC}$, remains constant throughout the year, and $P_{el,net}$, which represents the amount of electricity produced and sold to the grid, is lower during the cooling operation than during the heating operation. This is because the entire ADS subsystem, which is off during the heating operation, must be powered.

The constant average first-law efficiency of the ORC is 10.6%, owing to the steady working condition of the module.

The ADS, whose COP is 0.700 under nominal conditions (temperature of the feed hot water equal to 100 °C, the temperature of chilled water equal to 7 °C, and temperature of the cooling water equal to 22 °C), operates during the year with an average COP of 0.660: this is due to fluctuations in operating conditions (in terms of the feed water temperature, cooling water temperature, load), which move the module from the optimum point of operation.

7.2. Environmental and Economic Data Analysis

In this section, the main results of the economic and environmental analyses are presented and briefly discussed. The costs of the components are presented in Table 6, along with the cost function adopted. The total investment cost takes into account three different values of the depth of the geothermal well: the base case at 1500 m and the case of higher depth at 2000 and 2500 m by considering two production and one reinjection wells. Globally, the total investment cost (Z_{TOT}) ranges between 7.84 and 8.21 M€.

Table 6. Component plant costs.

Component	Cost (k€)
Z_{well}^{1500m} [94]	943
Z_{well}^{2000m} [94]	1122
Z_{well}^{2500m} [94]	1301
Z_{GHE1} [52]	53.97
Z_{GHE2} [52]	36.61
Z_{Pgeo} [52]	3.301
Z_{ORC} [52]	1500
Z_{ADS} [52]	882.0
Z_{BB} [78]	1600
Z_{CT} [78]	153.2
$Z_{Piping-DHCN}$ [78]	26.75
$Z_{Piping-DHCN}$ [78]	1784
$Z_{Substations}$ [94]	463.8
Z_{TOT}^{1500m}	7838
Z_{TOT}^{2000m}	8023
Z_{TOT}^{2500m}	8207

Table 7 presents the detailed yearly economic analysis results for each well depth and the different scenarios of the electricity market (with and without feed-in tariffs).

In the case of no feed-in tariffs, the total revenue R_{TOT} ranges from 1.09 to 1.18 M€. With a total operational and maintenance cost C_{TOT} of approximately 0.40 M€, the CFS ranges between 0.69 and 0.78 M€ per year.

All the economic indicators are negatively affected by the low CFS, and the system appears to be unprofitable.

Table 7. Main results of the economic analysis (on a yearly basis).

	R_{el} (M€)		R_{th} (M€)	R_{cool} (M€)	R_{TOT} (M€)		$C_{O\&M}$ (M€)	C_{Op} (M€)	C_{TOT} (M€)	CFS (M€)	
No Feed-in Tariffs											
deep	min	max			min	max				min	max
1500	0.2018	0.2919	0.8175	0.0694	1.089	1.179	0.3919	0.0072	0.3991	0.6897	0.7797
2000	0.2018	0.2919	0.8175	0.0694	1.089	1.179	0.3919	0.0072	0.3991	0.6897	0.7797
2500	0.2018	0.2919	0.8175	0.0694	1.089	1.179	0.3919	0.0072	0.3991	0.6897	0.7797
Feed-in Tariffs											
1500	0.6312		0.8175	0.0694	1.518		0.3919	0.0072	0.3991	1.119	
2000	0.6312		0.8175	0.0694	1.518		0.3919	0.0072	0.3991	1.119	
2500	0.6312		0.8175	0.0694	1.518		0.3919	0.0072	0.3991	1.119	

	SPB (y)		DPB (y)		NPV (M€)		PI (%)		IRR (%)	
No Feed-in Tariffs										
H	min	max	min	max	min	max	min	max	min	max
1500	10.05	11.37	14.31	17.22	0.757	1.879	9.653	23.97	6.071	7.640
2000	10.29	11.63	14.81	17.86	0.5723	1.695	7.134	21.12	5.787	7.336
2500	10.53	11.90	15.31	18.52	0.3880	1.510	4.728	18.40	5.512	7.041
Feed-in Tariffs										
1500	7.004		8.836		6.107		77.92		12.99	
2000	7.169		9.098		5.923		73.83		12.61	
2500	7.334		9.363		5.739		69.93		12.24	

The SPB period exhibits a minimum value of 10.0 years and a maximum value of 11.9 years. These are far higher than the acceptable range of 5.00 to 7.00 years for private investments.

With a discount rate of 5.00%, the DPB period increases (minimum value of 14.3 years and maximum value of 18.5 years).

The NPV exhibits low values ranging between 0.388 and 1.88 M€, corresponding to a PI ranging between 4.73% and 24.0%. Acceptable values of the PI are between 60.0% and 70.0%.

Finally, the IRR exhibits low values between 5.51% and 7.64%

The economic profitability is improved in the case of the feed-in tariffs for electricity selling.

In fact, given the constant revenues related to the thermal and cooling energy for air conditioning, R_{th}, and R_{cool}, as well as the constant operational and maintenance costs C_{TOT}, in correspondence of CFS value of 1.12 M€ the higher revenue related to power production R_{el} are equal to 0.631 M€.

Consequently, acceptable values of the SPB and DPB are obtained.

The SPB period ranges between 7.00 and 7.33 years, and the DPB period ranges between 8.84 and 9.36 years.

The NPV is increased to values between 5.74 and 6.11 M€, corresponding to PIs of 69.9% and 77.9%, respectively.

Finally, the IRR ranges between 12.2% and 13.0%.

Globally, as shown in Table 8, such a renewable system allows avoiding the employment of 27.2 GWh of primary energy and (depending on the reference scenario adopted for the energy and material outputs) avoiding 5.49×10^3 tons of CO_2 emissions.

Table 8. Saved primary energy and CO_2 emissions.

Saved Primary Energy	Avoided CO_2 Emissions
GWh	t of CO_2
27.16	5.490×10^3

The main results of the thermodynamic and economic analyses performed by varying the operation time of the DHCN are presented from Tables A5–A9 in Appendix A. In detail, the data about electricity production (R_{el}), the CFS, the SPB, the NPV, and the PIs in the case of feed-in tariffs (green) and the cases of the minimum and maximum selling prices of electricity are reported.

Longer operation time of the DHCN subsystem corresponds to higher economic profitability: all the economic indices are affected by lower DHCN operation time.

In fact, in the case of the longest time of operation, a slightly lower amount of electricity is sold to the grid because the auxiliaries are on, and the revenue related to electricity is lower as well.

On the other hand, a higher amount of thermal energy is available from the geothermal source, then avoiding the use of biomass and reducing the operational cost of the biomass boiler to supply the DHW.

In case of the shortest time of operation, a higher amount of electricity is available (auxiliaries are off) for the selling. At the same time, any load from DHW network is cover by the biomass boiler, and a higher operational cost related to boiler fuel is obtained.

The higher operational cost of the biomass boiler prevails on the higher revenue related to electricity when the time of operation is reduced and the global profitability decreases.

The revenues related to space heating and cooling are constant.

It is worth noticing that, despite the analysis was carried out with the greatest possible precision, the results strictly depend on the specific case study analyzed and on the considered market context; moreover, they are affected by unavoidable uncertainty given by

- weather conditions, which affect the electric and thermal loads, and the ORC operation (including the cooling tower);
- real availability of geothermal source, which can be assessed only after a specific exploration campaign;
- fluctuations of the electricity prices and variation of the feed-in tariffs system; and variations of prices related to district heating and cooling (even though they are characterized by more stability if compared to electricity one).

8. Conclusions

A thermodynamic, economic, and environmental analysis of a renewable polygeneration system connected to a DHCN was performed.

The system is designed for a suburban area of the metropolitan city of Naples (South of Italy) and it is powered by geothermal sources and biomass, producing electricity, thermal energy for space heating and cooling, and DHW production.

The entire dynamic simulation model (which includes an organic Rankine cycle module, an ADS, an auxiliary biomass boiler, geothermal heat exchangers, thermal storage tanks, the DHCN, residential systems of space conditioning including terminals, and a suitable model of the building envelope) was developed and implemented in AspenONE and TRNSYS environments.

The layout and the control strategy were implemented to match the appropriate operating temperature levels for each component and to prevent the temperature of the geothermal fluid reinjected into the wells from decreasing below 70.0 °C.

The overall component calibration and the economic analysis were based on manufacturer data and market surveys.

The economical returns are estimated by considering the electric power sold to the national grid. Differently, the thermal energy for heating and cooling is employed to satisfy the thermal yearly load of the whole district.

An analysis was performed with different depths of geothermal wells where the source is available considering a geothermal temperature gradient of 0.1 °C/m.

An economic analysis was performed for two different electricity purchase scenarios: with and without feed-in tariffs.

The system, whose investment cost ranges between 7.84 and 8.02 M€, is economically feasible only under feed-in tariff conditions.

In fact, without feed-in tariffs, economic indicators suggest that the system is unprofitable: the minimum SPB period is 10.1 years (corresponding to a DPB period of 14.3 years if a discount rate of 5.00% is applied), and the maximum NPV is 1.88 M€ (corresponding to a maximum PI of 24.0%).

Finally, the maximum IRR is 7.64%.

Conversely, if feed-in tariffs are considered, the economic indicators suggest that for the specific case study analyzed, the system is attractive.

The minimum SPB period is 7.00 years, corresponding to a DPB period of 8.84 years.

The maximum NPV is increased to 6.11 M€, corresponding to a PI of 77.9%. The maximum IRR is 13.0%.

The system allows avoiding exploitation of 27.2 GWh of primary energy yearly, corresponding to 5.49×10^3 tons of CO_2 emissions avoided yearly.

The keys to making the small-medium scale systems powered with geothermal sources feasible and attractive are the polygeneration and the load-sharing, which aim at maximizing the source exploitation by diversifying the energy and material vectors produced and by obtaining loads and requests that are distributed over time to the greatest extent possible.

In future studies, different typologies of final users and different energy vectors could be considered using the sources availability 24 h/24 h to feed the plant.

Author Contributions: Conceptualization, F.C., A.M., E.M., C.R. and L.V.; methodology, F.C., A.M., E.M., C.R. and L.V.; software, F.C., A.M. and E.M.; formal analysis, F.C., A.M. and E.M.; investigation, F.C., A.M. and E.M.; resources, C.R. and L.V.; data curation, F.C., A.M. and E.M.; writing—original draft preparation, F.C., A.M. and E.M.; writing—review and editing, F.C., A.M. and E.M.; visualization, F.C., A.M. and E.M.; supervision, C.R. and L.V.; project administration, C.R. and L.V.; funding acquisition, C.R. and L.V.; All authors have read and agreed to the published version of the manuscript.

Funding: This research was funded by the GeoGrid project POR Campania FESR 2014/2020 CUP B43D18000230007.

Acknowledgments: The authors gratefully acknowledge the financial support of provided through the GeoGrid project POR Campania FESR 2014/2020 CUP B43D18000230007. Nicola Massarotti gratefully acknowledges the local program of the University of Napoli "Parthenope" for the support of individual research.

Conflicts of Interest: The authors declare no conflict of interest.

Nomenclature

A	Area (m^2)
a	discount rate (-)
AF	annuity factor
BOP	Balance of the Plant
C	Cost (€, k€, M€)
c	unit price (€/specific unit of measurement)
CFS	cash-flow statement (€, k€, M€)
COP	coefficient of performance (-)
d	Diameter (mm)
DPB	discounted payback (y)
E	Energy (kWh, MWh, kJ, MJ)
EM_{CO_2}	avoided CO_2 emissions (-,%)
EF	emission factor (kg_{CO_2}/MWh_{el}, t_{CO_2}/TJ)
H	geothermal well depth (m)
L	Length (m)
LHV	lower heating value (kJ/kg, kWh/m^3, MWh/t)
$\dot{m}$	mass flow rate (kg/s)
N	service life (y)
NPV	net present value (M€)

IRR	internal rate of return (-)
$\dot{P}$	Power (kW)
p	pressure (bar)
PE	primary energy (MWh)
PI	profit index (-)
E	thermal energy (kWh, MWh, kJ, MJ)
R	Revenue (€, k€, M€)
SPB	simple payback (y)
T	Temperature (°C)
t	ton
U	Transmittance (W/(m^{2}K))
V	Volume (m^3)
Z	investment cost (€, k€, M€)

Superscripts and Acronyms

a	air
amb	ambient
ADS	adsorption chiller
AUX	auxiliary
AVG	average
biom	biomass
boil	boiler
BB	biomass boiler
BOP	Balance Of Plant
chil	electric chillers
CT	cooling tower
ChW	chilled water
CW	cooling water
cool	cooling
CO_2	carbon dioxide
D	diverter
DHCN	district heating and cooling network
DHW	domestic hot water
DHWN	domestic hot water network
el	electric
F	timeslot of electricity market
f_i	yearly fraction of -ith timeslot of electricity market
fc	fan coil
geo	geothermic/geothermal
GF	geothermal fluid
GHE	geothermal heat exchanger
grid	power plant grid
h	hour
heat	heating
hot	hot
HP	traditional heat pump
HW	hot water
in	inlet
is	isentropic
LMTD	logarithm mean temperature difference (K)
LOSS	losses
max	maximum
min	minimum
M	mixer
nat, gas	natural gas

net	net
nom	nominal
NTU	number thermal unit
O & M	operation and maintenance
Op	operative
out	outlet
ORC	organic Rankine cycle
P	pump
piping	piping
PS	proposed system
ref	reference scenario
req	required
RES	renewable energy sources
RS	reference system
s	shell
sens	sensible
set	setpoint
SUPPLY	supplied
SUMM	summer
th	thermal
TOT	total
TK	thermal storage tank
w	water
wf	working fluid
WINT	winter
Greek Letters	
ε	heat-exchanger efficiency (-)
η	isentropic efficiency/first-law efficiency (-)

Appendix A

Appendix A.1. Input Parameters

Table A1. Lengths and diameters of the main line and of the pipes of the subdistrict.

Network Piping		
Segment	**Length (m)**	**Diameter (mm)**
MAIN	90	300
A1	80	250
A2	80	250
A3	70	175
A4	80	250
A5	50	200
A6	80	200
A7	70	200
A8	50	175
A9	70	175
B1	100	250
B2	100	250
B3	60	200

Table A2. Main thermodynamic and design input parameters.

Parameter	Value
m_{geo} (kg/s)	40.00
m_{HW_ORC} (kg/s)	44.15
m_{HW} (kg/s)	268.3
m_{ChW_ADS} (kg/s)	268.3
m_{HW_DHCN} (kg/s)	268.3
p_1 (bar)	8.000
T_1 (°C)	150.0
T_2 (°C)	122.0
T_3 (°C)	>70.00
T_4 (°C) *	115
T_5 (°C) *	140
T_6 (°C) *	20.0
T_7 (°C) *	45.00
T_{10} (°C) *	40.0
T_8 (°C) *	78.0
$\dot{P}_{el,ORC}$ (kW) **	500.0
$\dot{P}_{el,2}$ (kW) **	13.4
$\dot{P}_{el,3}$ (kW) **	13.40
$\dot{P}_{el,4}$ (kW) **	13.4
$\dot{P}_{el,5}$ (kW) **	13.4
COP_{ADS}(-) *	0.70
$P_{th,BB}$ (kW) **	2000.0
$P_{nom.ADS}$ (kW) **	7000.00
V_{TK1} (m³)	160.0
V_{TK2} (m³)	160.0
V_{TK3} (m³)	60.0

* Under nominal conditions. ** Rated power.

Table A3. Geothermal heat exchanger output parameters.

Parameter	Value
GHE1	
A_{GHE1} (m²)	175.9
Tube number (plain) (-)	248.0
L_{GHE1} (m)	6.0
Number of tube passes (-)	1
Number of tube passes (-)	1
$d_{s,GHE1}$ (mm)	457.2
Number of segmental baffles (-)	10
$\dot{P}_{th,GHE1}$ (kW) *	4698

Table A3. *Cont.*

Parameter	Value
ε_{GHE1} (-) *	0.7851
T_{LMTD} (°C) *	8.53
F_{GHE1} (-)	1.02
GHE2	
A_{GHE2} (m^2)	106.6
Tube number (plain) (-)	296.0
L_{GHE2} (m)	6.096
Number of tube passes (-)	1.000
Number of tube passes (-)	1.000
$d_{s,GHE2}$ (mm)	498.5
Number of segmental baffles (-)	10.00
$\dot{P}_{th,GHE2}$ (kW) *	9659
ε_{GHE2} (-) *	0.6365
T_{LMTD} (°C) *	36.69

* Under nominal conditions.

Table A4. Main ORC output parameters.

Parameter	Value
Evaporation pressure (bar)	19.35
Condensation pressure (bar)	3.432
Evaporation temperature (°C)	120.1
Condensation temperature (°C)	50.00
$\dot{m}_{R245fa}$ (kg/s)	20.04
Evaporator heat exchanger area (m^2)	207.7
Condenser heat exchanger area (m^2)	214.2

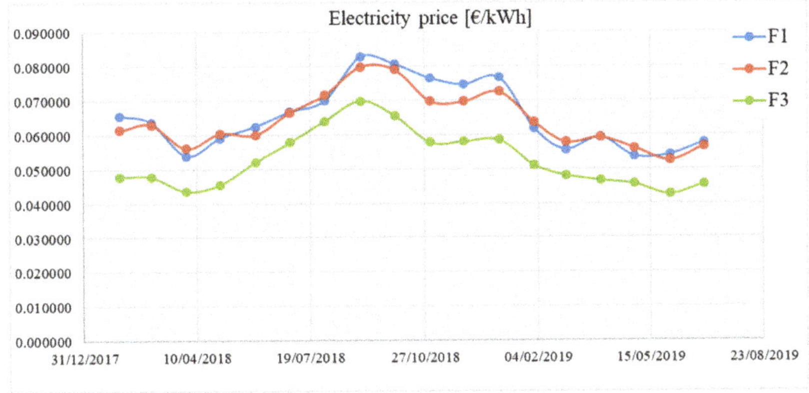

Figure A1. Electricity price from January 2018 to August 2019 for every partition (F1, F2, and F3) of the time-dependent tariff.

Appendix A.2. Cost Functions

In the following are the cost functions used in the economic analysis; it is also indicated the reference the function was taken from.

The cost of the well z_{well}, which is divided into three different segments, is a function of the diameter and depth of each segment. In this work, two production well and one reinjection well were supposed.

$$z_{well} = 2 \left(37.974 \times e^{(0.0079 \times D_{well,1})} \times H_{well,1} + 37.6 \times e^{(0.0039 \times D_{well,2})} \times H_{well,2} \right.$$
$$\left. + 37.6 \times e^{(0.0039 \times D_{well,3})} \times H_{well,3} \right.$$
$$D_{well,1} = 0.600 \text{ m}; \; H_{well,1} = 10.0 \text{ m}$$
$$D_{well,2} = 0.600 \text{ m}; \; H_{well,2} = 10.0 \text{ m} \tag{A1}$$
$$D_{well,3} = 0.600 \text{ m}; \; H_{well,3} = 10.0 \text{ m}$$

$$z_{geopump} = 107.26 \times \dot{P}_{pumping}^{0.7176} \tag{A2}$$

$$z_{ORC} = 3000 \times \dot{P}_{ORC} \tag{A3}$$

$$z_{HE} = 150 \times \left(\frac{A_{HE}}{0.093} \right)^{0.78} \tag{A4}$$

$$z_{ADS} = 126 \times \dot{Q}_{ADS} \tag{A5}$$

$$z_{biomass,boil} = 8000 \times \dot{Q}_{biomass,boil[kW]} \tag{A6}$$

$$z_{TK} = 167.19 \times V_{TK} \tag{A7}$$

$$z_{pump} = 389 \times \ln(\dot{m} * 3.6) - 289 \tag{A8}$$

$$Z_{CT} = 8815 \times \ln(\dot{P}_{cool,CT} \times 3600) - 174 \tag{A9}$$

$$z_{substations} = 1850/\text{unit} \tag{A10}$$

$$z_{piping} = \left(349.63 + 2472.37 \times 2472.37 \times D_{pipe} \right) \times L_{pipe} \tag{A11}$$

Appendix A.3. Main Results of Thermodynamic Analysis

Table A5. Main results of the thermodynamic analysis (stop time of the DHCN at 20:00).

Thermal/Electrical Energy	Jan	Feb	Mar	Apr	May	June	July	Aug	Sept	Oct	Nov	Dec	Year
E_{geo} (MWh)	6309	5726	5739	3776	3884	3833	4307	4195	3839	3863	4710	6214	56,394
$E_{TK1,in}$ (MWh)	2486	2261	1877	24	16.55	95.5	490.9	362.6	117.2	13.4	990	2392	11,125
$E_{TK1,out}$ (MWh)	2475	2257	1866	7.753	2.644	81.49	481.1	347.1	98.24	0.000	995	2376	10,987
$E_{TK2,in}$ (MWh)	0.000	0.000	0.000	4.159	0.9846	51.63	403.6	270.9	62.57	0.000	0.000	0.000	793.8
$E_{TK2,out}$ (MWh)	0.000	0.000	0.000	3.746	0.1533	50.45	402.2	269.2	61.37	0.000	0.000	0.000	787.2
$E_{heat,ADS}$ (MWh)	0.000	0.000	0.000	6.04	1.797	81.6	600.3	408.4	97.41	0.000	0.000	0.000	1196
$E_{cool,ADS}$ (MWh)	0.000	0.000	0.000	4.195	1.053	51.73	403.7	271	62.65	0.000	0.000	0.000	794.3
$E_{req,WINT,DHCN}$ (MWh)	2518	2294	1860	0.000	0.000	0.000	0.000	0.000	0.000	0.000	997.7	2402	10,072
$E_{req,SUMM,DHCN}$ (MWh)	0.000	0.000	0.000	0.000	0.000	48.8	399.8	265.5	57.58	0.000	0.000	0.000	771.6
$E_{SUPPLIED,DHCN}$ (MWh)	2527	2304	1871	0.000	0.000	50.36	402.2	269.2	61.29	0.000	1003	2411	10,900
$E_{TK3,in}$ (MWh)	331.8	309.8	373.7	358.7	361.1	346.5	313.2	330.1	329.2	343.1	336.1	334.2	4068
$E_{TK3,out}$ (MWh)	327.3	303.6	365.9	357.8	360.2	345.4	312.2	329.1	328.4	342.3	332.1	326.4	4031
$E_{req,DHW}$ (MWh)	468.1	428.7	476.4	457.9	465.3	440.4	445.3	438.4	422.5	440	433.5	458.2	5375
$E_{SUPPLIED,DHW}$ (MWh)	468.4	429	476.5	457.9	465.4	440.4	445.1	438.2	422.4	440.1	433.5	458.5	5376
$E_{AUX,DHCN}$ (MWh)	56.76	49.17	8.054	0.033	0.000	1.923	120.27	65.5	2.905	0	8.823	40.42	353.9
$E_{AUX,DHW}$ (MWh)	141.1	125.37	110.55	100.15	105.21	94.96	132.95	109.08	94.04	97.79	101.43	132.11	1344.7
$E_{LOSS,DHCN}$ (MWh)	7.368	7.296	8.894	0.5708	0.3691	1.386	2.476	3.402	3.615	1.212	2.889	6.17	45.65
$E_{LOSS,DHW}$ (MWh)	1.252	1.249	1.42	1.31	1.197	0.9611	0.7958	0.6574	0.5994	0.6895	0.8208	1.053	12.01
$E_{el,ORC}$ (MWh)	372	336	372	360	372	360	372	372	360	372	360	372	4380
$E_{el,net}$ (MWh)	327.4	295.9	328.8	312.3	322.8	312.1	322.1	322.1	312.3	330	318.5	328.2	3832
COP_{AVG} (-)	0.000	0.000	0.000	0.6944	0.5861	0.634	0.6724	0.6634	0.6432	0.000	0.000	0.000	0.6643
η_{ORC} (%)	10.64	10.64	10.64	10.64	10.64	10.64	10.64	10.64	10.64	10.64	10.64	10.64	10.64

Table A6. Main results of the thermodynamic analysis (stop time of the DHCN at 22:00).

Thermal/Electrical Energy	Jan	Feb	Mar	Apr	May	June	July	Aug	Sept	Oct	Nov	Dec	Year
E_{geo} (MWh)	6309	5726	5739	3776	3884	3833	4307	4195	3839	3863	4710	6214	56,394
$E_{TK1,in}$ (MWh)	2486	2261	1877	24.00	16.55	95.50	490.9	362.6	117.2	13.40	990.0	2392	11,125
$E_{TK1,out}$ (MWh)	2475	2257	1866	7.753	2.644	81.49	481.1	347.1	98.24	0.000	995	2376	10,987
$E_{TK2,in}$ (MWh)	0.000	0.000	0.000	4.159	0.9846	51.63	403.6	270.9	62.57	0.000	0.000	0.000	793.8
$E_{TK2,out}$ (MWh)	0.000	0.000	0.000	3.746	0.1533	50.45	402.2	269.2	61.37	0.000	0.000	0.000	787.2
$E_{heat,ADS}$ (MWh)	0.000	0.000	0.000	6.04	1.797	81.6	600.3	408.4	97.41	0.000	0.000	0.000	1196
$E_{cool,ADS}$ (MWh)	0.000	0.000	0.000	4.195	1.053	51.73	403.7	271	62.65	0.000	0.000	0.000	794.3
$E_{req,WINT,DHCN}$ (MWh)	2518	2294	1860	0.000	0.000	0.000	0.000	0.000	0.000	0.000	997.7	2402	10,072
$E_{req,SUMM,DHCN}$ (MWh)	0.000	0.000	0.000	0.000	0.000	48.8	399.8	265.5	57.58	0.000	0.000	0.000	771.6
$E_{SUPPLIED,DHCN}$ (MWh)	2527	2304	1871	0.000	0.000	50.36	402.2	269.2	61.29	0.000	1003	2411	10,900
$E_{TK3,in}$ (MWh)	331.8	309.8	373.7	358.7	361.1	346.5	313.2	330.1	329.2	343.1	336.1	334.2	4068
$E_{TK3,out}$ (MWh)	327.3	303.6	365.9	357.8	360.2	345.4	312.2	329.1	328.4	342.3	332.1	326.4	4031
$E_{req,DHW}$ (MWh)	468.1	428.7	476.4	457.9	465.3	440.4	445.3	438.4	422.5	440	433.5	458.2	5375
$E_{SUPPLIED,DHW}$ (MWh)	469.4	429.9	477.8	459.2	466.5	441.4	446.1	439.1	423.1	440.7	434.3	459.3	5387
$E_{AUX,DHCN}$ (MWh)	50.49	44.19	6.964	0.000	0.000	1.612	118.73	65.72	2.466	0.000	8.032	34.43	332.6
$E_{AUX,DHW}$ (MWh)	54.66	41.74	25.09	21.77	24.85	18.69	50.94	30.63	19.50	22.66	25.02	42.59	378.1
$E_{LOSS,DHCN}$ (MWh)	7.469	7.404	8.934	0.5708	0.3691	1.386	2.481	3.405	3.615	1.212	2.907	6.250	46.00
$E_{LOSS,DHW}$ (MWh)	1.252	1.249	1.420	1.310	1.197	0.9611	0.7957	0.6574	0.5994	0.6896	0.8208	1.053	12.01
$E_{el,ORC}$ (MWh)	372.0	336.0	372.0	360.0	372.0	360.0	372.0	372.0	360.0	372.0	360.0	372.0	4380
$E_{el,net}$ (MWh)	327.2	295.7	328.6	311.3	321.7	311.1	321.1	321.1	311.4	329.8	318.3	328.0	3825
COP_{AVG} (-)	0.000	0.000	0.000	0.6917	0.586	0.6335	0.6717	0.6627	0.6425	0.000	0.000	0.000	0.6636
η_{ORC} (%)	10.64	10.64	10.64	10.64	10.64	10.64	10.64	10.64	10.64	10.64	10.64	10.64	10.64

Appendix A.4. Main Results of Economic Analysis

Table A7. Main results of the economic analysis (on a yearly basis; stop time of the DHCN at 20:00).

	R_{el} (M€)		R_{th} (M€)	R_{cool} (M€)	R_{TOT} (M€)		$C_{O\&M}$ (M€)	C_{Op} (M€)	C_{TOT} (M€)	CFS (M€)	
				No Feed-in Tariffs							
deep	min	max			min	max				min	max
1500	0.2022	0.2924	0.8175	0.0694	1.0892	1.1794	0.3919	0.0275	0.4195	0.6697	0.7600
2000	0.2022	0.2924	0.8175	0.0694	1.0892	1.1794	0.3919	0.0275	0.4195	0.6697	0.7600
2500	0.2022	0.2924	0.8175	0.0694	1.0892	1.1794	0.3919	0.0275	0.4195	0.6697	0.7600
				Feed-in Tariffs							
1500	0.6324		0.8175	0.0694	1.519		0.3919	0.0275	0.4195	1.100	
2000	0.6324		0.8175	0.0694	1.519		0.3919	0.0275	0.4195	1.100	
2500	0.6324		0.8175	0.0694	1.519		0.3919	0.0275	0.4195	1.100	

	SPB (y)		DPB (y)		NPV (M€)		PI (%)		IRR (%)	
					No Feed-in Tariffs					
deep	min	max	min	max	min	max	min	max	min	max
1500	10.31	11.70	14.86	18.04	0.508	1.632	6.480	20.83	5.711	7.303
2000	10.56	11.98	15.38	18.73	0.3236	1.448	4.034	18.05	5.431	7.004
2500	10.80	12.25	15.91	19.44	0.1393	1.264	1.697	15.40	5.652	6.713
					Feed-in Tariffs					
1500	7.127		9.030		5.869		74.87		12.71	
2000	7.294		9.298		5.684		70.85		12.33	
2500	7.462		9.570		5.500		67.02		11.97	

Table A8. Main results of the economic analysis (on a yearly basis; stop time of the DHCN at 22:00).

	R_{el} (M€)		R_{th} (M€)	R_{cool} (M€)	R_{TOT} (M€)		$C_{O\&M}$ (M€)	C_{Op} (M€)	C_{TOT} (M€)	CFS (M€)	
					No feed-in tariffs						
deep	min	max			min	max				min	max
1500	0.2020	0.2922	0.8175	0.0694	1.0890	1.1791	0.3919	0.0115	0.4034	0.6856	0.7757
2000	0.2020	0.2922	0.8175	0.0694	1.0890	1.1791	0.3919	0.0115	0.4034	0.6856	0.7757
2500	0.2020	0.2922	0.8175	0.0694	1.0890	1.1791	0.3919	0.0115	0.4034	0.6856	0.7757
					Feed-in tariffs						
1500	0.6318		0.8175	0.0694	1.519	0.3919		0.0115	0.4034	1.115	
2000	0.6318		0.8175	0.0694	1.519	0.3919		0.0115	0.4034	1.115	
2500	0.6318		0.8175	0.0694	1.519	0.3919		0.0115	0.4034	1.115	
	SPB (y)		DPB (y)		NPV (M€)		PI (%)			IRR (%)	
					No feed-in tariffs						
deep	min	max	min	max	min	max	min	max		min	max
1500	10.11	11.43	14.42	17.38	0.705	1.829	8.997	23.33		5.998	7.572
2000	10.34	11.70	14.92	18.03	0.5209	1.644	6.493	20.50		5.714	7.269
2500	10.58	11.97	15.43	18.71	0.3366	1.460	4.101	17.79		5.441	6.795
					Feed-in tariffs						
1500	7.028		8.873		6.061		77.32			12.93	
2000	7.193		9.136		5.877		73.25			12.55	
2500	7.358		9.402		5.692		69.36			12.18	

Table A9. Main results of the environmental analysis (on a yearly basis).

Case	Saved PE (GWh)	Avoided CO_2 Emissions (t)
stop time of the DHCN at 22:00	27.17	5494
stop time of the DHCN at 20:00	27.17	5495

References

1. Dincer, I.; Acar, C. A review on clean energy solutions for better sustainability. *Int. J. Energy Res.* **2015**, *39*, 585–606. [CrossRef]
2. Rhodes, C.J. The 2015 Paris Climate Change Conference: Cop21. *Sci. Prog.* **2016**, *99*, 97–104. [CrossRef] [PubMed]
3. UNEP. Sustainable Innovation Forum 2015. 2014. Available online: http://www.cop21paris.org/ (accessed on 1 October 2019).
4. European Commission. Low-Carbon Economy 2050. 2018. Available online: https://ec.europa.eu/clima/policies/strategies/2050_en (accessed on 1 October 2019).
5. Mendes, G.; Ioakimidis, C.; Ferrao, P. On the planning and analysis of Integrated Community Energy Systems: A review and survey of available tools. *Renew. Sustain. Energy Rev.* **2011**, *15*, 4836–4854. [CrossRef]
6. Lucarelli, A.; Berg, P.O. City branding: A state-of-the-art review of the research domain. *J. Place Manag. Dev.* **2011**, *4*, 9–27. [CrossRef]
7. Andersson, I. Placing place branding: An analysis of an emerging research field in human geography. *Geogr. Tidsskr. J. Geogr.* **2014**, *114*, 143–155. [CrossRef]
8. Lund, H.; Vad Mathiesen, B.; Connolly, D.; Østergaard, P.A. Renewable energy systems—A smart energy systems approach to the choice and modelling of 100% renewable solutions. *Chem. Eng. Trans.* **2014**, *39*, 1–6.
9. Lund, H.; Duić, N.; Østergaard, P.A.; Mathiesen, B.V. Future district heating systems and technologies: On the role of smart energy systems and 4th generation district heating. *Energy* **2018**, *165*, 614–619. [CrossRef]
10. Serra, L.M.; Lozano, M.A.; Ramos, J.; Enzinas, A.V.; Nebra, S.A. Polygeneration and efficient use of natural sources. *Energy* **2009**, *34*, 575–586. [CrossRef]
11. Jana, K.; Ray, A.; Majoumerd, M.M.; Assadi, M.; De, S. Polygeneration as a future sustainable energy solution—A comprehensive review. *Appl. Energy* **2017**, *202*, 88–111. [CrossRef]
12. Busch, H.; Anderberg, S. Green Attraction—Transnational Municipal Climate Networks and Green City Branding. *J. Manag. Sustain.* **2015**, *5*, 1. [CrossRef]

13. Andersson, I. *Geographies of Place Branding-Researching Through Small and Medium Sized Cities*; Department of Human Geography, Stockholm University: Stockholm, Sweden, 2015. Available online: https://www.diva-portal.org/ (accessed on 17 May 2020).

14. Gustavsson, E.; Elander, I. Cocky and climate smart? Climate change mitigation and place-branding in three Swedish towns. *Local Environ.* **2012**, *17*, 769–782. [CrossRef]

15. Kristjansdottir, R.; Busch, H. Towards a Neutral North—The Urban Low Carbon Transitions of Akureyri, Iceland. *Sustainability* **2019**, *11*, 2014. [CrossRef]

16. Giubilato, E.; Bleicher, A. Neither risky technology nor renewable electricity: Contested frames in the development of geothermal energy in Germany. *Energy Res. Soc. Sci.* **2019**, *47*, 46–55. [CrossRef]

17. Ceglia, F.; Esposito, P.; Marrasso, E.; Sasso, M. From smart energy community to smart energy municipalities: Literature review, agendas and pathways. *J. Clean. Prod.* **2020**, *254*, 120118. [CrossRef]

18. Bertani, R. Geothermal Power Generation in the World 2010–2014 Update Report. In Proceedings of the World Geothermal Congress 2015, Melbourne, Australia, 19–25 April 2015. Available online: https://www.geothermal-energy.org/ (accessed on 3 September 2020).

19. International Energy Agency. Renewables. Available online: https://www.iea.org/topics/renewables/ (accessed on 10 September 2019).

20. Geothermal Capacity. Retrieved from: BP—Statistical Review of World Energy—June 2018. Available online: https://www.bp.com/content/dam/bp/businesssites/en/global/corporate/pdfs/energy-economics/statistical-review/bp-stats-review-2018-full-report.pdf (accessed on 10 September 2019).

21. Lund, J.W.; Toth, A.N. Direct Utilization of Geothermal Energy 2020 Worldwide Review. In Proceedings of the World Geothermal Congress 2020, Reykjavik, Iceland, 26 April–2 May 2020. Available online: https://www.geothermal-energy.org/ (accessed on 3 September 2020).

22. Østergaard, P.A.; Mathiesen, B.V.; Möller, B.; Lund, H. A renewable energy scenario for Aalborg Municipality based on low-temperature geothermal heat, wind power and biomass. *Energy* **2010**, *35*, 4892–4901. [CrossRef]

23. Weinand, J.; Kleinebrahm, M.; McKenna, R.; Mainzer, K.; Fichtner, W. Developing a combinatorial optimisation approach to design district heating networks based on deep geothermal energy. *Appl. Energy* **2019**, *251*, 113367. [CrossRef]

24. Sun, F.; Zhao, X.; Chen, X.; Fu, L.; Liu, L. New configurations of district heating system based on natural gas and deep geothermal energy for higher energy efficiency in northern China. *Appl. Therm. Eng.* **2019**, *151*, 439–450. [CrossRef]

25. Rubio-Maya, C.; Díaz, V.M.A.; Martínez, E.P.; Belman-Flores, J.M. Cascade utilization of low and meiudm enthalpy geothermal resources—A review. *Renew. Sustain. Energy Rev.* **2015**, *52*, 689–716. [CrossRef]

26. Moya, D.; Aldás, C.; Kaparaju, P. Geothermal energy: Power plant technology and direct heat applications. *Renew. Sustain. Energy Rev.* **2018**, *94*, 889–901. [CrossRef]

27. Intergovernmental Panel on Climate Change. *Renewable Energy Sources and Climate Change Mitigation: Special Report of the Intergovernmental Panel on Climate Change*; Cambridge University Press: Cambridge, UK, 2012. Available online: https://www.ipcc.ch (accessed on 17 May 2020).

28. Farinelli, U. Renewable energy policies in Italy. *Energy Sustain. Dev.* **2004**, *8*, 58–66. [CrossRef]

29. Manzella, A.; Bonciani, R.; Allansdottir, A.; Botteghi, S.; Donato, A.; Giamberini, S.; Lenzi, A.; Paci, M.; Pellizzone, A.; Scrocca, D. Environmental and social aspects of geothermal energy in Italy. *Geothermics* **2018**, *72*, 232–248. [CrossRef]

30. TERNA. Manager of the Italian High and Medium Voltage Electricity Grid. Available online: https://www.terna.it (accessed on 10 September 2019).

31. GSE—Gestore Servizi Energetici. Manager of Italian Energy Services. Available online: https://www.gse.it/ (accessed on 10 September 2019).

32. Quoilin, S.; van den Broek, M.; Declaye, S.; Dewallef, P.; Lemort, V. Techno-economic survey of Organic Rankine cycle (ORC) systemas. *Renew. Sustain. Energy Rev.* **2013**, *22*, 168–186. [CrossRef]

33. Larjola, J.; Uusitalo, A.; Turunen-Saaresti, T. Background and summary of commercial ORC development and exploitation. In Proceedings of the First International Seminar on ORC Power Systems, Delft, The Netherlands, 22–23 September 2011.

34. Walraven, D.; Laenen, B.; D'Haeseleer, W. Comparison of thermodynamic cycles for power production from low-temperature geothermal heat sources. *Energy Convers. Manag.* **2013**, *66*, 220–233. [CrossRef]

35. Ahmed, A.; Esmaeil, K.K.; Irfan, M.A.; Al-Mufadi, F.A. Design methodology of organic Rankine cycle for waste heat recovery in cement plants. *Appl. Therm. Eng.* **2018**, *129*, 421–430. [CrossRef]

36. Bertrand, G.L.; Lambrinos, G.; Frangoudakis, A.; Papadakis, G. Low-grade heat conversion into power using organic Rankine cycles—A review of various applications. *Renew. Sustain. Energy Rev.* **2011**, *15*, 3963–3979.

37. Lenzen, M. Current State of Development of Electricity-Generating Technologies: A Literature Review. *Energies* **2010**, *3*, 462–591. [CrossRef]

38. Tartière, T.; Astolfi, M. A World Overview of the Organic Rankine Cycle Market. *Energy Procedia* **2017**, *129*, 2–9. [CrossRef]

39. Astolfi, M.; Romano, M.C.; Bombarda, P.; Macchi, E. Binary ORC (organic Rankine cycles) power plants for the exploitation of mediumelow temperature geothermal sources: Part A: Thermodynamic optimization. *Energy Res. Soc. Sci.* **2014**, *66*, 423–434.

40. Turboden. Available online: https://www.turboden.com/ (accessed on 13 August 2020).

41. Exergy. Available online: http://www.exergy-orc.com/ (accessed on 13 August 2020).

42. Rank. Available online: https://www.rank-orc.com/ (accessed on 13 August 2020).

43. Knowledge Center Organic Rankine Cycle. Available online: http://www.kcorc.org (accessed on 10 September 2019).

44. Hajabdollahi, Z.; Hajabdollahi, F.; Tehrani, M.; Hajabdollahi, H. Thermo-economic environmental optimization of Organic Rankine Cycle for diesel waste heat recovery. *Energy* **2013**, *63*, 142–151. [CrossRef]

45. Vera, D.; Jurado, F.; Carpio, J.; Kamel, S. Biomass gasification coupled to an EFGT-ORC combined system to maximize the electrical energy generation: A case applied to the olive oil industry. *Energy* **2018**, *144*, 41–53. [CrossRef]

46. Kim, K.H.; Han, C.H. A Review on Solar Collector and Solar Organic Rankine Cycle (ORC) Systems. *J. Autom. Control. Eng.* **2015**, *3*, 66–73. [CrossRef]

47. Pikra, G.; Salim, A.; Prawara, B.; Purwanto, A.; Admono, T.; Eddy, Z. Development of Small Scale Concentrated Solar Power Plant Using Organic Rankine Cycle for Isolated Region in Indonesia. *Energy Procedia* **2013**, *32*, 122–128. [CrossRef]

48. Sami, S. Analysis of Nanofluids Behavior in Concentrated Solar Power Collectors with Organic Rankine Cycle. *Appl. Syst. Innov.* **2019**, *2*, 22. [CrossRef]

49. Boyaghchi, F.A.; Chavoshi, M.; Sabeti, V. Multi-generation system incorporated with PEM electrolyzer and dual ORC based on biomass gasification waste heat recovery: Exergetic, economic and environmental impact optimizations. *Energy* **2018**, *145*, 38–51. [CrossRef]

50. Calise, F.; Di Fraia, S.; Macaluso, A.; Massarotti, N.; Vanoli, L. A geothermal energy system for wastewater sludge drying and electricity production in a small island. *Energy* **2018**, *163*, 130–143. [CrossRef]

51. Di Fraia, S.; Macaluso, A.; Massarotti, N.; Vanoli, L. Energy, exergy and economic analysis of a novel geothermal energy system for wastewater and sludge treatment. *Energy Convers. Manag.* **2019**, *195*, 533–547. [CrossRef]

52. Calise, F.; D'Accadia, M.D.; Macaluso, A.; Vanoli, L.; Piacentino, A. A novel solar-geothermal trigeneration system integrating water desalination: Design, dynamic simulation and economic assessment. *Energy* **2016**, *115*, 1533–1547. [CrossRef]

53. Calise, F.; D'Accadia, M.D.; Macaluso, A.; Piacentino, A.; Vanoli, L. Exergetic and exergoeconomic analysis of a novel hybrid solar-geothermal polygeneration system producing energy and water. *Energy Convers. Manag.* **2016**, *115*, 200–220. [CrossRef]

54. Calise, F.; Macaluso, A.; Piacentino, A.; Vanoli, L. A novel hybrid polygeneration system supplying energy and desalinated water by renewable sources in Pantelleria Island. *Energy* **2017**, *137*, 1086–1106. [CrossRef]

55. Györke, G.; Deiters, U.K.; Groniewsky, A.; Lassu, I.; Imre, A. Novel classification of pure working fluids for Organic Rankine Cycle. *Energy* **2018**, *145*, 288–300. [CrossRef]

56. Invernizzi, C.; Bombarda, P. Thermodynamic performance of selected HCFS for geothermal applications. *Energy* **1997**, *22*, 887–895. [CrossRef]

57. Liu, X.; Zhang, Y.; Shen, J. System performance optimization of ORC-based geo-plant with R245fa under different geothermal water inlet temperatures. *Geothermics* **2017**, *66*, 134–142. [CrossRef]

58. Muhammad, U.; Imran, M.; Lee, N.H.; Park, B.S. Design and experimental investigation of a 1kW organic Rankine cycle system using R245fa as working fluid for low-grade waste heat recovery from steam. *Energy Convers. Manag.* **2015**, *103*, 1089–1100. [CrossRef]

59. Ghaebi, H.; Farhang, B.; Parikhani, T.; Rostamzadeh, H. Energy, exergy and exergoeconomic analysis of a cogeneration system for power and hydrogen production purpose based on TRR method and using low grade geothermal source. *Geothermics* **2018**, *71*, 132–145. [CrossRef]

60. Calise, F.; Capuozzo, C.; Vanoli, L. Design and parametric optimization of an organic rankine cycle powered by solar energy. *Am. J. Eng. Appl. Sci.* **2013**, *6*, 178–204. [CrossRef]

61. Wang, J.; Zhao, L.; Wang, X. A comparative study of pure and zeotropic mixtures in low-temperature solar Rankine cycle. *Appl. Energy* **2010**, *87*, 3366–3373. [CrossRef]

62. Suwa, A.; Sando, A. Geothermal energy and community sustainability. In Proceedings of the Grand Renewable Energy 2018 Pacific Conference, Yokohama, Japan, 17–22 June 2018.

63. Cook, D.; Fazeli, R.; Davíðsdóttir, B. The need for integrated valuation tools to support decision-making—The case of cultural ecosystem services sourced from geothermal areas. *Ecosyst. Serv.* **2019**, *37*, 100923. [CrossRef]

64. Cook, D.; Davíðsdóttir, B.; Kristófersson, D.M. An ecosystem services perspective for classifying and valuing the environmental impacts of geothermal power projects. *Energy Sustain. Dev.* **2017**, *40*, 126–138. [CrossRef]

65. University of Wisconsin—Madison Solar Energy Laboratory. *TRNSYS, a Transient Simulation Program*; University of Wisconsin—Madison Solar Energy Laboratory: Madison, WI, USA, 1975.

66. Aspen Technology. *Aspen Plus Engineering Software*; Aspen Technology: Bedford, MA, USA, 2019.

67. Barbier, E.; Bellani, S.; Musmeci, F. The Italian geothermal database. In Proceedings of the World Geothermal Congress, Kyusgu-Tohoku, Janpan, 28 May–10 June 2000.

68. Trumpy, E.; Botteghi, S.; Caiozzi, F.; Donato, A.; Gola, G.; Montanari, D.; Pluymaekers, M.; Santilano, A.; Van Wees, J.; Manzella, A. Geothermal potential assessment for a low carbon strategy: A new systematic approach applied in southern Italy. *Energy* **2016**, *103*, 167–181. [CrossRef]

69. Calamai, A.; Cataldi, R.; Locardi, E.; Praturlon, A. (1977) Distribuzione delle anomalie geotermiche nella fascia preappenninica tosco-laziale (Italia) [Distribution of geothermal anomalies in the Tuscany-Latium pre-Apennine belt (Italy)]. In Proceedings of the Simp Intern sobre Energia Geotérmica en America Latina, Ciudad de Guatemala, Guatemala, 16–23 October 1976; pp. 189–229.

70. Cataldi, R.; Mongelli, F.; Squarci, P.; Taffi, L.; Zito, G.; Calore, C. Geothermal ranking of Italian territory. *Geothermics* **1995**, *24*, 115–129. [CrossRef]

71. Mormone, A.; Tramelli, A.; Di Vito, M.A.; Piochi, M.; Troise, C.; De Natale, G. Secondary hydrothermal minerals in buried rocks at the Campi Flegrei caldera, Italy: A possible tool to understand the rock-physics and to assess the state of the volcanic system. *Period. Mineral.* **2011**, *80*, 385.

72. De Vivo, B.; Belkin, H.E.; Barbieri, M.; Chelini, W.; Lattanzi, P.; Lima, A.; Tolomeo, L. The campi flegrei (Italy) geothermal system: A fluid inclusion study of the mofete and San Vito fields. *J. Volcanol. Geotherm. Res.* **1989**, *36*, 303–326. [CrossRef]

73. Carlino, S.; Somma, R.; Troise, C.; De Natale, G. The geothermal exploration of Campanian volcanoes: Historical review and future development. *Renew. Sustain. Energy Rev.* **2012**, *16*, 1004–1030. [CrossRef]

74. Corrado, G.; De Lorenzo, S.; Mongelli, F.; Tramacere, A.; Zito, G. Surface heat flow density at the phlegrean fields caldera (SOUTHERN ITALY). *Geothermics* **1998**, *27*, 469–484. [CrossRef]

75. Carlino, S.; Cubellis, E.; Delizia, I.; Luongo, G. History of Ischia Harbour (Southern Italy), in Macro-engineering Seawater in UniEue Environments. *Environ. Sci. Eng.* **2010**, 27–57. [CrossRef]

76. Carella, R.; Verdiani, G.; Palmerini, C.G.; Stefani, C.G. Geothermal activity in Italy: Present status and future prospects. *Geothermics* **1985**, *14*, 247–254. [CrossRef]

77. AGIP. *Geologia e Geofisica del Sistema Geotermico dei Campi Flegrei, Technical Report*; Settore Esplor e Ric Geoterm-Metodol per l'Esplor Geotermica: San Donato Milanese, Italy, 1987; pp. 1–23.

78. Carotenuto, A.; Figaj, R.D.; Vanoli, L. A novel solar-geothermal district heating, cooling and domestic hot water system: Dynamic simulation and energy-economic analysis. *Energy* **2017**, *141*, 2652–2669. [CrossRef]

79. SketchUp, Trimble, Inc. Available online: https://www.sketchup.com/ (accessed on 11 September 2019).

80. Ballarini, I.; Corgnati, S.P.; Corrado, A. TABULA—Building Typology Brochure—Italy. Fascicolo Sulla Tipologia Edilizia Italiana. Politecnico di Torino—Dip. Energia. TEBE. 2014. Available online: https://episcope.eu/ (accessed on 17 May 2020).

81. Ahmed, K.; Pylsy, P.; Kurnitski, J. Hourly consumption profiles of domestic hot water for different occupant groups in dwellings. *Sol. Energy* **2016**, *137*, 516–530. [CrossRef]

82. Prestazioni energetiche degli edifice—Parte 2: Determinazione del fabbisogno di energia primaria e dei rendimenti per la climatizzazione invernale e per la produzione di acqua calda sanitariaEnergy (Performance of Buildings Part 2: Evaluation of Primary Energy Need and System Efficiencies for Space Heating and Domestic Hot Water Production); UNI TS 11300-2. 2008. Available online: http://store.uni.com/catalogo/uni-ts-11300-2-2008 (accessed on 3 September 2020).

83. Meteotest. *Meteonorm Handbook, Parts I, II and III*; Meteotest: Bern, Switzerland, 2003. Available online: http://www.meteotest.ch (accessed on 3 September 2020).

84. Grassiani, M. Siliceous scaling aspects of geothermal power generation using binary cycle heat recovery. In Proceedings of the Transactions-Geothermal Resources Council; 2000; pp. 475–5933. Available online: https://www.geothermal-energy.org/pdf/IGAstandard/WGC/2000/R0549.PDF (accessed on 3 September 2020).

85. Franco, A.; Villani, M. Optimal design of binary cycle power plants for water-dominated, medium-temperature geothermal fields. *Geothermics* **2009**, *38*, 379–391. [CrossRef]

86. Shah, R.K.; London, A.L. *Laminar Flow Forced Convection in Ducts: A Source Book for Compact Heat Exchanger Analytical Data*; Academic Press: Cambridge, MA, USA, 2014; ISBN 1483191303.

87. Bejan, A.; Kraus, A.D. *Handbook, Heat Transfer*; John Wiley & Sons: Hoboken, NJ, USA, 2003.

88. Calise, F.; D'Accadia, M.D.; Figaj, R.; Vanoli, L. A novel solar-assisted heat pump driven by photovotaic/thermal collectors: Dynamic simulation and thermoeconomic optimization. *Energy* **2016**, *95*, 346–366. [CrossRef]

89. ISPRA—Istituto Superiore per la Protezione Ambientale (Italian Institute for Environmental Protection). Available online: http://www.isprambiente.gov.it (accessed on 11 September 2019).

90. Calise, F.; D'Accadia, M.D.; Vicidomini, M.; Scarpellino, M. Design and simulation of a prototype of a small-scale solar CHP system based on evacuated flat-plate solar collectors and Organic Rankine Cycle. *Energy Convers. Manag.* **2015**, *90*, 347–363. [CrossRef]

91. Hendricks, A.M.; Wagner, J.E.; Volk, T.; Newman, D.H.; Brown, T.R. A cost-effective evaluation of biomass district heating in rural communities. *Appl. Energy* **2016**, *162*, 561–569. [CrossRef]

92. Calise, F.; Palombo, A.; Vanoli, L. Maximization of primary energy savings of solar heating and cooling systems by transient simulations and computer design of experiments. *Appl. Energy* **2010**, *87*, 524–540. [CrossRef]

93. Ministero dello Sviluppo Economico. Ministry of Economic Development. Available online: https://www.mise.gov.it (accessed on 10 September 2019).

94. Casasso, A.; Gallina, M.; Rivoire, M.; Sethi, R. 2048 Studio di Fattibilità per un Impianto Geotermico a Circuito Aperto in un Complesso Residenziale (Feasibility Study for an Open Circuit Geothermal Plant in a Residential Context). Master's Thesis, Politecnico di Torino, Torino, Italy, March 2018. Available online: http://webthesis.biblio.polito.it/id/eprint/7409 (accessed on 3 September 2020).

MDPI

St. Alban-Anlage 66

4052 Basel

Switzerland

Tel. +41 61 683 77 34

Fax +41 61 302 89 18

www.mdpi.com

Energies Editorial Office

E-mail: energies@mdpi.com

www.mdpi.com/journal/energies